Lecture Notes
in Control and Information Sciences 366

Editors: M. Thoma, M. Morari

Francesco Bullo, Kenji Fujimoto (Eds.)

Lagrangian and Hamiltonian Methods for Nonlinear Control 2006

Proceedings from the 3rd IFAC Workshop, Nagoya, Japan, July 2006

Springer

Editors

Francesco Bullo

Mechanical Engineering Department
University of California at Santa Barbara
2338 Engineering Bldg II
Santa Barbara, CA 93106-5070
USA
Email: bullo@engineering.ucsb.edu

Kenji Fujimoto

Department of Mechanical Science and Engineering
Nagoya University
Furo-cho, Chikusa-ku, Nagoya, 464-8603
Japan
Email: k.fujimoto@ieee.org

Library of Congress Control Number: 2007931367

ISSN print edition: 0170-8643
ISSN electronic edition: 1610-7411
ISBN-10 3-540-73889-4 Springer Berlin Heidelberg New York
ISBN-13 978-3-540-73889-3 Springer Berlin Heidelberg New York

Springer is a part of Springer Science+Business Media
springer.com

Typesetting: by the authors and SPS using a Springer LaTeX macro package

Printed on acid-free paper SPIN: 11898436 89/SPS 5 4 3 2 1 0

Preface

This proceedings volume documents the 3rd IFAC Workshop on Lagrangian and Hamiltonian Methods in Nonlinear Control (LHMNLC'06) that was held in Nagoya, Japan, on July 19-21, 2006. The first workshop in this series was chaired and organized by Professors N. E. Leonard and R. Ortega, and was held in Princeton, USA, in March 2000. The second one was chaired and organized by Professors A. Astolfi, F. Gordillo and A. J. van der Schaft, and was held in Seville, Spain, in April 2003.

A vibrant synergy is documented between areas such as nonlinear control and optimal control theory, differential and Riemannian geometry, Lagrangian and Hamiltonian mechanics, nonsmooth optimization, and dynamical systems. The articles in this volume focus on technological areas including not only control of mechanical systems, but also geometric optimization, networked control, control of chemical processes, robotic locomotion, quantum systems, multi-agent systems, and robotic grasping and telemanipulation. Novel scientific contribution are proposed in a wide variety of techniques including synchronization, control Lyapunov functions, energy and power-based control, optimization algorithms, fault-tolerant control, geometric reduction theory, and iterative learning control, to name a few.

Financial support for the workshop was provided by the 21st Century COE Program (Tokyo Institute of Technology) "Innovation of Creative Engineering through the Development of Advanced Robotics," the Suzuki Foundation, the Daiko Foundation and the University of Nagoya. We also would like to thank all the participants to the workshop, all the members of the national and international organizing committees, the IFAC Secretariat, the IFAC Publications Committee, and the Springer-Verlag review board for the LNCIS series.

Santa Barbara, USA, and Nagoya, Japan,
June 2007

Francesco Bullo
Kenji Fujimoto

List of Contributors

José Ángel Acosta
Depto. de Ingeniería de Sistemas y Automática
Escuela Superior de Ingenieros
Camino de los Descubrimientos s/n.
41092 Sevilla, Spain
jaar@esi.us.es

Frank Allgöwer
Institute for Systems Theory and Automatic Control (IST)
University of Stuttgart
Pfaffenwaldring 9
70550 Stuttgart, Germany
allgower@ist.uni-stuttgart.de

Aaron D. Ames
California Institute of Technology
1200 E. California Boulevard
Pasadena, CA 91125, USA
ames@cds.caltech.edu

Javier Aracil
University of Seville
Escuela Superior de Ingenieros.
Camino de los Descubrimientos s/n.
Sevilla, 41092, Spain
aracil@esi.us.es

Suguru Arimoto
Ritsumeikan University
1-1-1 Nojihigashi
Kusatsu, Shiga, 525-8577, Japan
arimoto@se.ritsumei.ac.jp

Alessandro Astolfi
Dept. Electrical & Electronic Eng.
Imperial College London
Exhibition Road
London, SW7 2AZ, UK
and
DISP, University of Rome
Tor Vergata, Via del Politecnico 1
00133 Rome, Italy
a.astolfi@imperial.ac.uk

Karl J. Åström
Department of Automatic Control
LTH. Lund University
Box 118
SE-221 00 Lund, Sweden
kja@control.lth.se

Ji-Hun Bae
Pohang Institute of Intelligent Robotics
San 31, Hyoja-Dong, Nam-Gu
Pohang
Gyeongbuk 790-784, S. Korea
joseph@postech.ac.kr

Ravi Banavar
Systems and Control Engineering
I. I. T. - Bombay
101 A, ACRE Building
Powai, Mumbai, 400 076, India
banavar@iitb.ac.in

Luca Bassi
CASY-DEIS, University of Bologna
viale Risorgimento 2
40136 Bologna, Italy
lbassi@deis.unibo.it

Carles Batlle
MA4 and IOC, EPSEVG
UPC
Av. V. Balaguer s/n
Vilanova i la Geltrú 08800, Spain
carles.batlle@upc.edu

Claudio Bonivento
CASY-DEIS, University of Bologna
Viale Risorgimento 2
40136 Bologna, Italy
claudio.bonivento@unibo.it

Bernard Bonnard
Institut de Mathématiques de Bourgogne (UMR CNRS 5584)
9 avenue A. Savary
F-21078 Dijon, France
Bernard.Bonnard@u-bourgogne.fr

Francesco Bullo
Mechanical Engineering, University of California at Santa Barbara
2338 Engineering Bldg II
Santa Barbara, CA 93106, USA
bullo@engineering.ucsb.edu

Jean-Baptiste Caillau
ENSEEIHT-IRIT
(UMR CNRS 5505)
2 rue Camichel
F-31071 Toulouse, France
caillau@n7.fr

Nikhil Chopra
Coordinated Science Laboratory
University of Illinois at Urbana-Champaign
1308 W. Main St.
Urbana, IL 61801, USA
nchopra@uiuc.edu

Monique Chyba
University of Hawaii at Manoa
2565 Mc Carthy Mall
Hawaii, HI 96822, USA
mchyba@math.hawaii.edu

Alessandro de Rinaldis
Laboratoire des Signaux et Systèmes
CNRS-SUPELEC
Gif–sur–Yvette 91192, France
alessandro.de-rinaldis@renault.com

Arnau Dòria-Cerezo
DEE and IOC, EPSEVG
UPC
Av. V. Balaguer s/n
Vilanova i la Geltrú 08800, Spain
arnau.doria@upc.edu

Gunther Dirr
Institute of Mathematics
University of Würzburg
Am Hubland
97074 Würzburg, Germany
dirr@mathematik.uni-wuerzburg.de

Romain Dujol
ENSEEIHT-IRIT
(UMR CNRS 5505)
2 rue Camichel
F-31071 Toulouse, France
dujol@n7.fr

Damien Eberard
Institute for Mathematics and Computer
Science, University of Groningen
P.O. Box 8009700 AV Groningen
The Netherlands
eberard@math.rug.nl

Gerardo Espinosa-Pérez
DEPFI–UNAM, Apartado Postal
70-256, 04510 México D.F.
Mexico
gerardoe@servidor.unam.mx

Cesare Fantuzzi
DISMI - University of Modena and Reggio Emilia
Via G. Amendola 2
Morselli Building
Reggio Emilia, Italy, 42100
cesare.fantuzzi@unimore.it

Robert Fuentes
Raytheon Missile Systems
1151 E. Hermans Road Bldg. 805
Tucson, AZ 85706, USA
rjfuentes@ieee.org

Kenji Fujimoto
Department of Mechanical Science and Engineering, Graduate School of Engineering, Nagoya University, Furo-cho, Chikusa-ku
Nagoya, 464-8603, Japan
k.fujimoto@ieee.org

Eloísa García–Canseco
Laboratoire des Signaux et Systèmes
SUPELEC, Plateau de Moulon,
91192, Gif–sur–Yvette, France
garcia@lss.supelec.fr

Luca Gentili
CASY-DEIS, University of Bologna
Viale Risorgimento, 2
40136 Bologna, Italy
lgentili@deis.unibo.it

Francisco Gordillo
University of Seville
Escuela Superior de Ingenieros
Camino de los Descubrimientos s/n.
Sevilla, 41092, Spain
gordillo@esi.us.es

Robert D. Gregg
University of Illinois at Urbana-Champaign
1406 W. Green Street
Urbana, IL 61801, USA
rgregg@uiuc.edu

Manuel Guerra
ISEG, Technical University of Lisbon
R. do Quelhas, 6
Lisboa, 1200-781, Portugal
mguerra@iseg.utl.pt

Thomas Haberkorn
University of Hawaii at Manoa
2565 Mc Carthy Mall
Hawaii, HI 96822, USA
haberkor@math.hawaii.edu

Uwe Helmke
Institute of Mathematics
University of Würzburg
Am Hubland
97074 Würzburg, Germany
helmke@mathematik.uni-wuerzburg.de

Gregory Hicks
General Dynamics Advanced Information Systems
14700 Lee Road
Chantilly, VA 20151, USA
gregory.hicks@gd-ais.com

Sang-Ho Hyon
Computational Brain Project
ICORP, JST, and ATR
Computational Neuroscience Laboratories, 2-2-2 Hikaridai
Kyoto, 619-0288, Japan
sangho@atr.jp

Dimitri Jeltsema
Delft Center for Systems and Control
Delft University of Technology
Mekelweg 2, 2628 CD Delft
The Netherlands
d.jeltsema@tudelft.nl

Xiaohong Jiao
Yanshan University
Qinhuangdao 066004, China
jiaoxh@tsinghua.edu.cn

Jørgen K. Johnsen
Institute for Systems Theory and Automatic Control (IST)
University of Stuttgart
Pfaffenwaldring 9
70550 Stuttgart, Germany
johnsen@ist.uni-stuttgart.de

Velimir Jurdjevic
University of Toronto
Toronto, Canada
jurdj@math.utoronto.ca

P.S. Krishnaprasad
Department of Electrical and Computer Engineering
University of Maryland, College Park, MD 20742-3285, USA
krishna@isr.umd.edu

Christian Lageman
Institute of Mathematics
University of Würzburg
Am Hubland
97074 Würzburg, Germany
lageman@mathematik.uni-wuerzburg.de

Dina Shona Laila
Johannes Kepler University Linz
Altenberger Strasse 69
A-4040 Linz, Austria
dina-shona.laila@jku.at

Naomi Ehrich Leonard
Princeton University
Mechanical and Aerospace Engineering
Princeton, NJ 08544, USA
naomi@princeton.edu

Zhiwei Luo
Kobe University / RIKEN BMC
1-1 Rokkodai, Nada-ku
Kobe, Hyogo 657-8501, Japan
luo@gold.kobe-u.ac.jp

Alessandro Macchelli
CASY-DEIS, University of Bologna
viale Risorgimento 2
40136 Bologna, Italy
amacchelli@deis.unibo.it

Jerrold E. Marsden
Control and Dynamical Systems
California Institute of Technology
Pasadena, CA 91125, USA
marsden@cds.caltech.edu

Bernhard Maschke
Laboratoire d'Automatique et de Génie des Procédés
UMR CNRS 5007
Université Lyon 1
F-69622 Villeurbanne cedex, France
maschke@lagep.univ-lyon1.fr

Claudio Melchiorri
CASY-DEIS, University of Bologna
viale Risorgimento 2,
40136 Bologna, Italy
cmelchiorri@deis.unibo.it

Benjamin Nabet
Princeton University
Mechanical and Aerospace Engineering
Princeton, NJ 08544, USA
bnabet@princeton.edu

Gou Nishida
RIKEN (The Institute of Physical and Chemical Research)
2271-130 Anagahora, Shimoshidami
Moriyama-ku
Nagoya, Aichi 463-0003, Japan
nishida@bmc.riken.jp

Romeo Ortega
Laboratoire des Signaux et Systèmes
CNRS-SUPELEC
Gif–sur–Yvette 91192, France
romeo.ortega@lss.supelec.fr

Jason Osborne
North Carolina State University
245 Harrelson Hall
Raleigh, NC 27695, USA
jmosborn@ncsu.edu

Andrea Paoli
CASY-DEIS, University of Bologna
Viale Risorgimento 2
40136 Bologna, Italy
apaoli@deis.unibo.it

Laura Poggiolini
Dipartimento di Matematica
Applicata "G. Sansone"
Università degli Studi di Firenze
Via di Santa Marta 3, I-50139
Firenze, Italy
laura.poggiolini@math.unifi.it

Tsuyoshi Sagami
Tokyo Institute of Technology
2-12-1-S5-24
Meguro-ku, Tokyo, 152-8550, Japan
sagami@sc.ctrl.titech.ac.jp

Satoru Sakai
Chiba University
Department of Electronics and Mechanical Engineering
Yayoi-cho, Inage-ku, Chiba-shi
Chiba 263-0043, Japan
satorusakai@faculty.chiba-u.jp

Mitsuji Sampei
Tokyo Institute of Technology
2-12-1-S5-24
Meguro-ku, Tokyo, 152-8550, Japan
sampei@ctrl.titech.ac.jp

Andrey Sarychev
DiMaD, University of Florence
v. C.Lombroso 6/17
Firenze, 50134, Italy
asarychev@unifi.it

Shankar Sastry
University of California, Berkeley
253 Cory Hall
Berkeley, CA 94720, USA
sastry@eecs.berkeley.edu

Satoshi Satoh
Mechanical Science and Engineering
Graduate School of Engineering
Nagoya University, Furo-cho,
Chikusa-ku, Nagoya 464-8603, Japan
s_satou@nuem.nagoya-u.ac.jp

Jacquelien M.A. Scherpen
University of Groningen
Faculty of Mathematics and Natural Sciences
ITM Nijenborgh 4 9747 AG
Groningen, The Netherlands
j.m.a.scherpen@rug.nl

Sonia Schirmer
University of Cambridge
Dept of Applied Maths & Theoretical Physics
Wilberforce Road, Cambridge, CB3 0WA, UK
sgs29@cam.ac.uk

Cristian Secchi
DISMI - University of Modena and Reggio Emilia, Morselli Building
Via G. Amendola 2
Reggio Emilia, 42100, Italy
cristian.secchi@unimore.it

Tielong Shen
Sophia University
Tokyo 102-8554, Japan
tetu-sin@sophia.ac.jp

Ryan N. Smith
University of Hawaii at Manoa
2540 Dole Street, Holmes Hall 402
Hawaii, HI 96822, USA
ryan@math.hawaii.edu

Mark W. Spong
Coordinated Science Laboratory
University of Illinois at
Urbana-Champaign
1308 W. Main St.
Urbana, IL 61801, USA
mspong@uiuc.edu

Gianna Stefani
Dipartimento di Matematica
Applicata "G. Sansone"
Università degli Studi di Firenze
Via di Santa Marta 3, I-50139
Firenze, Italy
gianna.stefani@unifi.it

Stefano Stramigioli
Impact Institute
University of Twente
P.O. Box 217, 7500 AE
Enschede, The Netherlands
s.stramigioli@ieee.org

Yuanzhang Sun
Tsinghua University
Beijing 100084, China
yzsun@tsinghua.edu.cn

Arjan van der Schaft
Institute for Mathematics and
Computer Science, University of
Groningen, P.O. Box 800 9700 AV
Groningen, The Netherlands
A.J.van.der.Schaft@math.rug.nl

Giuseppe Viola
Dipartimento di Informatica, Sistemi
e Produzione, Università di Roma
"Tor Vergata", Via del Politecnico 1
00133 Roma, Italy
viola@disp.uniroma2.it

Eric D.B. Wendel
Sensis Corporation
1700 Dell Avenue
Campbell, CA 95008, USA
eric.wendel@sensis.com

George Wilkens
University of Hawaii at Manoa
2565 Mc Carthy Mall
Hawaii, HI 96822, USA
grw@math.hawaii.edu

Masaki Yamakita
Tokyo Institute of Technology /
RIKEN BMC
2-12-1 Ohokayama, Meguro-ku
Tokyo 152-8552, Japan
yamakita@ac.ctrl.titech.ac.jp

Morio Yoshida
RIKEN Bio-Mimetic Control Research
Center
2271-130 Anagahora Shimoshidami
Moriyama-ku
Nagoya, Aichi, 463-0003, Japan
yoshida@bmc.riken.jp

Hiroaki Yoshimura
Department of Mechanical Engineering, Waseda
University, Ohkubo, Shinjuku, Tokyo
169-8555, Japan
yoshimura@waseda.jp

Jason Zimmerman
University of Toronto
Toronto, Canada
a.n.j@sympatico.ca

Contents

A Differential-Geometric Approach for Bernstein's Degrees-of-Freedom Problem

Suguru Arimoto

Department of Robotics, Ritsumeikan University, 1-1-1 Nojihigashi, Kusatsu, Shiga, 525-8577 Japan
arimoto@se.ritsumei.ac.jp
Nagoya, RIKEN

Summary. This article challenges Bernstein's problem of redundant degrees of freedom (DOF) that remains unsolved from the control-theoretic point of view as well as from the standpoint of both neuro-physiology and robotics. Firstly, a rather simpler but mysterious control problem of movements of human-like multi-joint reaching with excess DOFs is analyzed from Newtonian mechanics and differential geometry. Secondly, another illustrative control problem that seems to be sophisticated and complicated is tackled, which is to find a sensory coordinated control signal for 3-Dimensional stable grasping and object manipulation by a pair of robot fingers with multiple joints under the effect of gravity and nonholonomic constraints. In each illustrative control problem, it is possible to find a simple control signal that renders each corresponding closed-loop dynamics stable on its corresponding equilibrium-point manifold. It is claimed, however, that convergences of solutions of closed-loop dynamics to an equilibrium-point manifold can not be analyzed by using Lyapunov's direct method, because a Lyapunov-like energy form can not be positive definite due to redundancy of DOFs. Instead, a novel definition called "stability on a manifold" based upon the concept of Riemannian distance on the constraint manifold is introduced in both illustrative problems and used in the analysis of convergence of solution trajectories. It is also shown that finiteness of Riemannian metrics plays an important role in evaluation of the performance of control in both problems.

1 Introduction

This paper is concerned with one of unsolved problems posed more than a half century ago by A.N. Bernstein as the Degrees-of-Freedom problem[14, 15], particularly, in case of human or robotic multi-joint movements of reaching as shown in Fig.1. The problem in case of Fig.1 is how to generate a joint motion so as to transfer the endpoint of an upper limb with four joints (shoulder, elbow, wrist, and finger MP joint) to a given target point $\boldsymbol{x}_d = (x_d, y_d)$ in the two-dimensional horizontal plane. Since the objective task $\boldsymbol{x}_d$ is given in the task space $\boldsymbol{x} \in X(= R^2)$ and the joint coordinates $q = (q_1, q_2, q_3, q_4)^{\mathrm{T}}$ are of four-dimension, there exists an infinite number of inverses q_d that realize $\boldsymbol{x}(q_d) = \boldsymbol{x}_d$

F. Bullo et al. (Eds.): Lag. & Hamil. Methods for Nonlin. Ctrl., LNCIS 366, pp. 1–28, 2007.
springerlink.com

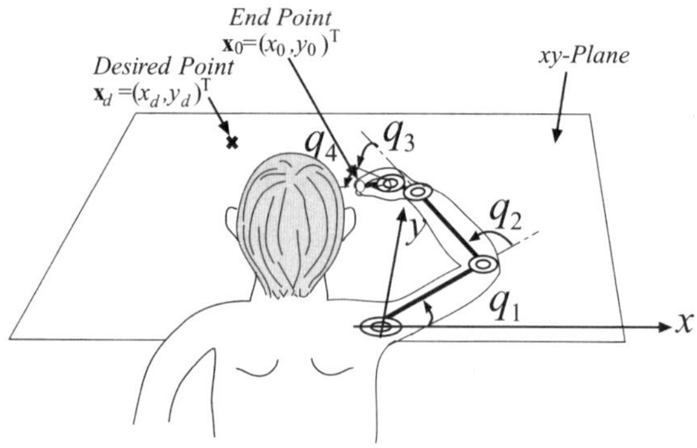

Fig. 1. "Reaching" by means of a surplus DOF system of arm-hand dynamics

and hence the problem of obtaining inverse kinematics from the task description space X to the 4-dimensional joint space becomes ill-posed. Under this circumstances, however, it is necessary to generate joint motions $q(t)$ starting from a given initial point $\boldsymbol{x}(0) = (x(0), y(0))$ in X with some initial posture $q(0) = (q_1(0), \cdots, q_4(0))^{\mathrm{T}}$ and leading the endpoint trajectory $\boldsymbol{x}(t)$ to reach the target $\boldsymbol{x}_d$ as $t \to \infty$. In order to get rid of such ill-posedness, many methods have been proposed as surveyed in a special issue of the journal (see [16]) and a book specially dedicated to problems of DOF redundancy (see [17]). Most of them are based on an idea of introducing some extra and artificial performance index for determining uniquely an appropriate joint space trajectory by minimizing it. In fact, examples of such performance index in robotics research are the followings: kinetic energy, quadratic norm of joint control torque, manipulability index, virtual fatigue function, etc. Most of proposed methods have been explicitly or implicitly based on the Jacobian pseudoinverse approach for planning an optimized joint velocity trajectory $\dot{q}(t) = J^{+}(q)\dot{\boldsymbol{x}}_d(t)$ together with an extra term $(I - J^{+}(q)J(q))\boldsymbol{v}$, where $\boldsymbol{v}$ is determined by optimizing the performance index, $J(q)$ stands for the Jacobian matrix of task coordinates $\boldsymbol{x}$ in joint coordinates q, and $J^{+}(q)$ the pseudoinverse of $J(q)$. In the history of robot control the idea of use of the pseudoinverse for generation of joint trajectories for redundant robots was initiated by [18]. However, it is impossible to calculate $J^{+}(q_d)$ in advance because q_d is undetermined. Therefore, it is recommended that a control signal u to be exerted through joint actuators of the robot is designed as

$$\begin{aligned} u = &\, H(q)J^{+}(\ddot{\boldsymbol{x}}_d - \dot{J}\dot{q}) + h(q, \dot{q}) \\ &+ \{I - J^{+}(q)J(q)\}\boldsymbol{v} + g(q) \end{aligned} \tag{1}$$

where $H(q)$ stands for the inertia matrix, $h(q, \dot{q})$ denotes direct compensation for the coriolis and centrifugal forces, and $g(q)$ compensation for the gravity term. On the other hand, there is the vast literature of research works concerned with even a simple multi-joint reaching motion or pointing movement in physiology, which are, for example, [19], [20], and [21]. Most of them published in physiological journals are more or less affected by the equilibrium point (EP) hypothesis of motor control found by [22], observed by [23], and argued explicitly by [24] as spring-like motion of a group of muscles. This EP hypothesis can be interpreted in the control-theoretic language such that for a specified arm endpoint $\boldsymbol{x}_d$ in cartesian space the central part of control for reaching it must be composed of the term $J^{\mathrm{T}}(q)K(\boldsymbol{x}-\boldsymbol{x}_d)$ which corresponds to assuming introduction of an artificial potential $\Delta\boldsymbol{x}^{\mathrm{T}}K\Delta\boldsymbol{x}/2$ in the external world (task space), where $\Delta\boldsymbol{x}=\boldsymbol{x}-\boldsymbol{x}_d$. This idea is also known in robotics as PD feedback with damping shaping first proposed by [25]. In the case of redundant multi-joint reaching, however, there arises the same problem of ill-posedness of inverse kinematics. Thus, in parallel with the vast literature ([16, 17]) in robotics research, many methods for elimination of redundancy of DOFs have been proposed in the physiological literature. Most of them are samely based on introduction of extra performance criterion to be optimized, which are in the following: squared norm of joint jerks (rate of change of acceleration), energy, effort applied during movement, minimum-torque, and minimum torque-change and some other cost functions though some of them are not used for redundancy resolution. Nevertheless, even all performance indices that lead successfully to unique determination of the inverse kinematics are not well-grounded physiologically and none of physiological evidence or principle that associates such a performance index to generation of human movements could be found.

In this article, we resolve this ill-posedness of inverse kinematics without considering any kind of inverse problems and without introducing any type of artificial performance index for the multi-joint reaching problems posed above. Instead, we use a surprisingly simpler sensory feedback scheme described as

$$u=-C\dot{q}-J^{\mathrm{T}}(q)k\Delta\boldsymbol{x} \tag{2}$$

or

$$u=-C_0\dot{q}-J^{\mathrm{T}}(q)(c\dot{\boldsymbol{x}}+k\Delta\boldsymbol{x}) \tag{3}$$

where $\Delta\boldsymbol{x}=\boldsymbol{x}-\boldsymbol{x}_d$, k a single stiffness parameter, C and C_0 a positive diagonal matrix corresponding to joint damping factors, and c a single positive damping parameter. It is proved theoretically that adequate choices of damping gain matrices C and stiffness parameter k render the closed-loop system dynamics convergent as time elapses, that is, $\boldsymbol{x}(t)\to\boldsymbol{x}_d$ and $\dot{q}(t)\to 0$ as $t\to\infty$. However, owing to the joint redundancy, the convergence in task space does not directly imply the convergence of joint variables $q(t)$ to some posture. In the paper, by introducing a novel concept named "stability on a manifold" it is shown that $q(t)$ remains in a specified region in joint space so that there does not arise any unexpected self-motion inherent to redundant systems (see [26]). In other words, the control scheme of eq.(2) suggests that the problem of elimination of joint redundancy need not be solved but can be ignored in control of the dynamics. Or it can be said that a natural physical principle for economies of skilled motions like the principle of least action in Newtonian mechanics may work in elimination of redundancy. Another concept named "transferability to a submanifold" is also introduced for discussing the asymptotic convergence in a case of middle-range reaching. These two concepts were originally and very recently defined in cases of control of multi-fingered hands with joint redundancy (see [27, 28]).

The latter part of the paper discusses how to construct a mathematical model of 3-D object grasping and manipulation by a pair of 3-D robot fingers with multi-joints. A noteworthy difference of modeling of 3-D object grasping from that of 2-D object is that the instantaneous axis of rotation of the object is fixed in the latter case but it is time-varying in the former case. Hence, dynamics of the overall fingers-object system is subject to non-holonomic constraints regarding a 3-D orthogonal matrix consisting of three mutually orthogonal unit-vectors fixed at the object. A further difference arises due to the physical assumption that spinning around the opposing axis between the two contact points does no more arise, which induces another nonholonomic constraint. It is shown that Lagrange's equation of motion of the overall system can be derived from Hamilton's principle without violating the causality that governs the nonholonomic constraints. Then, a simple control signal constructed on the basis of finger-thumb opposable forces and an object-mass estimator is proposed and shown to accomplish stable grasping in a dynamic sense without using object information or external sensing. This is called "blind grasping" if in addition the overall closed-loop dynamics converge to a state of force/torque balance. Stability and asymptotic stability (in this paper, this is called "transferability") under the gravity effect and the nonholonomic constraints are proved on the basis of principle of least action and Morse theory.

2 Closed-Loop Dynamics of Multi-joint Reaching Movement

Lagrange's equation of motion of a multi-joint system whose motion is confined to a plane as shown in Fig.1 is described by the formula (see [29])

$$H(q)\ddot{q} + \left\{ \frac{1}{2}\dot{H}(q) + S(q, \dot{q}) \right\} \dot{q} = u \tag{4}$$

where $q = (q_1, q_2, q_3, q_4)^{\mathrm{T}}$ denotes the vector of joint angles, $H(q)$ the inertia matrix, and $S(q, \dot{q})\dot{q}$ the gyroscopic force term including centrifugal and Coriolis forces. It is well known that the inertia matrix $H(q)$ is symmetric and positive definite and there exist a positive constant h_m together with a positive definite constant diagonal matrix H_0 such that

$$h_m H_0 \leq H(q) \leq H_0 \tag{5}$$

for any q. It should be also noted that $S(q, \dot{q})$ is skew symmetric and linear and homogeneous in $\dot{q}$. Any entry of $H(q)$ and $S(q, \dot{q})$ is constant or a sinusoidal function of components of q.

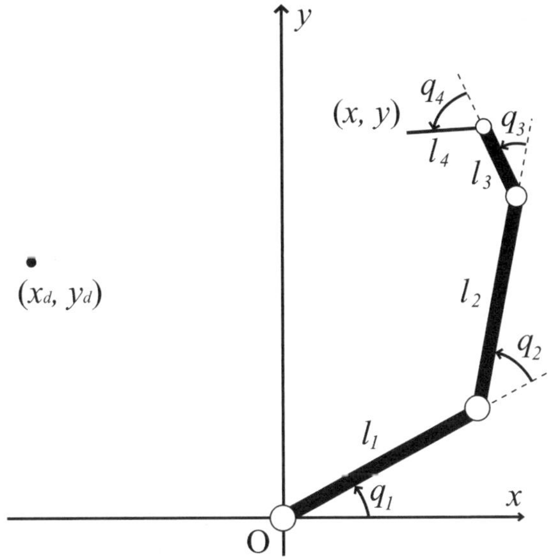

Fig. 2. Initial posture of the arm-hand system with four joints. The point (x_d, y_d) denotes the target for the robot endpoint.

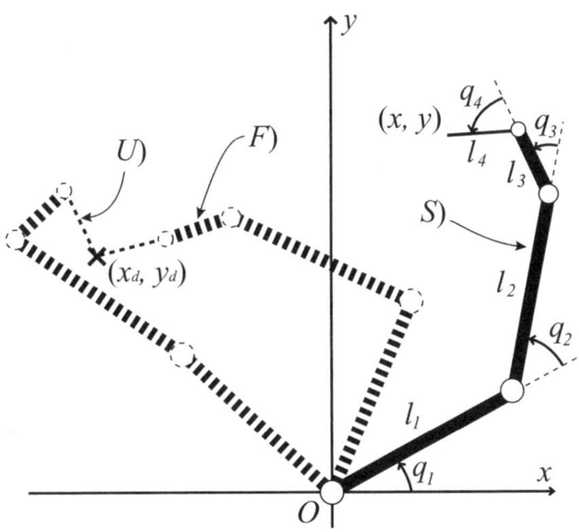

Fig. 3. S) Starting posture, F) Final posture, U) Unreasonable posture

For a given specified target position $\boldsymbol{x}_d = (x_d, y_d)^{\mathrm{T}}$ as shown in Fig.2, if the control input of eq.(3) is used at joint actuators then the closed-loop equation of motion of the system can be expressed as

$$H(q)\ddot{q} + \left\{ \frac{1}{2}\dot{H}(q) + S(q, \dot{q}) + C_0 \right\} \dot{q} + J^{\mathrm{T}}(q)\left\{ c\dot{\boldsymbol{x}} + k\Delta\boldsymbol{x} \right\} = 0 \tag{6}$$

which follows from substitution of eq.(3) into eq.(4). Since $\dot{\boldsymbol{x}} = J(q)\dot{q}$, the inner product of eq.(6) with $\dot{q}$ is reduced to

$$\frac{\mathrm{d}}{\mathrm{d}t}E = -\dot{q}^{\mathrm{T}} C_0 \dot{q} - c\|\dot{\boldsymbol{x}}\|^2 \tag{7}$$

where E stands for the total energy, i.e.,

$$E(q, \dot{q}) = \frac{1}{2}\left\{ \dot{q}^{\mathrm{T}} H(q)\dot{q} + k\|\Delta\boldsymbol{x}\|^2 \right\} \tag{8}$$

Evidently the first term of this quantity E stands for the kinetic energy of the system. The second term is called an artificial potential in this paper that appears due to addition of control signal $-J^{\mathrm{T}}(q)k\Delta\boldsymbol{x}$ based on the error $\Delta\boldsymbol{x}$ expressed in cartesian space. As it is well known in robot control (see [29]), the relation of eq.(7) denotes passivity of the closed-loop dynamics of eq.(6). It also reminds us of Lyapunov's stability analysis, since it shows that the derivative of a scalar function E in time t is negative semi-definite. However, it should be noted that the scalar function $E(q,\, \dot{q})$ is not positive definite with respect to the state vector $(q,\, \dot{q}) \in R^8$. In fact, E includes only a quadratic term of two-dimensional vector $\Delta\boldsymbol{x}$ except the kinetic energy as a positive definite quadratic function of $\dot{q}$. Therefore, it is natural and reasonable to introduce a manifold of 2-dimension defined as

$$M_2 = \{(q, \dot{q}) : E(q, \dot{q}) = 0 \ (\dot{q} = 0, \boldsymbol{x}(q) = \boldsymbol{x}_d)\}$$

which is called the zero space in the literature of robotics research (for example, [30]). Next, consider a posture $(q^0,\, 0)$ with still state (i.e., $\dot{q} = 0$) whose endpoint is located at $\boldsymbol{x}_d$, i.e., $\boldsymbol{x}(q^0) = \boldsymbol{x}_d$ and hence $(q^0,\, 0) \in M_2$, and analyze stability of motion of the closed-loop dynamics in a neighborhood of this equilibrium state. This equilibrium state in R^8 is called in this paper the reference equilibrium state.

3 Riemannian Metrics and Stability on a Manifold

According to terminologies of differential geometry, the set of all generalized position vector $q = (q_1, \cdots, q_n)^{\mathrm{T}}$ is called the n-dimensional configuration space, a set of all $\dot{q} = (\dot{q}_1, \cdots, \dot{q}_n)$ at a given position q is called a tangent space, and the set of all tangent spaces for all q is called the tangent bundle. For a given target endpoint $\boldsymbol{x}_d$ in task space the set of all vectors q satisfying $\boldsymbol{x}(q) = \boldsymbol{x}_d$ constitutes a two-dimensional manifold:

$$M_2(\boldsymbol{x}_d) = \{q : \boldsymbol{x}(\boldsymbol{q}) = \boldsymbol{x}_d\} \tag{9}$$

When an initial position $q = q(0)$ with $\dot{q}(0) = 0$, the reaching movement can be expressed by a trajectory of $q(t)$ in the configuration space that is a solution to the closed-loop equation of motion expressed by eq.(6) with satisfying initial conditions $q(0)$ and $\dot{q}(0) = 0$. Then, the total length of movements of $q(t)$ meaured by the metric along the trajectory $q(t)$

$$R(q(0)) = \int_{q(0)\in M_2(\boldsymbol{x}(0))}^{q(\infty)\in M_2(\boldsymbol{x}_d)} \sqrt{\frac{1}{2}\sum_{i,j=1}^{n} h_{ij}(q)\mathrm{d}q_i \mathrm{d}q_j} \tag{10}$$

can be rewritten in the form

$$\begin{aligned} R(q(0)) &= \int_0^\infty \sqrt{\frac{1}{2}\sum h_{ij}(q(t))\dot{q}_i(t)\dot{q}_j(t)}\,\mathrm{d}t \\ &= \int_0^\infty \sqrt{\frac{1}{2}\dot{q}^{\mathrm{T}}(t)H(q(t))\dot{q}(t)}\,\mathrm{d}t \end{aligned} \tag{11}$$

where $H(q) = (h_{ij}(q))$, provided that $\dot{q}(t) \in L^1(0, \infty)$.

It is also interesting to note that the original equation of motion of the robot arm depicted in Fig.1 and expressed in eq.(4) can be recast into the following form by multiplying eq.(4) with $H^{-1}(q)$:

$$\frac{\mathrm{d}^2}{\mathrm{d}t^2}q_i + \sum_{m,n} \Gamma^i_{nm} \frac{\mathrm{d}q_m}{\mathrm{d}t}\frac{\mathrm{d}q_n}{\mathrm{d}t} = \sum_j h^{ij} u_j \tag{12}$$

where h^{ij} denotes (i,j)-entry of $H^{-1}(q)$ and Γ^i_{nm} Christoffel's symbol. Notwithstanding such a mathematically simple expression of eq.(12), it loses physical meanings of the problem. In fact the finiteness of the Riemannian metrics such as eqs.(10) and (11) should be proved by gaining a physical insight into the expression of motion equation via eqs.(4) and (6). In this expository arricle the proof of finiteness of $R(q(0))$ is omitted (it can be found in a separate paper [34]). In differential geometry, eq.(4) is written in the form

$$\sum_{j=1}^{4} h_{ij}(q)\ddot{q}_j + \sum_{j,k=1}^{4} \Gamma_{ijk}(q)\dot{q}_j\dot{q}_k = u_i \tag{13}$$

by using another Christoffel's symbol called the second kind. When the external torque u_i is zero, a solution trajectory $q(t)$ to eq.(12) or eq.(13) starting from a given initial position to a target position $q(1)$ is called the geodesics. Then, the Riemannian distance between two points q^0 on the configuration R^4 can be defined as

$$\begin{aligned} &R(q^0, q^1) \\ &= \min_{q(t)} \int_0^1 \sqrt{\frac{1}{2}\sum_{i,j} h_{ij}(q(t))\dot{q}_i(t)\dot{q}_j(t)}\,\mathrm{d}t \end{aligned} \tag{14}$$

where minimization is taken over all curves $q(t)$ parametrized by $t \in [0,1]$ in such a way as $q(0) = q^0$ and $q(1) = q^1$. The optimal curve that attains $R(q^0, q^1)$ is shown to satisfy eq.(12) or equivalently eq.(13) when $u = 0$. It should be remarked at this stage that

1) The Riemannian distance $R(q^0, q^1)$ is invariant under any transformation $t = g(s)$ satisfying $\mathrm{d}t/\mathrm{d}s > 0$ and $\mathrm{T} = g^{-1}(1)$ and $0 = g^{-1}(0)$, because it follows that

$$\begin{aligned}
&\int_0^{\mathrm{T}} \sqrt{\sum_{i,j} \frac{1}{2} h_{ij}(\tilde{q}(s)) \frac{\mathrm{d}\tilde{q}_i}{\mathrm{d}s} \frac{\mathrm{d}\tilde{q}_j}{\mathrm{d}s}}\, \mathrm{d}s \\
&= \int_0^{\mathrm{T}} \sqrt{\sum_{i,j} \frac{1}{2} h_{ij}(q(t)) \cdot \left(\frac{\mathrm{d}q_i}{\mathrm{d}t} \frac{\mathrm{d}t}{\mathrm{d}s} \right) \cdot \left(\frac{\mathrm{d}q_i}{\mathrm{d}t} \frac{\mathrm{d}t}{\mathrm{d}s} \right)}\, \mathrm{d}s \\
&= \int_0^{1} \sqrt{\sum_{i,j} \frac{1}{2} h_{ij}(q(t)) \frac{\mathrm{d}q_i}{\mathrm{d}t} \frac{\mathrm{d}q_j}{\mathrm{d}t}}\, \mathrm{d}t
\end{aligned} \tag{15}$$

where $\tilde{q}(s) = q(g(s))$.

2) If the curve $q(t)$ for $t \in [0,1]$ minimizes the interal of eq.(14), then the quantity

$$L(q, \dot{q}) = \sqrt{\frac{1}{2} \sum_{i,j} h_{ij}(q) \dot{q}_i \dot{q}_j} \tag{16}$$

is constant for all $t \in [0,1]$. This follows from taking an inner product between $\dot{q}(t)$ and eq.(4) with $u = 0$, which yields

$$\begin{aligned}
0 &= \dot{q}^{\mathrm{T}} \left\{ H\ddot{q} + \left(\frac{1}{2}\dot{H} + S \right) \dot{q} \right\} \\
&= \frac{\mathrm{d}}{\mathrm{d}t} \left\{ \frac{1}{2} \dot{q}^{\mathrm{T}} H(q) \dot{q} \right\}
\end{aligned} \tag{17}$$

This shows $(1/2)\dot{q}^{\mathrm{T}} H(q)\dot{q} = \text{const}$.

It is now necessary to introduce the concept of neighborhoods of the reference equilibrium state $(q^0,\, 0) \in M_2$ in R^8, which are conveniently defined with positive parameters $\rho > 0$ and $r_0 > 0$ as

$$\begin{aligned}
&N^8(\rho, r_0) \\
&\quad = \left\{ (q, \dot{q}) : E(q, \dot{q}) \leq \rho^2 \text{ and } R(q, q^0) \leq r_0 \right\}
\end{aligned}$$

The necessity of imposing the inequality condition $R(q, q^0) \leq r_0$ comes from avoiding arise of possible movements such as self-motion ([26]) due to redundancy of DOFs far from the original posture. In fact, for the given endpoint $\boldsymbol{x}_d$ with the reference state as shown by the mark F) in Fig.3, one possible state with the posture marked by S) in Fig.3 may be inside $N^8(\rho, r_0)$ but another state

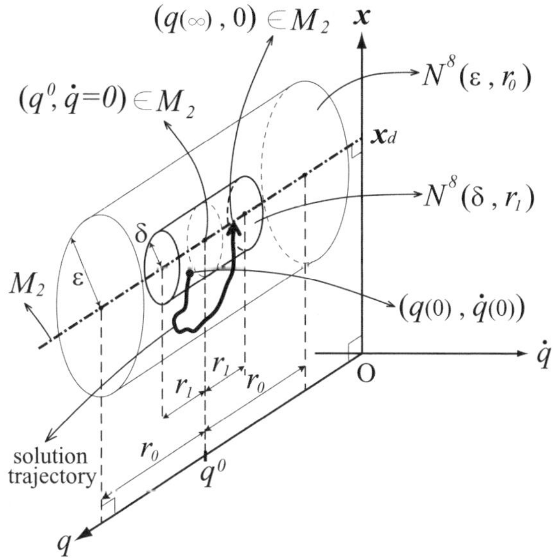

Fig. 4. Definitions of "stability on a manifold" and "transferability to a submanifold"

marked by U) must be excluded from the neighborhood $N^8(\rho,\, r_0)$ by choosing $r_0 > 0$ appropriately, because the overall posture of U) is by far deviated from that of the original reference equilibrium state $(q^0,\, 0)$. Further, it is necessary to assume that the reference equilibrium state $(q^0,\, 0)$ is considerably distant from the posture that has singularity of Jacobian matrix $J(q)$, which happens if and only if $q_2 = q_3 = q_4 = 0$.

We are now in a position to define the concept of stability of the reference equilibrium state lying on the manifold M_2.

Definition 1. If for an arbitrarily given $\varepsilon > 0$ there exist a constant $\delta > 0$ depending on ε and another constant $r_1 > 0$ independent of ε and less than r_0 such that a solution trajectory $(q(t),\, \dot{q}(t))$ of the closed-loop dynamics of eq.(6) starting from any initial state $(q(0),\, \dot{q}(0))$ inside $N^8(\delta(\varepsilon),\, r_1)$ remains in $N^8(\varepsilon,\, r_0)$, then the reference equilibrium state $(q^0,\, 0)$ is called stable on a manifold (see Fig.4).

Definition 2. If for a reference equilibrium state $(q^0,\, 0) \in R^8$ there exist constants $\varepsilon_1 > 0$ and $r_1 > 0$ $(r_1 < r_0)$ such that any solution of the closed-loop dynamics of eq.(6) starting from an arbitrary initial state in $N^8(\varepsilon_1,\, r_1)$ remains in $N^8(\varepsilon_1,\, r_0)$ and converges asymptotically as $t \to \infty$ to some point on $M_2 \cap N^8(\varepsilon_1,\, r_0)$, then the neighborhood $N^8(\varepsilon_1,\, r_1)$ of the reference equilibrium state $(q^0,\, 0)$ is said to be transferable to a subset of M_2 including q^0.

This definition means that, even if a still state $(q^0,\, 0) \in M_2$ of the multi-joint system is forced to move instantly to a different state $(q(0),\, \dot{q}(0))$ in a neighborhood of $(q^0,\, 0)$ by being exerted from some external disturbance, the sensory feedback control of eq.(3) assures that the system's state soon recovers to another still state $(q^{\infty}, 0) \in M_2 \cap N^8(\varepsilon_1, r_0)$ whose endpoint attains at the original point $\boldsymbol{x}(q^{\infty}) = \boldsymbol{x}_d$ though the convergent posture q^{∞} possibly differs from the original one q^0 but remains within $R(q, q^0) \leq r_0$.

4 Middle-Range Reaching

According to the energy balance law expressed by eq.(7), the total energy $E(t)(= E(q(t), \dot{q}(t)))$ is decreasing with increasing t as far as $\dot{q} \neq 0$ for an arbitrary positive definite damping gain matrix C_0. However, the motion profile $\boldsymbol{x}(t)$ of the endpoint is quite sensitive to choice for $c_i > 0$ for $i = 1, \cdots, 4$ where $C_0 = \mathrm{diag}(c_1, \cdots, c_4)$ though for a broad range of choice for c_i $(i = 1, \cdots, 4)$ the endpoint $\boldsymbol{x}(t)$ eventually converges to the target if the stiffness k is chosen adequately. For example, we show several endpoint trajectories for different k in Fig.5 obtained by computer simulation based on a human model shown in Table 1 in the case that the initial point $\boldsymbol{x}(0)$ in task space $X(= R^2)$, its corresponding

Table 1. Lengths of upper arm (l_1), lower arm (l_2), palm (l_3), and index finger (l_4) together with corresponding link masses and inertia moments. The data are taken from a 5 years old child.

link1 length (m)	l_1	0.175
link2 length (m)	l_2	0.170
link3 length (m)	l_3	0.0600
link4 length (m)	l_4	0.0600
link1 cylinder radius (m)	r_1	0.0247
link2 cylinder radius (m)	r_2	0.0223
link3 cuboid height (m)	h_3	0.0600
link3 cuboid depth (m)	d_3	0.0210
link4 cylinder radius (m)	r_4	0.00509
link1 mass (kg)	m_1	0.335
link2 mass (kg)	m_2	0.266
link3 mass (kg)	m_3	0.0756
link4 mass (kg)	m_4	0.00488
link1 inertia moment (kgm^2)	I_1	9.07×10^{-4}
link2 inertia moment (kgm^2)	I_2	6.73×10^{-4}
link3 inertia moment (kgm^2)	I_3	2.55×10^{-5}
link4 inertia moment (kgm^2)	I_4	1.50×10^{-6}

initial posture $q(0)$, and the target $\boldsymbol{x}_d$ in task space X are set as in Table 2, where damping gains in eq.(3) are chosen as

$$\begin{aligned} &c = 0.0\,(\mathrm{Ns/m}), \\ &c_1 = c_2 = c_3 = c_4 = 0.0025\,(\mathrm{Nms}) \end{aligned} \tag{18}$$

The endpoint trajectory $\boldsymbol{x}(t)$ starting from the initial posture shown as S) in Fig.3 is going to approach the target but overruns and oscillates around the target $\boldsymbol{x}_d$, but in a long run it converges to the target and stops with the final posture shown as U) in Fig.3 when $k = 10.0\,(\text{N/m})$ is chosen.

Table 2. Initial conditions and the target for middle-range reaching

Arm posture	: $q_1(0)=28.15^\circ$, $q_2(0)=51.14^\circ$,
	: $q_3(0)=36.40^\circ$, $q_4(0)=67.82^\circ$
Endpoint position (m)	: $\boldsymbol{x}(0) = (0.10,\ 0.30)^{\mathrm{T}}$
Target position (m)	: $\boldsymbol{x}_d = (-0.20,\ 0.20)^{\mathrm{T}}$
Distance (m)	: $\lVert \Delta\boldsymbol{x}(0) \rVert = 0.3162$

Table 3. Numerical values of inertia matrix $H(q(t))$ at $t = 0$ in the case of middle-range reaching (Unit: kgm^2)

$$H(q(0)) = \begin{bmatrix} 2.776\times10^{-2} & 9.707\times10^{-3} & 4.639\times10^{-4} & -2.021\times10^{-5} \\ 9.707\times10^{-3} & 5.730\times10^{-3} & 4.679\times10^{-4} & 3.091\times10^{-6} \\ 4.639\times10^{-4} & 4.679\times10^{-4} & 1.236\times10^{-4} & 9.210\times10^{-6} \\ -2.021\times10^{-5} & 3.091\times10^{-6} & 9.210\times10^{-6} & 5.892\times10^{-6} \end{bmatrix}$$

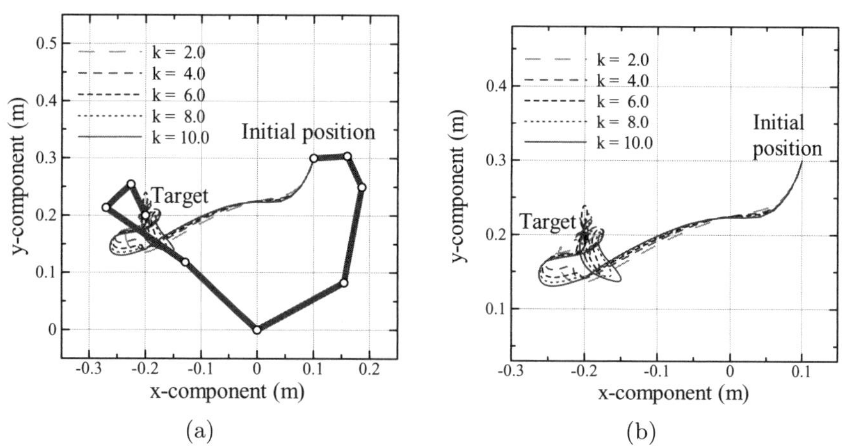

Fig. 5. Endpoint trajectories of multi-joint reaching movements when inadequate damping factors are selected

According to [31], human skilled multi-joint reaching is characterized as follows:

a) The profile of the endpoint trajectory in task space X becomes closely rectilinear,

b) the velocity profile of it in X becomes symmetric and bell-shaped,

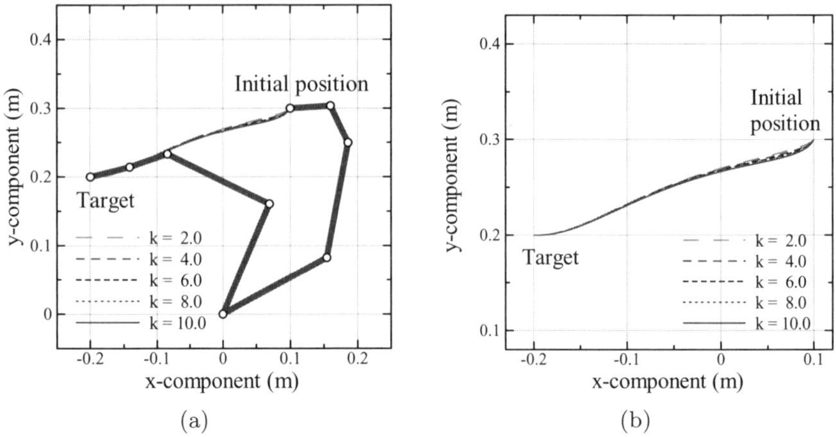

Fig. 6. Endpoint trajectories $\boldsymbol{x}(t)$ in the case of middle-range reaching movements

c) the acceleration profile has double peaks,

d) but each profile of time histories of joint angles $q_i(t)$ and angular velocities $\dot{q}_i(t)$ may differ for $i = 1, 2, \cdots, 4$.

We are now in a position to answer to the question whether it is possible to find a set of adequate damping factors c and c_i $(i = 1, \cdots, 4)$ together with an adequate stiffness parameter $k > 0$ so that the sensory feedback of eq.(2) or eq.(3) leads to the skilled motion of reaching realizing an approximately rectilinear endpoint trajectory without incurring any noteworthy self-motion. We first select damping factors as follows:

$$
\begin{aligned}
&c = 0.0\,(\mathrm{Ns/m}) \\
&c_1 = 0.59, c_2 = 0.38, c_3 = 0.098, c_4 = 0.019\,(\mathrm{Nms})
\end{aligned}
\tag{19}
$$

Then, numerical solutions of the closed-loop dynamics of eq.(6) for different stiffness parameters give rise to transient responses of the endpoint trajectory $(x(t),\, y(t))$ in Fig.6. As shown in Figs.6, the endpoint trajectories become well approximately rectilinear and do not change much for different stiffness parameters, though the speed of convergence to the target depends on k. The best choice of k in this chosen set of damping factors given in eq.(19) must be around $k = 10.0\,(\mathrm{N/m})$.

Now let us discuss how to select such a good set of damping factors as in eq.(19). If one of the tightest (smallest) diagonal matrix H_0 satisfying eq.(5) is found, then it is possible to select C in such a way that

$$
C \geq 3.0 H_0^{1/2} \tag{20}
$$

Further, it should be noted that such a matrix H_0 can be selected as the smallest constant diagonal matrix satisfying

$$
H(q) \leq H_0 \text{ for all } q \text{ such that } \|\boldsymbol{x}(q) - \boldsymbol{x}_d\| \leq r \tag{21}
$$

where r denotes the euclidean distance between the starting endpoint point $\boldsymbol{x}(0)$ $(= (x(0), y(0))^{\mathrm{T}})$ and the target $\boldsymbol{x}_d$, because according to eq.(7) the endpoint should remain inside the circle $\|\boldsymbol{x}(t) - \boldsymbol{x}_d\| \leq r$ for any $t > 0$. In the case of middle-range reaching with $r = 31.62$ (cm) for a typical five-years old child with 1.07 (m) in height, the initial value of $H(q)$ with the posture shown in Fig.2 is evaluated as in Table 3. We evaluate a tighter bound H_0 starting from the data of $H(q(0))$ in Table 3 by adding each possible contribution of off-diagonal element $|H_{ij}(q(0))|$ to $H_{ii}(q(0))$ and $H_{jj}(q(0))$. By the same computer simulation, we find that such a choice of $C = 3.0H_0^{1/2}$ as numerically given in eq.(19) gives rise to

$$9.0C^{-1}H(q)C^{-1} < I_4 \tag{22}$$

during the transient process of reaching.

5 Virtual Spring/Damper Hypothesis

Since the endpoint trajectory in the case of middle-range reaching as shown in Fig.6 is not so close to the straight line, better damping factors were sought for so as to make the endpoint trajectory more closely rectilinear. However, finding out such a set of adequate damping factors for each motion was very laborious. Furthermore, according to physiological data on muscle contraction ([32]), the magnitudes c_1 and c_2 in eq.(19) corresponding to passive damping factors for shoulder and elbow joints are too big to be compared with actual viscosity of contractile muscles ([32, 33]). Therefore, we suggest another form of control signals introduced as in eq.(3). Apparently, this control signal plays a role of a parallel pair of mechanical damper and spring that draws the endpoint of the whole arm to the target as shown in Fig.7. Based on the closed-loop dynamics eq.(6), we conducted numerical simulations of middle-range reaching. For comparison of the effectiveness of control signal shown in eq.(3) with that of eq.(2), the same physical parameters of the arm model, initial posture, and target point as in the simulation of Fig.6 are set as given in Tables 1 and 2. As shown in Fig.8, endpoint trajectories in this case become rectilinear almost completely from the beginning to the final stage even in the case of middle-range reaching movements. In this case, the damping factors C_0 in eq.(3) are set as $C_0 = 0.1C$, that is,

$$C_0 = \mathrm{diag}(0.059,\ 0.038,\ 0.0098,\ 0.0019)\ (\mathrm{Nms}) \tag{23}$$

and the coefficient for dampings in task space c in eq.(3) is set as $c = 4.5$ (Ns/m). Each passive damping factor C_0 corresponding to each joint can be reduced to the range of 10 percent of C. Thus, from simulation results of Fig.6 and Fig.8, it is shown that the control signal of eq.(3) adding the damping effect in task space can lead to movements of more skilled multi-joint reaching than the control signals of eq.(2) in various situations.

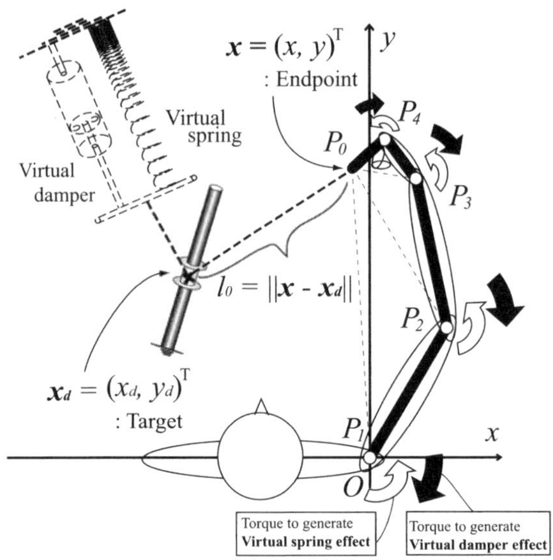

Fig. 7. Virtual spring/damper hypothesis

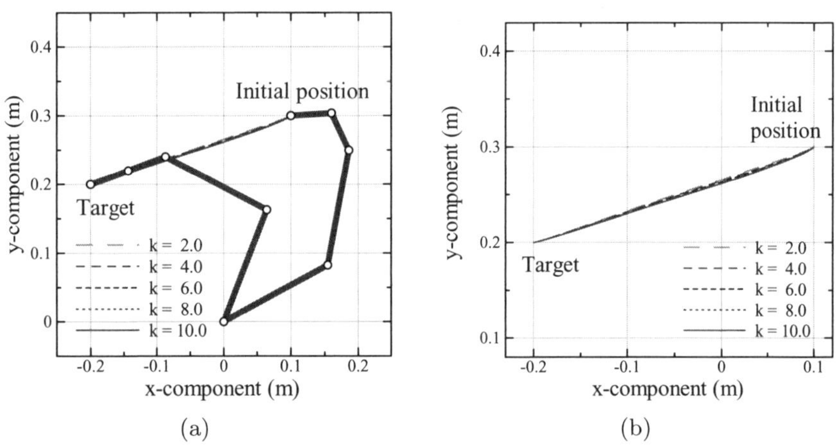

Fig. 8. Endpoint trajectories $\boldsymbol{x}(t)$ in case of middle reaching movements when damping factors based upon "Virtual spring/damper hypothesis" are used

6 Theoretical Proof of Transferability

As argued in Section 2, the Lyapunov-like relation of eq.(7) does neither directly imply the transferability of any neighborhood $N^8(\varepsilon_1, r_1)$ of a reference equilibrium state $(q^0, 0)$ nor its stability on a manifold. Similary to eq.(11), we introduce the quantity

$$X(q(0)) = \int_0^\infty \sqrt{k/2}\|\Delta \boldsymbol{x}(t)\|\mathrm{d}t \tag{24}$$

measured along the endpoint trajectory of the solution to eq.(6) starting from a given initial state $(q(0), \dot{q}(0))$. First, it is necessary to remark that the Lyapunov-like relation of eq.(7) implies only

$$0 \leq \int_0^\infty \left\{\dot{q}^{\mathrm{T}}(t)C_0\dot{q}(t) + c\|\dot{\boldsymbol{x}}(t)\|^2\right\} \mathrm{d}t \leq E(0) \tag{25}$$

that is, $\dot{q}(t) \in L^2(0,\infty)$ and $\dot{\boldsymbol{x}}(t) \in L^2(0,\infty)$, where we denote $E(q(t), \dot{q}(t))$ by $E(t)$ for brevity. It should be also noted that LaSalle's invariance theorem can not be applied for eq.(7) because $E(q, \dot{q})$ is not positive definite in the state $(q, \dot{q})$. However, it is possible to show that $\dot{q}(t) \to 0$ and $\dot{\boldsymbol{x}}(t) \to 0$ as $t \to \infty$ uniformly. In fact, eq.(7) implies $0 \leq E(t) \leq E(0)$ and shows that $\dot{q}(t)$ and $\Delta \boldsymbol{x}(t)$ are uniformly bounded. Hence, from eq.(6) $\ddot{q}(t)$ must be also uniformly bounded. This means that $\dot{q}(t)$ is uniformly continuous in t and also belongs to $L^2(0,\infty)$, which implies $\dot{q}(t) \to 0$ as $t \to \infty$ according to a well-known lemma (for example see Appendix C of the book ([29])). Similarly $\Delta \boldsymbol{x}(t) \to 0$ as $t \to \infty$. Nevertheless, this result does not imply the finiteness of metrics $R(q(0))$ defined by eq.(11) and $X(q(0))$ of eq.(24) for a given target $\boldsymbol{x} = \boldsymbol{x}_d$. Thus, it is crucial to gain a deep insight into the physical properties of Γ_{ijk} in eq.(13), or equivalently the second term $\{(1/2)\dot{H} + S\}\dot{q}$ of the left hand side of eq.(4). In particular, it is important to analyze physical interactions between the term $kJ^{\mathrm{T}}(q)\Delta \boldsymbol{x}$ exerted from the external spring-like force $k\Delta \boldsymbol{x}$ and the torque contributed from the inertia term in eq.(6). In the previous paper ([34]), we introduced another quadratic function defined as

$$\begin{aligned} W = & \frac{1}{2}\dot{q}^{\mathrm{T}}H(q)\dot{q} + \frac{k}{2}(1+\alpha\zeta)\|\Delta \boldsymbol{x}\|^2 \\ & + \alpha\sqrt{k}\Delta \boldsymbol{x}^{\mathrm{T}}(J^+(q))^{\mathrm{T}}H(q)\dot{q} \end{aligned} \tag{26}$$

with a positive parameter $1 \geq \alpha > 0$ and showed that $(1-\alpha)E \leq W \leq (1+\alpha)E$ and there exist a positive constant $\gamma(\alpha)$ dependening on $\alpha > 0$ such that

$$W(t) \leq e^{-\gamma(\alpha)t}W(0) \tag{27}$$

and

$$E(t) \leq \frac{e^{-\gamma(\alpha)t}}{1-\alpha}W(0) \leq \frac{1+\alpha}{1-\alpha}e^{-\gamma(\alpha)t}E(0) \tag{28}$$

where we set $\zeta = c/\sqrt{k}$. This shows that $E(t)$ converges to zero exponentially in t. Therefore, $\sqrt{E(t)}$ is also convergent to zero as $t \to \infty$ with exponent $\{-\gamma(\alpha)/2\}t$. Thus, it is now possible to prove the stability on a manifold and transferability of solution trajectories of eq.(6).

Proposition 1. For a given reference state $(q^0, 0)$ lying on M_2 with satisfying $\boldsymbol{x}(q^0) = \boldsymbol{x}_d$, then any solution trajectory of eq.(6) starting from an arbitrary initial state $(q(0), \dot{q}(0))$ belonging to a smaller neighborhood $N^8(\varepsilon_1, r_1)$ of $(q^0, 0)$ with some $\varepsilon_1 < \rho$ and $r_1 < r_0$ remains to stay in the neighborhood $N^8(\rho, r_0)$ of

$(q^0, 0)$. Moreover, the positional solution trajectory converges asymptotically to some point of M_2. That is, $N^8(\varepsilon_1, r_1)$ is transferable to M_2.

Proof. Since $\sqrt{E(t)}$ is exponentially convergent to zero according to eq.(28), we can set

$$\begin{cases} R(q(0)) \le d_R\sqrt{E(0)} \\ X(q(0)) \le d_R\sqrt{E(0)} \end{cases} \tag{29}$$

with a common constant $d_R > 0$. Then, we choose $r_1 > 0$ and $\delta(\varepsilon)$ in such a way that

$$r_1 < \frac{r_0}{2}, \quad \varepsilon_1 < \min\left\{\frac{r_0}{2d_R}, \rho\right\} \tag{30}$$

Consider a solution $(q(t), \dot{q}(t))$ of eq.(6) starting from an aribitrary state $(q(0), \dot{q}(0))$ inside the neighborhood $N^8(\varepsilon_1, r_1)$ of $(q^0, 0)$ and notice that at any position $q(T)$ along the position trajectory $q(t)$ of this solution it follows that

$$\begin{aligned} R(q(T), q^0) &\le R(q(T), q(0)) + R(q(0), q^0) \\ &\le R(q(T), q(0)) + r_1 \end{aligned} \tag{31}$$

Since the position trajectory $q(t)$ starting from $q(0)$ at $t = 0$ and reaching $q(T)$ at $t = T$ is a special curve in the configuration space, it follows from the meaning of Riemannian distance that

$$\begin{aligned} R(q(T), q(0)) &\le \int_0^T \sqrt{\frac{1}{2}\dot{q}^{\mathrm{T}} H(q(t))\dot{q}(t)}\mathrm{d}t \\ &\le \int_0^T \sqrt{E(t)}\mathrm{d}t \le d_R\sqrt{E(0)} \\ &= d_R\sqrt{E(q(0), \dot{q}(0))} \le d_R\varepsilon_1 \end{aligned} \tag{32}$$

Substituting this into eq.(31) and referring to eq.(30) conclude that

$$R(q(T), q^0) \le d_R\varepsilon_1 + r_1 < r_0 \tag{33}$$

At the same time, eq.(30) implies $E(q(t), \dot{q}(t)) \le E(0) = \varepsilon_1^2 < \rho^2$. This completes the proof.

In the above proof, if we choose $\delta(\varepsilon)$ for an arbitrarily given $\varepsilon > 0$ in such a way that

$$\delta(\varepsilon) = \min\left\{\frac{r_0}{2d_R}, \rho, \varepsilon\right\} \tag{34}$$

then it is possible to show that

Theorem 1. If $(q^0, 0) \in M_2$ and at any pose q satisfying $R(q, q^0) < r_0$ the Jacobian matrix $J(q) = \partial\boldsymbol{x}(q)/\partial q^{\mathrm{T}}$ is nondegenerate, then the reference equilibrium state $(q^0, 0)$ is stable on a manifold.

Theorem 2. Under the same assumption on q^0, there exists a pair of positive numbers (ε_1, r_1) such that the neighborhood $N^8(\varepsilon_1, r_1)$ of q^0 is transferable to a subset of M_2.

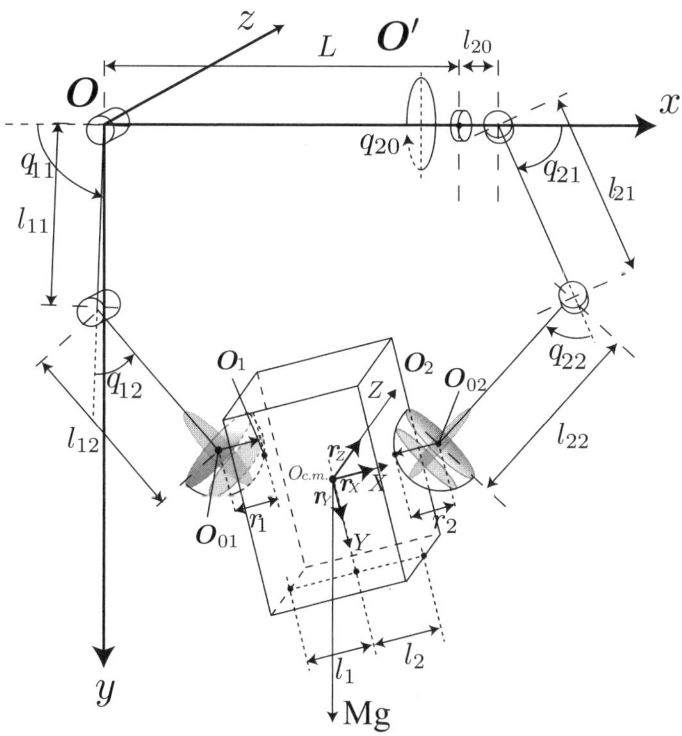

Fig. 9. The coordinates of the overall fingers-object system. The coordinates O_{xyz} is fixed at the frame.

7 3-D Grasping Under Nonholonomic Constraints

Next, we will discuss a problem of modeling physical interactions of a 3D-object by a pair of robot fingers with multi-joint whose ends are rigid and hemispherical (see Fig.9). Consider motion of a rigid object with parallel flat surfaces, which is grasped by a pair of robot fingers with 3 DOFs and 3 DOFs as shown in Fig.9. When the distance from the straight line $\overrightarrow{O_1O_2}$ (opposition axis) connecting two contact points between finger-ends and object surfaces to the vertical axis through the object mass center in the direction of gravity becomes large, there arises a spinning motion of the object around that opposition axis. This paper considers the problem of modeling of pinching in the situation that this spinning motion ceases after that the center of mass of the object came sufficiently close to a point beneath the opposing axis and there will no more arise such spinning due to dry friction between finger-ends and object surfaces. The cease of spinning, however, induces a non-holonomic constraint among rotational angular velocities ω_X, ω_Y and ω_Z around X, Y, and Z axis, respectively, where $O_{c.m.} - XYZ$ denotes the cartesian coordinates fixed at the object as shown in Fig.10. Since the opposing axis is expressed as $\overrightarrow{O_1O_2} = \boldsymbol{x}_1 - \boldsymbol{x}_2$ where $\boldsymbol{x}_i = (x_i, y_i, z_i)^{\mathrm{T}}$ denotes

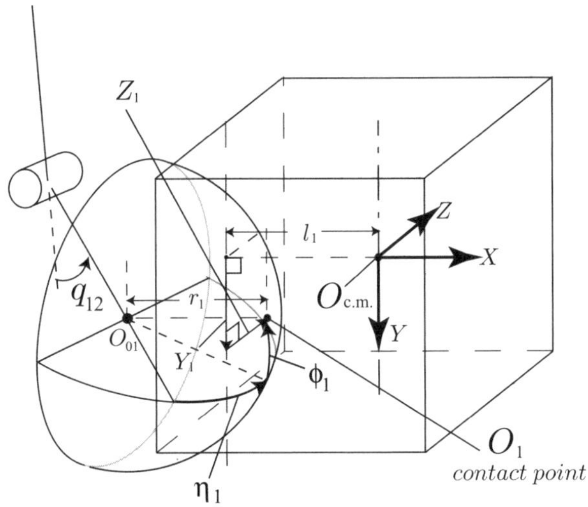

Fig. 10. Mutually orthogonal unit vectors $\boldsymbol{r}_X, \boldsymbol{r}_Y$ and $\boldsymbol{r}_Z$ express the rotational motion of the object. The pair (η_i, ϕ_i) for $(i = 1, 2)$ expresses the spherical coordinates of each hemispherical finger end.

the cartesian coordinates of contact point O_i (see Fig.10), the cease of spinning motion around the axis $\overrightarrow{O_1O_2}$ implies that the instantaneous axis of rotation of the object is orthogonal to $\boldsymbol{x}_1 - \boldsymbol{x}_2$, that is,

$$\boldsymbol{\omega}^{\mathrm{T}}(\boldsymbol{x}_1 - \boldsymbol{x}_2) = 0 \tag{35}$$

which is rewritten into the form

$$\omega_X = -\xi_y \omega_Y - \xi_z \omega_Z \tag{36}$$

$$\xi_y = \frac{y_1 - y_2}{x_1 - x_2}, \qquad \xi_z = \frac{z_1 - z_2}{x_1 - x_2} \tag{37}$$

where $\boldsymbol{\omega} = (\omega_X, \omega_Y, \omega_Z)^{\mathrm{T}}$. On the other hand, denote the cartesian coordinates of the object mass center $O_{c.m.}$ by $\boldsymbol{x} = (x, y, z)^{\mathrm{T}}$ based on the frame coordinates $O - xyz$ and three mutually orthogonal unit vectors fixed at the object frame by $\boldsymbol{r}_X, \boldsymbol{r}_Y$ and $\boldsymbol{r}_Z$. Then, the 3×3 rotation matrix

$$R(t) = (\boldsymbol{r}_X, \boldsymbol{r}_Y, \boldsymbol{r}_Z) \tag{38}$$

belongs to SO(3) and is subject to the first-order differential equation

$$\frac{\mathrm{d}}{\mathrm{d}t} R(t) = R(t) \Omega(t) \tag{39}$$

where

$$\Omega(t) = \begin{pmatrix} 0 & -\omega_Z & \omega_Y \\ \omega_Z & 0 & -\omega_X \\ -\omega_Y & \omega_X & 0 \end{pmatrix} \tag{40}$$

The equation (39) expresses another nonholonomic constraint on rotational motion of the object. Next, denote the position of the center of each hemispherical finger-end by $\boldsymbol{x}_{0i} = (x_{0i}, y_{0i}, z_{0i})^{\mathrm{T}}$. Then, it is possible to notice that (see Fig.10)

$$\boldsymbol{x}_i = \boldsymbol{x}_{0i} - (-1)^i r_i \boldsymbol{r}_X \tag{41}$$

$$\boldsymbol{x} = \boldsymbol{x}_{0i} - (-1)^i (r_i + l_i)\boldsymbol{r}_X - Y_i \boldsymbol{r}_Y - Z_i \boldsymbol{r}_Z \tag{42}$$

Since each contact point O_i can be expressed by the coordinates $((-1)^i l_i, Y_i, Z_i)$ based on the object frame $O_{c.m.}-XYZ$, taking an inner product between eq.(42) and $\boldsymbol{r}_X$ gives rise to

$$Q_i = -(r_i + l_i) - (-1)^i (\boldsymbol{x} - \boldsymbol{x}_{0i})^{\mathrm{T}} \boldsymbol{r}_X = 0, \quad i = 1, 2 \tag{43}$$

which express holonomic constraints of contacts between finger-ends and the object. The rolling constraints between finger-ends and object surfaces can be expressed by equalities of two contact point velocities expressed on the finger-end surfaces and on the object surfaces in terms of orthogonal coordinates $(\boldsymbol{r}_Y, \boldsymbol{r}_Z)$. In fact, they are expressed as

$$(-1)^i r_i \boldsymbol{r}_Y^{\mathrm{T}} \left\{ \dot{\boldsymbol{r}}_X - (\boldsymbol{e}_z \times \boldsymbol{r}_X)(\dot{q}_i^{\mathrm{T}} \boldsymbol{e}_i) - (\boldsymbol{e}_x \times \boldsymbol{r}_X)\dot{q}_{i0} \right\} = \frac{\mathrm{d}}{\mathrm{d}t} Y_i \tag{44}$$

$$(-1)^i r_i \boldsymbol{r}_Z^{\mathrm{T}} \left\{ \dot{\boldsymbol{r}}_X - (\boldsymbol{e}_z \times \boldsymbol{r}_X)(\dot{q}_i^{\mathrm{T}} \boldsymbol{e}_i) - (\boldsymbol{e}_x \times \boldsymbol{r}_X)\dot{q}_{i0} \right\} = \frac{\mathrm{d}}{\mathrm{d}t} Z_i \tag{45}$$

where $\boldsymbol{e}_1 = (1,1)^{\mathrm{T}}$, $\boldsymbol{e}_2 = (0,1,1)^{\mathrm{T}}$, $\boldsymbol{e}_z = (0,0,1)^{\mathrm{T}}$, $\boldsymbol{e}_x = (1,0,0)^{\mathrm{T}}$, and $\dot{q}_{10} = 0$ since the left finger has no joint in the x-axis. These four equalities expressing rolling constraints are non-holonomic, but they are Pfaffian since they can be expressed in linear homogeneous forms of velocity variables as follows:

$$-(-1)^i r_i \left\{ \dot{\theta} + r_{Zz}\dot{p}_i + r_{Zx}\dot{q}_{i0} \right\} = \dot{Y}_i, \quad i = 1, 2 \tag{46}$$

$$(-1)^i r_i \left\{ \dot{\phi} - r_{Yz}\dot{p}_i - r_{Yx}\dot{q}_{i0} \right\} = \dot{Z}_i, \quad i = 1, 2 \tag{47}$$

where $p_i = q_i^{\mathrm{T}} \boldsymbol{e}_i$ for $i = 1, 2$ and we conveniently express $\omega_Z = \dot{\theta}$ and $\omega_Y = \dot{\psi}$ (i.e., θ signifies an indefinite integral of ω_Z and ψ does that of ω_Y), and r_{Zz} denotes the z-component of $\boldsymbol{r}_Z$, and r_{Zx} and others have a similar meaning. Note that $\dot{\boldsymbol{r}}_X$ in eqs.(44) and (45) represents the angular velocities of the object surfaces (YZ-planes), $\dot{\boldsymbol{r}}_X = -\omega_Z \boldsymbol{r}_Y + \omega_Y \boldsymbol{r}_Z$ from eq.(40), $(\boldsymbol{e}_z \times \boldsymbol{r}_X)(\dot{q}_i^{\mathrm{T}} \boldsymbol{e}_i)$ expresses the velocity of the contact point O_i on the finger-end surface induced by the net angular velocity of all joints with the z-axis of finger i, and $(\boldsymbol{e}_x \times \boldsymbol{r}_X)\dot{q}_{i0}$ does a similar meaning. Thus, the rolling constraints can be expressed with accompany of Lagrange's multipliers λ_{Yi} for eq.(44) and λ_{Zi} for eq.(45) such that

$$\lambda_{Yi} \left\{ \boldsymbol{Y}_{qi}^{\mathrm{T}} \dot{q}_i + \boldsymbol{Y}_{\boldsymbol{x}i}^{\mathrm{T}} \dot{\boldsymbol{x}} + Y_{\theta i}\dot{\theta} + Y_{\psi i}\dot{\psi} \right\} = 0 \tag{48}$$

$$\lambda_{Zi} \left\{ \boldsymbol{Z}_{qi}^{\mathrm{T}} \dot{q}_i + \boldsymbol{Z}_{\boldsymbol{x}i}^{\mathrm{T}} \dot{\boldsymbol{x}} + Z_{\theta i}\dot{\theta} + Z_{\psi i}\dot{\psi} \right\} = 0 \tag{49}$$

where $\boldsymbol{Y}_{qi}$, $\boldsymbol{Z}_{qi}$, and others are shown in Table 4. Corresponding to eq.(43), we introduce the scalar quantity with multipliers f_1 and f_2 as follows:

$$Q = f_1 Q_1 + f_2 Q_2 = 0 \tag{50}$$

The Lagrangian for the overall fingers-object system can now be expressed by the scalar quantity

$$L = K - P + Q \tag{51}$$

where K denotes the total kinetic energy expressed as

$$\begin{aligned} K = & \frac{1}{2} \sum_{i=1,2} \dot{q}_i^{\mathrm{T}} H_i(q_i) \dot{q}_i + \frac{1}{2} M \left(\dot{x}^2 + \dot{y}^2 + \dot{z}^2 \right) \\ & + \frac{1}{2} (\omega_Z, \omega_Y) H_0 (\omega_Z, \omega_Y)^{\mathrm{T}} \end{aligned} \tag{52}$$

and P denotes the total potential energy expressed as

$$P = P_1(q_1) + P_2(q_2) - Mgy \tag{53}$$

where $H_i(q_i)$ stands for the inertia matrix for finger i, M the mass of the object, $P_i(q_i)$ the potential energy of finger i, g the gravity constant and H_0 is given in the following:

$$H_0 = \begin{pmatrix} \begin{matrix} I_{ZZ} + \xi_z^2 I_{XX} \\ -2\xi_z I_{ZX} \end{matrix} & \begin{matrix} I_{YZ} + \xi_y \xi_z I_{XX} \\ -\xi_z I_{YX} - \xi_y I_{XZ} \end{matrix} \\ & \\ \begin{matrix} I_{YZ} + \xi_y \xi_z I_{XX} \\ -\xi_z I_{YX} - \xi_y I_{XZ} \end{matrix} & \begin{matrix} I_{YY} + \xi_y^2 I_{XX} \\ -2\xi_y I_{YX} \end{matrix} \end{pmatrix} \tag{54}$$

provided that the inertia matrix of the object around its mass center $O_{c.m.}$ is expressed as

$$H = \begin{pmatrix} I_{XX} & I_{XY} & I_{XZ} \\ I_{XY} & I_{YY} & I_{YZ} \\ I_{XZ} & I_{YZ} & I_{ZZ} \end{pmatrix} \tag{55}$$

It should be noted that the kinetic energy of the object can be expressed as $K_0 = (1/2)\boldsymbol{\omega}_0^{\mathrm{T}} H_0 \boldsymbol{\omega}_0 = (1/2)\boldsymbol{\omega}^{\mathrm{T}} H \boldsymbol{\omega}$ under the nonholonomic constraint of eq.(35), where $\boldsymbol{\omega}_0 = (\omega_Z, \omega_Y)^{\mathrm{T}}$.

8 Lagrange's Equation of the Overall Fingers-Object System

By applying the variational principle

$$\int_{t_0}^{t_1} \left\{ \delta L - u_1^{\mathrm{T}} \delta q_1 - u_2^{\mathrm{T}} \delta q_2 \right\} \mathrm{d}t = \int_{t_0}^{t_1} \sum_{i=1,2} \left\{ \lambda_{Yi} \boldsymbol{F}_{Yi}^{\mathrm{T}} \delta \boldsymbol{X} + \lambda_{Zi} \boldsymbol{F}_{Zi}^{\mathrm{T}} \delta \boldsymbol{X} \right\} \mathrm{d}t \tag{56}$$

Table 4. Partial derivatives of holonomic constraints with respect to position variables

$$\frac{\partial Q}{\partial q_i} = f_i \left(\frac{\partial Q_i}{\partial q_i} \right) = (-1)^i f_i J_i^{\mathrm{T}}(q_i) \boldsymbol{r}_X, \quad J_i(q_i) = \frac{\partial \boldsymbol{x}_{0i}}{\partial q_i^{\mathrm{T}}}, \qquad i = 1, 2 \tag{T-1}$$

$$\frac{\partial Q}{\partial \boldsymbol{x}} = f_1 \left(\frac{\partial Q_1}{\partial \boldsymbol{x}} \right) + f_2 \left(\frac{\partial Q_2}{\partial \boldsymbol{x}} \right) = (f_1 - f_2) \boldsymbol{r}_X \tag{T-2}$$

$$\frac{\partial Q}{\partial \theta} = f_1 \frac{\partial Q_1}{\partial \theta} + f_2 \frac{\partial Q_2}{\partial \theta} = -f_1 Y_1 + f_2 Y_2 \tag{T-3}$$

$$\frac{\partial Q}{\partial \psi} = f_1 \frac{\partial Q_1}{\partial \psi} + f_2 \frac{\partial Q_2}{\partial \psi} = f_1 Z_1 - f_2 Z_2 \tag{T-4}$$

$$\boldsymbol{Y}_{qi} = J_i^{\mathrm{T}}(q_i) \boldsymbol{r}_Y + r_i \left\{ (-1)^i r_{Zz} \boldsymbol{e}_i + r_{Zx} \boldsymbol{e}_{0i} \right\} \tag{T-5}$$

$$\boldsymbol{Z}_{qi} = J_i^{\mathrm{T}}(q_i) \boldsymbol{r}_Z + r_i \left\{ (-1)^i r_{Yz} \boldsymbol{e}_i + r_{Yx} \boldsymbol{e}_{0i} \right\} \tag{T-6}$$

$$\boldsymbol{Y}_{\boldsymbol{x}i} = -\boldsymbol{r}_Y, \quad \boldsymbol{Z}_{\boldsymbol{x}i} = -\boldsymbol{r}_Z \tag{T-7}$$

$$Y_{\theta i} = \frac{\partial Y_i}{\partial \theta} + (-1)^i r_i = (-1)^i l_i - \xi_z Z_i \tag{T-8}$$

$$Y_{\psi i} = \frac{\partial Y_i}{\partial \psi} = -\xi_y Z_i, \quad Z_{\theta i} = \frac{\partial Z_i}{\partial \theta} = \xi_z Y_i \tag{T-9}$$

$$Z_{\psi i} = \frac{\partial Z_i}{\partial \psi} - (-1)^i r_i = (-1)^i l_i + \xi_y Y_i \tag{T-10}$$

to the Lagrangian L defined by eq. (51), we obtain

$$H\ddot{\boldsymbol{X}} + \left(\frac{1}{2} \dot{H} + S + C \right) \dot{\boldsymbol{X}} - A\boldsymbol{\lambda} + Mg\boldsymbol{e} = \boldsymbol{u} \tag{57}$$

where $\boldsymbol{e} = (0_6, 1, 0_3)^{\mathrm{T}}$, $\quad \boldsymbol{\lambda} = (f_1, f_2, \lambda_{Y1}, \lambda_{Y2}, \lambda_{Z1}, \lambda_{Z2})^{\mathrm{T}}$, $\boldsymbol{u} = (u_1^{\mathrm{T}}, u_2^{\mathrm{T}}, 0_5)^{\mathrm{T}}$,

$$\left\{ \begin{array}{l} \boldsymbol{X} = (q_1^{\mathrm{T}}, q_2^{\mathrm{T}}, \boldsymbol{x}^{\mathrm{T}}, \theta, \psi)^{\mathrm{T}}, \quad C = \mathrm{diag}(C_1, C_2, 0_5) \\ H = \begin{pmatrix} H_1 & 0_{2\times3} & 0_{2\times3} & 0_{2\times2} \\ 0_{3\times2} & H_2 & 0_{3\times3} & 0_{3\times2} \\ 0_{3\times2} & 0_{3\times3} & Mr^2 I_3 & 0_{3\times2} \\ 0_{2\times2} & 0_{2\times3} & 0_{2\times3} & H_0 \end{pmatrix} \\ S = \begin{pmatrix} S_1 & 0_{2\times3} & 0_{2\times3} & S_{14} \\ 0_{3\times2} & S_2 & 0_{3\times3} & S_{24} \\ 0_{3\times2} & 0_{3\times3} & 0_{3\times3} & 0_{3\times2} \\ -S_{14}^{\mathrm{T}} & -S_{24}^{\mathrm{T}} & 0_{2\times3} & S_0 \end{pmatrix} \end{array} \right. \tag{58}$$

and

$$A = \begin{pmatrix} -J_1^{\mathrm{T}} \boldsymbol{r}_X & 0_2 & \boldsymbol{Y}_{q1} & 0_2 & \boldsymbol{Z}_{q1} & 0_2 \\ 0_3 & J_2^{\mathrm{T}} \boldsymbol{r}_X & 0_3 & \boldsymbol{Y}_{q2} & 0_3 & \boldsymbol{Z}_{q2} \\ \boldsymbol{r}_X & -\boldsymbol{r}_X & -\boldsymbol{r}_Y & -\boldsymbol{r}_Y & -\boldsymbol{r}_Z & -\boldsymbol{r}_Z \\ -Y_1 & Y_2 & -l_1 - \xi_z Z_1 & l_2 - \xi_z Z_2 & -\xi_y Z_1 & -\xi_y Z_2 \\ Z_1 & -Z_2 & \xi_z Y_1 & \xi_z Y_2 & -l_1 + \xi_y Y_1 & l_2 + \xi_y Y_2 \end{pmatrix} \tag{59}$$

where $\boldsymbol{F}_{Y1}$ and $\boldsymbol{F}_{Z1}$ are defined as 10-dimensional vector-valued functions appearing in the third and fifth columns of matrix A and $\boldsymbol{F}_{Y2}$ and $\boldsymbol{F}_{Z2}$ the fourth and sixth columns respectively, and

$$S_0 = \begin{pmatrix} 0 & s_{12} \\ -s_{12} & 0 \end{pmatrix}, \quad S_{i4} = -\frac{1}{2}\left\{\frac{\partial \boldsymbol{\omega}_0^{\mathrm{T}} H_0}{\partial q_i}\right\}$$

$$s_{12} = \frac{1}{2}\left(\frac{\partial h_{11}}{\partial \psi} - \frac{\partial h_{12}}{\partial \theta}\right)\omega_Z + \frac{1}{2}\left(\frac{\partial h_{12}}{\partial \psi} - \frac{\partial h_{22}}{\partial \theta}\right)\omega_Y \tag{60}$$

where we denote the (i,j)-entry of H_0 by h_{ij}. Obviously from the variational principle of eq.(56), it follows that

$$\int_0^t \left\{\sum_{i=1,2} \dot{q}_i^{\mathrm{T}} u_i\right\} \mathrm{d}\tau = K(t) + P(t) - K(0) - P(0) \tag{61}$$

This relation can be utilized later in derivation of the passivity for a class of closed-loop dynamics when control signals u_i are designed in a form of function of state variables of robot fingers. Note that dynamics of the object exressed in eq.(57) can not be controlled directly from the control u_i $(i = 1, 2)$ but must be controlled indirectly through constraint forces f_i, λ_{Zi}, and λ_{Yi} $(i = 1, 2)$. It is also important to note that the position variables θ and ψ do not appear explicitly in eq.(59) which expresses the Lagrange equation for rotational motion of the object, because they (θ and ψ) do not appear in the right hand sides of (T-8) to (T-10). These variables (θ and ψ) should be determined on the basis of nonholonomic constraint expressed by eq.(39). Thus, it can be claimed that the overall system dynamics of eq.(57) together with the nonholonomic constraint of eq.(39) does not contradict the causality. Furthermore, it should be remarked that the use of Pfaffian constraints of eqs.(46) and (47) instead of nonholonomic expessions of eqs.(44) and (45) must be reasonable under the argument of "infinitesimal rotation" (see [35]) of the object and thereby partial derivatives of constraints $Q_i = 0$ $(i = 1, 2)$ and eqs.(46) and (47) in θ and ψ can be obtained through partial derivatives of $\boldsymbol{r}_X$, $\boldsymbol{r}_Y$, and $\boldsymbol{r}_Z$ in θ and ψ under the nonholonomic constraint of eq.(35), which are given as follows (see [36]):

$$\begin{cases} \dfrac{\partial \boldsymbol{r}_X}{\partial \theta} = \boldsymbol{r}_Y, \quad \dfrac{\partial \boldsymbol{r}_X}{\partial \psi} = -\boldsymbol{r}_Z \\ \dfrac{\partial \boldsymbol{r}_Y}{\partial \theta} = -\boldsymbol{r}_X - \xi_z \boldsymbol{r}_Z, \quad \dfrac{\partial \boldsymbol{r}_Y}{\partial \psi} = -\xi_y \boldsymbol{r}_Z \\ \dfrac{\partial \boldsymbol{r}_Z}{\partial \theta} = \xi_z \boldsymbol{r}_Y, \quad \dfrac{\partial \boldsymbol{r}_Z}{\partial \psi} = \boldsymbol{r}_X + \xi_y \boldsymbol{r}_Y \end{cases} \tag{62}$$

9 Control for Stable Blind Grasping

Even if the gravity affects motion of the object and fingers as shown in Fig.9, it is possible to construct a control signal for stable grasping in a dynamic sense without knowing object kinematics or using external sensing (visual or tactile). The signal is defined as

$$\begin{aligned} u_i = g_i(q_i) - C_i\dot{q}_i + \frac{(-1)^i f_d}{r_1+r_2} J_i^{\mathrm{T}}(q_i)(\boldsymbol{x}_{01} - \boldsymbol{x}_{02}) \\ -\frac{\hat{M}g}{2}\frac{\partial y_{0i}}{\partial q_i} - r_i\hat{N}_i\boldsymbol{e}_i - r_i\hat{N}_{0i}\boldsymbol{e}_{0i}, \quad i=1,2 \end{aligned} \tag{63}$$

where

$$\hat{M}(t) = \hat{M}(0) + \frac{g}{2\gamma_M}\sum_{i=1,2}(y_{0i}(t) - y_{0i}(0)) \tag{64}$$

$$\hat{N}_i(t) = \frac{r_i}{\gamma_i}\sum_{j=1}^{2}\{q_{ij}(t) - q_{ij}(0)\}, \quad \hat{N}_{02}(t) = \frac{r_2}{\gamma_0}(q_{20}(t) - q_{20}(0)) \tag{65}$$

Note that $\hat{M}(t)$ plays a role of estimator for the unknown object mass M. Signals $\hat{N}_i(t)$ $(i=1,2)$ and $\hat{N}_{02}$ are not any of estimators but play an important role in suppressing excess movements of rotation of finger joints together with damping terms $-C_i\dot{q}_i$.

Before discussing stability of closed-loop system that can be obtained by substituting eq. (63) into eq. (57), we need the following definitions:

$$\begin{cases} \Delta f_i = f_i - f_0 - (-1)^i \dfrac{Mg}{2} r_{Xy} \\ \Delta\lambda_{Yi} = \lambda_{Yi} + (-1)^i \dfrac{f_d(Y_1 - Y_2)}{r_1+r_2} - \dfrac{Mg}{2} r_{Yy} \\ \Delta\lambda_{Zi} = \lambda_{Zi} + (-1)^i \dfrac{f_d(Z_1 - Z_2)}{r_1+r_2} - \dfrac{Mg}{2} r_{Zy} \end{cases} \tag{66}$$

$$\Delta M = \hat{M} - M, \quad \Delta N_i = \hat{N}_i - N_i, \quad \Delta N_0 = \hat{N}_{02} - N_0 \tag{67}$$

$$N_i = (-1)^i\left\{\frac{Mg}{2}(r_{Zz} + r_{Yz}) - \frac{f_d}{r_1+r_2}(r_{Zz}(Y_1 - Y_2) - r_{Yz}(Z_1 - Z_2))\right\} \tag{68}$$

$$N_{02} = \frac{Mg}{2}(r_{Zx} + r_{Yx}) - \frac{f_d}{r_1+r_2}\{r_{Zx}(Y_1 - Y_2) - r_{Yz}(Z_1 - Z_2)\} \tag{69}$$

and describe the closed-loop equation in the following vector-matrix form:

$$H\ddot{\boldsymbol{X}} + \left(\frac{1}{2}\dot{H} + S + C\right)\dot{\boldsymbol{X}} - A\Delta\boldsymbol{\lambda} - B\Delta\boldsymbol{m} = 0 \tag{70}$$

where $\Delta\boldsymbol{\lambda} = (\Delta f_1, \Delta f_2, \Delta\lambda_{Y1}, \Delta\lambda_{Y2}, \Delta\lambda_{Z1}, \Delta\lambda_{Z2})^{\mathrm{T}}$, $\Delta\boldsymbol{m} = (\Delta Mg/2, \Delta N_1, \Delta N_2, \Delta N_0, r_\theta^{-1}S_Z, r_\psi^{-1}S_Y)^{\mathrm{T}}$, $\boldsymbol{e}_\theta = (1,0)^{\mathrm{T}}$, $\boldsymbol{e}_\psi = (0,1)^{\mathrm{T}}$, and

$$B = \begin{pmatrix} \dfrac{\partial y_{01}}{\partial q_1} & r_1 \boldsymbol{e}_1 & 0_2 & 0_2 & 0_2 & 0_2 \\ \dfrac{\partial y_{02}}{\partial q_2} & 0_3 & r_2 \boldsymbol{e}_2 & r_2 \boldsymbol{e}_{02} & 0_3 & 0_3 \\ & 0_{3\times 4} & & & 0_{3\times 2} & \\ & 0_{2\times 4} & & & r_\theta \boldsymbol{e}_\theta & r_\psi \boldsymbol{e}_\psi \end{pmatrix} \tag{71}$$

$$S_Z = \partial W/\partial \theta, \quad S_Y = \partial W/\partial \psi \tag{72}$$

and W stands for

$$\begin{aligned} W = & \frac{f_d}{2(r_1+r_2)}\left\{(Y_1-Y_2)^2+(Z_1-Z_2)^2\right\}+\frac{\gamma_M}{2}\Delta M^2 \\ & +\frac{Mg}{2}\Big[(Y_1+Y_2)r_{Yy}-\{(l_1+r_1)-(l_2+r_2)\}r_{Xy} \\ & +(Z_1+Z_2)r_{Zy}\Big]+\sum_{i=1,2}\frac{\gamma_i}{2}\hat{N}_i^2+\frac{\gamma_0}{2}\hat{N}_{02}^2 \end{aligned} \tag{73}$$

In derivation of the form of eq.(70) the relations

$$\frac{Mg}{2}J_i^{\mathrm{T}}(q_i)\left\{r_{Xy}\boldsymbol{r}_X+r_{Yy}\boldsymbol{r}_Y+r_{Zy}\boldsymbol{r}_Z\right\}=\frac{Mg}{2}\left(\frac{\partial y_{0i}}{\partial q_i}\right) \tag{74}$$

and

$$\begin{aligned} \sum_{i=1,2}\dot{q}_i^{\mathrm{T}}u_i = & \frac{\mathrm{d}}{\mathrm{d}t}\Bigg\{P_1+P_2-\frac{f_d}{2(r_1+r_2)}\|\boldsymbol{x}_{01}-\boldsymbol{x}_{02}\|^2 \\ & -\frac{\gamma_M}{2}\hat{M}^2-\sum_{i=1,2}\frac{\gamma_i}{2}\hat{N}_i^2-\frac{\gamma_0}{2}\hat{N}_0^2\Bigg\}+\sum_{i=1,2}\dot{q}_i^{\mathrm{T}}C_i\dot{q}_i \end{aligned} \tag{75}$$

play a crucial role. From these two relations it is possible to verify that

$$\frac{\mathrm{d}}{\mathrm{d}t}(K+W)=-\sum_{i=1,2}\dot{q}_i^{\mathrm{T}}C_i\dot{q}_i \tag{76}$$

At the same time it is important to note from eq.(74) that the third term of the right hand side of W in eq.(73) is equivalent to

$$\begin{aligned} & \frac{Mg}{2}\Big[(Y_1+Y_2)r_{Yy}-\{(l_1+r_1)-(l_2+r_2)\}r_{Xy}+(Z_1+Z_2)r_{Zy}\Big] \\ & \qquad = Mg\left\{y-\frac{y_{01}+y_{02}}{2}\right\} \end{aligned} \tag{77}$$

This means physically that the unknown potential energy $(-Mgy)$ is approximated by the quantity $\hat{M}g(y_{01}+y_{02})/2$ that can be calculated by using measured data on finger joint angles.

Now, consider minimization of the scalar function W in position variables $\boldsymbol{X}$ under constraints of eqs.(43), (46), and (47). Since $\boldsymbol{X}$ is of 10-dimension and there are two holonomic constraints, the following five variables

$$p_1 = q_{11} + q_{12}, \quad p_2 = q_{21} + q_{22}, \quad q_{20}, \quad \theta, \quad \psi \tag{78}$$

are independent and sufficient for minimization of W under the constraints. Then, it is possible to check that the Hessian matrix of W in p_1, p_2, q_{20}, θ, and ψ becomes positive definite in a broad region in $\boldsymbol{X} \in R^{10}$ provided that γ_M, γ_i, γ_0 and f_d are properly chosen. From this fact, it is possible to ascertain that there exists a critical point $\boldsymbol{X}^* \in R^{10}$ at which $W(\boldsymbol{X})$ attains the minimum $W(\boldsymbol{X}^*) = W_m$ under constraints of eqs.(43), (46), and (47). This critical position $\boldsymbol{X} = \boldsymbol{X}^*$ must satisfy the gradient equation:

$$A\Delta\boldsymbol{\lambda} + B\Delta\boldsymbol{m} = 0 \tag{79}$$

which is the same as the equation of eq.(70) when $\dot{\boldsymbol{X}} = 0$ and $\ddot{\boldsymbol{X}} = 0$. In other words, eq.(70) can be regarded as Lagrange's equation of motion with the Lagrangian $L = K - W$ and the external torque $C\dot{\boldsymbol{X}}$ and Pfaffian constraints appearing as a dissipation term. Thus, the critical position $\boldsymbol{X} = \boldsymbol{X}^*$ should be treated as an attractor in such a way that any closed-loop solution to eq.(70) starting from an arbitrary initial state $(\boldsymbol{X}(0), \dot{\boldsymbol{X}}(0))$ in a neighborhood of $(\boldsymbol{X}^*, 0)$ converges asymptotically to the critical state $(\boldsymbol{X}^*, 0)$.

10 Stability on a Constraint Manifold

The stability problem mentioned above can be treated in terms of differential geometry. Denote the position state vector by $\boldsymbol{X} = (q_1^{\mathrm{T}}, q_2^{\mathrm{T}}, \bar{\boldsymbol{x}}^{\mathrm{T}}, \theta, \psi)^{\mathrm{T}}$ and its corresponding time-derivative by $\dot{\boldsymbol{X}} = (\dot{q}_1^{\mathrm{T}}, \dot{q}_2^{\mathrm{T}}, \dot{\bar{\boldsymbol{x}}}^{\mathrm{T}}, \omega_Z, \omega_Y)^{\mathrm{T}}$. The set of all vectors $\boldsymbol{X}$ constituting a 10-dimensional vector space R^{10} is called the configuration space. This space is embedded in the state space $R^{20} = \{(\boldsymbol{X}, \dot{\boldsymbol{X}})\}$. Next, define the 8-dimensional space

$$M_8 = \{\boldsymbol{X} : Q_i = 0, i = 1, 2\}$$

which is called a constraint manifold. Further, at each point $\boldsymbol{X}$ on M_8 it is possible to define a tangent space TM_4 in such a way that

$$TM_4(\boldsymbol{X}) = \left\{\dot{\boldsymbol{X}} : \dot{Q}_i = 0, \text{ eq.(46)}, \text{ eq.(47)}, \ i = 1, 2\right\}$$

At this stage, it is possible to suppose formally a tangent boundle define as

$$T(\boldsymbol{X}, \dot{\boldsymbol{X}}) = \left\{(\boldsymbol{X}, \dot{\boldsymbol{X}}) : Q_i = 0, \dot{Q}_i = 0, \text{ eq.(46)}, \text{ and eq.(47) for } i = 1, 2\right\}$$

though it seems difficult to determine how many dimensions the set $T(\boldsymbol{X}, \dot{\boldsymbol{X}})$ has. Nevertheless, finiteness of the Riemannian metric

$$\int_0^\infty \sqrt{K}\mathrm{d}t = \int_0^\infty \sqrt{\sum_{i=1,2}\left\{\frac{1}{2}\dot{q}_i^{\mathrm{T}} H_i \dot{q}_i\right\} + \frac{M}{2}\|\dot{\boldsymbol{x}}\|^2 + \frac{1}{2}\boldsymbol{\omega}_0^{\mathrm{T}} H_0 \boldsymbol{\omega}_0 \ \mathrm{d}t} < +\infty \tag{80}$$

plays a crusial role in verification of the stability of an equilibrium position state $\boldsymbol{X}^*$ lying on M_8. It is also important to note that the integral

$$\int_0^\infty \sqrt{(W - W_m)}\,\mathrm{d}t < +\infty \tag{81}$$

is also finite, where W_m is the minimum of W under the constraints of $Q_i = 0$ eqs.(46) and (47) for $i = 1, 2$. Further, it is interesting to notice that eq. (80) implies that the locus of each rolling contact point on its corresponding object called E. Cartan's development ([37]) becomes finite in its total length over semi-infinite time interval $[0, \infty)$. Finally, if it is possible to define a neighborhood of the critical state $(\boldsymbol{X}^*, 0)$ by means of the Riemannian distance from $\boldsymbol{X}(0)$ to $\boldsymbol{X}^*$ on the manifold M_8 in such a way that

$$R(\boldsymbol{X}(0), \boldsymbol{X}^*) = \min_{T(\boldsymbol{X}, \dot{\boldsymbol{X}})} \int_0^t \sqrt{\sum_{i,j} h_{ij}(\boldsymbol{X})\dot{X}_i\dot{X}_j}\,\mathrm{d}t \tag{82}$$

then it is possible to discuss problems of stability of the equilibrium state corresponding to a certain state satisfying force/torque balance (i.e., stable grasp) in a rigorous mathematical argument and asymptotic stability (transferability). In illustrative example of Fig.9, the fingers-object system is not redundant in DOFs. In cases of redundant fingers-object systems, all equilibrium states constitute a manifold, too. Then, a Morse-theoretic treatment (see [38]) of critical points would become essential in relation to the problem of minimization of the artificial potential on a constraint manifold. Such an example of stability of 3-D grasp is treated recently in [39].

11 Conclusions

In this expository article a differential-geometric approach is presented for analyzing Lagrange's equations of the two control-theoretic problems of non-linear mechanical systems: 1) multi-joint point-to-point reaching movements under redundant DOFs and 2) 3-D object grasping and manipulation by means of a pair of robot fingers with multi-joints. It is shown that finiteness of Riemannian metrics naturally introduced on tangent bundles associated with a configuration space or a constraint manifold plays a key role in verification of stability and convergence of solution trajectories of the closed-loop dynamics.

References

1. M. R.Cutkovsky (1985) Robotic Grasping and Fine Manipulation. Kluwer Academic, Dordrecht, Netherlands,
2. K. B. Shimoga (1996) Robot grasp synthesis algorithms: A survey. Int. J. of Robotics Research,15,3:230–266

3. J. D. Crisman, C. Kanojia, I. Zeid (1996) Graspar: A flexible, easily controllable robotic hand. IEEE Robotics and Automation Magazine,June:32–38
4. A. M. Okamura, N. Smaby, M. R. Cutkovsky (2000) An overview of dexterous manipulation. Proc. of the 2000 IEEE Robotics and Automation, SanFrancisco, CA, April 25-28:255–262
5. R. M. Murray, Z. Li, S. S. Sastry (1994) A Mathematical Introduction to Robotic Manipulation. CRC Press, Boca Raton and Tokyo
6. S. Arimoto, P.T.A. Nguyen, H.-Y. Han, Z. Doulgeri (2000) Dynamics and control of a set of dual fingers with soft tips. Robotica,18,1:71–80
7. K. Tahara, M. Yamaguchi, P.T.A. Nguyen, H.-Y. Han, S. Arimoto (2000) Stable grasping and posture control for a pair of robot fingers with soft tips. Proc. of the Int. Conf. on Machine Automation(ICMA2000), Osaka, Japan, Sept. 27-29:33–38
8. J. P. LaSalle (1960) Some extensions of Lyapunov's second method. IRE Trans. on Circuit Theory,7:520–527
9. S. Arimoto, K. Tahara, M. Ymaguchi, P.T.A Nguyen, H.-Y. Han (2001) Principle of superposition for controlling pinch motions by means of robot fingers with soft tips. Robotica,19,1:21–28
10. S. Arimoto, P.T.A. Nguyen (2001) Principle of superposition for realizing dexterous pinching motion of a pair of robot fingers with soft-tips. IEICE Trans. on Fundamentals,E84-A,1:39–47
11. S. Arimoto, R. Ozawa, M. Yoshida (2005) Two-dimensional stable blind grasping under the gravity effect. Proc. of the 2005 IEEE Int. Conf. on Robotics and Automation:1208–1214
12. S. Arimoto (2004) Intelligent control of multi-fingered hands. Annual Review in Control,28,1,75–85
13. D.J. Montana (1992) Contact stability for Two-Fingered Grasps. IEEE Trans. on Robotics and Automation,8,4:421–430
14. N.A. Bernstein (1996) The Coordination and Regulation of Movements, Pergamon, London, 1967
15. N.A. Bernstein (translated from the Russian by M.L. Latash) (1996) On Dexterity and Its Development, Lawrence Erlbaum Associates, Inc.
16. J. Lenarcic (ed.) (1991) Special Issue on Redundant Robots. Laboratory Robotics and Automation,6,1
17. Y. Nakamura (1991) Advanced Robotics: Redundancy and Optimization, Addison-Wesley, Reading, MA
18. D.E. Whitney (1969) Resolved motion rate control of manipulators and human protheses, IEEE Trans. Man-Machine Syst.,MMS–10:47–53
19. A.G. Feldman (1986) Once more on the equilibrium-point hypothesis (λ model) for motor control, J. Mot. Behav.,18:17–54
20. T. Flash and N. Hogan (1985) The coordination of arm movements: An experimentally confirmed mathematical model, J. Neurosci.,5:1688–1703
21. E. Bizzi, N. Hogan, F.A. Mussa-Ivaldi, S. Giszter (1992) Does the nervous system use equilbrium-point control to guide single and multiple joint movements?, Behav. Brain Sci.,15:603–613
22. A.G. Feldman (1966) Functional tuning of the nervous system with control of movement or maintenance of steady posture. III. Mechanographic analysis of the execution by man of the simplest robot tasks, Biophysics, 11:766–775
23. E. Bizzi, A. Polit, and P. Morasso (1976) Mechanisms underlying achievement of final head position, J. Neurophysical, 39:435–444
24. N. Hogan (1984) An organizing principle for a class of volundary movements, J. Neurosci., 4:2745–2754

25. M. Takegaki and S. Arimoto (1981) A new feedback method for dynamic control of manipulators, Trans. of the ASME, Journal of Dynamic Systems, Measurement, and Control, 103, 2:119–125
26. H. Seraji (1989) Configuration control of redundant manipulators: Theory and implementation, IEEE Trans. on Robotics and Automation, 5, 4:472–490
27. S. Arimoto, M. Yoshida, and J.-H. Bae (2003) Dynamic force/torque balance of 2D polygonal objects by a pair of rolling contacts and sensory-motor coordination, J. of Robotic Systems, 20, 9:517–537
28. S. Arimoto (2004) Intelligent control of multi-fingered hands, Annual Review in Control, 28:75–85
29. S. Arimoto (1996) Control Theory of Nonlinear Mechanical Systems: A Passivity-based and Circuit-theoretic Approach, Oxford Univ. Press, Oxford, UK
30. J.M. Hollerbach and K.C. Suh (1987) Redundancy resolution of manipulators through torque optimization, IEEE J. of Robotics and Automation, RA–3, 4:308–316
31. M.L. Latash (1998) Neurophysiological Basis of Movement, Human Kinetics Pub., New York
32. J.M. Winters and L. Stark (1987) Muscle models: what is gained and what is lost by varying model complexity, Biological Cybernetics, 55, 6:403–420
33. N. Hogan (1985) The mechanics of multi-joint posture and movement control, Biological Cybernetics, 52, 5:315–331
34. S. Arimoto and M. Sekimoto (2006) Natural redundancy resolution for PTP control of multi-joint reaching movements: A virtual spring/damper hypothesis, Proc. of the 17th Int. Symp. on MTNS, Kyoto, Japan:1008–1017
35. H. Goldstein, C. Poole, and J. Safko (2002) Classical Mechanics – 3rd ed., Pearson Education Inc., Upper Saddle River, NJ
36. S. Arimoto, M. Yoshida, and J.-H. Bae (2006a) Modeling of 3-D object manipulation by multi-joint robot fingers under non-holonomic constraints and stable blind grasping, Proc. of Asian Conf. on Multi-body Dynamics, Tokyo, Japan, Aug. 1–4
37. K. Nomizu (1978) Kinematics and differential geometry of submanifolds – Rolling a ball with a prescribed locus of contact, Tohoku Mathematical Journal, 30:623–637
38. B. Bullo and A.D. Lewis (2005) Geometric Control of Mechanical Systems, Springer, New York
39. S. Arimoto, M. Yoshida, and J.-H. Bae (2007) Modeling of 3-D object manipulation by multi-joint robot fingers under non-holonomic constraints and stable blind grasping, to be published in a special issue on ACMD '06 in JSME Technical Journal, Japanese Soc. of Mechanical Engineers, Tokyo, Japan

Nonsmooth Riemannian Optimization with Applications to Sphere Packing and Grasping

Gunther Dirr, Uwe Helmke*, and Christian Lageman

Institute of Mathematics, University of Würzburg,
Am Hubland, 97074 Würzburg, Germany

Summary. This paper presents a survey on Riemannian geometry methods for smooth and nonsmooth constrained optimization. Gradient and subgradient descent algorithms on a Riemannian manifold are discussed. We illustrate the methods by applications from robotics and multi antenna communication. Gradient descent algorithms for dextrous hand grasping and for sphere packing problems on Grassmann manifolds are presented respectively.

1 Introduction

Riemannian optimization is a rather recent branch of constrained optimization theory that attempts to analyze and develop optimization algorithms by fully exploiting the information that is inherent in the differential geometric structure of the constraint set. Despite a number of important recent contributions, see e.g. [9, 14, 22, 23, 25], the use of differential geometric tools in modern optimization textbooks is confined to standard Lagrange multiplier techniques and therefore implicit, at best.

In this paper we survey some basic Riemannian optimization techniques that have played an important role in recent applications to geometric mechanics. First we illustrate our methods with a nonsmooth Rayleigh cost function. Then we focus our discussion on two major applications. First we tackle the problem of computing an optimal sphere packing on a Grassmann manifold with respect to the chordal distance. Such packings are of interest for coding theory [2, 11, 16]. The second application deals with the computation of optimal force distributions for multi–fingered dextrous robot hands. Following previous work [5, 12], the task is reformulated as a semidefinite optimization problem for which a Riemannian gradient algorithm yields good convergence results.

* Partially supported by the German Research Foundation under grant KONNEW HE 1858/10-1.

F. Bullo et al. (Eds.): Lag. & Hamil. Methods for Nonlin. Ctrl., LNCIS 366, pp. 29–45, 2007.
springerlink.com

2 Riemannian Optimization

2.1 Riemannian Geometry

To enhance the clarity of our subsequent discussion of Riemannian optimization algorithms, we begin with a brief summary of well known definitions and results from differential geometry; for further details we refer to any modern textbook on Riemannian geometry, e.g. [10]. Let M denote an n-dimensional Riemannian manifold with Riemannian metric $<,>$ and let df denote the differential of a smooth function f on M. In the sequel we will often write, as usual, $Xf :=$ dfX for the Lie derivative of f with respect to a smooth vector field X. A **Riemannian or Levi-Civita connection** ∇ on M assigns to each pair of vector fields X, Y on M a vector field $\nabla_X Y$ such that for all smooth functions f on M:

(i) $(X, Y) \mapsto \nabla_X Y$ is bilinear in X, Y.
(ii) $\nabla_{fX} Y = f \nabla_X Y$.
(iii) $\nabla_X (fY) = f \nabla_X Y + (\mathrm{dfX})\mathrm{Y}$.
(iv) $\nabla_X Y - \nabla_Y X = [X, Y]$.
(v) $X < Y, Z > = < \nabla_X Y, Z > + < Y, \nabla_X Z >$.

It is easy to see that these conditions (i)–(v) uniquely specify the Levi-Civita connection. Recalling the definition of a geodesic, we note, that for any smooth curve $x : I \to M$, $\dot{x}(t) \neq 0$, there exists locally a smooth vector field $X : M \to TM$ such that $\dot{x}(t) = X(x(t))$. By inspection, it is then easily seen that

$$\nabla_{\dot{x}} \dot{x} := (\nabla_X X)(x(t))$$

is independent of the choice of the vector field extension X. We call $\nabla_{\dot{x}} \dot{x}$ the **geodesic operator**. A smooth curve $\gamma : I \to M$ is called a **geodesic**, if

$$\nabla_{\dot{\gamma}} \dot{\gamma} = 0$$

holds for all $t \in I$. It is a well known and easily established fact from the theory of ordinary differential equations, that geodesics always exist. However, such geodesics may only be defined on a sufficiently small time interval I around 0; if geodesics exist for all $t \in \mathbb{R}$ and all initial conditions, then M is called **geodesically complete**. Any compact manifold is geodesically complete. In the sequel, we will mostly deal with geodesically complete Riemannian manifolds. The **Riemannian exponential map** is then a well-defined smooth map

$$\exp : TM \to M$$

on the tangent bundle, given by $\exp_x(v) = \gamma(1)$, where γ is the unique geodesic through x with initial velocity v. It is easily seen that this implies $\exp_x(tv) = \gamma(t)$ for all t where γ is defined.

We now come to the definitions of gradient and Hessian in Riemannian geometry. The **Riemannian gradient** of a smooth function $f : M \to \mathbb{R}$ on a

Riemannian manifold M is the uniquely defined smooth vector field grad f on M such that

$$\mathrm{df(x)v} = < \mathrm{grad\, f(x), v} >_{\mathrm{x}}$$

holds for all $x \in M$, $v \in T_x M$.

The definition of the Hessian is slightly more complicated and it depends on the use of geodesics. The **Riemannian Hessian** of a smooth function $f : M \to \mathbb{R}$ at $x \in M$ is the quadratic form on the tangent space

$$\mathrm{H}_f(x) : T_x M \times T_x M \to \mathbb{R},$$

defined for each tangent vector $\xi \in T_x M$ by

$$\mathrm{H}_f(x)(\xi, \xi) := (f \circ \gamma)''(0).$$

Here $\gamma : I \to M$ is a geodesic, with $\gamma(0) = x$ and $\dot{\gamma}(0) = \xi$. The **Riemannian Hesse operator** is the uniquely determined linear map

$$\mathbf{H}_f(x) : T_x M \to T_x M$$

satisfying

$$\mathrm{H}_f(x)(\xi, \eta) = < \mathbf{H}_f(x)\xi, \eta >_x$$

for all tangent vectors $\xi, \eta \in T_x M$.

There is an equivalent characterization in terms of the Levi-Civita connection.

Proposition 1. *Let $x \in M$, $\xi, \eta \in T_a M$ be tangent vectors and X, Y be smooth vectors fields on M with $X(x) = \xi, Y(x) = \eta$. Then the Riemannian Hesse form and Hesse operator, respectively, are given as*

$$\mathrm{H}_f(x)(\xi, \eta) = X(Yf)(x) - (\nabla_X Y)f(x)$$
$$\mathbf{H}_f(x)\xi = (\nabla_X \mathrm{grad}\, f)(x).$$

In particularly, for any smooth curve $x : I \to M$

$$\nabla_{\dot{x}} \mathrm{grad}\, f(x) = \mathbf{H}_f(x)\dot{x}.$$

As an example we consider the Rayleigh quotient function $f \colon S^n \to \mathbb{R}, f(x) = \frac{1}{2}x^\top Ax$, on the Euclidean n-sphere $S^n := \{x \in \mathbb{R}^{n+1} | \|x\|^2 = 1\}$, where A is a symmetric $(n+1) \times (n+1)$-matrix. Endow S^n with the Riemannian metric induced by the standard Euclidean inner product on $\mathbb{R}^{n+1}$. Then Riemannian gradient and Hessian are respectively

$$\mathrm{grad}\, f(x) = (I - xx^\top)Ax$$

$$\mathbf{H}_f(x) = (A - x^\top AxI)|_{T_x S^n}.$$

2.2 Gradient Algorithms

Gradient-descent algorithms are the most basic optimization methods on a manifold. Despite their simplicity they are often useful to obtain a quick improvement of a previous estimate to an optimum. In the sequel, let $f\colon M \to \mathbb{R}$ denote a smooth function with compact sublevel sets. By this assumption the global minimum of f on M exists. We will assume that M is a real analytic manifold that carries a real analytic Riemannian metric $< \cdot, \cdot >$. Let $||v|| := \sqrt{< v, v >}$ denote the associated norms on the tangent spaces $T_x M$. We can use continuous-time or discrete-time gradient-like methods to search for minima of f.

In the continuous-time case a **gradient-like descent flow** for minimizing f is given by a differential equation

$$\dot{x} = X(x)$$

on M such that each $x \in M$ with $X(x) = 0$ is a critical point[1] of f and for any compact subset $K \subset M$ there exists $\varepsilon_K > 0$ for which the **angle condition**

$$\langle X(x), \operatorname{grad} f(x) \rangle \leq -\varepsilon_K \|X(x)\| \|\operatorname{grad} f(x)\|$$

holds for all $x \in K$. Obviously, the **gradient descent flow**

$$\dot{x} = -\operatorname{grad} f(x)$$

is included in this class (with $\varepsilon_K = 1$). Standard arguments from Lyapunov theory show the convergence of the solutions to the set of critical points [14]. However, pointwise convergence to single equilibrium points holds only under additional regularity assumptions on the function f. For example, a result of Łojasiewicz [20] establishes convergence of trajectories of gradient flows[1] to single equilibrium points for arbitrary real analytic functions f. This can be extended to gradient-like descent flows, see [1, 3] and for more general results [17, 18].

Theorem 1. *The ω-limit sets of the solutions of a gradient-like descent flow are contained in the set of critical points of f. If f is analytic, then each solution either converges to a single point or has empty ω-limit set.*

The pointwise convergence fails in the general smooth case. If the angle condition is relaxed to $\langle \operatorname{grad} f(x), X(x) \rangle < 0$ for points $x \in M$ with $X(x) \neq 0$, $\operatorname{grad} f(x) \neq 0$, then the pointwise convergence fails even for analytic f as the example $f(x, y) = (x^2 + y^2 - 1)^2$,

$$X(x, y) = f(x, y) \left(\begin{pmatrix} -y \\ x \end{pmatrix} - f(x, y) \begin{pmatrix} x \\ y \end{pmatrix} \right)$$

in $\mathbb{R}^2$ shows.

We now discuss discrete-time gradient-like algorithms. Let $x_0 \in M$ be any initial point and α_t denote a sequence of non-negative numbers. The **Riemannian gradient-like descent algorithm** is given by an iteration

[1] This corrects inexact statements in [8].

$$x_{t+1} = \exp_{x_t}(\alpha_t v_t),$$

where the **descent directions** v_t are assumed to satisfy an **angle condition**

$$\langle v_t, \operatorname{grad} f(x_t)\rangle \leq -\varepsilon \|v_t\| \|\operatorname{grad} f(x_t)\|$$

with $\varepsilon > 0$ a constant depending only on x_0. Here, we normalize the descent directions such that $\|v_t\| = \|\operatorname{grad} f(x_t)\|$. The **step-sizes** α_t are often not chosen as constant but rather by sophisticated step-size selection schemes to ensure **descent of the iterates**, i.e. $f(x_{t+1}) < f(x_t)$, and convergence to the set of critical points. One example of such a step-size selection scheme is the so-called Armijo step-size rule[2]

$$\alpha_t = \max\{\tau\mu^l \mid f(x_t) - f\left(\exp_{x_t}\left(\tau\mu^l v_t\right)\right) \geq -\sigma\tau\mu^l df(x_t)(v_t)\}$$

with $\sigma, \mu \in (0,1), \tau > 0$ constants independent of t. The special case of

$$v_t = -\operatorname{grad} f(x_t)$$

is simply called **gradient descent**. We have the following convergence results, cf. [1, 17, 25].

Theorem 2. *For the Armijo step-size rule, the gradient-like descent iteration converges to the set of critical points. If f is analytic, then it converges to a single critical point.*

Note, that some arguments from [25] for the first statement can be directly extended from gradient to gradient-like descent.

The gradient descent scheme can be extended to nonsmooth, Lipschitz continuous functions as well. In this case, Clarke's concept of a subgradient on a Banach space [6] needs to be generalized to arbitrary Riemannian manifolds, which is straightforward to do. For our purposes it is sufficient to confine to the following more restrictive setting, which helps us to formulate the basic concepts in an easily accessible form. Thus we assume, that $f\colon M \to \mathbb{R}$ is defined as the pointwise maximum

$$f(x) = \max_{i=1,\ldots,m} f_i(x)$$

of smooth functions $f_i : M \to \mathbb{R}$, $i = 1, \ldots, m$ with compact sublevel sets.

Define

$$I_f(x) = \{i \in \{1, \ldots, m\} \mid f(x) = f_i(x)\}.$$

Here, the subgradient of f is

$$\operatorname{grad} f(x) = \operatorname{co}\{\operatorname{grad} f_i(x) \mid i \in I_f(x)\}$$

where co denotes the convex hull. See [18, 19] for the discussion of general Lipschitz continuous cost functions.

[2] This adjusts the incorrect formula for the Armijo rule in [8].

The continuous-time subgradient algorithm is given by the **subgradient differential inclusion**

$$\dot{x} \in -\operatorname{grad} f(x). \tag{1}$$

Solutions of (1) are defined as absolutely continuous functions, which satisfy the differential inclusion almost everywhere. Upper semicontinuity of $\operatorname{grad} f(x)$ ensures the existence of solutions, which are though not necessarily unique [4].

Theorem 3. *For any $x_0 \in M$, a solution of (1) starting in x_0 exists and all solutions converge to the set of critical points of f, i.e. to the set $\{x \in M \mid 0 \in \operatorname{grad} f(x)\}$.*

Theorem 3 can be proved by extending the chain-rule for the max-function from [21] to the Riemannian subgradient by local charts. Thus for a solution γ we have $-\frac{d}{dt} f(\gamma(t)) \geq \min\{||v||^2 \mid v \in \operatorname{grad} f(\gamma(t))\}$ for all points t where $f(\gamma(t))$ and $\gamma(t)$ are differentiable. Absolute continuity of $f(\gamma(t))$ and upper semicontinuity of $\operatorname{grad} f$ then yield the desired result.

Similar to gradient-like flows one can construct discrete-time versions of the subgradient algorithm. However, not every $v \in -\operatorname{grad} f(x)$ a step size selection scheme like the Armijo rule would be well-defined. Hence, we define the **Riemannian subgradient descent** by

$$x_{t+1} = \exp_{x_t}(\alpha_t v_t)),$$

with the **descent directions** v_t satisfying the **descent condition**

$$\forall w \in \operatorname{grad} f(x_t) : \langle v_t, w \rangle < 0.$$

Note, that under mild conditions on f, the set $\operatorname{grad} f(x_t)$ consists generically only of one element. In general, a suitable choice for v_t is $v_t = -\pi_0(\operatorname{grad} f(x_t))$, where π_0 denotes the minimum-distance projection of 0 to $\operatorname{grad} f(x_t)$ in $T_{x_t}M$. We use this descent direction in the sequel.

The discrete subgradient descent allows the possibility of converging to non-critical points, even for step size selection schemes like the Armijo rule. However, if for all t we have $-v_t \in \operatorname{grad} f(x_t)$ and the step sizes satisfy $\sum_{t=1}^{\infty} \alpha_t = \infty$ and $\alpha_t \to 0$, then it can be shown that the subgradient descent converges to the set of critical points [18, 24]. Note that in [24] only the Euclidean case is considered. Interestingly, the α_t can be chosen independently of the cost function f, for example the harmonic step sizes $\alpha_t = t^{-1}$ can be used. However, such general step sizes have often bad convergence properties. Thus a careful adaption to the actual cost function is usually necessary.

3 Applications

3.1 Rayleigh Quotients on the Sphere S^{n-1}

We consider the maximum of m Rayleigh quotients on the sphere S^{n-1}, i.e.

$$f(x) = \max_{i=1,\ldots,m} f_i(x) = \max_{i=1,\ldots,m} \frac{1}{2} x^\top A_i x \tag{2}$$

$A_i \in \mathbb{R}^{n \times n}$ symmetric. Our aim is to find a minimum of f.

The sphere S^{n-1} is treated as a submanifold of $\mathbb{R}^n$ with the standard Riemannian metric. As mentioned in Subsection 2.1 the gradient of $\frac{1}{2}x^\top A_i x$ on S^{n-1} with respect to this metric is $(I - xx^\top)A_i x$. Thus we have

$$\operatorname{grad} f(x) = \operatorname{co}\{A_i x \mid i \in I_f(x)\} - 2f(x)x.$$

To calculate the directions $v_t = -\pi_0(\operatorname{grad} f(x_t))$ we have to solve the optimization problem

$$\min_{t_i \geq 0, \sum t_i = 1} \left\| \sum_{i \in I_f(x_t)} t_i A_i x_t - 2f(x_t)x_t \right\|^2 .$$

Straightforward calculations show that this problem is equivalent to

$$\min_{t_i \geq 0, \sum t_i = 1} \left\| \sum_{i \in I_f(x_t)} t_i A_i x_t \right\|^2 .$$

This problem can be solved by e.g. quadratic programming methods. Note, that generically we will not meet points where several $x_t^\top A_i x_t$ achieve the maximum simultaneously, thus this optimization problem will not increase the computational costs of this algorithm very much. Given $v \in T_x S^{n-1}$, $\|v\| = 1$, the geodesic $\gamma(s)$ at x in direction v is easily calculated as $\gamma(s) = \cos(s)x + \sin(s)v$. This gives the subgradient iteration

$$x_{t+1} = \cos(\alpha_t \|v_t\|)x_t - \sin(\alpha_t \|v_t\|)\|v_t\|^{-1}v_t.$$

The the minimum of the cost function is achieved for the step size

$$\alpha_t = \operatorname*{argmin}_{\alpha \geq 0}\{f(\cos(\alpha\|v_t\|)x_t + \sin(\alpha\|v_t\|)\|v_t\|^{-1}v_t\}.$$

Solving this problem is equivalent to

$$\min_{\alpha \geq 0} \max_{i=1,\dots,m} y(\alpha)^\top \begin{pmatrix} x_t^\top A_i x_t & x_t^\top A_i v_t \\ v_t^\top A_i x_t & v_t^\top A_i v_t \end{pmatrix} y(\alpha)$$

with $y(\alpha) = (\cos(\alpha\|v_t\|), \|v_t\|^{-1}\sin(\alpha\|v_t\|))^\top$, i.e. it is equivalent to our optimization problem on a one dimensional sphere. Note, that even for this "optimal" step size there is no general convergence result known. In fact, there are examples of convex functions on $\mathbb{R}^n$, where the convergence with such a step size fails [15].

In Figure 1 the results[3] of the above algorithm are plotted for 20 randomly chosen symmetric matrices in $\mathbb{R}^{3\times3}$. We used the Armijo step size rule with $\sigma = 10^{-4}$, $\tau = 1/2$, $\mu = 0.9$, and the harmonic step size rule with $\alpha_t = 1/(2t)$. The algorithm with Armijo step sizes was stopped if the μ^l fell below 2^{-16}. In our simulations the both step size rules yield convergence to similar values in many

[3] These adjust the incorrect simulations in [8] for this problem.

cases, although the harmonic rule did not show monotone decrease of the cost function $f(x_t)$. However, there were significant number of cases, where once the Armijo rule or once the harmonic step size produced substantially better results. This effect might be caused by convergence to non-optimal critical points or local minima.

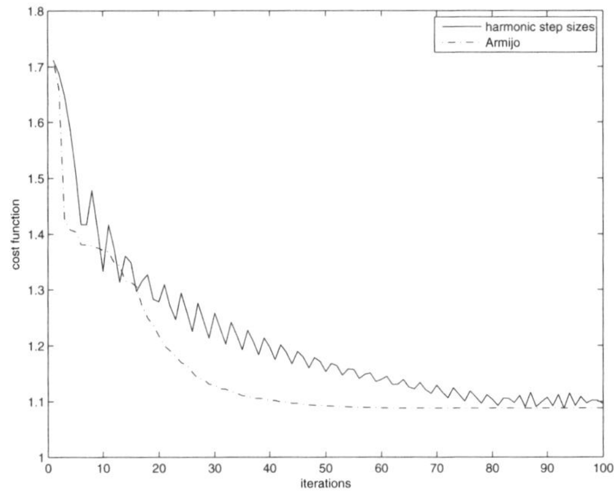

Fig. 1. Subgradient descent for the cost function (2) with 20 (3×3)-matrices

3.2 Sphere Packings on Grassmannians

Next, we discuss the packing of m spherical balls on a real Grassmann manifold with respect to the chordal distance. In [7], the packing problem on the real Grassmannian was approached by a family of smooth cost functions. In [2], the authors considered smooth exponential-type approximations of a nonsmooth cost function for the sphere packing problem on the complex Grassmannian. We present here a nonsmooth descent approach to this sphere packing problem. For a thorough discussion we refer the reader to [18, 19]. The real Grassmann manifold is the smooth manifold of k-dimensional linear subspaces of $\mathbb{R}^n$. Mapping a subspace to the associated orthogonal projector, we can identify the Grassmann manifold with the **Grassmannian**

$$\mathrm{Grass}(n,k) = \{P \in \mathbb{R}^{n\times n} \mid P^\top = P, P^2 = P, \mathrm{tr}\,(P) = k\},$$

see [14]. The tangent space of $\mathrm{Grass}(n,k)$ in a point P is given by

$$T_P\,\mathrm{Grass}(n,k) = \{[P,\Omega] \mid \Omega \in \mathfrak{so}(n)\}$$

with $\mathfrak{so}(n) = \{\Omega \in \mathbb{R}^{n\times n} \mid \Omega^\top = -\Omega\}$. We denote by K_P the kernel of the linear map $\mathfrak{so}(n) \to T_P\,\mathrm{Grass}(n,k), X \mapsto [P,X]$ and by $K_p^\perp$ the set

$$K_P^\perp := \{\Omega \in \mathfrak{so}(n) \mid \forall \Theta \in K_P \colon \mathrm{tr}\,(\Omega^\top \Theta) = 0\}.$$

Any $\Omega \in \mathfrak{so}(n)$ is uniquely decomposed into

$$\Omega = \Omega_P + \Omega_P^\perp$$

with $\Omega_P \in K_P$, $\Omega_P^\perp \in K_p^\perp$. We equip the Grassmannian with the **normal Riemannian metric** [14]

$$\langle [P,\Omega],[P,\Theta]\rangle := -\mathrm{tr}\,(\Omega_P^\perp \Theta_P^\perp), \quad [P,\Omega],[P,\Theta] \in T_P\,\mathrm{Grass}(n,k).$$

The geodesics on $\mathrm{Grass}(n,k)$ with respect to the normal metric have the form

$$\gamma(t) = \exp(t[\eta,P])P\exp(t[\eta,P])^\top, \eta \in T_P\,\mathrm{Grass}(n,k),$$

where exp denotes the matrix exponential function [13].

On $\mathrm{Grass}(n,k)$ the **chordal distance** $\mathrm{dist}(P,Q)$ is defined by

$$\mathrm{dist}(P,Q) = \sqrt{2}\|P-Q\|_F.$$

It can be shown that this definition is equivalent to the definition via the principal angles between the subspaces, see [7].

The **sphere packing problem** consists of finding m identical, non-overlapping balls $B_r(P_i) = \{Q \in \mathrm{Grass}(n,k) \mid \mathrm{dist}(P,Q) < r\}$, $i = 1,\dots,m$, such that their radius is maximized. This problem is equivalent to maximizing

$$\min_{i<j} \mathrm{dist}(P_i,P_j)^2$$

for the chordal distance on $\mathrm{Grass}(n,k)$. However, this is equivalent to finding the minima of the cost function

$$f(P_1,\dots,P_m) = \max_{i<j} \mathrm{tr}\,(P_iP_j)$$

on the m-fold product of $\mathrm{Grass}(n,k)$. The gradient of the function $h(P,Q) = \mathrm{tr}\,(PQ)$ on $\mathrm{Grass}(n,k) \times \mathrm{Grass}(n,k)$ with respect to the product of the normal Riemannian metric is given by

$$\mathrm{grad}\,h(P,Q) = ([P,[P,Q]],[Q,[Q,P]]).$$

For the m-fold product space of the Grassmannian equipped with the product metric, we get the following formula for the subgradient of f,

$$\mathrm{grad}\,f(P_1,\dots,P_m) = \mathrm{co}\{(\dots,[P_i,[P_i,P_j]],\dots,[P_j,[P_j,P_i]],\dots) \mid \\ i<j, \mathrm{tr}\,(P_iP_j) = f(P_1,\dots,P_j)\}.$$

The subgradient iteration gives a sequence of m-tuples $(P_1^t,\dots,P_m^t)$ of elements of $\mathrm{Grass}(n,k)$ which are updated by the rule

$$P_i^{t+1} = \exp(\alpha_t[\eta_i^t,P_i^t])P_i^t\exp(\alpha_t[\eta_i^t,P_i^t])^\top$$

where $(\eta_1^t,\dots,\eta_m^t) = -\pi_0(\mathrm{grad}\,f(P_1^t,\dots,P_m^t))/\|\pi_0(\mathrm{grad}\,f(P_1^t,\dots,P_m^t))\|$ is the steepest descent direction and α_t a suitable step size. Note, that we normalized the descent direction to length 1. In our experience, this ad hoc modification of the Armijo rule produces better results for the sphere packing problem. For alternative implementations of the algorithm we refer to [19].

In Figure 2 the results of our algorithm for packing 10 points in Grass(16, 8) are shown. The initial points on Grass(16, 8) were constructed by the following method: For each initial point, we chose a $\Theta \in SO(n)$ with $\Theta = \exp(10(A - A^\top)/2)$ for $A \in \mathbb{R}^{n\times n}$ with entries randomly distributed in $[-1, 1]$. Each initial point was then calculated by

$$\Theta^\top \begin{pmatrix} I_k & 0 \\ 0 & 0 \end{pmatrix} \Theta.$$

The diagram compares the minimal squared chordal distance for the Armijo rule, $\sigma = 10^{-4}$, $\tau = 1$, $\mu = 1/2$, and for the harmonic step size selection $\alpha_t = (0.3t + 1)^{-1}$. The algorithm with Armijo step sizes was stopped if μ^l fell below 2^{-16}. For the choice of initial points by the construction above, the harmonic step size showed slightly better convergence properties. However, in simulations with different constructions of initial configurations, especially for bad initial configurations with respect to the cost function, the harmonic step size performed worse than the Armijo step size. Hence, we cannot give a conclusive result which step size rule is better for general initial values.

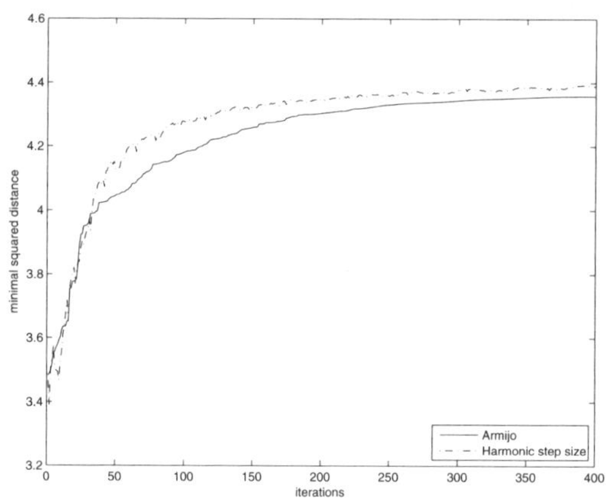

Fig. 2. Subgradient descent for sphere packing problem for 10 points on Grass(16, 8)

3.3 Sphere Packing on the Lagrange Grassmannian

The next example is the sphere packing problem on the **Lagrange Grassmann manifold**, i.e. the manifold of Lagrangian subspaces of $\mathbb{R}^{2n}$. This case has apparently not been discussed in the literature before. A subspace V of $\mathbb{R}^{2n}$ is called **Lagrangian** if for all $v, w \in V$ the equation

$$v^\top J w = 0, \text{ with } J = \begin{pmatrix} 0 & I_n \\ -I_n & 0 \end{pmatrix}$$

holds. Naturally, the Lagrange Grassmannian is a submanifold of the Grassmann manifold of n-dimensional subspaces of $\mathbb{R}^{2n}$. Further note that a symmetric projection matrix P onto a Lagrangian subspaces must satisfy $PJP = 0$. Therefore we can identify the Lagrange Grassmannian with

$$\mathrm{LGrass}(n) = \{P \in \mathbb{R}^{2n \times 2n} \mid P^\top = P, P^2 = P, PJP = 0, \mathrm{tr}\,(P) = n\}.$$

The tangent spaces of $\mathrm{LGrass}(n)$ at a point P is given by

$$T_P \,\mathrm{LGrass}(n) = \{[P, \Omega] \mid \Omega \in \mathfrak{osp}(2n)\}$$

with $\mathfrak{osp}(2n)$ the skew-symmetric Hamiltonian matrices, i.e.

$$\mathfrak{osp}(2n) = \{\Omega \in \mathbb{R}^{2n \times 2n} \mid \Omega^\top = -\Omega, \Omega J - J\Omega = 0\}.$$

We define the **normal metric** as for the Grassmannian. Setting K_P the kernel of the map $\mathfrak{osp}(2n) \to T_P \,\mathrm{LGrass}(n), X \mapsto [P, X]$ and $K_P^\perp = \{\Omega \in \mathfrak{osp}(2n) \mid \forall \Theta \in K_P\colon \mathrm{tr}\,(\Omega^\top \Theta) = 0\}$ we decompose every $\Omega \in \mathfrak{osp}(2n)$ as $\Omega = \Omega_P + \Omega_P^\perp$, $\Omega_P \in K_P$, $\Omega_P^\perp \in K_P^\perp$. The normal metric is again given by

$$\langle [P, \Omega], [P, \Theta] \rangle = -\mathrm{tr}\,(\Omega_P^\perp \Theta_P^\perp), \quad [P, \Omega], [P, \Theta] \in T_P \,\mathrm{LGrass}(n).$$

It can be shown that this is just the restriction of the normal metric on $\mathrm{Grass}(2n, n)$ to $\mathrm{LGrass}(n)$ [13]. The geodesics are again given by the formula

$$\gamma(t) = \exp(t[\eta, P]) P \exp(t[\eta, P])^\top, \eta \in T_P \,\mathrm{LGrass}(n).$$

As a submanifold of $\mathrm{Grass}(2n, n)$, the chordal distance is well defined on $\mathrm{LGrass}(n)$. The sphere packing problem on $\mathrm{LGrass}(n)$ with respect to the chordal distance, is again equivalent to minimizing the cost function

$$f(P_1, \ldots, P_m) = \max_{i<j} \mathrm{tr}\,(P_i P_j)$$

on the m-fold product space of the Lagrange Grassmannian. Obviously, this is just the restriction of the cost function of Subsection 3.2 to the submanifold $\mathrm{LGrass}(n) \times \ldots \times \mathrm{LGrass}(n)$. Analogous arguments for the general Grassmannian show that the subgradient of f is given by

$$\begin{aligned} \mathrm{grad}\, f(P_1, \ldots, P_m) = \mathrm{co}\{(\ldots, [P_i, [P_i, P_j]], \ldots, [P_j, [P_j, P_i]], \ldots) \mid \\ i < j, \mathrm{tr}\,(P_i P_j) = f(P_1, \ldots, P_j)\}. \end{aligned}$$

Furthermore, we get as the subgradient descent iteration the sequence of m-tuples $(P_1^t, \ldots, P_m^t)$ which are calculated by

$$P_i^{t+1} = \exp(\alpha_t[\eta_i^t, P_i^t])P_i^t \exp(\alpha_t[\eta_i^t, P_i^t])^\top$$

where $(\eta_1^t, \ldots, \eta_m^t) = -\pi_0(\operatorname{grad} f(P_1^t, \ldots, P_m^t))/\|\pi_0(\operatorname{grad} f(P_1^t, \ldots, P_m^t))\|$ is the steepest descent direction and α_t a suitable step size. The steepest descent direction is again normalized to length 1.

Thus the subgradient descent iteration for the sphere packing problem on the Lagrange Grassmannian with respect to the chordal distance is just the descent algorithm on the Grassmannian with an initial tuple $(P_1^0, \ldots, P_m^0)$ an element of the m-fold product of the Lagrange Grassmannian.

Figure 3 shows the results of our algorithm for packing 10 points in LGrass(8). The initial points were chosen on LGrass(8) with a similar method as for Grass(16, 8). The figure compares the minimal squared chordal distance for the Armijo rule, $\sigma = 10^{-4}$, $\tau = 1$, $\mu = 1/2$, and for the harmonic step size selection $\alpha_t = (0.3t + 1)^{-1}$. Again, the algorithm with Armijo step sizes was stopped if μ^l fell below 2^{-16}. The simulations behaved similar as for Grass(16, 8). Although we did not experiment with different constructions for the initial points, it is expected that the performance of the algorithm with harmonic step sizes degrades again significantly for constructions which produce bad initial configurations. Again, we cannot make a conclusive statement which step size selection is better in general.

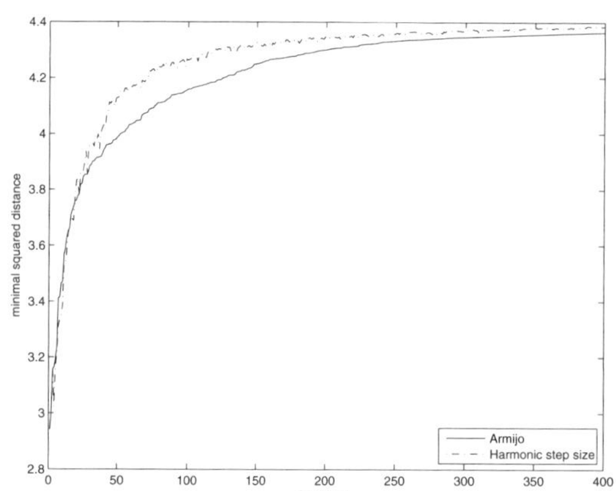

Fig. 3. Subgradient descent for sphere packing problem for 10 points on LGrass(8)

3.4 Grasping Force Optimization

The last problem we consider is taken from robotics, i.e. the computing of optimal forces in multi-finger dextrous hand grasping. The goal here is to achieve balancing of the external forces together with minimizing contact forces without violating nonlinear Coulomb friction constraints. In early work, linear and nonlinear programming techniques have been used but these approaches led to ill-conditioned problems and did not provide real time implementations. In [5], real time implementations have been achieved by reformulating the problem as a semidefinite optimization task on affinely constrained cones of positive definite matrices. A cost function is minimized that is the sum of a linear cost involving the total finger forces and a barrier term that prevents slippage at the finger contacts during the algorithm. Improvements of such algorithms have been developed in [12] where a damped Newton algorithm is proposed with locally quadratic convergence rates. Here, we consider extensions that allow the treatment of nonsmooth convex costs.

Specifically, we consider the case of semidefinite optimization over an affinely constrained cone of symmetric matrices. Let $\mathcal{S}(n)$ denote the vector space of $n \times n$ block diagonal matrices $S = \operatorname{diag}(S_1, \cdots, S_N)$, where $S_i = S_i^\top$ are real symmetric matrices of fixed sizes $r_1 > 0, \cdots, r_N > 0, r_1 + ... + r_N = n$. In the application to grasping we take $r_1 = ... = r_N = 2$. Let

$$\mathcal{P} = \{P \in \mathcal{S}(n) \quad | \quad P = P^\top > 0\}$$

denote the cone of block-diagonal positive definite symmetric matrices. In the grasping application the total sum of the finger forces is given by a linear trace function $\mathrm{tr(WP)}$, while a natural barrier term for preventing slippage is $-\log\det P$. External forces acting on the object are modelled by linear trace constraints $\mathrm{tr(A_jP) = b_j, j = 1, \cdots, m}$, where $A_1, \cdots, A_m \in \mathcal{S}(n)$ are linearly independent symmetric matrices with $\mathrm{tr(A_iA_j)} = \delta_{\mathrm{ij}}$ and $b_1, \cdots, b_m$ are scalars. In the grasping problem $m = 6$.

Thus the total grasping force optimization problem is equivalent to the constrained optimization task for the smooth, strictly convex function

$$f : \mathcal{C} \to \mathbb{R}, \quad f(P) = \mathrm{tr(WP)} - \log\det \mathrm{P}$$

on the constraint set

$$\mathcal{C} = \{P \in \mathcal{P} \quad | \quad \mathrm{tr(A_jP) = b_j, j = 1, \cdots, m}\}.$$

Here $W \in \mathcal{P}$ is assumed to be positive definite. We describe now different approaches to this problem via Riemannian optimization methods.

In a first step we compute the gradient for the restriction of the non-constant Riemannian metric $\langle \cdot, \cdot \rangle_P$,

$$< \xi, \eta >_P := \mathrm{tr(P^{-1}\xi P^{-1}\eta)}, \quad \xi, \eta \in \mathrm{T_P}\mathcal{P}$$

on $\mathcal{P}$ to $\mathcal{C}$. The orthogonal projection of the set $\mathcal{S}(n)$ of symmetric matrices onto the tangent space $T_P\mathcal{C}$ with respect to the inner product $< \cdot, \cdot >_P$ on $\mathcal{S}(n)$ is easily seen to be

$$\Pi_P : S(n) \to T_P\mathcal{C}, \quad \xi \mapsto \xi - \sum_{i=1}^{m} \lambda_j P A_j P,$$

where $\lambda = (\lambda_1, \cdots, \lambda_m)^\top$ is uniquely defined by

$$\lambda = \Gamma^{-1}\tau, \quad \Gamma := (\mathrm{trPA_iPA_j})_{\mathrm{i,j=1}}^{\mathrm{m}} = \left(\gamma^{\mathrm{ij}}\right)_{\mathrm{i,j=1}}^{\mathrm{m}}$$

and $\tau = (\mathrm{tr}(\mathrm{A}_1\xi), \cdots, \mathrm{tr}(\mathrm{A_m}\xi))^\top$.

The associated Riemannian gradient is easily computable as follows.

Theorem 4. *Endow $\mathcal{C}$ with the Riemannian metric defined by $< \cdot, \cdot >_P$. The Riemannian gradient* grad $f(P)$ *of f on $\mathcal{C}$ is*

$$PWP - P - \sum_{i,j=1}^{m} \gamma^{ij}(\mathrm{tr}(\mathrm{A_iPWP}) - \mathrm{b_i})\mathrm{PA_jP}.$$

Note that grad $f(P)$ coincides with the Newton direction with respect to the standard, constant metric $< \xi, \eta >= \mathrm{tr}\,(\xi\eta)$ [12]. Using the strict convexity of the cost function one then derives the following global convergence result for a variant of the gradient method, which uses the straight lines $P + t\eta$ as an approximation of the geodesics with respect to $< \cdot, \cdot >_P$; [12]. Thus we are not implementing the Riemannian gradient descent algorithm, but an approximation to it that is were easily computable.

Theorem 5. *For any $P \in \mathcal{C}$ let $\nu(P) = \|\mathrm{grad}\, f(P)\|_P$ and*

$$\alpha(P) := \frac{1 + 2\nu(P) - \sqrt{1 + 4\nu(P)}}{2\nu^2(P)} \in [0, 1].$$

For any initial condition $P_0 \in \mathcal{C}$, the gradient algorithm

$$P_{t+1} = P_t - \alpha(P_t)\left(P_tWP_t - P_t - \sum_{j=1}^{m} \lambda_j P_t A_j P_t\right)$$

converges quadratically fast to the unique global optimum P_* of f on $\mathcal{C}$. Here $\lambda_j = \lambda_j(P_t)$ is defined as above.

A difficulty of the above gradient algorithm is that it requires the computation of feasible initial point $P_0 \in \mathcal{C}$ to start the algorithm. Often, the computation of such feasible points is at least as difficult as the subsequent entire optimization process and the development of efficient numerical optimization algorithms that do not rely on the prior knowledge of a feasible point is an open research problem. Therefore, we do not consider here any discrete-time algorithms that combine optimization steps with the computation of feasible. However, it is relatively easy to develop continuous-time algorithms that achieve this purpose.

Theorem 6. *Let $\mu = \mu(P) = (\mu_1, \cdots, \mu_m)^\top$, $\mu = \Gamma^{-1}\sigma$, $\Gamma := (\mathrm{trPA_iPA_j})_{\mathrm{i,j=1}}^{\mathrm{m}}$ and $\sigma = (\mathrm{tr}(\mathrm{A_1(PWP - 2P)) + b_1}, \cdots, \mathrm{tr}(\mathrm{A_m(PWP - 2P)) + b_m})^\top$. Then the differential equation*

$$\dot{P} = P - PWP + \sum_{j=1}^{m} \mu_j P A_j P$$

converges from any initial condition $P_0 \in \mathcal{P}$ exponentially fast to the global minimum $P_{\min} \in \mathcal{C}$ of $f : \mathcal{C} \to \mathbb{R}$.

The proof of this result is a straightforward consequence of LaSalle's invariance principle and the fact that

$$\frac{d}{dt}\mathrm{tr}(\mathrm{A_jP}) = -\mathrm{tr}(\mathrm{A_jP}) + \mathrm{b_j}, \quad \mathrm{j} = 1, \cdots, \mathrm{m}$$

holds for any solution $P(t)$ of the differential equation.

In some applications, the total sum of finger forces may not be the right cost to impose on the problem. Thus one might replace the sum by a maximum. Such issues arise e.g. if dealing with the minimization of the maximal admissible forces at the fingers. This leads to the following subgradient optimization and is discussed now. Let $W = \mathrm{diag}\,(W_1, \ldots, W_m) \in \mathcal{P}$ and $\widehat{W_k} = \mathrm{diag}\,(0, \ldots, W_k, \ldots, 0)$. Then we are faced with a constrained optimization task for the nonsmooth, strictly convex function on $\mathcal{C}$

$$F(P) = \max_{k=1,\ldots,m} \left(\mathrm{tr}(\widehat{\mathrm{W}}_\mathrm{k}\mathrm{P})\right) - \log\det P.$$

Thus $F(P) = \max(F_1(P), ..., F_N(P))$ and $F_k(P) = \mathrm{tr}(\widehat{\mathrm{W}}_\mathrm{k}\mathrm{P}) - \log\det P$. The gradient of F_r on $\mathcal{C}$ for the Riemannian metric $\langle\cdot,\cdot\rangle_P$ is

$$\mathrm{gradF_r(P) = PW_{(r)}P - P - \sum_{i,j=1}^{m} \gamma^{ij}(tr(A_iPW_{(r)}P) - b_i)PA_jP}.$$

For the continuous-time subgradient descent method

$$\dot{P} \in -\mathrm{grad}\, F(P),$$

we get the following differential inclusion.

Theorem 7. *Let $P(t)$ be a solution of*

$$\dot{P} \in -\,\mathrm{co}\{P\widehat{W}_k P - P - \sum_{j=1}^{m} \lambda_{kj} P A_j P \mid k \in I_F(x)\}$$

with $\lambda_k = (\lambda_{k1}, \ldots, \lambda_{km})^\top$, $\lambda_k = \lambda_k(P) = \Gamma^{-1}\sigma_k$, $\Gamma = (\mathrm{tr(PA_iPA_j)})_{i,j=1}^{m}$, $\sigma_k = (\mathrm{tr(A_1P\widehat{W}_kP) - b_1}, \ldots, \mathrm{tr(A_mP\widehat{W}_kP) - b_m})$. If the initial condition $P(0)$ is contained in $\mathcal{C}$ then $F(P(t))$ is decreasing and $P(t)$ converges to the global minimum of $F(P)$.

Instead of using subgradient descent methods for the nonsmooth function F one can also apply optimization methods to appropriate smooth approximations of F. For example, one may consider, for any integer $q \geq 1$, the smooth strictly convex functions on $\mathcal{C}$

$$\Phi_q(P) := \left(\sum_{k=1}^{m} \left(\mathrm{tr}(\widehat{\mathrm{W}}_\mathrm{k}\mathrm{P})\right)^q\right)^{1/q} - \log\det P.$$

Note, that for $q = 1$ this coincides with the previously discussed cost function. We obtain the following convergence result.

Theorem 8. *Let*

$$\Omega := \left(\sum_{k=1}^{m} \left(\mathrm{tr}(\widehat{\mathrm{W}}_{\mathrm{k}}\mathrm{P})\right)^{q}\right)^{\frac{1-q}{q}} \sum_{k=1}^{m} \left(\mathrm{tr}(\widehat{\mathrm{W}}_{\mathrm{k}}\mathrm{P})\right)^{q-1} \widehat{W}_k$$

$\mu = \mu(P) = (\mu_1, \cdots, \mu_m)^\top$, $\mu = \Gamma^{-1}\sigma$, $\Gamma := (\mathrm{trPA_iPA_j})_{\mathrm{i,j=1}}^{\mathrm{m}}$ *and* $\sigma = (\mathrm{tr}(\mathrm{A}_1(\mathrm{P}\Omega\mathrm{P} - 2\mathrm{P})) + \mathrm{b}_1, \cdots, \mathrm{tr}(\mathrm{A}_{\mathrm{m}}(\mathrm{P}\Omega\mathrm{P} - 2\mathrm{P})) + \mathrm{b}_{\mathrm{m}})^\top$. *Then the differential equation*

$$\dot{P} = P - P\Omega P + \sum_{j=1}^{m} \mu_j P A_j P$$

converges from any initial condition $P_0 \in \mathcal{P}$ *exponentially fast to the global minimum* $P^q_{\min} \in \mathcal{C}$ *of* $\Phi_q : \mathcal{C} \to \mathbb{R}$. *For* $q \to \infty$ *the sequence of minima* $P^q_{\min} \in \mathcal{C}$ *converges to the global minimum of* F *on* $\mathcal{C}$.

This result is shown by similar arguments as Theorem 6.

References

1. P.-A. Absil, R. Mahony, and B. Andrews. Convergence of iterates of descent methods for analytic cost functions. *SIAM J. Optimization*, 16(2):531-547, 2005.
2. D. Agrawal, T. J. Richardson, and R. L. Urbanke. Multiple-antenna signal constellations for fading channels. *IEEE Trans. Information Theory*, 47(6):2618–2626, 2001.
3. B. Andrews. Monotone quantities and unique limits for evolving convex hypersurfaces. *International Mathematics Research Notices*, 20:1002-1031, 1997.
4. J.-P. Aubin and A. Cellina. *Differential inclusions. Set-valued maps and viability theory.* Springer-Verlag, Berlin, 1984.
5. M. Buss, L. Hashimoto, and J. Moore. Dextrous hand grasping force optimization. *IEEE Trans. Robotics and Automation*, 12:267–269, 1996.
6. F. Clarke, Y. Ledyaev, R. Stern, and P. Wolenski. *Nonsmooth analysis and control theory.* Springer-Verlag, New York, 1998.
7. J. H. Conway, R. H. Hardin, and N. J. A. Sloane. Packing lines, planes, etc.: Packings in Grassmannian spaces. *Experimental Mathematics*, 5:139–159, 1996.
8. G. Dirr, U. Helmke, and C. Lageman. Some applications of Riemannian optimization to mechanics. *3rd Workshop on Lagrangian and Hamiltonian Methods for Nonlinear Control*, Nagoya, Japan, 2006.
9. D. Gabay. Minimizing a differentiable function over a differentiable manifold. *J. Optimization Theory and Appl.*, 37:177–219, 1982.
10. S. Gallot, D. Hulin, and J. Lafontaine. *Riemannian Geometry.* Springer-Verlag, Berlin, 2004.
11. G. Han and J. Rosenthal. Geometrical and numerical design of structured unitary space time constellations. *IEEE Trans. Information Theory*, 52(8), 2006.
12. U. Helmke, K. Hüper, and J. Moore. Quadratically convergent algorithms for optimal dextrous hand grasping. *IEEE Trans. Robotics and Automation*, 18:138–146, 2002.
13. U. Helmke, K. Hüper, and J. Trumpf. Newton's method on Grassmann manifolds. Preprint, 2006.

14. U. Helmke and J. Moore. *Optimization and Dynamical Systems.* Springer-Publ., London, 1994.
15. J.-B. Hiriart-Urruty and C. Lemaréchal. *Convex Analysis and Minimization Algorithms. Part 1: Fundamentals.* Springer-Verlag, Berlin, 1993.
16. B. M. Hochwald and T. L. Marzetta. Unitary space-time modulation for multiple-antenna communications in Rayleigh flat fading. *IEEE Trans. Information Theory,* 46(2):543–564, 2000.
17. C. Lageman. Pointwise convergence of gradient-like systems. submitted to *Mathematische Nachrichten*
18. C. Lageman. Convergence of gradient-like dynamical systems and optimizations algorithms. PhD Thesis, Submitted, 2007.
19. C. Lageman and U. Helmke. Optimization methods for the sphere packing problem on Grassmannians. In F. Lamnabhi-Lagarrigue, A. Loria, E. Panteley, and S. Laghrouche, editors, *Taming Heterogeneity and Complexity of Embedded Control: CTS-HYCON Workshop on Nonlinear and Hybrid Control, Paris IV Sorbonne, 10-12 July 2006*, 2007.
20. S. Łojasiewicz. Sur les trajectoires du gradient d'une fonction analytique. In *Seminari di Geometria 1982-1983*, pages 115–117, Università di Bologna, 1984.
21. B. E. Paden and S. S. Sastry. A calculus for computing Filippov's differential inclusion with application to the variable structure control of robot manipulators. *IEEE Trans. Circuits and Systems*, CAS-34(1), 1987.
22. M. Shub. Some remarks on dynamical systems and numerical analysis. In J. L. L. Lara-Carrero, editor, *Dynamical Systems and Partial Differential Equations: Proc. VII ELAM*, pages 69–92. U. Simon Bolivar, Caracas, 1986.
23. S. Smith. *Geometric optimization methods for adaptive filtering.* PhD thesis, Harvard University, 1993.
24. A. R. Teel. Lyapunov methods in nonsmooth optimization. Part I: Quasi-Newton algorithms for Lipschitz regular functions. In *Proc. 39th IEEE Conference on Decision and Control, 2000*, volume 1, pages 112–117, 2000.
25. C. Udriste. *Convex functions and optimization methods on Riemannian manifolds.* Kluwer Academic Publ., Dordrecht, 1994.

Synchronization of Networked Lagrangian Systems*

Mark W. Spong and Nikhil Chopra

Coordinated Science Laboratory
University of Illinois at Urbana-Champaign
1308 West Main Street
Urbana, IL 61801, USA
{mspong,nchopra}@uiuc.edu

Summary. In this paper we study the problem of synchronization of networked Lagrangian systems. This problem arises in applications such as the control of multiple robots, formation control of UAV's, mobile sensor networks, Kuramoto oscillators, and teleoperation. We assume that the agents exchange position and velocity information over a network described by an interconnection graph. We consider only linear interconnections on balanced, directed graphs but similar results can also be derived for nonlinear interconnections on undirected graphs. Our results exploit the well-known passivity property of Lagrangian mechanical systems. We demonstrate synchronization on both fixed and switching graphs. In the case of fixed graphs, the passivity property allows us, in addition, to treat the practically important problem of time delay in communication.

1 Introduction and Motivation

In this paper we study the problem of synchronization of multiple Lagrangian systems that are interconnected over a network. This problem arises in several application domains, including the control of multiple robots, formation control of UAV's, mobile sensor networks, Kuramoto oscillators, and teleoperation. We treat this problem in the context of passivity-based control [20] exploiting the well-known passivity property of Lagrangian mechanical systems [27]. The passivity property allows us, in addition, to treat the practically important problem of time delay in communication. Most of the results in this paper are from [2] and are presented in detail in [5, 4]. Therefore, in this paper, we will present an abridged exposition of the main ideas and refer the reader to the aforementioned references for many of the details.

The problem of communication and control in multi-agent systems is currently a topic of intense research and there are many results available in the literature. We can mention here only a small portion of the literature in this area. For

* This research was partially supported by the Office of Naval Research under Grant N00014-02-1-0011, N00014-05-1-0186, and by the National Science Foundation under Grants ECS-0122412 and INT-0128656.

F. Bullo et al. (Eds.): Lag. & Hamil. Methods for Nonlin. Ctrl., LNCIS 366, pp. 47–59, 2007.
springerlink.com

example, group coordination and formation stability problems are addressed in [1, 14, 7, 16, 15, 30, 6]. Agreement, group consensus and oscillator synchronization problems have been studied in [3, 9, 10, 11, 13, 17, 23, 26, 19, 25, 28, 31]. The effect of communication delay on some of the aforementioned agreement protocols has been studied in [19, 29, 31, 32]. Passivity-based approaches have been used in [21, 1, 22]. The assumption of passivity is a natural one for the types of problems considered. Many of the existing results in the literature, in fact, model the agents as velocity-controlled particles; in other words, as first-order integrators, which are the simplest type of passive systems.

2 Background

2.1 Passivity

In this paper we view the multi-agent coordination problem from an input-output perspective. In such a framework multiple agents exchange output information with their neighbors and construct control strategies, using their own outputs and the outputs received from their neighbors, to achieve the coordination objective. To set the background and notation for what follows, consider a control affine nonlinear system of the form

$$\Sigma \begin{cases} \dot{x} = f(x) + g(x)u \\ y = h(x) \end{cases} \tag{1}$$

where $x \in R^n$, $u \in R^m$, and $y \in R^m$. The functions $f(.) \in R^n$, $g(.) \in R^{n \times m}$, and $h(.) \in R^m$ are assumed to be sufficiently smooth. The admissible inputs are taken to be piecewise continuous and locally square integrable and we note that the dimensions of the input and output are the same. We assume, for simplicity, that $f(0) = 0$ and $h(0) = 0$.

Definition 1. *The nonlinear system Σ is said to be passive if there exists a C^1 storage function $V(x) \geq 0$, $V(0) = 0$ and a function $S(x) \geq 0$ such that for all $t \geq 0$:*

$$V(x(t)) - V(x(0)) = \int_0^t u^T(s)y(s)ds - \int_0^t S(x(s))ds \tag{2}$$

The system Σ is strictly passive if $S(x) > 0$ and lossless if $S(x) = 0$.

An important characterization of passive systems is the following result, due to Moylan [18]:

Theorem 1. *The following statements are equivalent.*

1. *The system (1) is passive*
2. *There exists a C^1 scalar storage function $V : R^n \to R$, $V(x) \geq 0$, $V(0) = 0$, and $S(x) \geq 0$ such that*

$$\begin{aligned} L_f V(x) &= -S(x) \\ L_g V(x) &= \ h^T(x) \end{aligned} \tag{3}$$

where $L_f V(x)$ and $L_g V(x)$ are the Lie derivatives of V with respect to f and g, respectively.

2.2 Lagrangian Systems

Following [27], the Euler-Lagrange equations of motion for an n-degree-of-freedom mechanical system is given as

$$M(q)\ddot{q} + C(q,\dot{q})\dot{q} + g(q) = \tau$$

where $q \in R^n$ is the vector of generalized configuration coordinates, $\tau \in R^n$ is the vector of generalized forces acting on the system, $M(q)$ is the $n \times n$ symmetric, positive definite inertia matrix, $C(q,\dot{q})\dot{q}$ is the matrix of Coriolis and centripetal torques and $g(q)$ is the vector of gravitational torques. Although the above equations of motion are coupled and nonlinear, they exhibit certain fundamental properties due to their Lagrangian dynamic structure, the most important of which is the well-known skew-symmetry of the matrix $\dot{M} - 2C$.

The skew-symmetry property implies passivity of the Euler-Lagrange dynamics from input τ to output $\dot{q}$. More generally, we can induce passivity in a Lagrangian system with different choice of output and a preliminary feedback control. This is know as *Feedback Passivation* [24]. To this end, we choose a preliminary control input as

$$\tau = -M(q)\lambda\dot{q} - C(q,\dot{q})\lambda q + g(q) + u$$

where λ is a positive diagonal matrix and u is an additional control input. Setting $r = \dot{q} + \lambda q$ we obtain the system

$$\begin{aligned} \dot{q} &= -\lambda q + r \\ M(q)\dot{r} + C(q,\dot{q})r &= u \end{aligned}$$

which is of the form

$$\dot{x} = f(x) + g(x)u$$

with state vector $x = (q, r)$ and vector fields f and g given by

$$f(x) = \begin{pmatrix} -\lambda q + r \\ -M(q)^{-1}C(q,\dot{q})r \end{pmatrix} \; ; \; g(x) = \begin{pmatrix} 0 \\ M(q)^{-1} \end{pmatrix}$$

It is easily shown, using the skew-symmetry property, that the above system is passive with input u, output $y = h(x) = r$, and storage function

$$V(x) = r^T M(q) r$$

This system will be the one considered in the remainder of the paper.

2.3 Graph Theory and Communication Topology

Information exchange between agents can be represented as a graph. We give here some basic terminology and definitions from graph theory [8] sufficient to follow the subsequent development.

Definition 2. *By an (information)* **graph** $\mathcal{G}$ *we mean a finite set* $\mathcal{V}(\mathcal{G}) = \{v_i, \ldots, v_N\}$, *whose elements are called* **nodes** *or* **vertices**, *together with set* $\mathcal{E}(\mathcal{G}) \subset \mathcal{V} \times \mathcal{V}$, *whose elements are called* **edges**. *An edge is therefore an ordered pair of distinct vertices.*

If, for all $(v_i, v_j) \in \mathcal{E}(\mathcal{G})$, *the edge* $(v_j, v_i) \in \mathcal{E}(\mathcal{G})$ *then the graph is said to be* **undirected**. *Otherwise, it is called a* **directed graph**.

An edge (v_i, v_j) *is said to be* **incoming with respect to** v_j *and* **outgoing with respect to** v_i *and can be represented as an arrow with vertex* v_i *as its tail and vertex* v_j *as its head.*

The **in-degree** *of a vertex* $v \in \mathcal{G}$ *is the number of edges that have this vertex as a head. Similarly, the* **out-degree** *of a vertex* $v \in \mathcal{G}$ *is the number of edges that have this vertex as the tail.*

If the in-degree equals the out-degree for all vertices $v \in \mathcal{V}(\mathcal{G})$, *then the graph is said to be* **balanced**.

A **path** *of length* r *in a directed graph is a sequence* $v_0, \ldots, v_r$ *of* $r+1$ *distinct vertices such that for every* $i \in \{0, \ldots, r-1\}$, (v_i, v_{i+1}) *is an edge.*

A **weak path** *is a sequence* $v_0, \ldots, v_r$ *of* $r+1$ *distinct vertices such that for each* $i \in \{0, \ldots, r-1\}$ *either* (v_i, v_{i+1}) *or* (v_{i+1}, v_i) *is an edge.*

A directed graph is **strongly connected** *if any two vertices can be joined by a path and is* **weakly connected** *if any two vertices can be joined by a weak path.*

An important property of balanced graphs is the following [8],

Lemma 1. *Let* $\mathcal{G}$ *be a directed graph and suppose it is balanced. Then* $\mathcal{G}$ *is strongly connected if and only if it is weakly connected.*

3 Synchronization

In this section we present our main results on synchronization of Lagrangian systems. Consider N Lagrangian systems, each with passive dynamics of the form

$$\begin{aligned} \dot{x}_i &= f_i(x_i) + g_i(x_i)u_i \\ y_i &= h_i(x_i) \end{aligned} \tag{4}$$

for $i = 1, \ldots, N$.

Definition 3. *1) The agents are said to state synchronize (or synchronize) if*

$$\lim_{t\to\infty} |x_i(t) - x_j(t)| = 0 \quad \forall i, j = 1, \ldots, N \tag{5}$$

2) The agents are said to output synchronize if

$$\lim_{t\to\infty} |y_i(t) - y_j(t)| = 0 \quad \forall i, j = 1, \ldots, N \tag{6}$$

In general, of course, output synchronization is a weaker property than state synchronization. In the particular case of the Lagrangian system (4), with output

$y = r$, output synchronization is equivalent to state synchronization. To see this note that

$$\begin{aligned} r_j(t) - r_i(t) &= (\dot{q}_j(t) + \lambda q_j(t)) - (\dot{q}_i(t) + \lambda q_i(t)) \\ &= \dot{e}_{ij}(t) + \lambda e_{ij}(t) \end{aligned} \tag{7}$$

where $e_{ij}(t) = q_j(t) - q_i(t)$. Equation (7) represents an exponentially stable linear system with input $r_j(t) - r_i(t)$. As shown, for example, in [20], it follows that if $r_j - r_i$ is an $\mathcal{L}^2$ signal that converges asymptotically to zero, then

$$\lim_{t\to\infty} |e_{ij}(t)| = 0 \quad \forall i, j = 1, \dots, N$$

from which state synchronization follows immediately. Therefore, in the remainder of this paper we will consider the problem of output synchronization of the networked systems (4).

Our main assumption in this section is the following:

A1: The agents are weakly connected pointwise in time and form a balanced graph with respect to information exchange. Hence the information graph is also strongly connected from Lemma 1.

We first analyze the case when the communication topology is fixed, i.e., the information graph does not change with time. Suppose that the agents are coupled together using the control

$$u_i = \sum_{j\in\mathcal{N}_i} K(y_j - y_i), \quad i = 1, \dots, N \tag{8}$$

where K is a positive constant and $\mathcal{N}_i$ is the set of agents transmitting their outputs to the ith agent.

Theorem 2. *Consider the dynamical system described by* (4) *with the control* (8). *Then under assumption A1, all solutions of the nonlinear system* (4),(8) *are bounded and the agents output synchronize.*

Proof: Consider a storage function for the N agent system as

$$V = 2(V_1 + \cdots + V_N) \tag{9}$$

where V_i is the storage function for agent i. Using Theorem (1) and the control law (8), the derivative of V along trajectories of the system is

$$\begin{aligned} \dot{V} &= 2\textstyle\sum_{i=1}^N (L_{f_i} V_i + L_{g_i} V_i u_i) \\ &= = 2\textstyle\sum_{i=1}^N (-S_i(x_i) + y_i^T u_i) \\ &= -2\textstyle\sum_{i=1}^N S_i(x_i) + 2\sum_{i=1}^N \sum_{j\in\mathcal{N}_i} y_i^T K(y_j - y_i) \end{aligned} \tag{10}$$

As the information exchange graph is balanced, it can be shown, after some computation [2], that

$$\dot{V} \leq -K \sum_{i=1}^{N} \sum_{j \in \mathcal{N}_i} (y_i - y_j)^T (y_i - y_j) \leq 0 \tag{11}$$

and, hence, all signals are bounded. Lasalle's Invariance Principle [12] and strong connectivity of the network can then be invoked to prove output synchronization of (4).

The Case of Switching Topology

We next consider the case when the graph topology is not constant, such as in nearest neighbor scenarios. In the case of linear coupling, it turns out that output synchronization is still guaranteed for switching graphs under fairly mild assumptions.

Consider now the coupling control law

$$u_i(t) = \sum_{j \in \mathcal{N}_i(t)} K(t)(y_j - y_i), \quad i = 1, \ldots, N \tag{12}$$

In this equation we allow both the gain $K(t) > 0$ and the set $\mathcal{N}_i$ of neighbors of agent i to be time-dependent. We assume that the gain $K(t)$ is piecewise continuous and satisfies

$$K_l \leq K(t) \leq K_b, \quad K_l, K_b > 0 \;\; \forall i, j \tag{13}$$

Theorem 3. *Consider the dynamical system described by* (4)*, coupled together using the control law* (12) *together with the assumption A1. Suppose, in addition, that there are at most finitely many switches in the graph topology on any finite time interval. Then all solution trajectories are bounded and the agents output synchronize.*

A more interesting scenario occurs when the agents are allowed to lose connectivity at every instant, but maintain connectivity in an average sense to be made precise. Let $t_{ij}(e), t_{ij}(d)$ denote the time instances at which the information link or the edge (i, j) is established and broken respectively. In the subsequent analysis we require that

$$t_{ij}(d) - t_{ij}(e) \geq \delta_{ij} > 0 \;\; \forall i, j \in \mathcal{E} \tag{14}$$

The above assumption is similar to the dwell time assumption in Theorem 5 of [9], but is less stringent as it does not require the dwell time to be the same for every agent, but implies it it to be uniformly bounded away from zero. The notion of joint connectivity was introduced in [9] (see also [31]), and has been used by several authors including [17, 23, 26]. The agents are said to be jointly connected across the time interval $[t, t+T]$, $T > 0$ if the agents are weakly connected across the union $\cup_{\sigma \in [t,t+T]} \mathcal{E}(\mathcal{G}(\sigma))$. Note that the above assumption (14) implies that there are only finitely many distinct graphs in this union for each T. Under a modified assumption

A2: The agents form a balanced information graph pointwise in time, and there exists an infinite sequence of bounded, non-overlapping time intervals across which the agents are jointly connected,

we can now prove the following:

Theorem 4. *Consider the dynamical system described by* (4), *coupled together using the control described by* (12) *together with the assumption A2. Then all solution trajectories are bounded and the agents output synchronize.*

Proof. We note that a proof for this result implies Theorem 3 as a corollary. The proof basically proceeds by demonstrating that

$$V = 2(V_1 + \ldots + V_N) \tag{15}$$

is a common storage function for the switching system [33]. Using the pointwise balanced nature of the information graph, as in Theorem 2, the derivative of (15) can be written as

$$\dot{V}(t) = -2\sum_{i=1}^{N} S_i(x_i) - K(t)\sum_{i=1}^{N}\sum_{j\in\mathcal{N}_i(t)} (y_i - y_j)^T(y_i - y_j) \leq 0 \tag{16}$$

implying that all trajectories of the nonlinear system described by (4) and (12) are bounded. Also, $\lim_{t\to\infty} V(t)$ exists and is finite since V is lower bounded by zero. Consider a infinite sequence $V(t_i)$ $i = 1, \ldots,$ where the times t_i approach infinity as i approaches infinity. Then, using Cauchy's convergence criteria, $\forall\epsilon \geq 0$ $\exists M > 0$ s.t $\forall k > l \geq M$

$$|V(t_l) - V(t_k)| \leq \epsilon \tag{17}$$

As the above statement is true $\forall k, l > M$, in particular we choose k, l such that the time interval $[t_k, t_l]$ encompasses some time interval across which the agents are jointly connected. Existence of such a time interval is guaranteed by assumption A2. Therefore, from (16),

$$\begin{aligned}
&|\sum_{i=1}^{N}\sum_{j\in\mathcal{N}_i(t_k,t_l)} \int_{t_l}^{t_k} K(t)(y_i - y_j)^T(y_i - y_j)dt| \leq \epsilon \\
&\Rightarrow |\sum_{j\in\mathcal{N}_i(t_k,t_l)} \int_{t_l}^{t_k} K(t)(y_i - y_j)^T(y_i - y_j)dt| \leq \epsilon \;\; \forall i \\
&\Rightarrow |\sum_{j\in\mathcal{N}_i(t_k,t_l)} \int_{t_{ij}(e)}^{t_{ij}(d)} K(t)(y_i - y_j)^T(y_i - y_j)dt| \leq \epsilon \;\; \forall i
\end{aligned}$$

As ϵ is arbitrary this implies

$$\lim_{t\to\infty} \int_{t}^{t+\delta_{ij}} ||y_i - y_j||^2 dt = 0, \;\; \forall j \in \mathcal{N}_i(t, t+T), \;\; \forall i$$

As $x_i \in \mathcal{L}_\infty, \ \ \forall i$, $\dot{y}_i \in \mathcal{L}_\infty, \ \ \forall i$, and hence the above limit implies that

$$\lim_{t\to\infty} ||y_i - y_j|| = 0 \ \ \forall j \in \mathcal{N}_i(t, t+T) \ \ \forall i$$

Weak connectivity of the network across the time interval $[t, t+T]$ then guarantees output synchronization.

Remark 1

1. If the graph is undirected, the above results are still valid in the case that different gains (or weights) $K_{ij}(t)$ couple the agents, provided $K_{ij}(t) = K_{ji}(t) \ \ \forall i, j$.

3.1 Output Synchronization with Time Delay

In this section, we study the problem of output synchronization when there are communication delays in the network. The delays are assumed to be constant and bounded, T_{ij} denotes the communication delay from the ith agent to the jth agent. We do not assume T_{ij} to be necessarily equal to T_{ji}.

Definition 4. *The agents are said to output synchronize if*

$$\lim_{t\to\infty} |y_i(t - T_{ij}) - y_j(t)| = 0 \ \ \forall i, j \tag{18}$$

where T_{ij} is the sum of the delays along the path from the ith agent to the jth agent.

As before, for the class of systems considered here, output synchronization in the above sense implies state synchronization

$$\lim_{t\to\infty} |x_i(t - T_{ij}) - x_j(t)| = 0 \ \ \forall i, j \tag{19}$$

Let the agents be coupled together using the control

$$u_i(t) = \sum_{j\in\mathcal{N}_i} K(y_j(t - T_{ji}) - y_i) \tag{20}$$

where $K > 0$ is a constant.

A3: In addition to assumption A1, we assume that there exists a unique path between any two nodes.

It is to be noted that assumption A3 is needed to ensure uniqueness of $T_{ij} \ \forall j \neq i$.

Theorem 5. *Consider the dynamical system described by* (4) *with the control law* (20)*. Then under the assumption A3, all trajectories of the nonlinear system* (4), (20) *are bounded and the agents output synchronize in the sense of* (18).

Proof: Consider a storage function for the N-agent system as

$$V = K\sum_{i=1}^{N}\sum_{j\in\mathcal{N}_i}\int_{t-T_{ji}}^{t} y_j^T(\tau)y_j(\tau)d\tau + 2(V_1 + \ldots + V_N) \tag{21}$$

The derivative of V along trajectories of the system is given as

$$\dot{V} = K\sum_{i=1}^{N}\sum_{j\in\mathcal{N}_i}(y_j^T y_j - y_j(t-T_{ji})^T y_j(t-T_{ji})) + 2\sum_{i=1}^{N}(L_{f_i}V_i + L_{g_i}V_i u_i)$$

Using Theorem (1), the derivative reduces to

$$\begin{aligned}\dot{V} &= K\sum_{i=1}^{N}\sum_{j\in\mathcal{N}_i}(y_j^T y_j - y_j(t-T_{ji})^T y_j(t-T_{ji})) + 2\sum_{i=1}^{N}(-S_i(x_i) + y_i^T u_i)\\ &= K\sum_{i=1}^{N}\sum_{j\in\mathcal{N}_i)}(y_j^T y_j - y_j(t-T_{ji})^T y_j(t-T_{ji}))\\ &-2\sum_{i=1}^{N}S_i(x_i) + 2\sum_{i=1}^{N}\sum_{j\in\mathcal{N}_i}y_i^T K(y_j(t-T_{ji}) - y_i)\end{aligned}$$

As the graph is balanced, it can be shown that

$$\begin{aligned}\dot{V} &\leq -K\sum_{i=1}^{N}\sum_{j\in\mathcal{N}_i}(y_j(t-T_{ji}) - y_i)^T(y_j(t-T_{ji}) - y_i)\\ &\leq 0\end{aligned}$$

Therefore every solution is bounded and $\lim_{t\to\infty} V(t)$ exists and is finite. Barbalat's Lemma [12] can be used to show that

$$\lim_{t\to\infty}|y_j(t-T_{ji}) - y_i| = 0 \quad \forall j \in \mathcal{N}_i \quad i = 1,\ldots,N$$

Strong connectivity of the interconnection graph then implies output synchronization of (4) in the sense of (18). (See [2] for the details.)

Example 1: As a simple example, consider four point masses with dynamics

$$m_i\ddot{q}_i = \tau_i \ ; \ i = 1,\ldots,4 \tag{22}$$

It was shown recently in [14] that the flocking and consensus problem for second order drift free systems is nontrivial, and mimicking the linear consensus protocols, for example [19], can potentially make the system unstable for certain choices of the coupling gains. We apply the proposed methodology to ensure asymptotic convergence of the position and velocity tracking errors to the origin. Following our above derivation, applying the preliminary feedback

$$\tau_i = -m_i\lambda\dot{q}_i + u_i \ ; \ i = 1,\ldots,4 \tag{23}$$

the systems reduce to

$$\begin{aligned} \dot{q}_i &= -\lambda q_i + r_i \\ m_i \dot{r}_i &= u_i \; ; \; i = 1, \ldots, 4 \end{aligned} \tag{24}$$

where $r_i = \dot{q}_i + \lambda q_i$. Let the communication structure among the agents be described by a ring topology as shown in Figure 1.

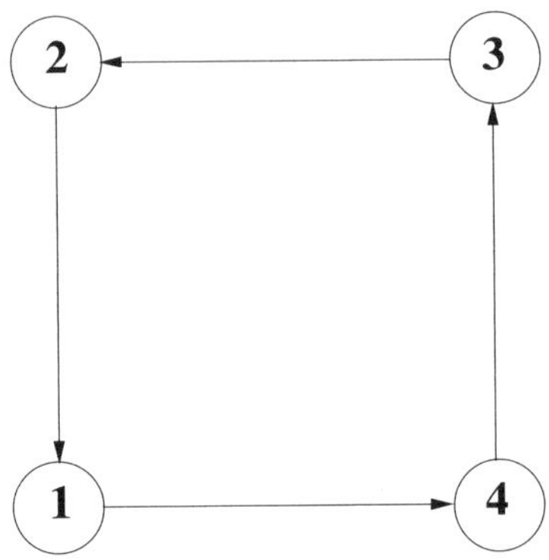

Fig. 1. The agents of Example 1 communicate in a ring topology

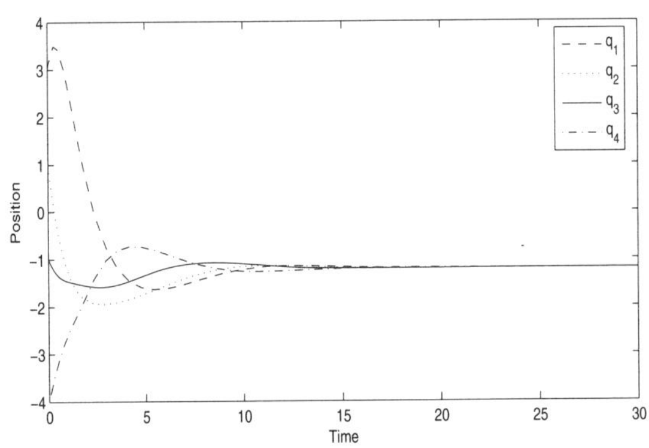

Fig. 2. Simulation with $m_1 = 1$, $m_2 = 2$, $m_3 = 4$, $m_4 = 4.5$, $K = 1$, $\lambda = 1$. The four agents synchronize in spite of the communication delay.

Then the closed loop dynamics are given as

$$\begin{aligned} \dot{q}_1 &= -q_1 + r_1 \; ; \; m_1 \dot{r}_1 = K(r_2(t - T_{21}) - r_1) \\ \dot{q}_2 &= -q_2 + r_2 \; ; \; m_2 \dot{r}_2 = K(r_3(t - T_{32}) - r_2) \\ \dot{q}_3 &= -q_3 + r_3 \; ; \; m_3 \dot{r}_3 = K(r_4(t - T_{43}) - r_3) \\ \dot{q}_4 &= -q_4 + r_4 \; ; \; m_4 \dot{r}_4 = K(r_1(t - T_{14}) - r_4) \end{aligned} \tag{25}$$

where T_{ij} denotes the delay from the ith agent to the jth agent. In this simulation $T_{21} = .1s$, $T_{32} = .2s$, $T_{43} = 1s$, $T_{14} = 1.5s$ and as seen in Figure 2, the agents synchronize asymptotically.

4 Conclusions

In this paper the multi-agent coordination and control problem for Lagrangian systems was treated from an input-output perspective. Given agents with nonlinear control affine, input/output passive, open loop stable dynamics, control laws were developed to achieve output synchronization. In the specific case where the coupling control is linear in the outputs, it was shown that output synchronization is robust to arbitrary constant (bounded) time delays.

It is easy to extend the results in this paper to underactuated systems provided that the zero-dynamics are asymptotically stable, i.e. provided that the system is minimum phase (See [21] for some results along this line.) However, since underactuated Lagrangian systems are typically only weakly minimum phase, the extension of the present results to underactuated systems is still open.

The problem of input saturation is also an important problem to consider. For example, in a mobile robot or UAV application, the agents spatial and/or velocity separation may initially be large, which may lead to large control effort. This problem, and the general problem of nonlinear coupling among the agents, is treated in [5], where we show that output synchronization is still possible provided that the interconnection graph is undirected.

References

1. Murat Arcak. Passivity as a design tool for group coordination. In *Proc. ACC*, Minneapolis, MN, June 2006.
2. N. Chopra. *Output Synchronization of Networked Passive Systems*. Ph.d. thesis, University of Illinois at Urbana-Champaign, Department of Industrial and Enterprise Systems Engineering, 2006.
3. N. Chopra and M. W. Spong. On synchronization of kuramoto oscillators. In *Proceedings of the Joint 44th IEEE Conference on Decision and Control and European Control Conference (CDC-ECC'05)*, pages 3916–3922, Seville, Spain, December 2005.
4. N. Chopra and M. W. Spong. Output synchronization of networked passive systems. Preprint, 2006.
5. N. Chopra and M. W. Spong. Output synchronization of nonlinear systems with time delay in communication. To Appear, IEEE Conference on Decision and Control, 2006.
6. J. Cortés, S. Martínez, T. Karatas, and F. Bullo. Coverage control for mobile sensing networks. *IEEE Transactions on Robotics and Automation*, 20(2):243–255, 2004.
7. J.A. Fax and R.M. Murray. Information flow and cooperative control of vehicle formations. *IEEE Transactions on Automatic Control*, 49(4):1465–1476, 2004.
8. C. Godsil and G. Royle. Algebraic graph theory. In *Springer Graduate Texts in Mathematics No. 207*. Springer, 2001.
9. A. Jadbabaie, J. Lin, and A. S. Morse. Coordination of groups of mobile autonomous agents using nearest neighbor rules. *IEEE Transactions on Automatic Control*, 48:988–1001, June 2003.

10. A. Jadbabaie, N. Motee, and M. Barahona. On the stability of the kuramoto model of coupled nonlinear oscillators. In *Proceedings of the American Control Conference*, pages 4296–4301, 2004.
11. E. W. Justh and P. S. Krishnaprasad. A simple control law for UAV formation flying. Technical 2002-38, Institute for Systems Research, 2002.
12. H. K. Khalil. Nonlinear systems. In *Upper Saddle River*. Prentice Hall, New Jersey, 2002.
13. Y. Kuramoto. In international symposium on mathematical problems in theoretical physics. In *Lecture Notes in Physics*, volume 39. Springer, New York, 1975.
14. D. J. Lee and M. W. Spong. Stable flocking of multiple inertial agents on balanced graphs. *IEEE Transactions on Automatic Control*, submitted, 2005.
15. N. E. Leonard and E. Fiorelli. Virtual leaders, artificial potentials and coordinated control of groups. In *Proceedings of the 40th IEEE Conference on Decision and Control*, pages 2968–2973, Orlando, Florida, 2001.
16. J. A. Marshall, M. E. Broucke, and B. A. Francis. Formation of vehicles in cyclic pursuit. *IEEE Transactions on Automatic Control*, 49(11):1963–1974, 2004.
17. L. Moreau. Stability of multi-agent systems with time-dependent communication links. *IEEE Transactions on Automatic Control*, 50(2):169–182, 2005.
18. P. J. Moylan. Implications of passivity for a class of nonlinear systems. *IEEE Tran. Automat. Contr.*, 19:373–381, August 1974.
19. R. Olfati-Saber and R.M. Murray. Consensus problems in networks of dynamic agents with switching topology and time-delays. *IEEE Transactions on Automatic Control*, 49:1520–1533, September 2004.
20. R. Ortega and M. W. Spong. Adaptive motion control of rigid robots: a tutorial. *Automatica*, 25:877–888, 1989.
21. A. Y. Pogromsky. Passivity based design of synchronizing systems. *Int. J. of Bifurcation and Chaos*, 8:295–319, 1998.
22. A. Y. Pogromsky and H. Nijmeijer. Cooperative oscillatory behavior of mutually coupled dynamical systems. *IEEE Transactions on Circuits and Systems 1*, 48:152–162, 2001.
23. W. Ren and R. W. Beard. Consensus seeking in multi-agent systems using dynamically changing interaction topologies. *IEEE Transactions on Automatic Control*, 5(5):655–661, May 2005.
24. R. Sepulchre, M. Janković, and P. Kokotović. *Constructive Nonlinear Control*. Springer-Verlag, London, 1997.
25. R. Sepulchre, D. Paley, and N. Leonard. Collection motion and oscillator synchronization. In V. Kumar, N. Leonard, and A. S. Morse, editors, *Cooperative Control, Lecture Notes in Control and Information Sciences*, volume 309. Springer Verlag, 2004.
26. J.-J. E. Slotine and W. Wang. A study of synchronization and group cooperation using partial contraction theory. In V. Kumar, N. E. Leonard, and A. S. Morse, editors, *Cooperative Control*, volume 309 of *Lecture Notes in Control and Information Sciences*, 2004.
27. M. W. Spong, S. Hutchinson, and M. Vidyasagar. *Robot Modeling and Control*. John Wiley & Sons, Inc., New York, 2006.
28. S. H. Strogatz. From Kuramoto to Crawford: exploring the onset of synchronization in populations of coupled oscillators. *Physica D*, 143:1–20, 2000.
29. H. G. Tanner and D. K. Christodoulakis. State synchronization in local-interaction networks is robust with respect to time delays, December 2005.

30. H. G. Tanner, G. J. Pappas, and V. Kumar. Leader-to-formation stability. *IEEE Transactions on Robotics and Automation*, 20(3):443–455, 2004.
31. J. N. Tsitsiklis, D. P. Bertsekas, and M. Athans. Distributed asynchronous deterministic and stochastic gradient optimization algorithms. *IEEE Transactions on Automatic Control*, 31(9):803–812, 1986.
32. W. Wang and J.-J. E. Slotine. Contraction analysis of time-delayed communications and group cooperation. *IEEE Transactions on Automatic Control*, 51(4):712–717, 2006.
33. J. Zhao and David J. Hill. Dissipative theory for switched systems. In *Proceedings of the Joint 44th IEEE Conference on Decision and Control and European Control Conference (CDC-ECC'05)*, pages 7003–7008, Seville, Spain, December 2005.

An Algorithm to Discretize One-Dimensional Distributed Port Hamiltonian Systems

Luca Bassi, Alessandro Macchelli, and Claudio Melchiorri

CASY–DEIS, University of Bologna (Italy)
{lbassi,amacchelli,cmelchiorri}@deis.unibo.it

1 Introduction

A key issue when dealing with distributed parameter systems is the solution of the set of partial differential equations that compose the system model. Even if we restrict our analysis to the linear case, it is often impossible to find a closed-form solution for this kind of equations, especially in control applications, where the forcing action on the system generates time-varying boundary conditions. Therefore, numerical solvers play a key role as support tools for the analysis of this kind of systems.

The algorithm presented here aims to provide a discretization procedure for lossless distributed port-Hamiltonian systems with spatial domain $\mathcal{Z} \equiv [0, L] \subset \mathbb{R}$. This work has been inspired by [6], where a spatial discretization procedure based on a particular type of mixed finite element and able to provide a finite dimensional input/output approximated system has been introduced. More in details, from the analysis of the geometric structure of an infinite dimensional port Hamiltonian system (i.e. of its Stokes–Dirac structure, [15]), a finite dimensional approximation still in port Hamiltonian form can be obtained. In this work, an extension to a wider class of distributed port Hamiltonian systems is illustrated. More precisely, it is possible to approximate infinite dimensional systems with higher order derivatives in the spatial variable.

In finite dimensions, a Dirac structure is a *particular* linear subset $\mathbb{D}$ of the space of power variables $\mathcal{F} \times \mathcal{E}$ in which $\mathcal{F}$ and $\mathcal{E}$ are finite dimensional vector spaces in duality representing the space of flows and efforts respectively. Main property is that, given $(f, e) \in \mathbb{D}$, the associated power $\langle f, e\rangle$ is equal to 0, where $\langle , \rangle$ is the duality product, [3, 14]. Then, this subset is able to mathematically describe the flows of energy within the system and between the system itself and the environment. The same property holds in the infinite dimensional case, with the difference that also the space of boundary terms $\mathcal{W}$ has to be taken into account. The Stokes–Dirac structure is, then, a subspace of $\mathcal{F} \times \mathcal{E} \times \mathcal{W}$ characterized by an energy conservation property which allows to relate the variation of internal energy with the boundary power flow.

In [15], flows and efforts of the infinite dimensional system were elements of spaces of differential forms [4] and the corresponding discretization method

F. Bullo et al. (Eds.): Lag. & Hamil. Methods for Nonlin. Ctrl., LNCIS 366, pp. 61–73, 2007.
springerlink.com

[6] was therefore accomplished through an approximation of these forms via Whitney forms [1], the differential form counterpart of the test functions used in the finite element method. The approach followed here is on the other hand based on the formulation described in [8, 13], which assumes that the power variables are vector-valued *smooth* functions with compact support. Even if the *physical* meaning of flow and effort is lost, this approach simplifies the definition of Stokes–Dirac structures in the case of higher order derivatives in the spatial variables and, at the same time, brings the port Hamiltonian framework closer to the classical theory of infinite dimensional systems [7], for which well-established results on analysis and control exist, [2].

At first, the analysis will be focused on linear mathematical models where only first-order spatial derivatives appear, then the extension of the procedure in case of higher-order derivatives will be briefly outlined. It is worth of mention that the introduction of a limited amount of non-linearity in the model does not preclude the application of this method, as briefly discussed in [11].

2 Problem Description and Notation

As discussed in [15, 8, 13], the classical notion of power-conserving structure, i.e. of Dirac structure, the heart of the port-Hamiltonian framework [14], needs to be reviewed in infinite dimensions in order to take into account the power flow through the boundary of the system. The result is the so-called Stokes-Dirac structure which can be defined as follows, under the hypothesis of one-dimensional spatial domain $\mathcal{Z} = [0, L]$, $L > 0$.

Denote by $\mathcal{F}$ the space of flows, which is the set of vector-valued smooth functions defined on $\mathcal{Z}$ with compact support and assume for simplicity that the space of efforts is $\mathcal{E} \equiv \mathcal{F}$. Then, a possible definition of Stokes–Dirac structure $\mathbb{D}$ is:

$$\mathbb{D} := \left\{(f, e, w) \in \mathcal{F} \times \mathcal{E} \times \mathcal{F} \mid f = -\mathcal{J}e,\ w = \mathcal{B}_{\mathcal{J}}(e)\right\} \tag{1}$$

where $\mathcal{J}$ is a skew-adjoint matrix constant differential operator given by:

$$\mathcal{J} := -\sum_{h=0}^{N} A_h \frac{\partial^h}{\partial z^h} \tag{2}$$

with $A_h = (-1)^{h+1} A_h^{\mathrm{T}}$ and $N \geq 1$ the order of the operator. Moreover, $\mathcal{B}_{\mathcal{J}}$ is an operator induced by $\mathcal{J}$ which defines the boundary variables w and, consequently, the space of boundary terms $\mathcal{W}$. In the case of one-dimensional spatial domain, this space results into a (finite dimensional) linear space of proper dimension. These objects can be easily determined by integrating $e^{\mathrm{T}} f$ on $\mathcal{Z}$, which provides the variation of internal energy, and applying the Stokes theorem. Roughly speaking, $\mathcal{W}$ is given by the restriction of the effort e and of its derivatives up to the $(N-1)$-th order on the boundary $\partial\mathcal{Z} \equiv \{0, L\}$ of the spatial domain.

The fundamental hypothesis of the algorithm is the possibility to define a power port on $\partial\mathcal{Z}$, i.e. a pair flow/effort whose *combination* should provide the power flowing inside the system. In other words, it is required to partition the boundary variables in couples of power conjugate inputs and outputs. Since the definition (1) of Stokes–Dirac structure developed around the differential operator (2) does not automatically guarantee such a property, in the following sections some fairly general sufficient conditions to check whether the procedure can be applied or not are illustrated. Finally, for the sake of compactness, matrices $r \times s$ containing only zeros will be written as $0_{r,s}$ and I_n will represent a $n \times n$ identity matrix.

3 Discretization of First-Order Structures

Let us start by setting $N = 1$ in (2). From (1), the relation between internal flows and efforts is given by:

$$f = A_0 e + A_1 \frac{\partial e}{\partial z} \tag{3}$$

where it has been assumed that $f = \left(f^1, \ldots, f^n\right)^{\mathrm{T}}$, $e = \left(e^1, \ldots, e^n\right)^{\mathrm{T}}$ and A_0 and A_1 are $n \times n$ matrices. If $\phi = f - A_0 e$ are the *modified flows*, (3) can be written as:

$$\phi = A_1 \frac{\partial e}{\partial z} \tag{4}$$

Intuitively, this means that the efforts linked algebraically with the flows, i.e. the efforts upon which A_0 acts, have the same nature and thus it is useful lumping them together. Now, it is necessary to check whether the problem is well-posed, in other words whether the differential operator $A_1 \frac{\partial}{\partial z}$ allows the definition of a power port on the boundary.

Assumption 3.1. *A_1 has even rank equal to $2l$.*

Assumption 3.2. *There exists a rotation matrix R and an $l \times l$ matrix $A_{1/2}$ such that A_1 may be written in the form:*

$$\tilde{A}_1 = R^{\mathrm{T}} A_1 R = \begin{pmatrix} 0_{l,l} & 0_{l,n-2l} & A_{1/2} \\ 0_{n-2l,l} & 0_{n-2l,n-2l} & 0_{n-2l,l} \\ A_{1/2}^{\mathrm{T}} & 0_{l,n-2l} & 0_{l,l} \end{pmatrix} \tag{5}$$

As it turns out, assuming the block structure (5) for the A_1 matrix is not particularly restrictive. In many important cases [15, 12, 10], the system equations are already in the desired form, thus $R = I_n$, or this can be achieved by means of a simple row/column permutation. Differently, when the equations are not expressed in term of power variables, but of their linear combinations, the discretization procedure calls for a change of variables in order to let the underlying port structure appear. Since A_1 is symmetric, it is always possible to find a matrix R_1 composed by n orthogonal eigenvectors such that $R_1^{\mathrm{T}} A_1 R_1 = \mathrm{diag}(\lambda_1, \ldots, \lambda_n)$. As prescribed by Assumption 3.1, the number of non-zero eigenvalues is $2l$. Then, let $\Lambda^+ := \{\lambda_i : \lambda_i > 0\}$ and $\Lambda^- := \{\lambda_i : \lambda_i < 0\}$.

Since, from Assumption 3.2, we have that $\Lambda^+ = -\Lambda^- = \Lambda$, it is possible to isolate a power port quite easily. Let us choose R_1 such that:

$$A_1' = R_1^{\mathrm{T}} A_1 R_1 = \begin{pmatrix} \Lambda & 0_{l,n-2l} & 0_{l,l} \\ 0_{n-2l,l} & 0_{n-2l,n-2l} & 0_{n-2l,l} \\ 0_{l,l} & 0_{l,n-2l} & -\Lambda \end{pmatrix}$$

Then, the desired change of variables is given by $R = R_1 R_2$, where

$$R_2 = \begin{pmatrix} \frac{1}{\sqrt{2}} I_l & 0_{l,n-2l} & \frac{1}{\sqrt{2}} I_l \\ 0_{n-2l,l} & I_{n-2l} & 0_{n-2l,l} \\ -\frac{1}{\sqrt{2}} I_l & 0_{l,n-2l} & \frac{1}{\sqrt{2}} I_l \end{pmatrix} \text{ and } \tilde{A}_1 = \begin{pmatrix} 0_{l,l} & 0_{l,n-2l} & \Lambda \\ 0_{n-2l,l} & 0_{n-2l,n-2l} & 0_{n-2l,l} \\ \Lambda & 0_{l,n-2l} & 0_{l,l} \end{pmatrix}$$

Since the eigenvalues of A_1 define the slopes of the characteristic curves of the linear partial differential equation $\partial_t u = A_1 \partial_z u$, i.e. the curves where the solution remains constant, the constraint $\Lambda^+ = -\Lambda^-$ requires the system to have two families of real characteristics, or two classes of wave-like solutions, which is a general consequence of power conservation. It is possible to conclude, then, that the well-posedness of the discretization problem is strictly related, as expected, to the energetic properties of the system. The remaining part of this section will therefore refer to first order Stokes–Dirac structures for which A_1 in (3) is in the form (5).

Denote by $\mathcal{Z}_{ab} := \{z : z \in [a,b]\}$ a small portion of the one-dimensional domain $\mathcal{Z}$, with $0 \le a < b \le L$. For every $i = 1, \dots, n$, the flows f^i and ϕ^i are approximated as:

$$f^i(t,z) = f^i_{ab}(t)\omega^i_{ab}(z) \qquad \phi^i(t,z) = \phi^i_{ab}(t)\omega^i_{ab}(z) \tag{6}$$

with $\int_{\mathcal{Z}_{ab}} \omega^i_{ab} = 1$, whereas the efforts e^i are treated as

$$e^i(t,z) = e^i_a(t)\omega^i_a(z) + e^i_b(t)\omega^i_b(z) \tag{7}$$

with $\omega^i_a(a) = \omega^i_b(b) = 1$ and $\omega^i_a(b) = \omega^i_b(a) = 0$. Clearly, it is possible to rewrite (4) as:

$$\phi^i_{ab}\omega^i_{ab} = \sum_{j=1}^{n} A_1^{ij} \left(e^j_a \frac{\partial \omega^j_a}{\partial z} + e^j_b \frac{\partial \omega^j_b}{\partial z} \right) \tag{8}$$

Assumption 3.3 (compatibility of forms, [6]). *The approximating functions ω^i_{ab}, ω^i_a, ω^i_b should be chosen in such a way that for every e^i_a and e^i_b is possible to find a group of quantities ϕ^i_{ab} that satisfies (4).*

This assumption provides additional constraints for the ω functions (6) and (7). In particular, for every i and j such that $A_1^{ij} \neq 0$:

$$\begin{aligned} -\frac{\partial \omega^j_a}{\partial z} &= \frac{\partial \omega^j_b}{\partial z} = \omega^i_{ab} \\ \frac{\partial}{\partial z}(\omega^j_a + \omega^j_b) &= 0 \qquad \Rightarrow \omega^j_a + \omega^j_b = 1 \\ \int_{\mathcal{Z}_{ab}} (\omega^j_a + \omega^j_b)\omega^j_{ab} &= \int_{\mathcal{Z}_{ab}} \omega^j_{ab} = 1 \end{aligned} \tag{9}$$

Once the conditions (9) have been taken into account, (8) can be written as:

$$\phi_{ab} = A_1(e_b - e_a) \tag{10}$$

Assumptions 3.1 and 3.2 guarantee that it is possible to divide the boundary variables into two components, i.e. a flow and an effort, so that a port can be defined. The boundary variables f_a^B, f_b^B, e_a^B and e_b^B are simply the restriction along the boundary of the efforts, and therefore can be equated to the discretized efforts e_a and e_b as follows:

$$\begin{aligned} f_a^B &= \left(e_a^1, \dots, e_a^l\right)^{\mathrm{T}} & f_b^B &= \left(e_b^1, \dots, e_b^l\right)^{\mathrm{T}} \\ e_a^B &= \left(e_a^{n-l+1}, \dots, e_a^n\right)^{\mathrm{T}} & e_b^B &= \left(e_b^{n-l+1}, \dots, e_b^n\right)^{\mathrm{T}} \end{aligned} \tag{11}$$

Taking into account (5), the power balance associated to the small portion of the system with spatial domain $\mathcal{Z}_{ab}$ is given by the sum of the variation of internal energy and the power flow through the boundary in $z = a$ and $z = b$:

$$P_{ab}^{net} = \int_{\mathcal{Z}_{ab}} \sum_{i=1}^{n} e^i f^i + {f_a^B}^{\mathrm{T}} A_{1/2} e_a^B - {f_b^B}^{\mathrm{T}} A_{1/2} e_b^B$$

¿From the fundamental energy conservation property of the Stokes–Dirac structures, it is clear that $P_{ab}^{net} = 0$. From (6) and (7) and since A_0 is skew-symmetric, we get:

$$P_{ab}^{net} = \sum_{j\in\sigma} f_{ab}^j \left(\alpha_{ab}^j e_a^j + (1-\alpha_{ab}^j) e_b^j\right) + {f_a^B}^{\mathrm{T}} A_{1/2} e_a^B - {f_b^B}^{\mathrm{T}} A_{1/2} e_b^B \tag{12}$$

where $\sigma = \{1, \dots, l, n-l+1, \dots, n\}$ and

$$\alpha_{ab}^j = \begin{cases} \int_{\mathcal{Z}_{ab}} \omega_{ab}^j \omega_a^j & \text{if } j \in \sigma \\ 0 & \text{otherwise} \end{cases}$$

Among the coefficients α_{ab}^j, at most l are independent. In fact, given two pairs of functions ω_{ab}^i, ω_a^i and ω_{ab}^j, ω_a^j for which $A_1^{ij} \neq 0$, we have that:

$$\int_{\mathcal{Z}_{ab}} \left(\omega_{ab}^i \omega_a^i + \omega_{ab}^j \omega_a^j\right) = -\int_{\mathcal{Z}_{ab}} \left(\frac{\partial \omega_a^j}{\partial z}\omega_a^i + \frac{\partial \omega_a^i}{\partial z}\omega_a^j\right) = -\int_{\mathcal{Z}_{ab}} \frac{\partial}{\partial z}\left(\omega_a^i \omega_a^j\right) = 1$$

and, consequently:

$$\alpha_{ab}^j = 1 - \alpha_{ab}^i =: \alpha_{ba}^i, \quad \forall i,j : A_1^{ij} \neq 0 \tag{13}$$

From the power balance relation (12), it is possible to define the finite dimensional efforts e_{ab}^j which are dual to f_{ab}^j for every $j \in \sigma$:

$$\begin{pmatrix} \left(e_{ab}^1, \dots, e_{ab}^l\right)^{\mathrm{T}} \\ \left(e_{ab}^{n-l+1}, \dots, e_{ab}^n\right)^{\mathrm{T}} \end{pmatrix} = \begin{pmatrix} \bar{\alpha}_{ab,1} & 0_{l,l} \\ 0_{l,l} & \bar{\alpha}_{ab,2} \end{pmatrix} \begin{pmatrix} f_a^B \\ e_a^B \end{pmatrix} + \begin{pmatrix} \bar{\alpha}_{ba,1} & 0_{l,l} \\ 0_{l,l} & \bar{\alpha}_{ba,2} \end{pmatrix} \begin{pmatrix} f_b^B \\ e_b^B \end{pmatrix} \tag{14}$$

where $\bar{\alpha}_{ab,1} := \text{diag}(\alpha_{ab}^1, \ldots, \alpha_{ab}^l)$, $\bar{\alpha}_{ab,2} := \text{diag}(\alpha_{ab}^{n-l+1}, \ldots, \alpha_{ab}^n)$, $\bar{\alpha}_{ba,1} := \text{diag}(\alpha_{ba}^1, \ldots, \alpha_{ba}^l)$ and $\bar{\alpha}_{ba,2} := \text{diag}(\alpha_{ba}^{n-l+1}, \ldots, \alpha_{ba}^n)$. Finally, it is necessary to derive an equation linking the discretized flows f_{ab} and the discretized modified flows ϕ_{ab}. This is possible under the following assumption.

Assumption 3.4. *The functions ω_{ab}^i must be consistent for each couple of variables that are linked through an algebraic relation, i.e.:*

$$\forall A_0^{ij}, A_0^{ik} \neq 0 \Rightarrow \omega_{ab}^j(z) = \omega_{ab}^k(z)$$

Since $\phi = f - A_0$, taking into account (6), (7), the properties of the ω functions (9) and the definition of the boundary power variables (11), we have:

$$\begin{aligned}
\phi_{ab}^i \omega_{ab}^i &= f_{ab}^i \omega_{ab}^i - \sum_{k=1}^n A_0^{ik} \left(e_a^k \omega_a^k + e_b^k \omega_b^k \right) \\
\phi_{ab}^i \omega_{ab}^i \omega_{ab}^k &= f_{ab}^i \omega_{ab}^i \omega_{ab}^k - \sum_{k=1}^n A_0^{ik} \left(e_a^k \omega_a^k \omega_{ab}^k + e_b^k \omega_b^k \omega_{ab}^k \right) \\
\left(f_{ab}^i - \phi_{ab}^i \right) \int_{\mathcal{Z}_{ab}} \omega_{ab}^i \omega_{ab}^k &= \sum_{k=1}^n A_0^{ik} \underbrace{\left(e_a^k \overbrace{\int_{\mathcal{Z}_{ab}} \omega_a^k \omega_{ab}^k}^{\alpha_{ab}^k} + e_b^k \overbrace{\int_{\mathcal{Z}_{ab}} \omega_b^k \omega_{ab}^k}^{\alpha_{ba}^k} \right)}_{e_{ab}^k}
\end{aligned}$$

$$f_{ab} = \phi_{ab} + \frac{A_0}{\bar{\alpha}} e_{ab} \quad \Rightarrow \quad f_{ab} = \begin{pmatrix} A_{1/2}(e_b^B - e_a^B) \\ 0_{n-2l,1} \\ A_{1/2}^{\mathrm{T}}(f_b^B - f_a^B) \end{pmatrix} + \bar{\alpha}^{-1} A_0 e_{ab} \tag{15}$$

where $\bar{\alpha} := \text{diag}(\int_{\mathcal{Z}_{ab}} \omega_{ab}^i \omega_{ab}^k)$, with $i, k = 1, \ldots, n$ and k such that $A_0^{ik} \neq 0$. Note that $\bar{\alpha}^{-1} A_0$ is still skew-symmetric since $\bar{\alpha}$ is a diagonal matrix and thanks to Assumption 3.4.

In order to get an explicit form of the finite-dimensional dynamics, it is necessary to choose an input and its dual output. This algorithm assumes that such choice is given by:

$$\begin{aligned}
u &= \left(e_a^{1B}, \ldots, e_a^{lB}, 0_{1,n-2l}, -f_b^{1B}, \ldots, -f_b^{lB} \right)^{\mathrm{T}} \\
y &= \left(f_a^{1B}, \ldots, f_a^{lB}, 0_{1,n-2l}, e_b^{1B}, \ldots, e_b^{lB} \right)^{\mathrm{T}}
\end{aligned} \tag{16}$$

Then, from (14), we have that:

$$\begin{aligned}
f_a^B &= \bar{\alpha}_{ab,1}^{-1} \left(e_{ab}^1, \ldots, e_{ab}^l \right)^{\mathrm{T}} - \bar{\alpha}_{ba,1} \bar{\alpha}_{ab,1}^{-1} f_b^B \\
e_b^B &= \bar{\alpha}_{ba,2}^{-1} \left(e_{ab}^{n-l+1}, \ldots, e_{ab}^n \right)^{\mathrm{T}} - \bar{\alpha}_{ab,2} \bar{\alpha}_{ba,2}^{-1} e_a^B
\end{aligned}$$

so that the following relations to be used with (15) are obtained:

$$A_{1/2}\left(e_b^B - e_a^B\right) = A_{1/2}\left[\bar{\alpha}_{ba,2}^{-1}\begin{pmatrix} e_{ab}^{n-l+1} \\ \vdots \\ e_{ab}^{n}\end{pmatrix} - \underbrace{\left(\bar{\alpha}_{ab,2}\bar{\alpha}_{ba,2}^{-1} + I_l\right)}_{\bar{\alpha}_{ba,2}^{-1}} e_a^B\right]$$

$$A_{1/2}^{\mathrm{T}}\left(f_b^B - f_a^B\right) = A_{1/2}^{\mathrm{T}}\left[-\bar{\alpha}_{ab,1}^{-1}\begin{pmatrix} e_{ab}^{1} \\ \vdots \\ e_{ab}^{l}\end{pmatrix} - \underbrace{\left(\bar{\alpha}_{ba,1}\bar{\alpha}_{ab,1}^{-1} + I_l\right)}_{\bar{\alpha}_{ab,1}^{-1}}(-f_b^B)\right]$$

It is now necessary to investigate if and under which conditions a finite-dimensional port-Hamiltonian system in input/output form [14] can be deduced from (15). Generally speaking, it is necessary to take into account the fact that more than a single effort variable may be related to any given flow. If, at first, the interdependencies among the α variables are considered, from (13) we have that $\alpha_{ab,1}^i = \alpha_{ba,2}^j$ for every i and j such that $A_{1/2}^{ij} \neq 0$. Therefore,

$$A_{1/2}\bar{\alpha}_{ba,2}^{-1} = A_{1/2}\bar{\alpha}_{ab,1}^{-1} \quad \text{and} \quad A_{1/2}^{\mathrm{T}}\bar{\alpha}_{ab,1}^{-1} = A_{1/2}^{\mathrm{T}}\bar{\alpha}_{ba,2}^{-1} \tag{17}$$

As for the input/output map, it is useful to rewrite (15) in the following form:

$$\begin{cases} -f_{ab} = J_\Delta e_{ab} + G_\Delta B_\Delta u \\ \quad\;\; y = (G_\Delta)^{\mathrm{T}} e_{ab} + D_\Delta u \end{cases} \tag{18}$$

where u and y have been defined in (16), while

$$J_\Delta = \begin{pmatrix} 0_{l,l} & 0_{l,n-2l} & A_{1/2}\alpha_{ba,2}^{-1} \\ 0_{n-2l,l} & 0_{n-2l,n-2l} & 0_{n-2l,l} \\ -A_{1/2}^{\mathrm{T}}\alpha_{ab,1}^{-1} & 0_{l,n-2l} & 0_{l,l} \end{pmatrix} - \bar{\alpha}^{-1}A_0$$

$$G_\Delta = \begin{pmatrix} \bar{\alpha}_{ab,1}^{-1} & 0_{l,n-2l} & 0_{l,l} \\ 0_{n-2l,l} & 0_{n-2l,n-2l} & 0_{n-2l,l} \\ 0_{l,l} & 0_{l,n-2l} & \bar{\alpha}_{ba,2}^{-1} \end{pmatrix} \qquad B_\Delta = \begin{pmatrix} A_{1/2} & 0_{l,n-2l} & 0_{l,l} \\ 0_{n-2l,l} & 0_{n-2l,n-2l} & 0_{n-2l,l} \\ 0_{l,l} & 0_{l,n-2l} & A_{1/2}^{\mathrm{T}} \end{pmatrix}$$

$$D_\Delta = \begin{pmatrix} 0_{l,l} & 0_{l,n-2l} & \bar{\alpha}_{ba,1}\bar{\alpha}_{ab,1}^{-1} \\ 0_{n-2l,l} & 0_{n-2l,n-2l} & 0_{n-2l,l} \\ \bar{\alpha}_{ab,2}\bar{\alpha}_{ba,2}^{-1} & 0_{l,n-2l} & 0_{l,l} \end{pmatrix}$$

Finally, since it is possible to prove through the use of (17) that J_Δ and $B_\Delta^{\mathrm{T}} D_\Delta$ are skew-symmetric, the power balance of the system is given by:

$$e_{ab}^{\mathrm{T}} f_{ab} + y^{\mathrm{T}} B_\Delta u = 0$$

where B_Δ represents the fact that multiple couplings occur between flows and efforts in order to generate the power flow through the boundary.

4 Extension to Higher Order Derivatives

Let us now consider a larger class of systems, by allowing $N \geq 2$ in (2). This involves the presence of higher order derivatives, and the necessity to consider as boundary variables also the restriction on the boundary of the efforts' derivatives up to the $(N-1)$-th order. The discretization problem may be solved by introducing a number of auxiliary variables so that it is possible to write the system equations in a form where only first-order derivatives appear. Flows and efforts can be redefined as follows:

$$\begin{aligned}\mathbf{f} &= \left(f^1, \dots, f^n, f^{1\int 1}, \dots, f^{n\int 1}, \dots, f^{1\int(N-1)}, \dots, f^{n\int(N-1)}\right)^{\mathrm{T}} \\ \mathbf{e} &= \left(e^1, \dots, e^n, e^{1\partial 1}, \dots, e^{n\partial 1}, \dots, e^{1\partial(N-1)}, \dots, e^{n\partial(N-1)}\right)^{\mathrm{T}}\end{aligned} \tag{19}$$

Then, assuming that in (2) A_N is non-singular, it can be shown [7] that relation $f = -\mathcal{J}e$ can be equivalently rewritten in the form:

$$\mathbf{f} = \mathbf{A}_0\mathbf{e} + \mathbf{A}_1\frac{\partial \mathbf{e}}{\partial z} \tag{20}$$

where:

$$\mathbf{A}_0 = \begin{pmatrix} A_0 & 0_{n,n} & \cdots & 0_{n,n} \\ 0_{n,n} & 0_{n,n} & \cdots & 0_{n,n} \\ \vdots & \vdots & \ddots & \vdots \\ 0_{n,n} & 0_{n,n} & \cdots & 0_{n,n} \end{pmatrix} \qquad \mathbf{A}_1 = \begin{pmatrix} A_1 & A_2 & \cdots & A_N \\ -A_2 & A_3 & \cdots & 0_{n,n} \\ \vdots & \vdots & \ddots & \vdots \\ (-1)^{N+1}A_N & 0_{n,n} & \cdots & 0_{n,n} \end{pmatrix}$$

Since the power balance of the system is preserved, provided that the additional ports are terminated properly, the discretization of the Stokes–Dirac structure may be carried out following the procedure described in the previous section, and then by imposing the correct termination on the ports of the finite element. It is easy to show that the correct power balance is restored with the following termination of the additional ports:

$$\begin{pmatrix} f^{\int 1} \\ f^{\int 2} \\ \vdots \\ f^{\int(N-1)} \end{pmatrix} = \begin{pmatrix} -A_2 & -A_3 & \cdots & -A_N \\ A_3 & -A_4 & \cdots & 0_{n,n} \\ \vdots & \vdots & \ddots & \vdots \\ (-1)^{h+1}A_N & 0_{n,n} & \cdots & 0_{n,n} \end{pmatrix} \begin{pmatrix} e^{\partial 1} \\ e^{\partial 2} \\ \vdots \\ e^{\partial(N-1)} \end{pmatrix}$$

An example of the procedure is reported in Sect. 6.2.

5 Energy Flow Discretization

The procedure described in the previous sections addresses the discretization of the Stokes-Dirac structure. However, a distributed port Hamiltonian system is complete only once the port behavior of this structure is specified. In infinite dimensions, if ξ denotes the energy variable, the Hamiltonian function $\mathcal{H}$ is defined as

$$\mathcal{H}(\xi) = \int_{\mathcal{Z}} H(\xi(z,t))\mathrm{dz} \tag{21}$$

where H is the energy density. The port Hamiltonian system with Stokes–Dirac structure (2) and Hamiltonian (21) follows, then, from the port behavior $f = -\dot{\xi}$ and $e = \delta_\xi \mathcal{H}$, where δ denotes the variational derivative [15, 13]:

$$\begin{cases} \dot{\xi} = \mathcal{J}\,\delta_\xi \mathcal{H} \\ w = B_{\mathcal{J}}(\delta_\xi \mathcal{H}) \end{cases}$$

In order to be able to terminate appropriately the ports of the finite element, it is necessary to have a lumped-parameter representation also for the energy part of the system. Energy variables ξ and flows f have the same geometric nature. This fact can be used in order to carry out an approximation in the form (6) by imposing that $\xi^i(z,t) = x^i(t)\omega^i_{ab}(z)$, with $x = \left(x^1, \ldots, x^n\right)^{\mathrm{T}}$ energy variable of the finite dimensional approximation. Then, if we assume that H in (21) is quadratic, i.e. $H = \frac{1}{2}\xi^{\mathrm{T}}\Sigma\,\xi$, the discretized energy function is again quadratic and given by $H_{ab}(x) = \frac{1}{2}x^{\mathrm{T}}S_{ab}\,x$, where $S^{ij}_{ab} = \int_{\mathcal{Z}_{ab}} \omega^i_{ab}\Sigma^{ij}\omega^j_{ab}$. The finite-dimensional model is completed once $f_{ab} = -\dot{x}$ and $e_{ab} = \frac{\partial H_{ab}}{\partial x}$.

6 Examples

6.1 The Timoshenko Beam

The Timoshenko beam model with spatial domain $\mathcal{Z} = [0, L]$ has a Stokes–Dirac structure characterized by the following differential relation, [5, 9]:

$$\begin{pmatrix} f^{p_t} \\ f^{p_r} \\ f^{\epsilon_t} \\ f^{\epsilon_r} \end{pmatrix} = \begin{pmatrix} 0 & 0 & -\partial_z & 0 \\ 0 & 0 & -1 & -\partial_z \\ -\partial_z & 1 & 0 & 0 \\ 0 & -\partial_z & 0 & 0 \end{pmatrix} \begin{pmatrix} e^{p_t} \\ e^{p_r} \\ e^{\epsilon_t} \\ e^{\epsilon_r} \end{pmatrix}$$

for which $(\phi^{p_t}, \phi^{p_r}, \phi^{\epsilon_t}, \phi^{\epsilon_r}) := (f^{p_t}, f^{p_r} + e^{\epsilon_t}, f^{\epsilon_t} - e^{p_r}, f^{\epsilon_r})$ are the *modified flows*. In this case, relation (10) becomes:

$$\begin{pmatrix} \phi^{p_t}_{ab} \\ \phi^{p_r}_{ab} \\ \phi^{\epsilon_t}_{ab} \\ \phi^{\epsilon_r}_{ab} \end{pmatrix} = \begin{pmatrix} 0 & 0 & -1 & 0 \\ 0 & 0 & 0 & -1 \\ -1 & 0 & 0 & 0 \\ 0 & -1 & 0 & 0 \end{pmatrix} \begin{pmatrix} e^{p_t}_b - e^{p_t}_a \\ e^{p_r}_b - e^{p_r}_a \\ e^{\epsilon_t}_b - e^{\epsilon_t}_a \\ e^{\epsilon_r}_b - e^{\epsilon_r}_a \end{pmatrix}$$

with the beam model that satisfies Assumption 3.2 without further permutations. The boundary variables, then, can be chosen as:

$$w = \left(f_a^{tB}, f_b^{tB}, f_a^{rB}, f_b^{rB}, e_a^{tB}, e_b^{tB}, e_a^{rB}, e_b^{rB}\right)^{\mathrm{T}}$$
$$\equiv \left(e_a^{p_t}, e_b^{p_t}, e_a^{p_r}, e_b^{p_r}, e_a^{\epsilon_t}, e_b^{\epsilon_t}, e_a^{\epsilon_r}, e_b^{\epsilon_r}\right)^{\mathrm{T}}$$

while the net power in the portion $\mathcal{Z}_{ab} = [a, b]$ of the beam is

$$P_{ab}^{net} = \int_{\mathcal{Z}_{ab}} \sum_i e^i f^i + f_b^{tB} e_b^{tB} - f_a^{tB} e_a^{tB} + f_b^{rB} e_b^{rB} - f_a^{rB} e_a^{rB}$$
$$= \sum_i f_{ab}^i \left(\alpha_{ab}^i e_a^i + \alpha_{ba}^i e_b^i\right) + f_b^{tB} e_b^{tB} - f_a^{tB} e_a^{tB} + f_b^{rB} e_b^{rB} - f_a^{rB} e_a^{rB}$$

$i \in \{p_t, p_r, \epsilon_t, \epsilon_r\}$, which leads to the definition of the distributed efforts $e_{ab}^{p_t} := \alpha_{ab}^t e_a^{p_t} + \alpha_{ba}^t e_b^{p_t}$, $e_{ab}^{p_r} := \alpha_{ab}^r e_a^{p_r} + \alpha_{ba}^r e_b^{p_r}$, $e_{ab}^{\epsilon_t} := \alpha_{ba}^t e_a^{\epsilon_t} + \alpha_{ab}^t e_b^{\epsilon_t}$ and $e_{ab}^{\epsilon_r} := \alpha_{ba}^r e_a^{\epsilon_r} + \alpha_{ab}^r e_b^{\epsilon_r}$. From Assumption 3.4, which relates f and ϕ, we have that $\phi_{ab}^{p_t} = f_{ab}^{p_t}$, $\phi_{ab}^{p_r} = f_{ab}^{p_r} + \frac{e_{ab}^{\epsilon_t}}{\alpha^{\epsilon_t p_r}}$, $\phi_{ab}^{\epsilon_t} = f_{ab}^{\epsilon_t} - \frac{e_{ab}^{p_r}}{\alpha^{\epsilon_t p_r}}$ and $\phi_{ab}^{\epsilon_r} = f_{ab}^{\epsilon_r}$, where $\alpha^{\epsilon_t p_r} = \int_{\mathcal{Z}_{ab}} \omega_{ab}^{\epsilon_t} \omega_{ab}^{p_r}$. The final step is to rearrange the previous results in order to obtain a finite dimensional Dirac structure in the form (18) provided that $u = (-e_a^{tB}, -e_a^{rB}, f_b^{tB}, f_b^{rB})^{\mathrm{T}}$ and $y = (f_a^{tB}, f_a^{rB}, e_b^{tB}, e_b^{rB})^{\mathrm{T}}$:

$$J_\Delta = \begin{pmatrix} 0 & 0 & \frac{1}{\alpha_{ab}^t} & 0 \\ 0 & 0 & \frac{1}{\alpha^{\epsilon_t p_r}} & \frac{1}{\alpha_{ab}^r} \\ -\frac{1}{\alpha_{ab}^t} & -\frac{1}{\alpha^{\epsilon_t p_r}} & 0 & 0 \\ 0 & -\frac{1}{\alpha_{ab}^r} & 0 & 0 \end{pmatrix} \qquad G_\Delta = \begin{pmatrix} \frac{1}{\alpha_{ab}^t} & 0 & 0 & 0 \\ 0 & \frac{1}{\alpha_{ab}^r} & 0 & 0 \\ 0 & 0 & \frac{1}{\alpha_{ab}^t} & 0 \\ 0 & 0 & 0 & \frac{1}{\alpha_{ab}^r} \end{pmatrix}$$

$$D_\Delta = \begin{pmatrix} 0 & 0 & -\frac{\alpha_{ba}^t}{\alpha_{ab}^t} & 0 \\ 0 & 0 & 0 & -\frac{\alpha_{ba}^r}{\alpha_{ab}^r} \\ \frac{\alpha_{ba}^t}{\alpha_{ab}^t} & 0 & 0 & 0 \\ 0 & \frac{\alpha_{ba}^r}{\alpha_{ab}^r} & 0 & 0 \end{pmatrix} \qquad B_\Delta = I_4$$

As far as concerns the $\xi = (p_t, p_r, \epsilon_t, \epsilon_r)^{\mathrm{T}}$, it is reasonable to use the same discretization technique (6), i.e. $p_t = P_{t,ab}\omega_{ab}^{p_t}$, $p_r = P_{r,ab}\omega_{ab}^{p_r}$, $\epsilon_t = E_{t,ab}\omega_{ab}^{\epsilon_t}$ and $\epsilon_r = E_{r,ab}\omega_{ab}^{\epsilon_r}$, which implies:

$$\frac{\mathrm{d}P_{t,ab}}{\mathrm{d}t} = -f_{ab}^{p_t} \quad \frac{\mathrm{d}P_{r,ab}}{\mathrm{d}t} = -f_{ab}^{p_r} \quad \frac{\mathrm{d}E_{ab,t}}{\mathrm{d}t} = -f_{ab}^{\epsilon_t} \quad \frac{\mathrm{d}E_{ab,r}}{\mathrm{d}t} = -f_{ab}^{\epsilon_r}$$

Finally, it is necessary to discretize the Hamiltonian function:

$$\mathcal{H}(p_t, p_r, \epsilon_t, \epsilon_r) = \frac{1}{2}\int_{\mathcal{Z}} \left(\frac{p_t^2}{\rho} + \frac{p_r^2}{I_\rho} + K\epsilon_t^2 + EI\epsilon_r^2\right) \mathrm{d}z$$

that can be written for the finite element as:

$$H_{ab}(P_{t,ab}, P_{r,ab}, E_{t,ab}, E_{r,ab}) = \frac{1}{2}\left(\frac{P_{t,ab}^2}{\rho_{ab}} + \frac{P_{r,ab}^2}{I_{\rho,ab}} + K_{ab}E_{t,ab}^2 + EI_{ab}E_{r,ab}^2\right)$$

where $\rho_{ab}^{-1} = \int_{\mathcal{Z}_{ab}} \left(\frac{\omega_{ab}^{p_t}}{\rho}\right)\omega_{ab}^{p_t}$, $I_{\rho,ab}^{-1} = \int_{\mathcal{Z}_{ab}} \left(\frac{\omega_{ab}^{p_r}}{I_\rho}\right)\omega_{ab}^{p_r}$, $K_{ab} = \int_{\mathcal{Z}_{ab}} \omega_{ab}^{\epsilon_t} K \omega_{ab}^{\epsilon_t}$ and $EI_{ab} = \int_{\mathcal{Z}_{ab}} \omega_{ab}^{\epsilon_r} EI \omega_{ab}^{\epsilon_r}$. This implies that the discretized effort variables e_{ab}^i are now equal to the gradient of the Hamiltonian function H_{ab}.

The discretized model for the entire beam follows from its finite element by partitioning the domain $\mathcal{Z} = [\ell_0, \ell_1] \cup \cdots \cup [\ell_{M-1}, \ell_M]$, repeating the previous steps for each of the M portions, and interconnecting the finite elements that are adjacent. The total discretized model has the ports $(f_{\ell_0}^{tB}, f_{\ell_0}^{rB}, e_{\ell_0}^{tB}, e_{\ell_0}^{rB})$ and $(f_{\ell_M}^{tB}, f_{\ell_M}^{rB}, e_{\ell_M}^{tB}, e_{\ell_M}^{rB})$. Its energy is given by:

$$\frac{1}{2}\sum_{k=1}^{M}\left(\frac{P_{t,\ell_{k-1},\ell_k}^2}{\rho_{\ell_{k-1},\ell_k}} + \frac{P_{r,\ell_{k-1},\ell_k}^2}{I_{\rho,\ell_{k-1},\ell_k}} + K_{\ell_{k-1},\ell_k}E_{t,\ell_{k-1},\ell_k}^2 + EI_{\ell_{k-1},\ell_k}E_{r,\ell_{k-1},\ell_k}^2\right)$$

6.2 A Generic Second-Order Structure

Let us consider the Stokes–Dirac structure (1) where in (2) we have $N = 2$, $A_0 = A_1 = 0$ and

$$A_2 = \begin{pmatrix} 0 & 1 \\ -1 & 0 \end{pmatrix}$$

As previously discussed, the finite dimensional approximation of the interconnection structure can be obtained once $f = -\mathcal{J}e$ is written in first-order form, where $f = (f_1, f_2)^{\mathrm{T}}$ and $e = (e_1, e_2)^{\mathrm{T}}$. This means introducing some additional variables, as specified by (19):

$$\mathbf{f} = \left(f^1, f^{1\int 1}, f^2, f^{2\int 1}\right)^{\mathrm{T}} \qquad \mathbf{e} = \left(e^1, e^{1\partial 1}, e^2, e^{2\partial 1}\right)^{\mathrm{T}}$$

Then, the equivalent system (20) is characterized by:

$$\mathbf{A}_0 = 0_{4,4} \quad \text{and} \quad \mathbf{A}_1 = \begin{pmatrix} 0 & 0 & 0 & 1 \\ 0 & 0 & -1 & 0 \\ 0 & -1 & 0 & 0 \\ 1 & 0 & 0 & 0 \end{pmatrix}$$

while the finite dimensional Dirac structure in the form (18) results to be

$$\begin{pmatrix} -f^1_{ab} \\ -f^2_{ab} \end{pmatrix} = \underbrace{\begin{pmatrix} 0 & -\frac{1}{\alpha^1_{ab}\alpha^2_{ba}} \\ \frac{1}{\alpha^1_{ab}\alpha^2_{ba}} & 0 \end{pmatrix}}_{J_\Delta} \begin{pmatrix} e^1_{ab} \\ e^2_{ab} \end{pmatrix} + \underbrace{\begin{pmatrix} \frac{1}{\alpha^1_{ab}\alpha^2_{ba}} & \frac{1}{\alpha^1_{ab}} & 0 & 0 \\ 0 & 0 & \frac{1}{\alpha^1_{ab}\alpha^2_{ba}} & -\frac{1}{\alpha^2_{ba}} \end{pmatrix}}_{G_\Delta B_\Delta} u$$

$$y = \underbrace{\begin{pmatrix} \frac{1}{\alpha^1_{ab}} & 0 \\ -\frac{1}{\alpha^1_{ab}\alpha^2_{ba}} & 0 \\ 0 & \frac{1}{\alpha^2_{ba}} \\ 0 & \frac{1}{\alpha^1_{ab}\alpha^2_{ba}} \end{pmatrix}}_{G^{\mathrm{T}}_\Delta} \begin{pmatrix} e^1_{ab} \\ e^2_{ab} \end{pmatrix} + \underbrace{\begin{pmatrix} 0 & 0 & \frac{\alpha^1_{ba}}{\alpha^1_{ab}} & 0 \\ 0 & 0 & -\frac{1}{\alpha^1_{ab}\alpha^2_{ba}} & \frac{\alpha^2_{ab}}{\alpha^2_{ba}} \\ -\frac{\alpha^2_{ab}}{\alpha^2_{ba}} & 0 & 0 & 0 \\ -\frac{1}{\alpha^1_{ab}\alpha^2_{ba}} & -\frac{\alpha^1_{ba}}{\alpha^1_{ab}} & 0 & 0 \end{pmatrix}}_{D_\Delta} u$$

with $u = \left(e^{1B}_a, e^{1\partial 1B}_a, -f^{1B}_b, -f^{1\partial 1B}_b\right)^{\mathrm{T}}$ and $y = \left(f^{1B}_a, f^{1\partial 1B}_a, e^{1B}_b, e^{1\partial 1B}_b\right)^{\mathrm{T}}$.

7 Conclusions

In this work, the discretization procedure for distributed port Hamiltonian systems introduced in [6] has been generalized in order to cope with a wider set of dynamical systems. The algorithm operates in two steps. Once the geometric structure which describes internal and boundary power flows (i.e. the Stokes–Dirac structure) is discretized, the port behavior of the energy storing elements is approximated. Main property of the simplified model is that it obeys the same energy balance relation. The procedure is described for distributed systems with Stokes–Dirac structure defined by a first order differential operator, while the extension to the higher order case is briefly outlined. A couple of examples have been included to better illustrate the main steps of the algorithm.

References

1. A. Bossavit. Differential forms and the computation of fields and forces in electromagnetism. *European Journal of Mechanics, B/Fluids,*, 10(5):474–488, 1991.
2. R. F. Curtain and H. J. Zwart. *An introduction to infinite dimensional linear systems theory.* Springler–Verlag, New York, 1995.
3. M. Dalsmo and A. J. van der Schaft. On representation and integrability of mathematical structures in energy-conserving physical systems. *SIAM J. Control and Optimization*, (37):54–91, 1999.
4. T. Frankel. *The Geometry of Physics. An Introduction.* Cambridge University Press, 2nd edition, 2003.
5. G. Golo, V. Talasila, and A. J. van der Schaft. A Hamiltonian formulation of the Timoshenko beam model. In *Proc. of Mechatronics 2002.* University of Twente, June 2002.
6. G. Golo, V. Talasila, A. J. van der Schaft, and B. M. Maschke. Hamiltonian discretization of boundary control systems. *Automatica*, 40(5):757–771, May 2004.
7. Y. Le Gorrec, H. Zwart, and B. J. Maschke. A semigroup approach to port Hamiltonian systems associated with linear skew symmetric operator. In *Proc. Sixteenth International Symposium on Mathematical Theory of Networks and Systems (MTNS2004), Leuven*, 2004.

8. A. Macchelli. *Port Hamiltonian systems. A unified approach for modeling and control finite and infinite dimensional physical systems.* PhD thesis, University of Bologna – DEIS, 2003. Available at `http://www-lar.deis.unibo.it/ woda/spider/e499.htm`.
9. A. Macchelli and C. Melchiorri. Modeling and control of the Timoshenko beam. The distributed port Hamiltonian approach. *SIAM Journal on Control and Optimization (SICON)*, 43(2):743–767, 2004.
10. A. Macchelli, C. Melchiorri, and L. Bassi. Port-based modelling and control of the Mindlin plate. In *Proc. 44th IEEE Conference on Decision and Control and European Control Conference 2005 (CDC-ECC'05)*, December 12–15 2005.
11. A. Macchelli, S. Stramigioli, and C. Melchiorri. Network modelling and simulation of robots with flexible links. A port-based approach. In *Proc. 17th International Symposium on Mathematical Theory of Networks and Systems (MTNS2006)*, 2006.
12. A. Macchelli, A. J. van der Schaft, and C. Melchiorri. Multi-variable port Hamiltonian model of piezoelectric material. In *Proc. IEEE/RSJ International Conference on Intelligent Robots and Systems, IROS'04*, 2004.
13. A. Macchelli, A. J. van der Schaft, and C. Melchiorri. Port Hamiltonian formulation of infinite dimensional systems. I. Modeling. In *Proc. 43th IEEE Conference on Decisions and Control (CDC04)*, December 14–17 2004.
14. A. J. van der Schaft. *L_2-Gain and Passivity Techniques in Nonlinear Control.* Communication and Control Engineering. Springer Verlag, 2000.
15. A. J. van der Schaft and B. M. Maschke. Hamiltonian formulation of distributed parameter systems with boundary energy flow. *Journal of Geometry and Physics*, 42(1–2):166–194, May 2002.

Virtual Lagrangian Construction Method for Infinite-Dimensional Systems with Homotopy Operators

Gou Nishida[1], Masaki Yamakita[2], and Zhiwei Luo[3]

[1] Environment Adaptive Robotic Systems Lab., Bio-Mimetic Control Research Center, RIKEN(The Institute of Physical and Chemical Research); 2271-130 Anagahora, Shimoshidami, Moriyama-ku, Nagoya, Aichi, 463-0003, Japan
nishida@bmc.riken.jp

[2] Department of Mechanical and Control Engineering, Tokyo Institute of Technology; 2-12-1 Ohokayama, Meguro-ku, Tokyo, 152-8552 Japan / RIKEN BMC
yamakita@ac.ctrl.titech.ac.jp

[3] Department of Computer and Systems Engineering, Kobe University; 1-1 Rokkodai, Nada-ku, Kobe, Hyogo, 657-8501 Japan / RIKEN BMC
luo@gold.kobe-u.ac.jp

Summary. This paper presents a general modeling method to construct a virtual Lagrangian for infinite-dimensional systems. If the system is self-adjoint in the sense of the Frétchet derivative, there exists some Lagrangian for a stationary condition of variational problems. A system having such a Lagrangian can be formulated as a field port-Lagrangian system by using a Stokes-Dirac structure on a variational complex. However, it is unknown whether any infinite-dimensional system can be expressed as an Euler-Lagrange equation. Then, we introduce a virtual Lagrangian with a homotopy operator. The virtual Lagrangian defines a self-adjoint subsystem, which is realized by a cancellation of non-self-adjoint error subsystems.

1 Introduction

From the viewpoint of control applications, model-based control has been one of significant strategies in robotics. One such control is the method of port-Hamiltonian systems, which has been developed as the general framework of passivity [13]. The port-Hamiltonian systems describe a pair of energy variables such that the product is equal to its power. The port-Hamiltonian system can be connected to each subsystem written as a port-Hamiltonian system through the boundary ports while preserving the total energy. Such a power-conserving property is characterized by a Dirac structure [14]. In particular, distributed parameter systems are written in such a formulation by using a Stokes-Dirac structure [1]. The Stokes-Dirac structure is defined using differential forms on a spatial domain with a boundary. The structure means that the change in the interior energy is equal to the power supplied to the system through its boundary.

F. Bullo et al. (Eds.): Lag. & Hamil. Methods for Nonlin. Ctrl., LNCIS 366, pp. 75–86, 2007.
springerlink.com

The internal energy variables can be stabilized by damping injection in terms of a boundary control.

On the other hand, many important physical systems are introduced from variational calculus of a Lagrangian. The correspondence between the field Euler-Lagrange equations and a *field port-Lagrangian system*, which is counterpart of distributed port-Hamiltonian systems, has been established by using a Stokes-Dirac structure on a variational complex [7]. If there exists at least one Lagrangian on the system, then the system can be written as a field port-Lagrangian system. However, it is not clear whether every system has a Lagrangian or not. The existence of a Lagrangian has been considered in mathematical physics [8, 9, 10]. However, in the topic of automatic control, we require the stability of a given unknown system regardless of its physical meaning, and it is necessary to elaborate a methodology for a more general case.

In this paper, a general modeling method is presented to construct a virtual Lagrangian for infinite-dimensional systems.

First, we introduce a unified notation for a higher order differential structure, called a jet bundle. To put the above more concretely, let us consider a situation in which a differential function $\triangle \in \mathscr{A}$ defines the Euler-Lagrange equation $\triangle = 0$ for a variational problem $\mathscr{L} = \int \mathcal{L}\, dx$, where $\mathscr{A}$ denotes the space of smooth functions $\mathcal{L}(x, u^{(v)})$ depending on x, u, and the derivatives of u up to some finite order v. If $\triangle$ is determined by a self-adjoint Fréchet derivative $\mathcal{D}_\triangle^* = \mathcal{D}_\triangle$, then $\triangle = \mathscr{E}(\mathcal{L})$, where $\mathscr{E}$ is an Euler operator that implies a variational derivative. Conversely, if $\triangle = \mathscr{E}(\mathcal{L})$ for some $\mathcal{L}$, then $\mathcal{D}_\triangle$ is self-adjoint. The conditions of the operator $\mathcal{D}_\triangle$ are often referred to as the *Helmholtz conditions*. In this case, the Lagrangian can be reconstructed with a homotopy operator [17]. For example, a homogeneous system of linear differential equations $\triangle[u] = 0$ comprises the Euler-Lagrange equations for a variational problem if and only if the matrix of the differential operators is self-adjoint, where the Lagrangian density is simply $1/2\,(u \cdot \triangle[u])$.

Next, for the case that the whole differential equation $\triangle$ cannot be reconstructed by homotopy operators, we will give a method constructing of a virtual Lagrangian on variational complexes with homotopy operators for a general infinite-dimensional system. The method can extract both a subsystem $\triangle^E$ associated with virtual Lagrangian $\mathscr{L}_v$ and an error term $\triangle^A$ from the original system. The word 'virtual' means that the subsystem can be compensated for the error term by a control. Since the time derivative of the Lagrangian can be considered as a power consisting of pair of generalized energy variables, this approach naturally clarifies the correspondence between infinite-dimensional systems and field port-Lagrangian systems.

2 Mathematical Preliminaries

This section introduces the required mathematical concepts and follows the references [17, 16].

Higher order partial derivatives of a smooth real-valued function $f(x) = f(x^1, \cdots, x^m)$ are denoted by $\partial_J f(x) = \partial^k f(x) \,/\, \partial x^{j_1} \partial x^{j_2} \cdots \partial x^{j_k}$, where $J = (j_1, \cdots, j_k)$ is a multi-index of order k and $x = (x^1, \cdots, x^m)$ are m independent variables. If $f\colon X \to U$ is a smooth function from $X \simeq \mathbb{R}^m$ to $U \simeq \mathbb{R}^l$, there are $m \cdot p_k$ numbers $u^\alpha_J = \partial_J f^\alpha(x)$, where $u^\alpha = (u^1, \cdots, u^l)$ are l dependent variables and $p_k := \binom{m+k-1}{k}$. Let $U_k := \mathbb{R}^{m \cdot p_k}$ be an Euclidean space of this dimension. The *v-th prolongation* $u^{(v)} = \mathrm{pr}^{(v)} f(x)\colon X \to U^{(v)}$ is defined by $u^\alpha_J = \partial_J f^\alpha(x) \subset U_k$, where $U^{(v)} := U \times U_1 \times \cdots \times U_v$. We introduce the *v-th jet space* $M^{(v)} = X \times U^{(v)}$. Let $\mathscr{A}$ be a space of smooth functions $P[u] := P(x, u^{(v)})$ called a *differential function*. Consider a fiber bundle (M, π, X), called a *jet bundle*, defined by the union $M := \bigcup_{(x,u) \in M} M^{(v)}|_{(x,u)}$, where x are horizontal variables and u are vertical variables. A smooth section of the bundle is defined by $u^\alpha = f^\alpha(x)$.

Let a 1-form du_J be the basic form on the fiber of M. du_J consists of the vertical variables u and all their derivatives u_J. Such a differential form $\hat{\omega} = P_J \, du_{J_1} \wedge \cdots \wedge du_{J_k}$ defined by du_J is called a *vertical k-form*, where $P_J \in \mathscr{A}$. A *vertical differential* $\hat{d}$ is defined for $\hat{\omega}$ in the same manner as the exterior differential of an ordinary differential form. As opposed to the vertical differential on the complex, a homotopy operator is defined as follows:

$$\hat{h}(\omega) = \int_0^1 \left[\left(\sum_J u_J \frac{\partial}{\partial u_J} \right) \rfloor \, \omega(\lambda u) \right] \lambda^{-1} \, d\lambda \,. \tag{1}$$

such that $\hat{h} \circ \hat{h} = 0$. Then, a *vertical complex* $\{\hat{\Lambda}^\bullet, \hat{d}\}$ can be defined. *Total derivative* $D_i = \partial \,/\, \partial x^i + u_{J,i}(\partial \,/\, \partial u_J)$ acts on the basic forms by $D_i \, du_J = d(D_i u_J) = du_{J,i}$.

Let us consider an equivalence relation on the space of vertical forms $\hat{\Lambda}^\bullet$, with $[\hat{\omega}] = \hat{\omega} + \operatorname{div} \hat{\eta}$, $\hat{\omega}, \hat{\eta} \in \hat{\Lambda}^k$. The equivalence class defines *functional k-forms* $\Lambda^k_* = \hat{\Lambda}^k \,/\, \operatorname{div}(\hat{\Lambda}^k)^p$. Note that derivations on a fiber are not always exact, but the equivalent class achieves the exactness of sequences. The natural projection from $\hat{\Lambda}^k$ to Λ^k_* is denoted by an integral sign $\int \hat{\omega} \, dx$ which stands for $[\hat{\omega}]$. This definition gives the integration by parts formula,

$$\int \hat{\psi} \wedge D_i \hat{\eta} \, dx = - \int (D_i \hat{\psi}) \wedge \hat{\eta} \, dx \,, \tag{2}$$

where $\hat{\psi} \in \hat{\Lambda}^k$, $\hat{\eta} \in \hat{\Lambda}^l$. Let $\omega = \int \hat{\omega} \, dx$ be a functional k-form corresponding to the vertical k-form $\hat{\omega}$. The *variational differential* of ω is the functional $(k+1)$-form corresponding to the vertical differential of ω: $\delta \omega = \int \hat{d} \hat{\omega} \, dx$. On vertically star-shaped domains, the variational differential determines an exact complex $\{\Lambda^\bullet_*, \delta\}$ called a *variational complex*. Let $\mathscr{L} = \int \mathcal{L}(x, u^{(v)}) \, dx$ be a Lagrangian density functional. The variational of $\mathscr{L}$ is the functional 1-form,

$$\delta \mathscr{L} = \int \left\{ \frac{\partial \mathcal{L}}{\partial u_J} \, du_J \right\} dx = \int \{ \mathscr{E}(\mathcal{L}) \cdot du \} \, dx \,, \tag{3}$$

where $\mathscr{E} = (-D)_J \cdot (\partial / \partial u_J)$ is the *Euler operator*. The relation $\mathscr{E}(\mathcal{L}) \equiv 0$ yields the Euler-Lagrange equations.

Let $P[u] = P(x, u^{(v)}) \in \mathscr{A}^r$ be an r-tuple of differential functions. The *Fréchet derivative* of P is the differential operator $\mathcal{D}_P : \mathscr{A}^l \to \mathscr{A}^r$ defined such that

$$\mathcal{D}_P(S) = \left.\frac{d}{d\epsilon}\right|_{\epsilon=0} P\big[u + \epsilon S[u]\big] \tag{4}$$

for any $S \in \mathscr{A}^l$. Then the Fréchet derivative of a general r-tuple $P = (P_1, \cdots, P_r)$ and the adjoint $\mathcal{D}_P^* : \mathscr{A}^r \to \mathscr{A}^l$ are

$$(\mathcal{D}_P)_{\mu\nu} = \sum_J \frac{\partial P_\mu}{\partial u_J^\nu} \cdot D_J\,, \quad (\mathcal{D}_P^*)_{\nu\mu} = \sum_J (-D)_J \cdot \frac{\partial P_\mu}{\partial u_J^\nu}\,, \tag{5}$$

respectively, where $\mu = 1, \cdots, r\,,\ \nu = 1, \cdots, l$. The operator $\mathcal{D}_P$ is *self-adjoint* if $\mathcal{D}_P^* = \mathcal{D}_P$.

The above framework gives us the following structure called a *field port-Lagrangian system* [7].

Theorem 1. *Consider a linear subspace $\mathbb{D}$ having a Stokes-Dirac structure on a variational complex such that*

$$\mathbb{D} = \left\{ (f,\, e) \left| \begin{bmatrix} f_p \\ f_r \\ f_q \end{bmatrix} = \begin{bmatrix} 0 & -I & D_i \\ I & 0 & 0 \\ D_i & 0 & 0 \end{bmatrix} \begin{bmatrix} e_p \\ e_r \\ e_q \end{bmatrix},\ \begin{bmatrix} f_b \\ e_b \end{bmatrix} = \begin{bmatrix} e_p|_{\partial X} \\ -e_q|_{\partial X} \end{bmatrix} \right.\right\}, \tag{6}$$

where $f = (f_p, f_q, f_r, f_b) \in \{\hat{\Lambda}_c^1(M)\}^3 \times \hat{\Lambda}^0(\partial M)$, $e = (e_p, e_q, e_r, e_b) \in \{\hat{\Lambda}^0(M)\}^3 \times \hat{\Lambda}^0(\partial M)$, D_i is the total differential operator as regards all spatial variables, the summation rule is used for the suffix i, and $I = \mathrm{id}_M$ is the identity operator. Regarding (6), define

$$f = \left(\frac{\partial \mathcal{L}}{\partial u^\alpha} - D_i \frac{\partial \mathcal{L}}{\partial u_i^\alpha},\, -\partial_0 \phi^\alpha,\, -\partial_0 \phi_i^\alpha \right), \quad e = \left(-\phi_0^\alpha,\, \frac{\partial \mathcal{L}}{\partial \phi^\alpha},\, \frac{\partial \mathcal{L}}{\partial \phi_i^\alpha} \right), \tag{7}$$

where $\phi^\alpha(x)$ are field quantities, α is an index of independent components of the field, x is a set of spatial coordinates $x^1, \cdots, x^m$ and a time coordinate $x^0 = t$, $\phi_\mu^\alpha = \partial \phi^\alpha / \partial x^\mu = \partial_\mu \phi^\alpha$ for $\mu = 0, \cdots, m$, $\hat{\Lambda}_c^1(M)$ is a space of vertical 1-forms with the coefficient in $\mathbb{Z}_2$, and we use the identification $\mathcal{T} = \iota \circ \hat{h}$ such that $\hat{h} : \hat{\Lambda}_c^1 \hookrightarrow \hat{\Lambda}^0$ and $\iota : \hat{\Lambda}^0 \overset{\cong}{\rightrightarrows} \mathscr{A}$. Then the first row of (6) corresponds to an Euler-Lagrange equation. The second and the third rows of (6) define an identity relation of higher order variables.

The representation (6) can be extended to a higher order structure and a multi-variable structure [7].

3 Decomposition of Self-adjoint Subsystems

A self-adjointness of Hamiltonian systems has been discussed in [11]. The study of nonlinear adjoint operators is a current topic [12]. The self-adjointness of Fréchet derivatives is an important property in the framework of variational problems [17, pp. 364].

Theorem 2. *Consider a differential function $\mathbb{A} \in \mathscr{A}^p$ defined on a vertically star-shaped domain $Z \subset M$. Then, $\mathbb{A} = \mathscr{E}(\mathcal{L})$ for some variational problem $\mathscr{L} = \int \mathcal{L}\, dx \in \Lambda^0_*$ if and only if the Fréchet derivative $\mathcal{D}_{\mathbb{A}}$ is self-adjoint $\mathcal{D}^*_{\mathbb{A}} = \mathcal{D}_{\mathbb{A}}$. In this case, a Lagrangian for $\mathbb{A}$ can be explicitly constructed using the homotopy formula*

$$\mathscr{L}[u] = \int \left\{ \int_0^1 u \cdot \mathbb{A}[\lambda u]\, d\lambda \right\} dx\,. \tag{8}$$

Based on this fact, we have the following result.

Theorem 3. *On a vertically star-shaped domain, an arbitrary system of differential equations $\mathbb{A} = 0$ can be uniquely written as the sum of a self-adjoint system $\mathbb{A}^E = 0$ and a non-self-adjoint system $\mathbb{A}^A = 0$ except for total divergence terms on $\hat{\Lambda}^0(M)$, that is,*

$$\mathbb{A} = \mathbb{A}^E \oplus \mathbb{A}^A\,, \tag{9}$$

where $\mathcal{D}^*_{\mathbb{A}^E} = \mathcal{D}_{\mathbb{A}^E}$ and $\mathcal{D}^*_{\mathbb{A}^A} \neq \mathcal{D}_{\mathbb{A}^A}$.

First, we need the following for the proof of *Theorem 3*. We shall attempt to connect the de Rham complex with the vertical complex with regard to a cohomology group. Let N be an oriented compact Riemannian n-dimensional manifold. The de Rham complex $\{\Omega^\bullet(N), d\}$ consists of spaces of differential forms $\Omega^\bullet(N)$ and an exterior differential $d^k : \Omega^k(N) \to \Omega^{k+1}(N)$. A dimension of the domain N of $\Omega^\bullet(N)$ is usually assumed to be finite. In this case, we can use the following useful relations [18, pp. 159-160]. One of them is the *Hodge decomposition*: $\Omega^k(N) = \mathbb{H}^k(N) \oplus d\Omega^{k-1}(N) \oplus d^\dagger \Omega^{k+1}(N)$, where $d^\dagger = (-1)^{n(k+1)+1} * d *$ is an adjoint operator of d. The other is the *Hodge theorem* that gives the existence of the isomorphism $H^k_{DR}(N) \to \mathbb{H}^k(N)$, where $\mathbb{H}^k$ is a space of *harmonic forms* that is a finite subspace of $\Omega^k(N)$. On the other hand, the total manifold M has an infinite-dimension, because the dimension of $\hat{\Lambda}^k(M)$ is still indefinite. Nevertheless, the space restricted to an appropriate finite dimension is used in a practice. As such, we cannot use the Hodge decomposition, because the Hodge star operator $*$ is no longer valid on the vertical complex of jet bundles. Accordingly, we consider the condition determining a common structure on both complexes as a morphism between complexes. The common structure means the cohomology groups. Indeed, let $\mathcal{G}$ be a category of Abelian groups and $\mathrm{Co}(\mathcal{G})$ be a category of complexes $\{A^\bullet, B^\bullet\}$ with a morphism $f^\bullet : \mathrm{Hom}(A^\bullet, B^\bullet)$ such that $'d^k \circ f^k = f^{k+1} \circ d^k$, where $d^k = \mathrm{Hom}(A^k, A^{k+1})$ and $'d^k = \mathrm{Hom}(B^k, B^{k+1})$.

Let $H^k(A^\bullet) = \mathrm{Ker}\, d^k \,/\, \mathrm{Im}\, d^{k-1}$ be a k-th cohomology group of $A^\bullet$. The cohomology group plays the role of a covariant functor $H^k(-) : \mathrm{Co}(\mathcal{G}) \rightsquigarrow \mathcal{G}$ in terms of category theory [19]. As a result, the above definition gives the identification between the de Rham complex $\{\Omega^\bullet(N), d\}$ and the vertical complex $\{\hat{\Lambda}^\bullet(M), \hat{d}\}$ on a star-shaped domain, because N and M are spaces having the same cohomology, where a star-shaped domain is contractible ($\simeq \mathbb{R}^n$) and the de Rham cohomology of $\mathbb{R}^n$ is trivial; that is, $H^k_{DR}(\mathbb{R}^n) = \{\mathbb{R}$, if $k = 0$; 0, if $k > 0\}$ from Poincaré's lemma [18].

Lemma 1. *Let us consider a vertically star-shaped region $Z \subset M$ on a v-th jet bundle. Then, for $k > 0$, an arbitrary vertical k-form can be uniquely written as*

$$\hat{\Lambda}^k(M) = \hat{d}\hat{\Lambda}^{k-1}(M) \oplus \hat{h}\hat{\Lambda}^{k+1}(M)\,. \tag{10}$$

Proof. The direct sum decomposition of de Rham complexes on a star-shaped domain: $\Omega^k(N) = d\Omega^{k-1}(N) \oplus h\Omega^{k+1}(N)$ has been given on an oriented compact Riemannian manifold N [15]. Consider the case that the degree of jet bundles is restricted to be finite. If we assume that the dimension of the fiber $U^{(v)}$ of M is equal to the dimension N, the cohomology of both N and M is the same, that is, trivial. □

Proof of Theorem 3. For any $\omega \in \hat{\Lambda}^k(M)$, *Lemma 1* gives the expression $\omega = \hat{d}\eta + \hat{h}\theta$, where $\eta \in \hat{\Lambda}^{k-1}(M)$ and $\theta \in \hat{\Lambda}^{k+1}(M)$. Let us consider a self-adjoint systems $\triangle = 0$ that are defined by the variation $\mathscr{K} = \int \omega\, dx$, $\omega = \triangle\, du$. In this case, the Lagrangian can be calculated by $\hat{h}\omega$. Actually, from the exactness of $\hat{\Lambda}^\bullet$, the following should be zero.

$$\begin{aligned}
\delta\mathscr{K} &= \int \hat{d}(\triangle) \wedge du\, dx = \int \mathcal{D}_\triangle(du) \wedge du\, dx \\
&= \int \frac{1}{2} du \wedge \left(\mathcal{D}^*_{\triangle^E} - \mathcal{D}_{\triangle^E}\right) du + \left(\mathcal{D}_\triangle - \mathcal{D}_{\triangle^E}\right)(du) \wedge du\, dx \\
&= \int \mathcal{D}_{\triangle^A}(du) \wedge du\, dx\,.
\end{aligned} \tag{11}$$

If there exists $\hat{d}\omega \in \hat{\Lambda}^2(M)$, then $\theta = \hat{h}\hat{d}\omega$ is an anti-exact form on $\hat{\Lambda}^1(M)$. However, according to the assumption, there is no such form θ on a self-adjoint system. Thus, $\hat{d}\hat{h}\omega = \omega$. Consequently, exact forms on $\hat{\Lambda}^1(M)$ are related to some Lagrangian density $\mathscr{L} = \int \mathcal{L} = \int \hat{h}\omega$; hence, the system $\triangle^E$ defined by $\hat{d}\mathcal{L} = 0$ is self-adjoint. On the other hand, anti-exact forms on $\hat{\Lambda}^1(M)$ are mapped to zero on $\hat{\Lambda}^0(M)$; hence such forms correspond to a non-self-adjoint system $\triangle^A$ □

Corollary 1. *If $\left[\triangle - (\hat{d}\hat{h} + \hat{h}\hat{d})\triangle\right] du = 0 \in \hat{\Lambda}^1$, then $\triangle^E = \hat{d}\hat{h}\triangle = \triangle - \hat{h}\hat{d}\triangle$ and $\triangle^A = \hat{h}\hat{d}\triangle = \triangle - \hat{d}\hat{h}\triangle$.*

Proof. If the cohomology is trivial, then $\triangle\, du = 0$ is equal to $(\hat{d}\hat{h} + \hat{h}\hat{d})\triangle\, du = 0$. Hence the proof follows from Theorem 3. □

4 Boundary Conditions and Reconstruction of Virtual Lagrangian

Identifying total divergence terms on the variational complex allows us to use the integration by parts formula (2). Eliminating the total divergence terms is equivalent to assuming that all boundary conditions are zero. However, the assumption is somewhat conservative. In this section, we will reduce the restriction of this assumption. First, we introduce a unified notation for the integration by parts formula with explicit treatment of boundary conditions. Then, the relation between boundary conditions and the self-adjointness of the Fréchet derivative is discussed. Next, we consider the minimum combination of fixed boundary conditions to define a one-to-one correspondence between a vertical derivative and an Euler operator.

Definition 1. *Let $\mathcal{J} = \{j_1, \cdots, j_k\}$ be a set of ordered indexes of partial differentials for any $k > 0$. Then $\mathcal{J}$ is a fixed combination of k indexes in J. For $\mathcal{J}_i \subset \mathcal{J}$, we have*

$$\mathcal{J} = \{\mathcal{J}_i, \bar{\mathcal{J}}_i\} = \{\{j_1, \cdots, j_i\}, \{j_{i+1}, \cdots, j_k\}\}\,. \tag{12}$$

Let us define a functional 1-form such that

$$\Xi_i = \int \left\{ (-D)_{\mathcal{J}_i} \frac{\partial \mathcal{L}}{\partial u_{\mathcal{J}}} \cdot D_{\bar{\mathcal{J}}_i}(du) \right\} dx\,. \tag{13}$$

By using this definition, we have the following relations.

Lemma 2

$$\Xi_0 = \int \{\hat{d}\mathcal{L}\}\, dx\,, \quad \Xi_k = \int \{\mathscr{E}(\mathcal{L})\, du\}\, dx\,. \tag{14}$$

Proof. The following relation can be considered:

$$\Xi_0 = \int \left\{ (-D)_0 \frac{\partial \mathcal{L}}{\partial u_{\mathcal{J}}} \cdot D_{\mathcal{J}}(du) \right\} dx = \int \left\{ \frac{\partial \mathcal{L}}{\partial u_{\mathcal{J}}}\, du_{\mathcal{J}} \right\} dx = \int \{\hat{d}\mathcal{L}\}\, dx\,, \tag{15}$$

$$\Xi_k = \int \left\{ (-D)_{\mathcal{J}} \frac{\partial \mathcal{L}}{\partial u_{\mathcal{J}}} \cdot D_0(du) \right\} dx = \int \{\mathscr{E}(\mathcal{L})\, du\}\, dx\,, \tag{16}$$

where $D_0 = \mathrm{id}_M$. Then the relations (14) have been given. □

Lemma 3. *If we consider the boundary conditions:*

$$\Phi_i = \int D_{j_i} \left\{ (-D)_{\mathcal{J}_{i-1}} \frac{\partial \mathcal{L}}{\partial u_{\mathcal{J}}} \cdot D_{\bar{\mathcal{J}}_i}(du) \right\} dx \tag{17}$$

explicitly, then we have $\Xi_{i-1} = \Xi_i + \Phi_i$.

Proof. The relation follows directly from *definition 1* and the integration by parts formula. □

Definition 2. *If the boundary conditions are fixed $\Phi_i = 0$ for any j_i, functional 1-forms can be integrated by parts as follows: $\varsigma(j_i) : \Lambda^1_* \ni \Xi_{i-1} \mapsto \Xi_i \in \Lambda^1_*$, where $i = 1, \cdots, k$ and ς is an invertible map.*

From the above definition, we have the following relation.

$$0 \longrightarrow \mathscr{L} \xrightarrow{\delta} \Xi_0 \xrightarrow{\varsigma(j_1)} \Xi_1 \xrightarrow{\varsigma(j_2)} \cdots \xrightarrow{\varsigma(j_k)} \Xi_k \xrightarrow{\delta} \mathscr{C} \longrightarrow \cdots, \tag{18}$$

where $\mathscr{C} \in \Lambda^2_*$. Thus, the integration by parts formula is repeatedly used from the variational derivative of the Lagrangian: Ξ_0 to the determination of the Euler-Lagrange equation: Ξ_k. Conversely, if we consider Lagrangian reconstruction, the information about the zero boundary conditions from Ξ_k to Ξ_0 is necessary.

Proposition 1. *Consider an integration by parts formulation $\sigma : \Xi_0 \to \Xi_k$, where $\sigma = \varsigma(j_k) \circ \cdots \circ \varsigma(j_1)$. If there exists at least one set of ordered indexes $\mathcal{J} = \{j_1, \cdots, j_k\}$ such that the boundary conditions are fixed by zero for any set of unordered indexes $J = (j_1, \cdots, j_k)$; that is, $^\exists\mathcal{J}$ s.t. $\Phi_i = 0$ $(i = 1, \cdots, k)$, then σ is one-to-one.*

Proof. The action of $\varsigma(j_i)$ can be written as follows:

$$\int \left\{ \frac{\partial \mathcal{L}}{\partial u_{\mathcal{J}}} \cdot D_{j_i}(du) \right\} dx + \int \left\{ D_{j_i} \left(\frac{\partial \mathcal{L}}{\partial u_{\mathcal{J}}} \right) du \right\} dx = \int D_{j_i} \left\{ \frac{\partial \mathcal{L}}{\partial u_{\mathcal{J}}} du \right\} dx\,. \tag{19}$$

The right-hand side of (19) is assumed to be a zero in the sequence (18). □

Next, we can consider the following relation, that is, a virtual Lagrangian construction with homotopy operators, by using the previous definitions.

Proposition 2. *If the system has a self-adjoint Fréchet derivative $\mathcal{D}_\Delta = \mathcal{D}^*_\Delta$, then $\Xi_0 = \Xi_k$.*

Proof. The proposition follows directly from *lemma 2*. □

This leads to a concrete calculation of homotopy operators.

Proposition 3. *Consider a system which has a self-adjoint Fréchet derivative $\mathcal{D}_\Delta = \mathcal{D}^*_\Delta$. Then $g = \hat{h} \circ \sigma^{-1}$.*

$$\begin{array}{ccccccccc} 0 \longrightarrow & \mathscr{L} & \xrightarrow{\delta} & \Xi_0 & \xrightarrow{\sigma} & \Xi_k & \xrightarrow{\delta} & \mathscr{C} & \longrightarrow \cdots \\ & & & \hat{h}\downarrow & \swarrow g & & & & \\ & 0 \longrightarrow & \mathscr{L}' & \xrightarrow{\delta} & \Xi'_0 & \xrightarrow{\sigma} & \Xi'_k & \xrightarrow{\delta} \mathscr{C}' & \longrightarrow \cdots \end{array} \tag{20}$$

where $\hat{h}$ is a homotopy operator on a vertical complex and δ is a variational derivative.

Proof. From *proposition 2*, there exists $\sigma = \varsigma(j_k) \circ \cdots \circ \varsigma(j_1)$, $j_i \in \mathcal{J}$ for any $J = (j_1, \cdots, j_k)$, which corresponds with the situation of *proposition 1*. □

Needless to say, *proposition 3* always holds on a variational complex. Finally, the next theorem gives the main result of the virtual Lagrangian reconstruction.

Theorem 4. *Let $Z \subset M$ be vertically star-shaped. Consider a decomposition $\triangle = \tilde{\triangle} + \alpha$, where $\tilde{\triangle} = \hat{d}\hat{h}\triangle$ and $\alpha \in \hat{\Lambda}^1$. If there exists σ in the sense of Proposition 3, $\hat{h}\triangle$ gives the virtual Lagrangian $\mathscr{L}_v$ which determines the subsystem $\tilde{\triangle}$ of $\triangle$.*

Proof. If there exists some Lagrangian $\mathscr{L}_v$ such that $\hat{d}\mathscr{L}_v \subset \triangle$, then we can write $\triangle = \ddot{d}\mathscr{L}_v \oplus \alpha$, where $\alpha \in \hat{\Lambda}^1$ defines a non-self-adjoint subsystem by *Theorem 3*. Then if the Lagrangian $\mathscr{L}_v$ is given once, the stationary condition of the variational $\mathscr{E}(\mathscr{L}_v) \equiv 0$ is decided uniquely. In other words, there does not exist any $\eta \in \hat{\Lambda}^0$ such that $\alpha = \hat{d}\eta$. □

Conversely, if the boundary conditions have not been explicitly given, the test of self-adjointness of Fréchet derivatives results in a failure although the real Lagrangian of the system can be reconstructed [17]. In this case, the setting of the free boundary conditions keeps the total divergence terms going in existence. Actually, such a total divergence term can be considered as the part of field port-Lagrangian systems. The next section shows an example of this situation.

5 Example

We consider the virtual Lagrangian reconstruction as an inverse problem of the calculus of variations and the transformation from the self-adjoint subsystem to field port-Lagrangian systems.

Example 1. Let $m = l = 1$. The differential equation

$$\triangle = -u_{tt} + u_{xxxx} + u^2 u_{xxx} + u_x^3 \tag{21}$$

which has Fréchet derivatives: $\mathcal{D}_\triangle = -D_t^2 + D_x^4 + u^2 D_x^3 + 3u_x^2 D_x + 2uu_{xxx}$ and $\mathcal{D}_\triangle^* = -D_t^2 + D_x^4 - u^2 D_x^3 - 3u_x^2 D_x - 12u_x u_{xx}$. It is clear that $\mathcal{D}_\triangle \neq \mathcal{D}_\triangle^*$. Next, we consider the virtual Lagrangian $\mathscr{L}_v$ of $\triangle$. Let $\omega = \triangle \cdot du$. We can consider a virtual Lagrangian as follows.

$$\mathscr{L}_v^1(\omega) = \int \hat{h}(\triangle \cdot du)\, dx\,. \tag{22}$$

However, (22) still has an error,

$$\zeta = \omega - \hat{d}\hat{h}(\omega) = \left(3u_x^3 + 6uu_x u_{xx} + u^2 u_{xxx}\right) du\,, \tag{23}$$

in comparison with the Euler-Lagrange expression. If (23) can be canceled out, (22) can be regarded as the new Lagrangian of the revised system, where

$$\begin{aligned} \hat{h}(\omega) &= -\frac{1}{2}uu_{tt} + \frac{1}{4}uu_x^3 + \frac{1}{4}u^2 u_{xxx} + \frac{1}{2}uu_{xxxx}\,, \\ \hat{d}\hat{h}(\omega) &= \left(-u_{tt} - 2u_x^3 - 6uu_x u_{xx} + u_{xxxx}\right) du\,. \end{aligned}$$

The second virtual Lagrangian for the compensated system $\psi = \omega - \zeta$ can be written as follows:

$$\mathscr{L}_v^2 = \int \hat{h}(\psi)\, dx = \int \left(-\frac{1}{2} u u_{tt} - \frac{1}{2} u u_x^3 - \frac{3}{2} u^2 u_x u_{xx} + \frac{1}{2} u u_{xxxx} \right) dx \,, \qquad (24)$$

$$\tilde{\Delta} = \hat{d}\hat{h}(\psi) = \left(-u_{tt} - 2u_x^3 - 6u u_x u_{xx} + u_{xxxx} \right) du \,. \qquad (25)$$

Here, note that $\hat{d}\hat{h}(\psi) = \hat{d}\hat{h}(\omega)$. Next, we rewrite the system $\tilde{\Delta} = u_t\, du_t + \left(-2u_x^3 - 6u u_x u_{xx} + u_{xxxx}\right) du$ in the field port-Lagrangian formulation.

$$\begin{aligned} \hat{d}\mathscr{L}_v^2 &= -\frac{1}{2} u\, du_{tt} + \frac{1}{2} u\, du_{xxxx} - \frac{3}{2} u^2 u_x\, du_{xx} - \frac{3}{2} (u u_x^2 + u^2 u_{xx})\, du_x \\ &\quad + \left(-\frac{1}{2} u_{tt} - \frac{1}{2} u_x^3 - 3u u_x u_{xx} + \frac{1}{2} u_{xxxx} \right) du \\ &= D_t\{u\}\, du_t + D_x \left[D_x^3 \left\{ \frac{1}{2} u \right\} - D_x \left\{ \frac{3}{2} u^2 u_x \right\} + \frac{3}{2} (u u_x^2 + u^2 u_{xx}) \right] du \\ &\quad + \left(-\frac{1}{2} u_x^3 - 3u u_x u_{xx} + \frac{1}{2} u_{xxxx} \right) du \,. \end{aligned} \qquad (26)$$

From the derivation of the virtual Lagrangian, energy variables are defined in two different ways. One of them is as follows:

$$\begin{aligned} f_p &= -\partial_t u_t \,, \quad e_p = -du_t \,, \\ f_r &= -\partial_t du \,, \quad e_r = -\frac{1}{2} u_x^3 - 3u u_x u_{xx} + \frac{1}{2} u_{xxxx} \,, \\ f_q &= -\partial_t du_x \,, \quad e_q = -D_x^3 \left\{ \frac{1}{2} u \right\} + D_x \left\{ \frac{3}{2} u^2 u_x \right\} - \frac{3}{2} (u u_x^2 + u^2 u_{xx}) \,. \end{aligned} \qquad (27)$$

In this case, we have a standard form of field port-Lagrangian systems,

$$\begin{bmatrix} f_p \\ f_r \\ f_q \end{bmatrix} = \begin{bmatrix} 0 & -I & D_x \\ I & 0 & 0 \\ D_x & 0 & 0 \end{bmatrix} \begin{bmatrix} e_p \\ e_r \\ e_q \end{bmatrix} , \quad \begin{bmatrix} f_b \\ e_b \end{bmatrix} = \begin{bmatrix} e_p|_{\partial Z} \\ -e_q|_{\partial Z} \end{bmatrix} . \qquad (28)$$

The other way, which includes a lot of energy variables unlike the previous definitions, is as follows:

$$\begin{aligned} f_p &= -\partial_t u_t \,, \quad e_p = -du_t \,, \\ f_r &= -\partial_t du \,, \quad e_r = -\frac{1}{2} u_x^3 - 3u u_x u_{xx} + \frac{1}{2} u_{xxxx} \,, \end{aligned}$$

$$\begin{aligned} f_{q1} &= -\partial_t du_x \,, \quad e_{q1} = -\frac{3}{2} (u u_x^2 + u^2 u_{xx}) \,, \\ f_{q2} &= -\partial_t du_{xx} \,, \quad e_{q2} = -\frac{3}{2} u^2 u_x \,, \\ f_{q4} &= -\partial_t du_{xxxx} \,, \quad e_{q4} = \frac{1}{2} u \,. \end{aligned} \qquad (29)$$

Hence, we obtain the higher order representation:

$$\begin{bmatrix} f_p \\ f_r \\ f_{q1} \\ f_{q2} \\ f_{q4} \end{bmatrix} = \begin{bmatrix} 0 & -I & D_x & -D_x^2 & -D_x^4 \\ I & 0 & 0 & 0 & 0 \\ D_x & 0 & 0 & 0 & 0 \\ D_x^2 & 0 & 0 & 0 & 0 \\ D_x^4 & 0 & 0 & 0 & 0 \end{bmatrix} \begin{bmatrix} e_p \\ e_r \\ e_{q1} \\ e_{q2} \\ e_{q4} \end{bmatrix},$$

$$\boldsymbol{f}_b = \Big[e_p|_{\partial Z},\, D_x e_p|_{\partial Z},\, e_p|_{\partial Z},\, D_x^3 e_p|_{\partial Z},\, D_x^2 e_p|_{\partial Z},\, D_x e_p|_{\partial Z},\, e_p|_{\partial Z} \Big]^\top,$$

$$\boldsymbol{e}_b = \Big[-e_{q1}|_{\partial Z},\, -e_{q2}|_{\partial Z},\, D_x e_{q2}|_{\partial Z},\, - e_{q4}|_{\partial Z},\, D_x e_{q4}|_{\partial Z},\, -D_x^2 e_{q4}|_{\partial Z},\, D_x^3 e_{q4}|_{\partial Z} \Big]^\top. \quad (30)$$

This is as different an expression from (28) as it looks. However, if we substitute the following zero boundary conditions into the boundary ports of (30), we set the same system.

$$\begin{aligned} &\Xi_0(-e_{q2}, D_x e_p) \xrightarrow{\varsigma(x)} \Xi_1(D_x e_{q2}, e_p), \\ &\Xi_0(-e_{q4}, D_x^3 e_p) \xrightarrow{\varsigma(x)} \Xi_1(D_x e_{q4}, D_x^2 e_p) \xrightarrow{\varsigma(x)} \cdots \\ &\qquad \cdots \longrightarrow \Xi_2(-D_x^2 e_{q4}, D_x e_p) \xrightarrow{\varsigma(x)} \Xi_3(D_x^3 e_{q4}, e_p), \end{aligned} \quad (31)$$

where $\Xi_i(e, f) = \int (e \cdot f)\, dx$. The above representations have the same error term ζ as that of the original system (21).

6 Conclusions

We presented the virtual Lagrangian construction method for non-self-adjoint infinite-dimensional systems. The key result of this study is that every non-self adjoint system includes a self-adjoint subsystem and a non-self-adjoint subsystem and the system decomposition is unique. If the control can compensate the non-self adjoint subsystem, the controlled system becomes a self-adjoint system. We can apply the formal field-port-Lagrangian representation to the self-adjoint subsystem. The non-self adjoint subsystem corresponds with a distributed energy structure. If the distributed energy structure has an internal network of unidirectional flows, the structure is equal to an irreversible energy structure [3]. In the present study, the concrete meaning of the virtual Lagrangian has not been mentioned. A more quantitative evaluation of such a subsystem should be elaborated for actual applications. We will consider this problem in our future work.

References

1. A.J. van der Schaft and B.M. Maschke, (2002), Hamiltonian formulation of distributed–parameter systems with boundary energy flow, Journal of Geometry and Physics, 42, 166–194

2. A. Macchelli, A.J. van der Schaft and C. Melchiorri, (2004), Port Hamiltonian Formulation of Infinite Dimensional Systems I. Modeling, Proc. 43rd IEEE Conference on Decision and Control, 3762–3767
3. D. Eberard and B.M. Maschke, (2004), An extension of port Hamiltonian systems with boundary energy flow to irreversible systems: the example of heat conduction, NOLCOS 2004, Stuttgart
4. G. Nishida and M. Yamakita, (2004), A Higher Order Stokes–Dirac Structure for Distributed–Parameter Port–Hamiltonian Systems, Proc. 2004 American Control Conf., 5004–5009
5. G. Nishida and M. Yamakita, (2004), Disturbance Structure Decomposition for Distributed–Parameter Port–Hamiltonian Systems, Proc. 43rd IEEE CDC, 2082–2087
6. D. Eberard, B.M. Maschke and A.J. van der Schaft, (2005), Conservative systems with port on contact manifolds, Proc. IFAC world cong. 2005
7. G. Nishida and M. Yamakita, (2005), Formal Distributed Port-Hamiltonian Representation of Field Equations, Proc. 44th IEEE CDC
8. J. Douglas, (1941), Solution of the inverse problem of the caluculus of variations, Trans. Amer. Math. Soc., 50, 71–128
9. L.A.Ibort, (1990), On the existence of local and global Lagrangians for ordinary differential equations J. Phys. A: Math. Gen., 23, 4779–4792
10. M. Crampin, W. Sarlet, E. Martínez, G.B. Gyrnes, G.E. Prince, (1994), Towards a geometrical understanding of Dauglas's solution of the inverse problem of the calculus of variations, Inverse Problems, 10, 245–260
11. A. van der Schaft and P.E. Crouch, (1987), Hamiltonian and self-adjoint control systems, Systems & Control Letters, 8, 289-295
12. K. Fujimoto, J.M.A. Scherpen and W.S. Gray, (2002), Hamiltonian realizations of nonlinear adjoint operators, Automatica, 38, 1769–1775
13. A.J. van der Schaft, (2000), L_2-Gain and Passivity Techniques in Nonlinear Control, Springer-Verlag, London
14. I. Dorfman, (1993), Dirac Structures and Integrability of Nonlinear Evolution Equations, John Wiley, Chichester
15. D.G.B. Edelen, (1985), Applied Exterior Calculus, John Wiley & Sons, Inc.
16. D.J. Saunders, (1989), The Geometry of Jet Bundles, Cambridge Univ. Press, Cambridge
17. P.J. Olver, (1993), Applications of Lie Groups to Differential Equations: Second Edition, Springer-Verlag, New York
18. S. Morita, (2001), Geometry of Differential Forms, A.M.S.
19. S. Mac Lane, (1998), Categories for the Working Mathematician, 2nd ed., Springer-Verlag

Direct Discrete-Time Design for Sampled-Data Hamiltonian Control Systems

Dina Shona Laila[1] and Alessandro Astolfi[2,3]

[1] Institute for Design and Control of Mechatronical Systems, Johannes Kepler University, A-4040 Linz, Austria
dina-shona.laila@jku.at
[2] Electrical & Electronic Engineering Department, Imperial College, Exhibition Road, London SW7 2AZ, UK
a.astolfi@imperial.ac.uk
[3] Dipartimento di Informatica, Sistemi e Produzione, University of Rome, Tor Vergata, 00133 Romc, Italy

1 Introduction

The success in a model-based direct discrete-time design for nonlinear sampled-data control systems depends on the availability of a good discrete-time plant model to use for the design. Unfortunately, even if the continuous-time model of a plant is known, we cannot in general compute the exact discrete-time model of the plant, since it requires computing an explicit analytic solution of a nonlinear differential equation. One way to solve the problem of finding a good model is by using an approximate model of the plant. A general framework for stabilization of sampled-data nonlinear systems via their approximate discrete-time models was presented in [11]. It is suggested that approximate discrete-time models can be obtained using various numerical algorithms, such as Runge-Kutta and multistep methods. Yet, to the best of the authors knowledge, almost all available results on this direction view the systems as *dissipative* systems, whereas for design purpose, there are many systems that are better modeled as Hamiltonian *conservative* systems.

The issues of constructing a discrete-time model for Hamiltonian conservative systems are in general more complicated than those for dissipative systems. The discretization is usually directed to preserving two important properties, symplectic mapping and Hamiltonian conservation. Many researchers have been putting a lot of effort to study the discretization of Hamiltonian conservative systems, focusing mainly on obtaining algorithms that are computationally robust to model the dynamics of a Hamiltonian systems for long time simulations (see [3, 9, 14, 15] and references therein). Unfortunately, although many of the algorithms are satisfactory for numerical modeling, they are often too complicated to use for control design purposes.

This paper presents a review of the result from [7] and its generalization to nonseparable, underactuated Hamiltonian systems. Our main results apply

F. Bullo et al. (Eds.): Lag. & Hamil. Methods for Nonlin. Ctrl., LNCIS 366, pp. 87–98, 2007.
springerlink.com

the idea of making the Lyapunov difference more negative than what is obtained when using an emulation controller. This approach helps avoiding loss of stability (due to the loss of Hamiltonian conservation under sampling) that inevitably occurs when we use emulation controller. For fully actuated systems, under certain conditions, this goal is achieved simply by replacing a *gradient* in continuous-time design with a *discrete gradient* in the discrete-time design using the coordinate increment discrete gradient [4]. Further, the generalization of our design obtained by modifying the discrete gradient to guarantee the decrease of the Lyapunov difference when applying our result to underactuated systems is presented. An example is provided to illustrate the usefulness of the results.

2 Preliminaries

2.1 Definitions and Notation

The set of real and natural numbers (including 0) are denoted respectively by $\mathbb{R}$ and $\mathbb{N}$. Given an arbitrary matrix G, we denote the transpose and the Moore-Penrose inverse of G by $G^\top$ and G^+, respectively. Moreover, $I_G := G(G^\top G)^{-1}G^\top = GG^+$ and $O_G := I - I_G$. If G is invertible, then $I_G = I$ and $O_G = O$, where I and O are the identity matrix and the zero matrix, respectively. For any function $H(i,j)$, we define $\nabla_i H(i,j) := \partial H(i,j)/\partial i$, and similar definition holds for the discrete gradient $\overline{\nabla}_i H(i,j)$ (more details in Subsection 2.3). Due to limited space the arguments of functions are often dropped whenever they are clear from the context.

Consider continuous-time Hamiltonian systems

$$\begin{bmatrix} \dot{q} \\ \dot{p} \end{bmatrix} = \begin{bmatrix} 0 & I_n \\ -I_n & 0 \end{bmatrix} \begin{bmatrix} \nabla_q H \\ \nabla_p H \end{bmatrix} + \begin{bmatrix} 0 \\ G(q) \end{bmatrix} u, \tag{1}$$

where $p \in \mathbb{R}^n$ and $q \in \mathbb{R}^n$ are the states, and $u \in \mathbb{R}^m$, $m \leq n$, is the control. The matrix $G \in \mathbb{R}^{n\times m}$. If $m = n$ the system is called fully actuated, whereas if $m < n$ it is called underactuated. The Hamiltonian function of the system is defined as the sum of the kinetic and the potential energy,

$$H(q,p) = K(q,p) + V(q) = \frac{1}{2} p^\top M^{-1}(q) p + V(q), \tag{2}$$

where $M(\cdot)$ is the symmetric inertia matrix. System (1) is called a *separable Hamiltonian system* if the matrix M is constant, i.e.

$$H(q,p) = K(p) + V(q) = \frac{1}{2} p^\top M^{-1} p + V(q). \tag{3}$$

2.2 Continuous-Time IDA-PBC Design

The design method we are developing in this paper is known as IDA-PBC design, which is well-known in the continuous-time context as a powerful tool to solve the stabilization problem for Hamiltonian systems [13]. Although IDA-PBC design is applicable to a broader class of systems (see [1]), it applies naturally to Hamiltonian systems due to the special structure of this class of systems. The idea is to construct a controller for system (1) so that the stabilization is achieved assigning a desired energy function

$$H_d(q,p) = K_d(p,q) + V_d(q) = \frac{1}{2}p^\top M_d^{-1}(q)p + V_d(q) \tag{4}$$

that has an isolated minimum at the desired equilibrium point $(q^e, 0)$ of the closed-loop system. The IDA-PBC design consists of two steps. First, design the energy shaping controller u_{es} to reshape the total energy of the system to obtain the target dynamic; second, design the damping injection controller u_{di} to achieve asymptotic stability. Hence, an IDA-PBC controller is of the form

$$u = u_{es}(q,p) + u_{di}(q,p). \tag{5}$$

The desired Hamiltonian dynamics take the form

$$\begin{bmatrix}\dot{q}\\ \dot{p}\end{bmatrix} = [J_d(q,p) - R_d(q,p)] \begin{bmatrix}\nabla_q H_d\\ \nabla_p H_d\end{bmatrix}, \tag{6}$$

where

$$J_d = -J_d^\top = \begin{bmatrix}0 & M^{-1}M_d\\ -M_dM^{-1} & J_2(q,p)\end{bmatrix}; \quad R_d = R_d^\top = \begin{bmatrix}0 & 0\\ 0 & GK_vG^\top\end{bmatrix},$$

with $K_v = K_v^\top > 0$ and $J_2(\cdot)$ skew symmetric. The controller u_{es} is obtained by solving

$$\begin{bmatrix}0 & I_n\\ -I_n & 0\end{bmatrix} \begin{bmatrix}\nabla_q H\\ \nabla_p H\end{bmatrix} + \begin{bmatrix}0\\ G(q)\end{bmatrix} u_{es} = \begin{bmatrix}0 & M^{-1}M_d\\ -M_dM^{-1} & J_2(q,p)\end{bmatrix} \begin{bmatrix}\nabla_q H_d\\ \nabla_p H_d\end{bmatrix}. \tag{7}$$

Note that for separable Hamiltonian systems, the parameter J_2 can be set to zero (see [13]). The first row of (7) is directly satisfied, and the second row can be written as

$$Gu_{es} = \nabla_q H - M_dM^{-1}\nabla_q H_d + J_2M_d^{-1}p. \tag{8}$$

If G is full column rank (not necessarily invertible), the following constraint must be satisfied:

$$G^\perp\{\nabla_q H - M_dM^{-1}\nabla_q H_d + J_2M_d^{-1}p\} = 0, \tag{9}$$

with $G^\perp$ a full rank left annihilator of G ($G^\perp G = 0$). If PDE (9) is solvable, u_{es} is obtained as

$$u_{es} = G^+(\nabla_q H - M_dM^{-1}\nabla_q H_d + J_2M_d^{-1}p). \tag{10}$$

The target dynamics when applying u_{es} also represent a Hamiltonian system. Hence the energy of the new dynamics is also conserved. Moreover, the damping injection controller u_{di} is constructed as

$$u_{di} = -K_vG^\top\nabla_p H_d\,, \quad k_v > 0 \tag{11}$$

which drives the state to the equilibrium point, provided that a detectability condition is satisfied.

2.3 Discrete-Time Model

As shown in Section 2.2, the objective of designing the energy shaping controller u_{es} is to shape the total energy of the system while keeping the closed-loop system conservative. For this, we are free to choose a method to construct the discrete-time model of the system that would lead to *almost* conserving the desired energy function with the discrete-time energy shaping controller. We choose to use the Euler model, which is not Hamiltonian conserving, but better preserves the Hamiltonian structure of the plant, i.e.

$$\begin{aligned} q(k+1) &= q(k) + T\dot{q}(k) \\ p(k+1) &= p(k) + T\dot{p}(k), \end{aligned} \tag{12}$$

with $q(k) := q(kT)$ and $p(k) := p(kT)$, $k \in \mathbb{N}$, and $T > 0$ the sampling period.

The formulae for constructing the IDA-PBC controller involve computing the gradient of the Hamiltonian. Recall that for a continuous-time function $H(x)$, $x := (q\ p)^\top$, the gradient is defined as

$$\nabla H(x) := \operatorname{grad} H(x) = \left[\frac{\partial H(x)}{\partial x_1}, \cdots, \frac{\partial H(x)}{\partial x_n}\right]^\top .$$

In numerical analysis, the discrete-time analogy of the gradient of a function is called a *discrete gradient.* There are various forms of discrete gradients, which are introduced based on the purpose of use and applications. In general, a discrete gradient is defined in [3] as follows.

Definition 1. *Let $H(x)$ be a differentiable function of $x \in \mathbb{R}^n$. Then $\overline{\nabla} H(x(k), x(k+1))$ is a discrete gradient of $H(x)$ if it is continuous and*

$$\begin{cases} \overline{\nabla} H(x(k), x(k+1))^\top [x(k+1) - x(k)] = H(x(k+1)) - H(x(k)) \\ \overline{\nabla} H(x(k), x(k)) = \nabla H(x) \ . \end{cases}$$ ■

In this paper, we use the coordinate increment discrete gradient introduced in [4]. We also use the standard definition of asymptotic stability (AS), semiglobal practical asymptotic stability (SPAS) and semiglobal practical asymptotic stability Lyapunov function [5, 11].

3 Main Results

To begin with, we present a simple example to illustrate the ideas that motivate our main results.

3.1 Motivating Example

Given the dynamic model of a nonlinear pendulum

$$\dot{q} = p, \quad \dot{p} = -\sin(q) + u. \tag{13}$$

Note that this system belongs to the class of separable Hamiltonian systems, as the Hamiltonian of this system is

$$H(q,p) = K(p) + V(q) = p^2/2 - \cos(q). \tag{14}$$

The equilibrium point to be stabilized is the origin. Choosing $M_d = I$, $V_d = -\cos(q) + \frac{k_1}{2}q^2 + 1$, $k_1 \geq 1$ and $J_2 = 0$, we assign the desired energy function

$$H_d(q,p) = K_d(p) + V_d(q) = \frac{1}{2}p^2 - \cos(q) + \frac{k_1}{2}q^2 + 1.$$

The energy shaping and the damping injection controllers for the system (13)

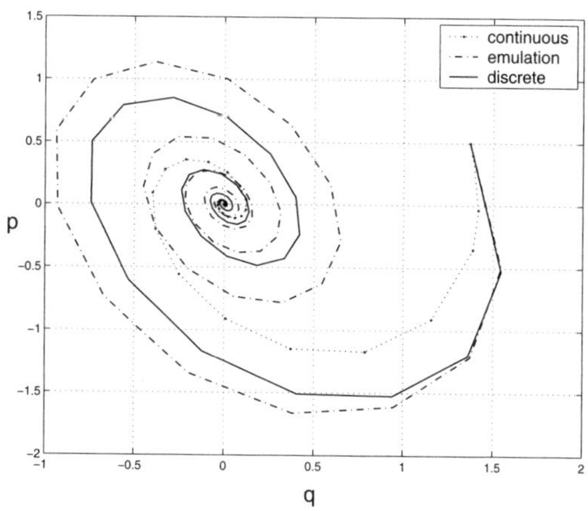

Fig. 1. States trajectory of the nonlinear pendulum

are obtained as

$$u_{es}(t) = \nabla_q H - M_d M^{-1} \nabla_q H_d = -k_1 q, \tag{15a}$$
$$u_{di}(t) = -k_v G^\top \nabla_p H_d = -k_v p, \tag{15b}$$

with $k_v > 0$. Applying $u(t) := u_{es}(t) + u_{di}(t)$ to (13), yields $\dot{H}_d(p,q) = -k_v p^2$, which, by LaSalle Invariance Principle, shows that $u(t)$ asymptotically stabilizes the system (13). Note that the emulation controller $u(k) := u_{es}(k) + u_{di}(k)$ obtained by sample and hold of the controller $u(t)$ is a SPAS controller for the plant (13) [6]. Now we replace the controller (15a) with the following discrete-time controller[1]

$$u_{es}^T(k) = \nabla_q H(q(k), p(k)) - M_d M^{-1} \overline{\nabla}_q H_d(k, k+1) \approx u_{es}(k) - \frac{k_1}{2} T p(k), \tag{16}$$

for T sufficiently small (so that $\sin(T\alpha) \approx T\alpha$ and $\cos(T\alpha) \approx 1$), and let $u_{di}^T(k) = u_{di}(k)$.

[1] $H_d(k, k+1) := H_d(q(k), p(k), q(k+1), p(k+1))$.

Taking the trajectory of the continuous-time system as reference, Figure 1 shows that applying the discrete-time controller $u^T(k) := u^T_{es}(k)+u^T_{di}(k)$ keeps the trajectory of the closed-loop system closer to the reference than using the emulation controller $u(k)$. In the simulation we have used the initial state $(q_\circ, p_\circ) = (\frac{\pi}{2} - 0.2, \frac{1}{2})$, $k_1 = 1$, $k_v = 1$ and $T = 0.35$. This phenomenon is partly explained by Figure 2, which displays the desired Hamiltonian function when applying only the energy shaping controller to the plant. Applying the controllers

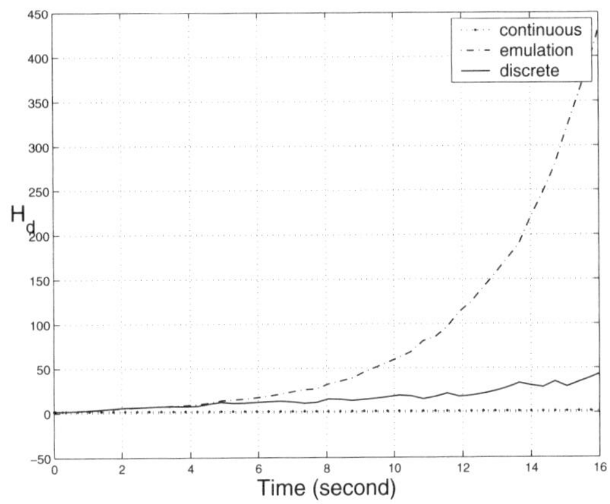

Fig. 2. Desired energy function H_d with $k_v = 0$

(16) and (15a) respectively to the Euler model of (13) and then computing the difference between $\Delta H_d^{u^T_{es}}$ and $\Delta H_d^{u_{es}}$, the corresponding desired energy difference with each controller, we obtain $\Delta H_d^{u^T_{es}} - \Delta H_d^{u_{es}} = -\frac{k_1}{2}T^2p^2 + O(T^3)$, which shows that $\Delta H_d^{u^T_{es}}$ is more negative than $\Delta H_d^{u_{es}}$ in a practical sense. With the same damping injection controller, we see in Figure 1 that the discrete-time controller $u^T(k)$ outperforms the emulation controller $u(k)$.

Remark 1. It is known that Euler approximation is not Hamiltonian conserving. To avoid confusion about the motivation of using this method in our construction we emphasize that IDA-PBC design does not involve Hamiltonian conservation as in the numerical analysis context and we need to distinguish these two different issues. Constructing u_{es} is not aimed to conserve the Hamiltonian of the system, but to transform the system to another Hamiltonian system by using feedback and reshaping the energy of the system (defining the desired Hamiltonian). Therefore, the use of Euler approximation in this context is justified. ■

3.2 Discrete-Time IDA-PBC Design

In this subsection we present our main results, namely the discrete-time IDA-PBC design. We consider a class of Hamiltonian systems (1) with Hamiltonian (2). The Euler model of system (1) is

$$
\begin{aligned}
q(k+1) &= q(k) + T\nabla_p H(q(k), p(k)) \\
p(k+1) &= p(k) - T\Big(\nabla_q H(q(k), p(k)) - Gu(k)\Big).
\end{aligned}
\tag{17}
$$

Suppose all conditions of the continuous-time design hold, and we have assigned the desired energy function (4) for the system. We are now ready to state our main results.

A. Fully actuated separable Hamiltonian systems. We consider the class of separable Hamiltonian systems, a considerably simple subclass of Hamiltonian systems that is still practical in linear system design. Suppose the system is fully actuated, i.e. the matrix G is full rank and invertible. We state the following theorem.

Theorem 1. *Consider the Euler model (17) of the separable Hamiltonian system (1) with Hamiltonian (3). The discrete-time controller* $u^T = u^T_{es} + u^T_{di}$ *where*

$$u^T_{es} = G^{+}\Big(\nabla_q V - M_d M^{-1}\overline{\nabla}_q V_d\Big) \tag{18a}$$

$$u^T_{di} = -K_v G^\top \nabla_p K_d(p) = -K_v G^\top M_d^{-1} p, \tag{18b}$$

with $K_v = K_v^\top > 0$ *and* $\overline{\nabla}_q V_d := \overline{\nabla}_q V_d(q(k), q(k+1))$ *the coordinate increment discrete gradient of* V_d*, is a SPAS controller for the Euler model (17). Moreover, the function*

$$W(q,p) = H_d(q,p) + \epsilon q^T p, \tag{19}$$

with $\epsilon > 0$ *sufficiently small is a SPAS Lyapunov function for the system (17), (18).* ■

Proof of Theorem 1: Suppose M_d and V_d have been obtained by assigning the desired energy function H_d for the system in the same way as in the continuous-time design. Comparing the formulae of the discrete-time controller (18) and the formulae of the continuous-time controller (10), (11), the only different is the term $\overline{\nabla}_q V_d$ that appears in (18a), replacing $\nabla_q V_d$ of (10). Since Euler method is used to build the discrete-time model of the plant, the coordinate increment discrete gradient can then be written as

$$\overline{\nabla}_q V_d = \nabla_q V_d + T L_V \dot{q} = \nabla_q V_d + T L_V M^{-1} p, \tag{20}$$

where L_V is an $n \times n$ matrix. Hence, the difference between the two controllers can be written as

$$
\begin{aligned}
u^T(k) - u(k) &= u^T_{es}(k) + u^T_{di}(k) - u_{es}(k) - u_{di}(k) \\
&= -G^{+} M_d M^{-1}\left(\overline{\nabla}_q V_d - \nabla_q H_d\right) = -TG^{+} M_d M^{-1} L_V M^{-1} p \\
&=: -TG^{+} M_d \overline{M} p =: T\tilde{u}(k),
\end{aligned}
$$

where we defined $\overline{M} := M^{-1} L_V M^{-1}$. Hence, the controller (18) takes form $u^T(k) := u(k) + T\tilde{u}(k)$.

We use the fact that $\lim_{T\to 0} u^T(kT) \to u(t)$, where $u(t)$ is the AS controller (5), to conclude practical asymptotic stability of the closed-loop system. The semiglobal property comes from the fact that T is also dependent on the set in which the initial states are defined. Moreover,

$$\begin{aligned}\Delta W(q,p) &= H_d(q(k+1),p(k+1)) + \epsilon p(k+1)^\top q(k+1)\\ &\quad - H_d(q(k),p(k)) - \epsilon p(k)^\top q(k)\\ &= -TK_v(G^\top M_d^{-1}p)^\top (G^\top M_d^{-1}p) - T^2 p^\top \overline{M} p + \epsilon T p^\top M^{-1} p\\ &\quad + \epsilon T (M_d M^{-1} \nabla_q V_d)^\top q - T^2 \epsilon (M_d M^{-1} L_V M^{-1} p)^\top q + O(T^3).\end{aligned}$$

Note that since $\nabla_q V_d(0) = 0$, using the Mean Value Theorem we can write $\nabla_q V_d(q) = \nabla_q V_d(q) - \nabla_q V_d(0) \leq (\nabla_{qq} V_d(q))^\top \theta q = \theta L_V^\top q$ with $\theta \in (0,1)$. Hence

$$\begin{aligned}\Delta W(q,p) &= -Tp^\top \mathcal{M}_1 p - \epsilon T (M_d M^{-1} \theta L_V q)^\top q\\ &\quad - T^2 \epsilon (M_d M^{-1} L_V M^{-1} p)^\top q + O(T^3)\\ &= -Tp^\top \mathcal{M}_1 p - \epsilon\theta T q^\top \mathcal{M}_2 q + O(T^3),\end{aligned}$$

with $\mathcal{M}_1 := (G^\top M_d^{-1})^\top (G^\top M_d^{-1}) + T\overline{M} - \epsilon M^{-1}$ and $\mathcal{M}_2 := (M_d M^{-1} L_V)^\top$ positive definite for $T > 0$ and $\epsilon > 0$ sufficiently small. Hence $\Delta W(q,p)$ is negative definite for $T > 0$ and $\epsilon > 0$ sufficiently small, and therefore W is a strict Lyapunov function for the closed-loop system (17), (18), and this completes the proof of the theorem. ■

Discussion 1: The proof of Theorem 1 provides the qualitative argument that the discrete-time controller (18) is a SPAS controller for the plant (17). In fact we observe in the motivating example in Subsection 3.1 that our discrete-time controller significantly outperforms the emulation controller which is also a SPAS controller for the same plant. To clarify this, we concentrate on the effect of the energy shaping controller, on which the two controllers differ.

The desired Hamiltonian difference for the discrete-time model is computed as

$$\begin{aligned}\Delta H_d &= H_d(q(k+1),p(k+1)) - H_d(q(k),p(k))\\ &= \frac{1}{2} p(k+1)^\top M_d^{-1} p(k+1) + V_d(q(k+1)) - \frac{1}{2} p(k)^\top M_d^{-1} p(k) - V_d(q(k)).\end{aligned}$$

By direct calculation and using the Mean Value Theorem, applying the controller (18a) to Euler model (17) yields

$$\Delta H_d^{u_{es}^T} = -T^2 p^\top M^{-1} L_V M^{-1} p + O(T^2), \tag{21}$$

with $O(T^2)$ sign indefinite. Replacing (18a) with the emulation of controller (10), we can compute $\Delta H_d^{u_{es}}$. Without loss of generality, due to full actuation, we assume that $V_d(q)$ is convex and at least twice differentiable with respect to q. Subtracting $\Delta H_d^{u_{es}}$ from $\Delta H_d^{u_{es}^T}$ we obtain

$$\Delta H_d^{u_{es}^T} - \Delta H_d^{u_{es}} = -T^2 p^\top \overline{M} p + O(T^3),$$

where $\overline{M} := M^{-1} L_V M^{-1}$ and $L_V := \nabla_{qq} V_d$ are positive semidefinite matrices. The fact that $\Delta H_d^{u_{es}^T}$ is more negative than $\Delta H_d^{u_{es}}$ explains why $u^T(k)$ performs better than $u(k)$.

B. Underactuated, nonseparable Hamiltonian systems. We have considered fully actuated systems when stating Theorem 1. However, most mechanical systems, for what Hamiltonian model is used, are underactuated. The next result, which is proved in [8], is a generalization of Theorem 1 to a more general class of underactuated Hamiltonian systems. We emphasize that for nonseparable Hamiltonian systems (1) the inertia matrices $M(\cdot)$ and $M_d(\cdot)$ are functions of q.

Theorem 2. *Consider the Euler model (17) of Hamiltonian system (1) with Hamiltonian (2). Then the discrete-time controller* $u^T = u_{es}^T + u_{di}^T$ *where*

$$u_{es}^T = G^+\Big(\nabla_q H - M_d M^{-1}\Big[\frac{\Delta H_d}{\Delta q}\Big] + J_2 M_d^{-1} p\Big) \tag{22a}$$

$$u_{di}^T = -K_v G^\top \nabla_p H_d = -K_v G^\top M_d^{-1} p, \tag{22b}$$

with $K_v = K_v^\top > 0$, $\left[\frac{\Delta H_d}{\Delta q}\right] := \nabla_q H_d + T\kappa L_V(q) M^{-1} p$, $\kappa > 0$ *and a matrix* L_V *such that* $\overline{M} := M_d^{-1} I_G M_d M^{-1} L_V M^{-1}$ *is positive semidefinite, is a SPAS controller for the Euler model (17).* ∎

Solving the PDEs in the construction of the controller (22) in Theorem 2 is in general very difficult due to the system that is nonseparable. To solve this problem, linearized model of the plant at the equilibrium point which is then in the form of separable Hamiltonian system may be used to design the controller. In this case, the problem reduces to the case of Theorem 1. However, while with the controller (22) the stabilization is achieved in a semiglobal practical sense, with linearization we can only obtain local stability. Nevertheless, this technique is useful when solving the stabilization problem of, for instance, a pendulum system where the swing up and the stabilization can be seen as two separate problems. The example presented in Section 4 shows that the Euler based IDA-PBC controller obtained using a linearized model of the plant results in a larger domain of attraction (DOA) than the DOA with emulated linearized controller.

Corollary 1. *Consider the Euler model (17) of the linearization of the Hamiltonian system (1). Then the discrete-time controller* $u^T = u_{es}^T + u_{di}^T$ *where* u_{es}^T *and* u_{di}^T *follow (18a) and (18b) respectively, is a locally asymptotically stabilizing controller for the Euler model (17) of the original nonlinear Hamiltonian system (1).* ∎

Remark 2. In Theorems 1 and 2 (as well as Corollary 1) we have solved the stabilization problem for the approximate model (17), whereas in application we are more interested in the stability of the sampled-data system, when implementing the controller (22) to stabilize the continuous-time plant (1). Using the results of [12, 10], we conclude SPAS of the sampled-data system from SPAS of the closed-loop approximate model, the consistency of the Euler model (17) with respect to the exact discrete-time model of (1), uniform boundedness of the controller (22) and uniform boundedness of the sampled-data solutions. ■

4 Example

We consider the stabilization of a cart and pendulum system [2], with

$$M = \begin{bmatrix} \alpha & \beta\cos(q_1) \\ \beta\cos(q_1) & \gamma \end{bmatrix}, \quad G = \begin{bmatrix} 0 \\ 1 \end{bmatrix} \tag{23}$$

and $V(q) = c\cos(q_1)$, where $\alpha := ml^2$, $\beta := ml$, $\gamma := M_s + m$ and $c := mgl$ with the cart mass $M_s = 0.14$kg, the pendulum bob mass $m = 0.44$kg, the pendulum length $l = 0.215$m and $g = 9.8\text{ms}^{-2}$. $q_1 = \theta$ is the pendulum angle from its upright position and $q_2 = s$ is the cart position. The control objective is to stabilize the continuous-time system at the origin. As the controllers to be designed are implemented digitally, the system is a sampled-data control system. Assuming that the swing up of the pendulum is taken care by a separate control, we use the linearization of the kinetic energy about the upright position of the pendulum for the design. Hence in the matrix M we use the approximation $\cos(q_1) \approx 1$.

Continuous-time controller design: A continuous-time IDA-PBC controller for system (23) is first designed using the construction provided in [13]. The desired inertia matrix is chosen in the form

$$M_d = \begin{bmatrix} a_1 & a_2 \\ a_2 & a_3 \end{bmatrix} \tag{24}$$

where $a_1 := \alpha$, $a_2 := \frac{\alpha\gamma}{\beta} + \nu$ and $a_3 := \frac{(\frac{\alpha\gamma}{\beta}+\nu)^2}{\alpha} + \nu$, and the desired potential energy in the form

$$V_d = \frac{c(\alpha\gamma - \beta^2)}{a_1\gamma - a_2\beta}\cos(q_1) + \left(q_2 + \frac{a_1\beta - a_2\gamma}{a_1\gamma - a_2\beta}q_1\right)^2 := -\sigma\cos(q_1) + (q_2 + \lambda q_1)^2 .$$

Following the design procedure given in [13], we obtain an IDA-PBC controller as follow,

$$\begin{aligned} u_{es} &= k_1\left[\left(\frac{M_s L^3}{\nu} + l^2\right)q_1 + lq_2\right] + \left(ml + M_s + m + \frac{\nu}{l}\right)g\sin(q_1), \\ u_{di} &= k_v\left(\frac{(M_s+m)l+\nu}{\nu m l^2}p_1 - \frac{1}{\nu}p_2\right). \end{aligned} \tag{25}$$

Discrete-time controller design: Applying Corollary 1, we obtain a discrete-time IDA-PBC controller as follows.

$$
\begin{aligned}
u_{es}^T(k) &= u_{es}(k) - TG^{+} M_d M^{-1} L_V M^{-1} p(k) \\
u_{di}^T(k) &= u_{di}(k)
\end{aligned}
\tag{26}
$$

where

$$
L_V = k_2 \begin{bmatrix} \sigma \sin(q_1) + k_1 \lambda^2 & k_1 \lambda \\ k_1 \lambda & k_1 \end{bmatrix}, \quad \sigma < 0, k_2 > 0.
$$

Both controllers (25) and (26) SPA stabilize the closed-loop linearized model and locally asymptotically stabilize the closed-loop nonlinear system. In this example we apply the controllers to control the original nonlinear plant. For initial conditions that are relatively close to the equilibrium and with fast sampling, both controllers (25) and (26) perform well in stabilizing the closed-loop system. However, as the initial condition gets larger and the sampling is slower, the controller (25) destabilizes the equilibrium point of the system whereas the controller (26) still performs very well. Figure (3) shows the simulation results for $(q_{10}, q_{20}, p_{10}, p_{20}) = (0.15, -0.5, 0.1, 0)$, $\nu = 0.1$, $k_1 = 10$, $k_v = 1$, $k_2 = 8$ and $T = 0.01$.

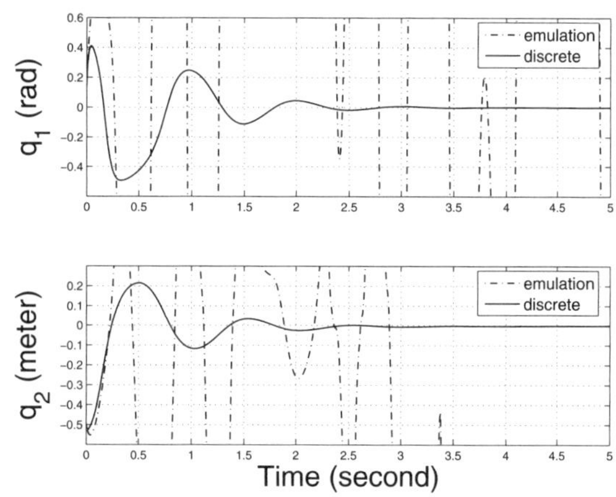

Fig. 3. Response of the pendulum on cart system

References

1. Acosta JA, Ortega R, Astolfi A (2005) Interconnection and damping assignment passivity-based control of mechanical systems with underactuation degree one. IEEE TAC 50: 1936–1955.
2. Bloch AM, Marsden JE, Sánchez de Alvarez G (1997) Feedback stabilization of relative equilibria of mechanical systems with symmetry. In: Alber M, Hu B, Rosenthal J (eds) Current and Future Directions in Applied Mathematics pp. 43–64, Birkhaüser.
3. Gonzales O (1996) Time integration and discrete Hamiltonian systems. Jour Nonlin Sci 6:449–467.
4. Itoh T, Abe K (1988) Hamiltonian-conserving discrete canonical equations based on variational different quotients. Jour. Comput. Physics 76: 85-102.

5. Khalil HK (1996) Nonlinear Control Systems 2nd Ed. Prentice Hall.
6. Laila DS, Nešić D, Teel AR (2002) Open and closed loop dissipation inequalities under sampling and controller emulation. European J. of Control 8: 109-125.
7. Laila DS, Astolfi A (2005) Discrete-time IDA-PBC design for separable of Hamiltonian systems. In: Proc. 16th IFAC World Congress, Prague.
8. Laila DS, Astolfi A (2006) Discrete-time IDA-PBC design for underactuated Hamiltonian control systems. In: Proc. 25th ACC, Minneapolis, Minnesota.
9. Marsden JE, West M (2001) Discrete mechanics and variational integrators. Acta Numerica pp. 357–514.
10. Nešić D., Teel AR, Sontag E (1999) Formulas relating $\mathcal{KL}$ stability estimates of discrete-time and sampled-data nonlinear systems. Syst. Contr. Lett. 38: 49–60.
11. Nešić D, Teel AR (2004) A framework for stabilization of nonlinear sampled-data systems based on their approximate discrete-time models. IEEE TAC 49: 1103-1122.
12. Nešić D, Laila DS (2002) A note on input-to-state stabilization for nonlinear sampled-data systems. IEEE TAC 47: 1153-1158.
13. Ortega R, Spong MW, Gomez-Estern F, Blankenstein G (2002) Stabilization of a class of underactuated mechanical systems via interconnection and damping assignment. IEEE TAC 47: 1218-1233.
14. Sanz-Serna JM, Calvo MP (1994) Numerical Hamiltonian Problems. Chapman & Hall.
15. Stuart AM, Humphries AR (1996) Dynamical Systems and Numerical Analysis. Cambridge Univ. Press, New York.

Kinematic Compensation in Port-Hamiltonian Telemanipulation

Cristian Secchi[1], Stefano Stramigioli[2], and Cesare Fantuzzi[1]

[1] DISMI, University of Modena and Reggio Emilia,
Viale Allegri 13, 42100 Reggio Emilia, Italy
secchi.cristian@unimore.it, fantuzzi.cesare@unimore.it

[2] IMPACT Institute, University of Twente
P.O. Box 217, 7500 AE Enschede, The Netherlands
S.Stramigioli@ieee.org

1 Introduction

A bilateral telemanipulator is a robotic system that allows the interaction with remote environments and it is composed by a controlled local robot (the master) and a controlled remote robot (the slave) interconnected through a communication channel. The motion imposed to the master by the human is transmitted to the slave which moves accordingly; when the slave interacts with a remote environment, the interaction force is fed back to the master side in order to improve the perception of the remote environment felt by the user. Passivity theory is a very suitable tool for the implementation of bilateral telemanipulation schemes over delayed communication channels. In passivity based telemanipulation, both master and slave are controlled by means of passive impedance controllers and master and slave sides are interconnected through a scattering based communication channel [1, 8] which allows an exchange of information which is passive independently of any constant communication delay. In this way, the overall telemanipulation system is passive and, consequently, its behavior is stable both in case of free motion and in case of interaction with any passive environment. In [12, 11], a generic framework for geometric telemanipulation of port-Hamiltonian systems [13] has been proposed; master and slave are interconnected through intrinsically passive port-Hamiltonian impedance controllers which allow to shape the energetic behavior of the robots and to achieve desired dynamic properties at master and slave sides. Local and remote sides are interconnected through a scattering based communication channel that allows a lossless exchange of energy.

When using scattering based communication channels, local and remote sides exchange only velocity and force information and this can cause the rise of a position error between master and slave which is particularly evident during interaction tasks [3, 6]. Thus, the human operator feels the remote environment as if it was at a position different from the real one and this kinematic mismatch degrades the performances of the overall system. In [14, 8] a physical

F. Bullo et al. (Eds.): Lag. & Hamil. Methods for Nonlin. Ctrl., LNCIS 366, pp. 99–110, 2007.
springerlink.com

interpretation of a scattering based communication channel is given and it is shown that the magnitude of the position error can be determined by measuring the interaction force between the slave and the environment and by the knowledge of the impedance of the scattering transformation. Several works addressed the problem of position tracking in telemanipulation, see for example [3, 6, 8].

In [2] a linear telemanipulator with a linear intrinsically passive controllers is considered and the controller is endowed with a virtual variable rest length spring that allows to compensate the position error without altering the impedance perceived by the operator. In order to preserve passivity, part of the energy injected into the system is used for changing the rest length of the spring; this energy deviation introduces a spurious effect which deteriorates the perception of the user and, furthermore, it can lead to slow transients in the compensation. In this contribution we exploit the port-Hamiltonian framework to extend the ideas reported in [2] to nonlinear telemanipulators. Furthermore, we show how it is possible to exploit the interconnection structure of port-Hamiltonian systems to implement the kinematic compensation without requiring any deviation of the energy introduced by the user and, consequently, avoiding any disturbing spurious effect and the limitation on the transient behavior of the compensation phase.

The rest of the document is organized as follows: in Sec. 2 some background on port-Hamiltonian systems and on port-Hamiltonian telemanipulation is given. In Sec. 3 we show how to modify the interconnection structure of the port-Hamiltonian controller at the slave side to store the energy that would be dissipated in the impedance matching process and in Sec. 4 we will further modify the slave controller for exploiting the stored energy for compensating the position error. In Sec. 5 some simulations to validate the obtained results are proposed and in Sec. 6 some conclusions are drawn.

2 Background

2.1 Port-Hamiltonian Systems

We can consider a port-Hamiltonian system as composed of a state manifold $\mathcal{X}$, a lower bounded energy function $H : \mathcal{X} \to \mathbb{R}$ corresponding to the internal energy, a network structure, represented by a skew-symmetric matrix, $D(x) = -D^T(x)$ whose graph has the mathematical structure of a Dirac structure, which a state dependent power continuous interconnection structure, and an interconnection port represented by a pair of dual power variables $(e, f) \in V^* \times V$ called effort and flow respectively. This port is used to interact energetically with the system: the power supplied through a port is equal to $e^T f$. We can furthermore split the interaction port in more sub-ports, each of which can be used to model different power flows. We will indicate with the subscript I the power ports by means of which the system interacts with the rest of the world, with the subscript C the

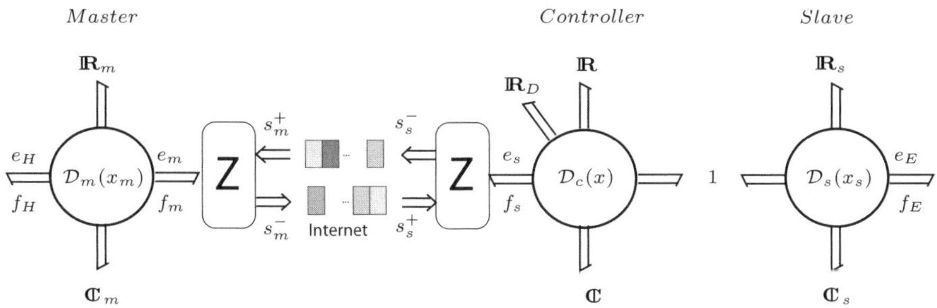

Fig. 1. The Port-Hamiltonian based Telemanipulation Scheme

power ports associated with the storage of energy and with the subscript R the power ports relative to power dissipation. Summarizing, we have:

$$\begin{pmatrix} e_I \\ f_C \\ e_R \end{pmatrix} = D(x) \begin{pmatrix} f_I \\ e_C \\ f_R \end{pmatrix} \tag{1}$$

where $D(x)$ is a skew symmetric matrix representing the Dirac structure. Loosely speaking, a port-Hamiltonian system is made up of a set of energy processing elements (energy storing, energy dissipating and sources of energy) that exchange energy by means of their power ports through a set of energy paths which form a power preserving interconnection that can be modeled as a Dirac structure. Using Eq.(1), it can be easily seen that port-Hamiltonian systems are passive because of the skew-symmetry of $D(x)$. A very broad class of physical systems, both linear and non linear, can be modeled within the port-Hamiltonian framework which can therefore be used to model telemanipulation systems endowed with nonlinear robots. For further information the reader is addressed to [13].

2.2 Port-Hamiltonian Based Bilateral Telemanipulation

The port-Hamiltonian based bilateral telemanipulation scheme is represented in Fig. 1 in a bond-graph notation. Both master and slave robots can be modeled as port-Hamiltonian systems. The slave is interconnected in a power preserving way to a port-Hamiltonian controller which acts as an intrinsically passive impedance controller [11]. In the considered scheme, we use only one impedance controller at the slave side, as recently proposed in [8], instead of two impedance controllers, as proposed in [12, 11]. Master and slave sides exchange power through a transmission line that is characterized, in general, by a non negligible communication delay. Each power port by means of which master and slave sides exchange power through the communication channel is characterized by an effort $e(t)$ and by a flow $f(t)$ and it can be equivalently represented by an incoming power wave $s^+(t)$ and an outgoing power wave $s^-(t)$ defined as

$$\begin{cases} s^+(t) = \frac{1}{\sqrt{2}} N^{-1}(e(t) + Zf(t)) \\ \\ s^-(t) = \frac{1}{\sqrt{2}} N^{-1}(e(t) - Zf(t)) \end{cases} \tag{2}$$

where $Z = NN > 0$ is the symmetric positive definite impedance of the scattering transformation. In order to get a passive exchange of energy independently of any constant communication delay, the power ports connected to the transmission line are decomposed into a pair of scattering variables which are transmitted along the channel [7, 12]. The controller is used to impose the impedance perceived by the user and it can be physically interpreted as a set of (possibly non linear, [9]) elastic, inertial and dissipative elements interconnected together. In order to avoid the wave reflection phenomenon [8, 12], which arises when using scattering based communication channels and which highly degrades performances, the controller is endowed with a dissipative element ($\mathbb{R}_D$) which implements the so called *impedance matching*. This element must be characterized by the following port behavior:

$$e_D = Zf_D \tag{3}$$

where (e_D, f_D) is the power port of the dissipative element, $f_D = f_s$ and Z is the impedance of the scattering transformation.

Since a port-Hamiltonian based bilateral telemanipulator is made up of passive subsystems interconnected in a power preserving way, the overall system is intrinsically passive and, therefore, characterized by a stable behavior. For further details the reader is addressed to [12, 11].

3 Impedance Matching Through a Storing Element

The role of the dissipating element added to the controller to implement impedance matching, is to absorb the energy content of the scattering wave that otherwise would be reflected back to the master side. In order to use the energy that is absorbed for compensating the position error, we would like to store, instead of dissipating, the energy absorbed during the impedance matching process. Thus, we have to replace the dissipating element with a storing element, that we call *tank*. The port behavior of an energy storing element is described by

$$\begin{cases} \dot{x}_T = f_C \\ \\ e_C = \frac{\partial H_T}{\partial x_T} \end{cases} \tag{4}$$

where (e_C, f_C) is the power port through which the element exchanges energy, x_C is the state and $H_T(\cdot)$ is a lower bounded function corresponding to the stored energy. If we simply replaced the dissipative element $\mathbb{R}_D$ with a storing element we wouldn't achieve the port behavior described by Eq.(3) and, consequently, we wouldn't match the impedance of the communication channel. This is mainly due to the fact that energy storing elements and energy dissipating elements process energy in two different ways: the first perform a reversible energy

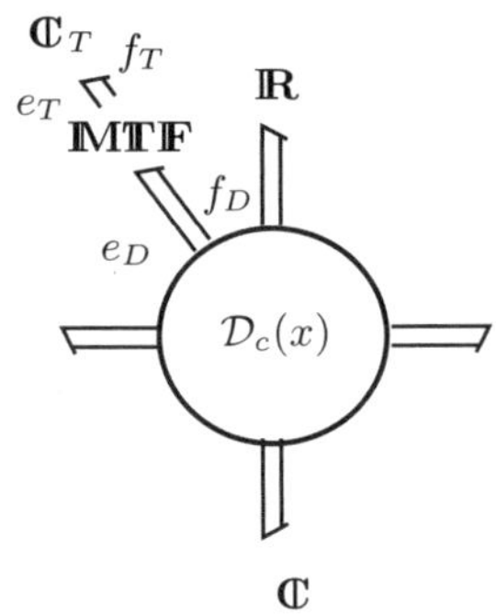

Fig. 2. The modified controller

transformation (i.e. energy can be both absorbed and released) while the latter perform an irreversible energy transformation (i.e. energy can travel only in one direction). In order to keep on implementing the required impedance matching, we need to force the port behavior reported in Eq.(3) and, therefore, we need to force the storing element to process energy irreversibly. The way in which an element processes energy is determined either by its constitutive equations or by the energetic interconnection that joins it to the rest of the system. Thus, we have to modify the interconnection structure that joins the tank to the rest of the controller in such a way that energy is always absorbed and that the port behavior reported in Eq.(3) is reproduced. In the following we will indicate with $diag(a_i)$, $i = 1, \ldots, n$ the $n \times n$ diagonal matrix whose diagonal entries are $a_1 \ldots a_n$. Furthermore, we will drop the dependence on time of efforts and flows to lighten the notation.

Proposition 1. *Consider the control structure reported in Fig. 2 where* **MTF** *represents the following modulation law:*

$$\begin{cases} f_T = m_b f_D \\ e_D = m_b^T e_T \end{cases} \qquad m_b = Z diag(\frac{f_{D_i}}{e_{T_i}})\ i = 1, \ldots, n \tag{5}$$

where f_{D_i} and e_{T_i} represent the ith component of the vectors f_D and e_T respectively. If $e_{T_i} \neq 0$, the tank element absorbs energy and the behavior implemented at the port (e_D, f_D) is that reported in Eq.(3).

Proof. We first prove that energy is always absorbed by the tank. The power flowing into the tank is given by $e_T^T f_T$. Using Eq.(5) we have that:

$$e_T^T f_T = e_T^T m_b f_D = e_T^T Z diag(\frac{f_{D_i}}{e_{T_i}}) f_D \tag{6}$$

since Z is symmetric and positive definite we can write:

$$e_T^T f_T = \underbrace{e_T^T diag(\frac{f_{D_i}}{e_{T_i}})}_{f_D^T} Z f_D = f_D^T Z f_D > 0 \tag{7}$$

Thus, power is always flowing into the tank and, therefore, the tank always absorbs energy. Furthermore we have that

$$e_D = m_b^T e_T = diag(\frac{f_{D_i}}{e_{T_i}}) Z^T e_T = Z \underbrace{diag(\frac{f_{D_i}}{e_{T_i}}) e_T}_{f_D} = Z f_D \tag{8}$$

which proves that the behavior implemented at the port (e_D, f_D) is that reported in Eq.(3)

Remark 1. In case $e_{T_i} = \frac{\partial H_T}{\partial x_{T_i}} = 0$, the modulation reported in Eq.(5) is not well defined. This situation can be avoided by choosing, for example, an energy tank characterized by a function $1/2 x_T^T K x_T$ with K diagonal positive definite and by precharging the tank by considering an initial configuration x_T such that $x_{T_i} > 0 \; \forall i$. In this way, since the tank always absorbs energy following Eq.(3), it is impossible that one of the components of e_T goes to zero.

Since their port behavior is the same, the user perceives the dissipative element used to implement the impedance matching proposed in [12] and the energy tank in the same way. Thus the energy storing process is completely hidden to the user that can keep on using the system without taking care of the modification made on the control structure and without feeling any spurious dynamic effect while charging the tank.

Finally, let us remark that the modulation reported in Eq.(5) is power preserving since it can be easily seen that $e_T^T f_T = e_D^T f_D$ and therefore we can embed it in the interconnection structure that joins all the energy processing elements of the controller achieving a new augmented power preserving interconnection structure $\mathcal{I}_{ca}(x, x_T)$. The controller keeps on being passive, which is our main objective, but it cannot be modeled as a port-Hamiltonian system anymore since its interconnection structure cannot be modeled as a Dirac structure, which cannot be modulated by power variables [13]. This is due to the fact that for modeling irreversible energy transfers towards storing elements, it is necessary to consider a nonlinear (e.g. depending on power variables) interconnection structure, see [4, 5] for further details.

4 The Transfer Interconnection Structure

In port-Hamiltonian telemanipulation, the controller always contains at least one elastic element which is used to implement the stiffness of the impedance perceived by the human operator [12, 11]. An elastic element is an energy storing element characterized by a power port (e, f) through which it exchanges energy and by a behavior described by

$$\begin{cases} \dot{x}(t) = f(t) \\ e(t) = \frac{\partial H}{\partial x} \end{cases} \tag{9}$$

where x is the state associated to the energy storing process, $H(x - l)$ is the lower bounded energy function, usually characterized by a global minimum configuration, and l is a constant configuration that models the rest length of the elastic element, namely an offset in its dynamic behavior. In order to consider variable rest length elements, we model l as a state variable and we endow the elastic element with an extra power port (e_L, f_L) through which it is possible to exchange energy for modifying l [9]. Thus the behavior of a variable rest length spring is represented by

$$\begin{cases} \dot{x}(t) = f(t) \\ \dot{l}(t) = f_L(t) \\ e(t) = \frac{\partial H}{\partial x} \\ e_L(t) = \frac{\partial H}{\partial l} \end{cases} \tag{10}$$

In [8] it is shown that, at steady state, the position mismatch that takes place between master and slave sides is given by:

$$\Delta = TZ^{-1}e_{env} \tag{11}$$

where e_{env} is the effort applied by the remote environment, T is the transmission delay and Z is the impedance of the scattering transformation. Thus, during contact tasks, the user perceives the remote environment as if it was in a different position. To give a proper perception, at steady state, for the same force on both sides the positions of master and slave should be the same. The idea for eliminating this error is to introduce an offset that compensates it by changing the rest length of the elastic element of the controller. Once the compensation has been done, the user can interact with the environment (e.g. sliding over a surface while applying a certain force) that is felt at its effective position at the master side.

To change the rest length, it is necessary to exchange energy via the length port (e_L, f_L). In order to preserve the passivity of the overall scheme we do not have to inject extra energy into the system for the compensation process and, therefore, we want to use the energy stored into the tank we described in Sec. 3. Thus, it is necessary to be able to drive energy from the tank to the length port and viceversa. In order to implement this energy exchange we interpose an interconnection structure, that we call *transfer interconnection structure*, that joins the length port and the tank port. The following result can be proven:

Proposition 2. *Let (e_T, f_T) and (e_L, f_L) be the tank port and the length port respectively and interconnect them with the interconnection structure $\mathcal{I}_T$ described by*

$$\begin{pmatrix} f_T \\ f_L \end{pmatrix} = \begin{pmatrix} 0 & -m_t \\ m_t & 0 \end{pmatrix} \begin{pmatrix} e_T \\ e_L \end{pmatrix} \tag{12}$$

where

$$m_t = diag(\gamma_i e_{L_i} e_{T_i}) \quad i = 1 \ldots n \tag{13}$$

If $\gamma_i > 0\ \forall i$ then energy is extracted from the tank and injected into the length port while if $\gamma_i < 0\ \forall i$ then energy is extracted from the length port and injected into the tank.

Proof. Using Eq.(12), we have that

$$e_L^T f_L = e_L^T diag(\gamma_i e_{L_i} e_{T_i}) e_T = \sum_{i=0}^{n} \gamma_i (e_{L_i})^2 (e_{T_i})^2 \tag{14}$$

and that

$$e_L^T f_L = e_L^T m_t e_T = -e_T^T f_T \tag{15}$$

Thus, from Eq.(14) and Eq.(15), we have that the transfer interconnection structure is power preserving and that if $\gamma_i > 0\ \forall i$, then $-e_T^T f_T = e_L^T f_L > 0$ which means that energy is extracted from the tank port and injected in the length port. Instead, if $\gamma_i < 0\ \forall i$ we have that $-e_T^T f_T = e_L^T f_L < 0$ which means that energy is extracted from the length port and injected into the tank port.

Thus, introducing the interconnection structure described in Eq.(12) between the tank port and the length port it is possible to drive the energy flow by properly setting the control parameter γ. The signs of the elements of γ determine the direction of the power transfer while their magnitudes can be used to boost the power transfer allowing us to achieve an energy transfer which is as fast as desired. In order to eliminate the position error introduced by the scattering based communication channel, we need to introduce an offset that compensates what reported in Eq.(11). Thus, the target rest length of the elastic element introduced in the controller has to be set to $l_T = -TZ^{-1}e_{env}$ the effort exchanged with the environment e_{env} can be measured by means of force sensors on the slave robot. For each component of the rest length, we need to properly set the control parameters γ_i depending on whether energy has to be extracted from or supplied to the length port to reach the target.

Using Eq.(10) it is possible to see that if $l_i < l_{T_i}$, then the ith component of the length has to increase and therefore, it is necessary to set $f_{L_i} > 0$. If $\frac{\partial H}{\partial l_i} > 0$, requiring $f_{L_i} > 0$ means requiring a positive power flow towards the length port and thus, in this case, it is necessary to set $\gamma_i > 0$. On the other hand, if $\frac{\partial H}{\partial l_i} < 0$, requiring $f_{L_i} > 0$ means requiring a negative power flow towards the length port or, equivalently, a positive power flow towards the tank; thus in this case it is necessary to set $\gamma_i < 0$. Similar considerations hold in case it is required $f_{L_i} < 0$. Thus, once the sign of the components of f_L is determined by comparing l with l_T, it is necessary to check the sign of the components of $\frac{\partial H}{\partial l}$ to determine the sign of the transfer parameter γ. Summarizing, the following tuning algorithm can be used:

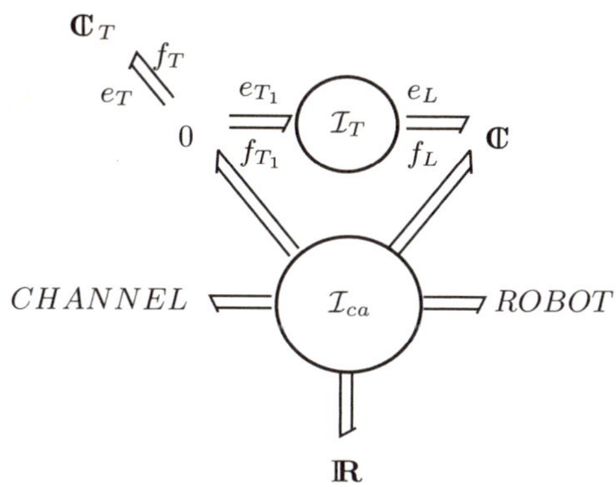

Fig. 3. The global port-Hamiltonian controller

- If $sign(l_{T_i} - l_i) = sign(\frac{\partial H}{\partial l_i})$ then $\gamma_i > 0$
- If $sign(l_{T_i} - l_i) \neq sign(\frac{\partial H}{\partial l_i})$ then $\gamma_i < 0$

Remark 2. The proposed tuning strategy can give rise to chattering phenomena which can degrade the perception felt by the user. This can be avoided using standard control techniques as boundary layers [10].

We can now put together the tank implementation described in Sec. 3 and the energy transfer mechanism described in Eq.(12) to get the overall controller which is represented in Fig. 3 in a bond graph notation. We can distinguish two power preserving interconnection structures: $\mathcal{I}_{ca}$ and $\mathcal{I}_T$ interconnected in a power preserving way. The first derives directly from the port-Hamiltonian impedance control which implements the dynamics perceived by the user augmented by the modulation reported in Eq.(5). The transfer structure $\mathcal{I}_T$ allows us to control the energy transfer between the tank $\mathbb{C}_T$ and the length port of the elastic element $\mathbb{C}$. When extracting energy from the tank, it is necessary to be careful. In fact, it is necessary to leave some energy into the tank to avoid the problems reported in Remark 1. This can be done by setting an energy threshold below which energy extraction is forbidden. It can happen that the energy stored in the tank is not sufficient for performing the desired rest length variation; in this case only a partial compensation is achieved. Nevertheless, while using the telemanipulator, more and more energy is stored in the tank and, since the energy injected in the elastic element is released back in the tank when the slave goes from contact to free motion, we will surely arrive at a point where there will always be a sufficient amount of energy for implementing the required compensation. Intuitively, the overall controller is passive since the energy used for implementing the kinematic compensation is that stored into the energy tank which comes from the impedance matching process. More formally, intrinsic passivity of the controller can be proven by recalling that a power preserving interconnection of two power preserving structures is still power preserving [5].

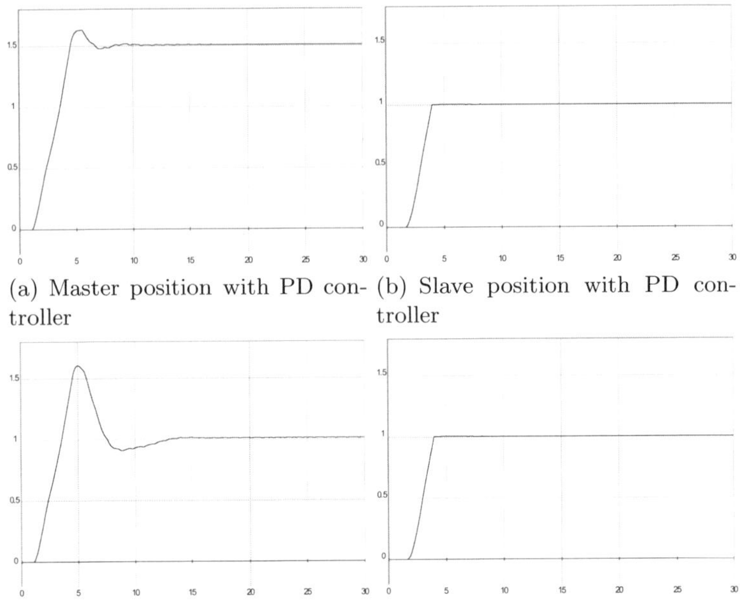

(a) Master position with PD controller (b) Slave position with PD controller

(c) Master position with the proposed controller (d) Slave position with the proposed controller

Fig. 4. Contact task

Thus, the interconnection of $\mathcal{I}_{ca}$ and $\mathcal{I}_T$ gives a controller made by a *nonlinear* power preserving interconnection of energy storing elements, energy dissipating elements and interaction ports. It can be easily seen that the interconnection structure can still be represented as a skew-symmetric matrix and that, therefore, the controller, even if it cannot be modeled as a port-Hamiltonian system, is intrinsically passive.

5 Simulations

We consider a simple one degree of freedom telemanipulator where master and slave are simple masses of 0.5 Kg. The slave is interconnected to a port-Hamiltonian impedance controller and local and remote sides are joined through a scattering based communication channel characterized by an impedance $Z = 1$ and by a transmission delay $T = 0.5$ s in both senses of communication. The port-Hamiltonian controller is a simple PD, physically equivalent to the parallel of a spring with stiffness $K = 200$ N/m and a damper with dissipation coefficient $b = 1$ $Nsec/m$ used to match the impedance of the communication channel.

We consider a contact task and we compare the behaviors of the system when using a PD controller and the proposed control scheme. The user applies a constant force $e_H = 1$ N and the slave gets in contact with a rigid visco-elastic wall characterized by a stiffness $K_w = 5000$ N/m and by a damping

$b_w = 40\ Ns/m$ at position $x = 1\ m$. Master and slave positions are reported in Fig. 4. The slave robot stops when it meets the wall and the interaction force is fed back to the master side. In case the PD controller is adopted (Fig. 4(a) and Fig. 4(b)) we can see that the interaction force is fed back to the master side through the transmission line and that it equilibrates the force applied by the user. Nevertheless, because of the scattering based nature of the communication channel, a significant position mismatch exists between master and slave position at the equilibrium. Thus, the user experiences a good force feedback but a bad kinematic feedback. If the proposed control scheme is used, the interaction force is still fed back to the master side and it equilibrates the force applied by the user implementing the same force feedback as in the case PD control is used. Furthermore, when the slave gets in touch with the environment, the rest length of the spring is changed to introduce an offset that compensates that introduced by the communication channel; the compensation is passivity preserving since it uses the energy that is stored in the tank through the impedance matching process. We can see in Fig. 4(c) and Fig. 4(d) that the position mismatch between master and slave drastically decreases giving thus to the user both a good force feedback and a good kinematic feedback.

6 Conclusions

In this contribution we have proposed a novel control algorithm for port-Hamiltonian based bilateral telemanipulation. It allows to compensate the position error introduced by the scattering based communication channel during contact tasks while preserving passivity of the overall scheme *without* requiring the direct intervention of the user. The controller that we have obtained is no more a port-Hamiltonian system but it is still passive and, therefore, the stable behavior of the overall telemanipulation system is still guaranteed. Future work aims at an experimental implementation of the proposed algorithm and at experimentally proving its usefulness in improving the perception of the remote environment. We will also focus on the use of the proposed control strategy for Internet based communication channels where the communication delay is variable and some packets can get lost.

References

1. R. Anderson and M. Spong. Bilateral control of teleoperators with time delay. *IEEE Transactions on Automatic Control*, 34(5):494–501, 1989.
2. P. Arcara and C. Melchiorri. Position drift compensation for a passivity-based telemanipulation control scheme. In *Proceedings to Mechatronics Conference*, Enschede, The Netherlands, June 2002.
3. N. Chopra, M.W. Spong, S. Hirche, and M. Buss. Bilateral teleoperation over the internet: the time varying delay problem. In *Proceedings of American Control Conference*, volume 1, Denver, Colorado, USA, June 2003.

4. D. Eberard, B. Maschke, and A.J. van der Schaft. Port contact systems for irreversible thermodynamical systems. In *Proceedings of IEEE Conference on Decision and Control*, Seville, Spain, December 2005.
5. D. Eberard, B. Maschke, and A.J. van der Schaft. Energy conserving formulation of RLC-circuits with linear resistors. In *Proceedings of the international symposium on mathematical theory of networks and systems*, Kyoto, Japan, July 2006.
6. D. Lee and M.W. Spong. Passive bilateral control of teleoperators under constant time-delay. In *Proceedings of IFAC world Congress*, Prague, Czech Republic, July 2005.
7. G. Niemeyer and J. Slotine. Stable adaptive teleoperation. *IEEE Journal of Oceanic Engineering*, 16(1):152–162, 1991.
8. J.-J. Niemeyer, G. ans Slotine. Telemanipulation with Time Delays. *International Journal of Robotics Research*, 23(9):873–890, September 2004.
9. C. Secchi. *Interactive Robotic Interfaces: a port-Hamiltonian Approach.* PhD thesis, University of Modena and Reggio Emilia, 2004. available at http://www.dismi.unimore.it/download/thesis.pdf.
10. J.-J. Slotine and W. Li. *Applied Nonlinear Control.* Prentice Hall, 1991.
11. S. Stramigioli, C. Secchi, A.J. van der Schaft, and C. Fantuzzi. Sampled data systems passivity and discrete port-hamiltonian systems. *IEEE Transactions on Robotics*, 21(4):574–587, 2005.
12. S. Stramigioli, A. van der Schaft, B. Maschke, and C. Melchiorri. Geometric scattering in robotic telemanipulation. *IEEE Transactions on Robotics and Automation*, 18(4), 2002.
13. A.J. van der Schaft. *L_2-Gain and Passivity Techniques in Nonlinear Control.* Communication and Control Engineering. Springer Verlag, 2000.
14. Y. Yokokohji, T. Tsujioka, and T. Yoshikawa. Bilateral control with time-varying delay including communication blackout. In *Proceedings of the 10th Symposyum on Haptic Interfaces for Virtual Environments and Teleoperator Systems*, 2002.

Interconnection and Damping Assignment Passivity-Based Control of a Four-Tank System

Jørgen K. Johnsen and Frank Allgöwer

Institute for Systems Theory and Automatic Control (IST),
University of Stuttgart, Germany
{johnsen, allgower}@ist.uni-stuttgart.de

Summary. Interconnection and damping assignment passivity-based control (IDA-PBC) is a recently developed method for nonlinear controller design. A majority of the examples of IDA-PBC found in literature are from the electro-mechanical domain. To show its applicability in other domains, in this work IDA-PBC is used to design a stabilizing controller for a process control example, the four-tank system. Both simulations and real experiments are provided for this example system.

1 Introduction

Interconnection and damping assignment passivity-based control (IDA-PBC) is a technique that regulates the behavior of nonlinear systems assigning a desired port-Hamiltonian structure to the closed loop. IDA-PBC was introduced by [3], and has been illustrated with examples such as mass-balance systems, electrical motors, magnetic levitation systems, the inverted pendulum etc. See [2] for a general survey of IDA-PBC with a list of references. See also [4].

Examples of IDA-PBC applications that are reported in the literature are mainly from the electrical and mechanical domains. However, the port-Hamiltonian framework is also applicable to systems and control problems from the chemical engineering domain and to process control. One of the few examples of IDA-PBC in process control is the averaging level control problem considered in [5]. In chemical engineering port-Hamiltonians are not the standard modeling framework – in many cases there are only identified models without any physical meaning of the states and the right hand side of the differential equations, in other cases energy, the key design variable in IDA-PBC, is not so easy to define. In these cases, IDA-PBC can still be used to find a closed-loop port-Hamiltonian system with an energy function which does not necessarily have physical meaning, but is suitable as a Lyapunov function guaranteeing stability. In this paper it will be shown how to use IDA-PBC for a typical example in process control, namely the four-tank system, and we will discuss issues related to its practical application.

It is well known that the main obstacle with IDA-PBC is the so-called matching equation for making the closed loop a port-Hamiltonian system. The matching equation consists of a set of PDE's which generally becomes increasingly

F. Bullo et al. (Eds.): Lag. & Hamil. Methods for Nonlin. Ctrl., LNCIS 366, pp. 111–122, 2007.
springerlink.com

difficult to solve as the system order increases. Examples in the literature are hence low-order, typically only two or three states.

The four-tank system considered in this paper is illustrated in Figure 1. It is a fourth order system, but when the input flows are directed only to the upper tanks, it consists of two decoupled two-tank subsystems. In this case the matching equation is easily solved and a control law for the two input flows can be determined. When part of the input flows are directed to the lower tanks, the two subsystems are interconnected through the inputs and the full fourth-order system must be considered when applying IDA-PBC. This makes the four-tank system a particularly suitable example for looking at what happens when two subsystems are interconnected. There are similarities in the structure of the interconnected and the decoupled system, and it is reasonable to assume that there are also similarities in the solution of the matching equation for the respective cases.

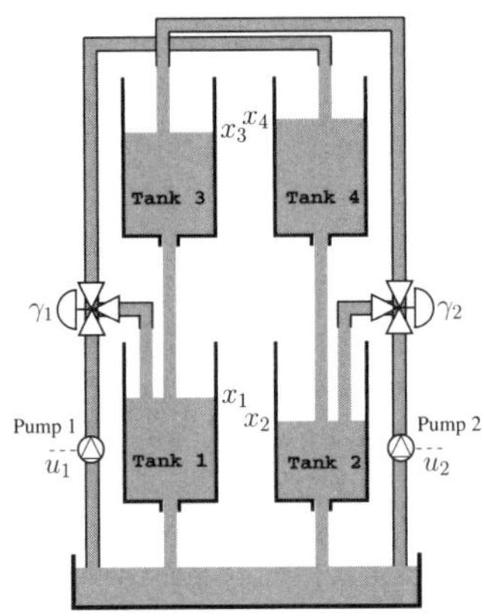

Fig. 1. The four-tank system

The general idea is that IDA-PBC based controller design for an interconnected system in some cases can be done in an incremental fashion by first finding solutions to smaller decoupled problems. These solutions can then be of help for finding a solution when the subsystems are interconnected in some way. Chemical plants are typically high-order systems consisting of several subsystems which are interconnected via pipes or in cascaded loops etc. Finding a solution of the matching equation for the whole system is difficult at best. A systematic way of designing controllers by solving the matching equations "bottom-up" will make it easier to apply IDA-PBC in process control.

The four-tank system is used as an example in a first attempt to understand and use these properties. More specifically, it is shown how a solution of the matching equation for the two-tank system can be used when trying to find a solution of the matching equation for the full system with interaction through the input.

There are only few experimental results with IDA-PBC available in the literature, most application examples are verified with simulations only. The IDA-PBC controller derived in this paper for the four-tank system is implemented on a lab experiment. It is shown how steady state offset due to modelling errors can be removed by using the internal model principle and adding error states to the dynamical system.

The paper is organized as follows. In Section 2 a short overview of the IDA-PBC methodology is given. In Section 3 we apply IDA-PBC to the four-tank system. Two cases will be discussed, the decoupled case where the matching equation is easy to solve, and the interconnected case in which the matching equation gets more complex. The solution in the decoupled case is shown to be useful in the sense that it helps in finding a solution for the interconnected case. In Section 4 the controller for the four-tank system is tested both with simulations and real life experiments. We conclude with Section 5.

2 Basic IDA-PBC Theory

Given an input affine system

$$\dot{x} = f(x) + g(x)u \tag{1}$$

with $x \in \mathbb{R}^n$, $u \in \mathbb{R}^m$. The idea of IDA-PBC is to make the closed loop with $u = \beta(x)$ an explicit port-Hamiltonian system of the form

$$\dot{x} = (J_d(x) - R_d(x))\nabla H_d(x), \tag{2}$$

where $J_d(x)$ is a skew-symmetrix matrix, $R_d(x)$ is a symmetric, positive semi-definite matrix and $H_d(x)$ is the Hamiltonian of the closed loop. $\nabla H_d(x) = \frac{\partial}{\partial x} H_d(x)$ is the gradient of $H_d(x)$. To make (1) equal (2), we need to find a solution of the *matching equation*

$$f(x) + g(x)\beta(x) = (J_d(x) - R_d(x))\nabla H_d(x), \tag{3}$$

using $J_d(x)$, $R_d(x)$ and $H_d(x)$ as design variables. The origin of the closed-loop system is stable if the desired Hamiltonian $H_d(x)$ is positive definite. This becomes clear when differentiating $H_d(x)$ along closed-loop trajectories:

$$\frac{\mathrm{d}}{\mathrm{d}t} H_d(x) = -\nabla^T H_d(x) R_d(x) \nabla H_d(x) \leq 0. \tag{4}$$

Thus $H_d(x) > 0$ serves as a Lyapunov function and the origin of the closed loop is stable. Asymptotic stability is guaranteed if $R_d(x)$ is strictly positive definite. If $R_d(x)$ and $H_d(x)$ are only locally positive definite, local stability is guaranteed.

There are different strategies for solving the matching equation. The approach used here is to split (3) into two parts; a fully actuated part and an un-actuated part. Let $g^{\perp}(x)$ denote a maximum rank left annihilator of $g(x)$ and $g^{\dagger}(x)$ a left inverse of $g(x)$, i.e. $g^{\perp}(x)g(x) = 0$ and $g^{\dagger}(x)g(x) = I$. Multiplying (3) from the left for one with $g^{\perp}(x)$ and as a second case with $g^{\dagger}(x)$ gives a PDE and an algebraic equation

$$g^{\perp} f = g^{\perp}(J_d - R_d)\nabla H_d, \tag{5}$$

$$\beta = g^{\dagger}\left\{(J_d - R_d)\nabla H_d - f\right\}, \tag{6}$$

where for simplicity, the arguments have been left out. Functions $J_d = -J_d^T$, $R_d \geq 0$ and H_d must be found such that (5) is satisfied and the Hamiltonian has an isolated minimum at the desired equilibrium point x^*. The following two conditions are imposed on the Hamiltonian:

- Necessary extremum assignment condition:

$$\nabla H_d(x^*) = 0. \tag{7}$$

- Sufficient minimum assignment condition:

$$\nabla^2 H_d(x^*) > 0. \tag{8}$$

The control law $u = \beta(x)$ can then be calculated directly from (6).

In the four-tank application below, the input matrix g is constant. We will further assume J_d and R_d to be constant matrices. Although possibly restrictive, this assumption makes (5) linear in the partial derivatives, and hence simpler to solve.

Remark 1. It is not necessary to require the system to be input affine, systems of the form $\dot{x} = f(x, u)$ may also be considered. However, the procedure for solving the matching equation will necessarily have to follow a different path.

3 The Four-Tank System

The four-tank system is a multi-input multi-output nonlinear system consisting of four tanks filled with water as illustrated in Figure 1. The water flows from the two upper tanks into the lower tanks and from there into a reservoir. There are two input flows, u_1 and u_2, and the goal is to stabilize the levels x_1 and x_2 of the lower tanks at the desired values. Two valves γ_1 and γ_2 are used to split each input flow in two directions which renders the system more or less interacting. The valve positions are not actuated, and are thus not considered as inputs to the system, but as parameters.

3.1 System Dynamics

Using Torricelli's law for the outflows of each tank, the dynamic model for the four tank system can be written as [1]

$$\dot{x} = f(x) + g(x)u, \tag{9}$$

where

$$f(x) = \begin{bmatrix} -\frac{a_1\sqrt{2gx_1}}{A_1} + \frac{a_3\sqrt{2gx_3}}{A_1} \\ -\frac{a_2\sqrt{2gx_2}}{A_2} + \frac{a_4\sqrt{2gx_4}}{A_2} \\ -\frac{a_3\sqrt{2gx_3}}{A_3} \\ -\frac{a_4\sqrt{2gx_4}}{A_4} \end{bmatrix}, \tag{10}$$

$$g(x) = \begin{bmatrix} \frac{\gamma_1}{A_1} & 0 \\ 0 & \frac{\gamma_2}{A_2} \\ 0 & \frac{1-\gamma_2}{A_3} \\ \frac{1-\gamma_1}{A_4} & 0 \end{bmatrix}, \tag{11}$$

and $x = [x_1, x_2, x_3, x_4]^T$, $u = [u_1, u_2]^T$. The state variables x_i, $i = 1, \ldots, 4$ represent the tank levels, u_i, $i = 1, 2$ the input flows, A_i, $i = 1, \ldots, 4$ the cross sections of the tanks, a_i, $i = 1, \ldots, 4$ the cross sections of the outlet holes and $g = 981$ cm/s^2 is the gravitational constant. The valve parameters γ_1, $\gamma_2 \in [0, 1]$ determine how much of the input flows are directed into the lower tanks. The parameter values of A_i and a_i for the lab experiment are given in Table 1. To avoid confusion with the gravitational constant g, we will in the following write $g(x)$ for the input matrix, even though it is independent of x.

Table 1. System parameter values

	A_i (cm^2)	a_i (cm^2)
$i = 1, 2$:	50.3	0.233
$i = 3, 4$:	28.3	0.127

When $\gamma_1 = \gamma_2 = 0$, all the water is entering in the upper tanks, and the overall system consists of two decoupled two-tank systems. Let's therefore first design a controller for this smaller problem.

3.2 Controller Design for the Two-Tank System

Consider a two-tank system with one input flow split between two tanks;

$$\begin{bmatrix} \dot{x}_1 \\ \dot{x}_2 \end{bmatrix} = \underbrace{\begin{bmatrix} -\frac{a_1\sqrt{2gx_1}}{A_1} + \frac{a_2\sqrt{2gx_2}}{A_1} \\ -\frac{a_2\sqrt{2gx_2}}{A_2} \end{bmatrix}}_{f(x)} + \underbrace{\begin{bmatrix} \frac{\gamma}{A_1} \\ \frac{1-\gamma}{A_2} \end{bmatrix}}_{g(x)} u. \tag{12}$$

The state variables x_1 and x_2 represent the water level in the lower and the upper tank respectively. The valve parameter $\gamma \in [0, 1]$ determines how much water is directed to the the different tanks; a value of zero directs all the water to the upper tank. Following the procedure outlined in Section 2, we will now design a controller which stabilizes the water level of the lower tank.

For the two-tank system, a left annihilator and a left inverse of $g(x)$ are given by

$$g^{\perp}(x) = \left[(1-\gamma)A_1,\, -\gamma A_2\right], \tag{13}$$
$$g^{\dagger}(x) = \left[A_1,\, A_2\right] \tag{14}$$

respectively.

The interconnection and damping matrices J_d and R_d are initially chosen to be constant, but otherwise free as

$$J_d - R_d = \begin{bmatrix} 0 & j \\ -j & 0 \end{bmatrix} - \begin{bmatrix} r_1 & r_3 \\ r_3 & r_2 \end{bmatrix}. \tag{15}$$

R_d is positive definite for $r_1, r_2 > 0$ and $r_1 r_2 - r_3^2 > 0$. The parameter j can take any constant value. Written out, the matching equation (5) to be solved becomes

$$\begin{aligned} -(1-\gamma)a_1\sqrt{2gx_1} + a_2\sqrt{2gx_2} &= \{-(1-\gamma)A_1 r_1 + \gamma A_2(j + r_3)\}\frac{\partial H_d}{\partial x_1} \\ &\quad + \{(1-\gamma)A_1(j - r_3) + \gamma A_2 r_2\}\frac{\partial H_d}{\partial x_2}. \end{aligned} \tag{16}$$

To simplify (16), we choose $j = -r_3$ and $r_3 = -A_2 r_2/(2A_1)$ to get

$$-(1-\gamma)a_1\sqrt{2gx_1} + a_2\sqrt{2gx_2} = -(1-\gamma)A_1 r_1\frac{\partial H_d}{\partial x_1} + A_2 r_2\frac{\partial H_d}{\partial x_2} \tag{17}$$

with a solution

$$H_d(x) = \frac{2a_1\sqrt{2g}}{3A_1 r_1}x_1^{3/2} + \frac{2a_2\sqrt{2g}}{3A_2 r_2}x_2^{3/2} + \Phi(z). \tag{18}$$

The first two terms on the right hand side are the particular solution. The homogeneous solution, $\Phi(z)$, where $z = \frac{x_1}{A_1 r_1} + \frac{(1-\gamma)x_2}{A_2 r_2} = p^T x$, is chosen such that the closed-loop system is stabilized, i.e. such that H_d is positive definite with a minimum at the desired equilibrium point x^*.

Remark 2. The gradient of the homogeneous solution is $\frac{\partial}{\partial x}\Phi = \frac{\partial z}{\partial x}\frac{\partial}{\partial z}\Phi = p\frac{\partial}{\partial z}\Phi$. Hence, (5) is satisfied for all$\Phi(p^T x)$ with p in the null space of $g^{\perp}(x)(J_d - R_d)$.

One possibility, yielding a simple controller, is to choose a quadratic function $\Phi(z) = 1/2q\tilde{z}^T\tilde{z} + l^T z$ where $\tilde{z} = z - z^*$ and $q > 0$ and l are scalars. For equilibrium assignment, i.e. $\nabla H_d(x^*) = 0$, we let $l = -a_1\sqrt{2gx_1^*} = -a_2\sqrt{2gx_2^*}/(1-\gamma)$. Note that since there is only one input, only one level can be independently stabilized. The Hessian of H_d is calculated as

$$\nabla^2 H_d = \begin{bmatrix} \frac{a_1\sqrt{2g}}{2A_1 r_1\sqrt{x_1}} & 0 \\ 0 & \frac{a_2\sqrt{2g}}{2A_2 r_2\sqrt{x_2}} \end{bmatrix} + qpp^T. \tag{19}$$

The Hessian is positive definite for all positive x and hence x^* is a unique minimum of $H_d(x)$.

The control law for the two-tank system is found from (6) as

$$\begin{aligned}\beta(x) &= \overbrace{g^\dagger(x)(J_d - R_d)p}^{=-1}(qp^T\tilde{x} + l) \\ &= -k^T\tilde{x} + u^*, \end{aligned} \tag{20}$$

where $\tilde{x} = x - x^*$, $k^T = qp^T = [\frac{q}{A_1 r_1}, \frac{(1-\gamma)q}{A_2 r_2}]$ and $u^* = -l$. Note that p in this case was normalized such that $g^\dagger(x)(J_d - R_d)p = -1$. The controller parameters are r_1, r_2 and q, the latter may be chosen equal to one without loss of generality. The closed loop is asymptotically stable if $r_1 > 0$ and $0 < r_2 < 4r_1(A_1/A_2)^2$, which makes the damping matrix R_d positive definite. To simplify slightly, define $k_1 = (A_1 r_1)^{-1}$ and $k_2 = (A_2 r_2)^{-1}$, the control law can then be written

$$u = \beta(x) = -k_1\tilde{x}_1 - (1-\gamma)k_2\tilde{x}_2 + u^*, \tag{21}$$

which is asymptotically stabilizing provided $k_1 > 0$ and $k_2 > k_1 A_2/(4A_1)$.

The simplifications in this subsection may look somewhat arbitrary. To summarize, two main simplifications were used: First we assumed J_d and R_d constant. Secondly, some of the elements of J_d and R_d were chosen such that (16) was simplified. Finally, with these simplification, a positive definite Hamiltonian was found. Had the Hamiltonian not been positive definite, one would have to try over again, using other simplifications, in an iterative fashion.

With the above calulations in mind, it is now time to look at the larger four-tank system.

3.3 Controller Design for the Four-Tank System

Before designing a controller for the four-tank system, it is instructive to write the open-loop system in a port-Hamiltonian form. To achieve zero steady-state offset for the water level in the two lower tanks, x_1 and x_2, the four-tank system described in Section 3.1 is extended with two error states:

$$\dot{x}_5 = k_{I1}\left(a_1\sqrt{2gx_1} - a_1\sqrt{2gx_1^*}\right), \tag{22}$$

$$\dot{x}_6 = k_{I2}\left(a_2\sqrt{2gx_2} - a_2\sqrt{2gx_2^*}\right). \tag{23}$$

If the extended system is stabilized, i.e. if $\dot{x}$ goes to zero, then x_1 and x_2 will converge to their desired values. Observe that the errors in the outflows of tank 1 and 2 are being integrated. The reason for choosing such an error term instead of the classical integrated state error is that the square root will show up in the gradient of the Hamiltonian, making it possible to keep the damping and interconnection matrices constant, similar to what was seen in the previous section.

Inspired by the Hamiltonian (18) for the two-tank system, and adding terms linear in in the error variables, define the Hamiltonian for the open-loop four-tank system as

$$H(x) = \sum_{i=1}^{4} \frac{2}{3} k_i a_i \sqrt{2g} x_i^{3/2} + k_1 a_1 \sqrt{2g x_1^*} x_5 + k_2 a_2 \sqrt{2g x_2^*} x_6. \tag{24}$$

With this Hamiltonian, the extended four-tank system can be written

$$\dot{x} = (J - R)\nabla H(x) + g(x)u, \tag{25}$$

where x is the extended state vector and

$$J - R = \begin{bmatrix} \frac{-1}{k_1 A_1} & 0 & \frac{1}{k_3 A_1} & 0 & 0 & 0 \\ 0 & \frac{-1}{k_2 A_2} & 0 & \frac{1}{k_4 A_2} & 0 & 0 \\ 0 & 0 & \frac{-1}{k_3 A_3} & 0 & 0 & 0 \\ 0 & 0 & 0 & \frac{-1}{k_4 A_4} & 0 & 0 \\ \frac{k_{I1}}{k_1} & 0 & 0 & 0 & \frac{-k_{I1}}{k_1} & 0 \\ 0 & \frac{k_{I2}}{k_2} & 0 & 0 & 0 & \frac{-k_{I2}}{k_2} \end{bmatrix}$$

$$g(x) = \begin{bmatrix} \frac{\gamma_1}{A_1} & 0 & 0 & \frac{1-\gamma_1}{A_4} & 0 & 0 \\ 0 & \frac{\gamma_2}{A_2} & \frac{1-\gamma_2}{A_3} & 0 & 0 & 0 \end{bmatrix}^T .$$

Similar to the two-tank system, the damping matrix R is positive definite if $k_1, k_2 > 0$, $k_3 > k_1 A_3/(4A_1)$, $k_4 > k_2 A_4/(4A_2)$ and additionally, the integral terms $k_{I1} < 4 - k_1/k_3$ and $k_{I2} < 4 - k_2/k_4$.

Remark 3. There are other ways of writing the system in a port-Hamiltonian form. The advantage of this particular choice is that $J-R$ is constant. A different approach would be e.g. to choose the Hamiltonian as the total volume of water, which would be physically more meaningful, and use nonlinear damping and interconnection matrices.

To solve the matching equation, we first choose a left annihilator and a left inverse of $g(x)$ as

$$g^{\perp}(x) = \begin{bmatrix} (1-\gamma_1)A_1 & 0 & 0 & 0 \\ 0 & (1-\gamma_2)A_2 & 0 & 0 \\ 0 & -\gamma_2 A_3 & 0 & 0 \\ -\gamma_1 A_4 & 0 & 0 & 0 \\ 0 & 0 & 1 & 0 \\ 0 & 0 & 0 & 1 \end{bmatrix}^T , \tag{26}$$

$$g^{\dagger}(x) = \begin{bmatrix} A_1 & 0 & 0 & A_4 & 0 & 0 \\ 0 & A_2 & A_3 & 0 & 0 & 0 \end{bmatrix} \tag{27}$$

respectively. If the interconnection and damping matrices are *kept unchanged*, i.e. $J_d - R_d = J - R$, a solution to (5) is given by

$$H_d(x) = H(x) + \frac{1}{2}\tilde{x}^T P^T Q P \tilde{x} + l^T P x, \tag{28}$$

where P gives a basis for the kernel of $g^\perp(J_d - R_d)$ (see Remark 2)and l is used to place the equilibrium point at x^*. One possible P is given by

$$P = \begin{bmatrix} \gamma_1 k_1 & (1-\gamma_2)k_1 \\ (1-\gamma_1)k_2 & \gamma_2 k_2 \\ 0 & (1-\gamma_2)k_3 \\ (1-\gamma_1)k_4 & 0 \\ \gamma_1 k_1 & (1-\gamma_2)k_1 \\ (1-\gamma_1)k_2 & \gamma_2 k_2 \end{bmatrix}^T . \tag{29}$$

Some calculations show that selecting

$$l = \begin{bmatrix} \gamma_1 & 1-\gamma_2 \\ 1-\gamma_1 & \gamma_2 \end{bmatrix}^{-1} \begin{bmatrix} a_1\sqrt{2gx_1^*} \\ a_2\sqrt{2gx_2^*} \end{bmatrix} \tag{30}$$

$$= -\begin{bmatrix} 0 & 1-\gamma_2 \\ 1-\gamma_1 & 0 \end{bmatrix}^{-1} \begin{bmatrix} a_3\sqrt{2gx_3^*} \\ a_4\sqrt{2gx_4^*} \end{bmatrix} \tag{31}$$

will make $\nabla H_d(x^*) = 0$.

The Hessian of the energy function,

$$\nabla^2 H_d(x) = \text{diag}\{\frac{k_i a_i \sqrt{2g}}{2\sqrt{x_i}}, 0, 0\} + P^T Q P, \tag{32}$$

can be shown to be positive definite if $Q > 0$, hence the system is asymptotically stable if $R_d > 0$.

Using (6) together with $J - R = J_d - R_d$, the control law is obtained as

$$\begin{aligned} \beta(x) &= \overbrace{g^\dagger(x)(J_d - R_d)P^T}^{=-I} (QP\tilde{x} + l) \\ &= -K\tilde{x} + u^*, \end{aligned} \tag{33}$$

where $K = QP$ and $u^* = -l$. The controller is a PI-type state feedback controller with a nonlinear integral term. The controller has tuning parameters k_1 to k_4, k_{I1}, k_{I2} and Q which has to be chosen such that the controller asymptotically stabilizes x_1 and x_2, that is $k_1, k_2 > 0$, $k_3 > k_1 A_3/(4A_1)$, $k_4 > k_2 A_4/(4A_2)$ and $k_{I1} < 4 - k_1/k_3$, $k_{I2} < 4 - k_2/k_4$ ($\Rightarrow R_d > 0$) and $Q > 0$ ($\Rightarrow H_d > 0$).

4 Simulations and Experiments

Simulations were performed with the model and experimental runs were conducted on the lab setup discussed in Section 3.1 with valve parameters $\gamma_1 = \gamma_2 = 0.6$.

To avoid integrator wind-up, the error states x_5 and x_6 were only integrated when none of the water pumps were saturated. If saturation occured in any pump, both derivatives $\dot{x}_5$ and $\dot{x}_6$ were simply set to zero.

The controller parameter values used in simulations and experiments, found by trial and error, are listed in Table 2. Since the system has a symmetry, the same values are used for similar parameters, i.e. for k_1 and k_2 etc.

Table 2. Controller parameter values

k_1, k_2	k_3, k_4	k_{I1}, k_{I2}	Q
10	5.0	0.13	$I_{2\times 2}$

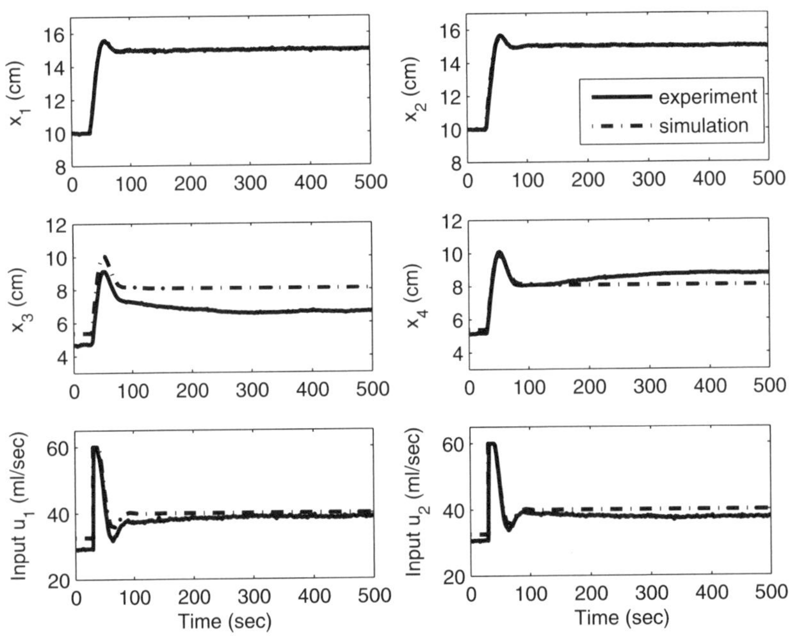

Fig. 2. Response of the four-tank system with controller given by (33). Solid line: Experimental result. Dashed line: Simulation result.

The result of applying a step in the desired values for the lower tanks from $x_1^* = x_2^* = 10$ cm to $x_1^* = x_2^* = 15$ cm at time $t = 25$ seconds is shown in Figure 2. There is evidently a small model/plant mismatch since the steady state levels of the upper tanks, x_3 and x_4, are different in the simulation (solid line) and in the experiment (dashed line). There is practically no difference in the dynamic response for the lower tank levels. The integral term makes sure that the lower tank levels converge to the desired value also in the experiment. Experiments with no integral action in the controller showed a steady-state offset (not shown). Saturation occurs during the transient phase, but causes no serious problems.

5 Discussion and Conclusion

We have shown how IDA-PBC can be used to design a controller with guaranteed stability properties for the four-tank system. Even though the controller in this case may be seen as very simple, the Hamiltonian, which acts as a Lyapunov function guaranteeing stability, is not trivial.

The controller design was done incrementally. First it was shown how to design a controller for the two-tank system, where it was easy to find a solution to the matching equation just by visual inspection. Going from the two-tank to the four-tank system was, due to the port-Hamiltonian setup, only a matter of extending everything in a natural way. The open- and closed-loop interconnection and damping matrices were chosen equal and constant, which made it easier to solve the matching equation. Terms were added to the closed-loop Hamiltonian to stabilize the desired equilibrium. The internal model principle was used when adding two output error states. This guaranteed no steady-state offset in the two lower levels even in the case of some parametric uncertainties.

To verify that the controller works, simulations were performed and the controller was implemented on a lab scale experiment. The performance of the controller was in both cases satisfactory, and even though there were some smaller discrepancies between simulations and experimental results, the convergence of the water levels in the lower tanks were close to identical.

The way IDA-PBC was used here is perhaps not revealing its main strength. It was used to stabilize the closed-loop system, which in itself is of course not unimportant. However, the interconnection and damping assignment part of IDA-PBC was not exploited. As mentioned above, the structure of the interconnection and damping matrices was left unchanged. This may actually be where IDA-PBC has one of its strengths; the possibility of coherently decomposing and rebuilding the internal structure of a system. However, no matter how attractive this may sound, it is not necessarily easy to determine what a "good" interconnection structure could or would look like. In most examples in the literature it is either left unchanged, like in this work, or changed in a way such that the matching equation for the respective problem is simplified. There is a need for further investigations to understand and use these properties.

References

1. K.H. Johansson. The quadruple-tank process: A multivariable laboratory process with adjustable zero. *IEEE Transactions on Control Systems Technology*, 8:456–465, 2000.

2. R. Ortega and E. Garcia-Canseco. Interconnection and damping assignment passivity-based control: A survey. *European Journal of Control*, 10(5):432–450, 2004.
3. R. Ortega, A. van der Schaft, B. Maschke, and G. Escobar. Energy shaping of port-controlled Hamiltonian systems by interconnection. In *Proceedings of the 38th IEEE Conference on Decision and Control*, volume 2, pages 1646–1651, 1999.
4. R. Ortega, A. van der Schaft, B. Maschke, and G. Escobar. Interconnection and damping assignment passivity-based control of port-controlled Hamiltonian systems. *Automatica*, 38(4):585–596, 2002.
5. D.G. Sbarbaro and R. Ortega. Averaging level control of multiple tanks, a passivity based approach. In *Proceedings of the 44th IEEE Conference on Decision and Control and the European Control Conference 2005*, pages 7384–7389, December 2005.

Towards Power-Based Control Strategies for a Class of Nonlinear Mechanical Systems

Alessandro de Rinaldis[1], Jacquelien M.A. Scherpen[2], and Romeo Ortega[1]

[1] Lab. des Signaux et Systèmes, CNRS-SUPELEC, Gif–sur–Yvette 91192, FR
alessandro.de-rinaldis@renault.com , romeo.ortega@lss.supelec.fr
[2] University of Groningen, Faculty of Mathematics and Natural Sciences, ITM
Nijenborgh 4 9747 AG Groningen, NL
J.M.A.Scherpen@rug.nl

Summary. In the present work we are interested on the derivation of power-based passivity properties for a certain class of non-linear mechanical systems. While for general mechanical systems, it is of common use to adopt a storage function related to the *system's energy* in order to show passivity and stabilize the system on a desired equilibrium point(e.g., IDA-PBC [1]), we want here to obtain similar properties related to the *system's power*. The motivation arises from the idea that in some engineering applications(satellite orbit motion, aircraft dynamic,etc...)seems more sensible to cope with the power flowing into the system instead of the energy that, for stabilization purposes, means to consider the systems's equilibrium the state for which the energy flow-rate(i.e.,system's power)achieve its minimum. In this respect, we recall first the power-based description for a certain class of (non)-linear mechanical systems given in [2] and then we give sufficient conditions to obtain power-based passivity properties, provided a suitable choice of port-variables. We conclude with the example of the inverted pendulum on the cart.

1 Introduction

In a previous work of the authors [2] an electrical interpretation of the motion equations of mechanical systems moving in a plane has been provided via the Brayton-Moser equations. In particular, it is proved that under certain generic assumptions the system's behavior derived from its Lagrangian function can be alternatively described through a power-based representation in an electrical fashion. It can be viewed as an extension of the well-known analogy mass/inductor, spring/capacitor and damper/resistor for linear mechanical systems to a larger class of (possibly) nonlinear systems. *The double pendulum* and *the inverted pendulum on the cart* are the illustrative examples which have been studied and electrically interpreted as nonlinear RLC circuits.

We are here interested on exploiting this power-based description for such mechanical system class in order to achieve a new passivity property using as port variables the external forces/torques and the linear/angular acceleration,and with the storage function being related to the system's power.

F. Bullo et al. (Eds.): Lag. & Hamil. Methods for Nonlin. Ctrl., LNCIS 366, pp. 123–133, 2007.
springerlink.com

In Section 2 we will first recall the fundamentals of Euler-Lagrange(EL) and Brayton-Moser(BM) equations in the standard form. Via the introduction of the pseudo-inductor the Brayton-Moser equations can be extended to a large class of non-linear mechanical systems, [2]. This is reviewed in Section 3, and followed by the presentation of the main result. Taking inspiration from [3] we provide a method to generate storage function candidates based on the power. We give sufficient conditions to show the power-based passivity properties. We conclude the paper in section 4 with the example of the *the inverted pendulum on the cart* for which our passivity conditions have a clear physical meaning.

2 Preliminaries

2.1 Euler-Lagrange Systems (EL)

The standard Euler-Lagrange equations (e.g., [1]) for an r degrees of freedom mechanical system with generalized coordinates $q \in \mathbb{R}^r$ and external forces $\tau \in \mathbb{R}^r$ are given by

$$\frac{d}{dt}\left(\frac{\partial \mathcal{L}(q,\dot{q})}{\partial \dot{q}}\right) - \frac{\partial \mathcal{L}(q,\dot{q})}{\partial q} = \tau \tag{1}$$

where

$$\mathcal{L}(q,\dot{q}) \triangleq \mathcal{T}(q,\dot{q}) - \mathcal{V}(q) \tag{2}$$

is the so-called Lagrangian function, $\mathcal{T}(q,\dot{q})$ is the kinetic energy which is of the form

$$\mathcal{T}(q,\dot{q}) = \frac{1}{2}\dot{q}^T D(q)\dot{q}, \tag{3}$$

where $D(q) \in \mathbb{R}^{r\times r}$ is a symmetric positive definite matrix, and $\mathcal{V}(q)$ is the potential function which is assumed to be bounded from below. Furthermore, dissipative elements can be included via the Rayleigh dissipation function as part of the external forces.

2.2 RLC-Circuits: The Brayton-Moser Equations (BM)

The electrical circuits considered in this paper are complete RLC-circuits in which all the elements can be nonlinear. The standard definitions of respectively inductance and capacitance matrices are given by

$$L(i_\rho) = \frac{\partial \phi_\rho(i_\rho)}{\partial i_\rho}\ ,\ C(v_\sigma) = \frac{\partial q_\sigma(v_\sigma)}{\partial v_\sigma}$$

where $i_\rho \in \mathbb{R}^r$ represents the currents flowing through the inductors and $\phi_\rho(i_\rho) \in \mathbb{R}^r$ is the related magnetic flux vector. On the other hand $v_\sigma \in \mathbb{R}^s$ defines the voltages across the capacitors and the vector $q_\sigma(v_\sigma) \in \mathbb{R}^s$ represents the charges stored in the capacitors. From [4] we know that the differential equations of such electrical circuits have the special form

$$Q(x)\dot{x} = \nabla P(x) \tag{4}$$

where $x = (i_\rho, v_\sigma) \in \mathbb{R}^{r+s}$, $\nabla = (\partial/\partial i_\rho, \partial/\partial v_\sigma)^T$, and

$$Q(x) = \begin{bmatrix} -L(i_\rho) & 0 \\ 0 & C(v_\sigma) \end{bmatrix}. \tag{5}$$

Furthermore the mixed potential function $P(x)$ which contains the interconnection and resistive structure of the circuit is defined as

$$P(x) = -F(i_\rho) + G(v_\sigma) + {i_\rho}^T \Lambda v_\sigma. \tag{6}$$

$F : \mathbb{R}^r \to \mathbb{R}$ and $G : \mathbb{R}^s \to \mathbb{R}$ being the current potential (content) related with the current-controlled resistors (R) and the voltage potential (co-content) related with the voltage-controlled resistors (i.e., conductors, G), respectively. More specifically, the content and co-content are defined by the integrals

$$\int_0^{i_\rho} \hat{v}_R(i'_\rho) di'_\rho, \quad \int_0^{v_\sigma} \hat{i}_G(v'_\sigma) dv'_\sigma,$$

where $\hat{v}_R(i_\rho)$ and $\hat{i}_G(v_\sigma)$ are the characteristic functions of the (current-controlled) resistors and conductors (voltage-controlled resistors), respectively. The $r \times s$ matrix Λ is given by the interconnection of the inductors and capacitors, and the elements of Λ are in $\{-1, 0, 1\}$.

2.3 Definitions

In order to introduce the electrical counter part of the position dependent mass we introduce the so-called *pseudo-inductor*. This is an inductor, but now relating the magnetic flux linkages to current and the voltage, which differs from the "usual" electrical case, i.e.,

$$\phi = f_\phi(x). \tag{7}$$

where $\phi \in \mathbb{R}^r$ is the flux related to the inductors. This definition lead to the following implicit relation between voltage and current

$$v_\rho = \frac{d\phi}{dt} = \frac{\partial f_\phi}{\partial i_\rho}\frac{di_\rho}{dt} + \frac{\partial f_\phi}{\partial v_\sigma}\frac{dv_\sigma}{dt}. \tag{8}$$

Now, define the *pseudo-inductance matrix* and the *co-pseudo-inductance matrix* as

$$\widetilde{L}(x) = \frac{\partial f_\phi}{\partial i_\rho} \quad , \quad \widetilde{M}(x) = \frac{\partial f_\phi}{\partial v_\sigma}$$

respectively, then (8) can be written as

$$v_\rho = \widetilde{L}(x)\frac{di_\rho}{dt} + \widetilde{M}(x)\frac{dv_\sigma}{dt}. \tag{9}$$

Similarly, we will consider a *capacitor* as a function relating the charge and the voltage, i.e.,

$$q_{\sigma j} = f_j^v(v_{\sigma j}) \ , \ j = 1, \dots, s. \tag{10}$$

By defining the non-negative capacitance matrix

$$C(v_\sigma) = diag\left[\frac{\partial f_j^v(v_{\sigma j})}{\partial v_{\sigma j}}\right] \ , \ j = 1, \dots, s,$$

we have from differentiation of (10) that

$$i_\sigma = C(v_\sigma)\frac{dv_\sigma}{dt}. \tag{11}$$

3 Power-Based Description for a Class of Mechanical Systems

In [2] the authors enlarged the class of mechanical systems for which an electrical interpretation can be provided replacing the generalized coordinates vector $(\dot{q}, q) \in \mathbb{R}^{2r}$ by the electrical states vector $(i_\rho, v_\sigma) \in \mathbb{R}^{r+s}$. In order to make the following relation a one-to-one mapping the equivalent circuit has to present a number of inductors r equal to the capacitors s. Moreover, all conservative forces acting on the masses should be (locally) invertible functions of its angular or linear position. The main result of [2] is as follows.

Theorem 1. *Consider the general Lagrangian function* (2). *Assume that:*

A1. (interconnection) $i_\rho = i_\sigma$, [1]

A2. (force-position link) $q_{\sigma j} = f_j^v(v_{\sigma j}) \in \mathcal{C}^1$ *with* $j = 1, \dots, r$ *is a set of invertible functions such that:*

- $\frac{\partial f_j^v(v_{\sigma j})}{\partial v_{\sigma j}} = C_j(v_{\sigma j})$,
- $f_j^q(q_{\sigma j}) = v_{\sigma j}$.

Then: the Euler-Lagrange (1) *equations can be rewritten in terms of the Brayton-Moser framework as follows*

$$\begin{bmatrix} -\widetilde{D}(v_\sigma) & -[\bar{D}(x) - \hat{D}(x)]C(v_\sigma) \\ 0 & C(v_\sigma) \end{bmatrix} \begin{bmatrix} \frac{di_\rho}{dt} \\ \frac{dv_\sigma}{dt} \end{bmatrix} = \begin{bmatrix} \frac{\partial P(x)}{\partial i_\rho} \\ \frac{\partial P(x)}{\partial v_\sigma} \end{bmatrix}$$

[1] Implying that $s = r$ and $\Lambda = I$. See Remark 4 of [2] for the physical implications.

with

$$P(x) = -F(i_\rho) + G(v_\sigma) + {i_\rho}^T v_\sigma,$$

being the mixed potential function $P(i_\rho, v_\sigma)$ *and where*

$$\hat{D}(i_\rho, v_\sigma) = \begin{bmatrix} \frac{1}{2}{i_\rho}^T \frac{\partial D(q_\rho)}{\partial q_{\rho_1}}|_{q_\rho = f^v(v_\sigma)} \\ \cdots \\ \frac{1}{2}{i_\rho}^T \frac{\partial D(q_\rho)}{\partial q_{\rho_r}}|_{q_\rho = f^v(v_\sigma)} \end{bmatrix} \tag{12}$$

$$C(v_\sigma) = diag\left[\frac{\partial f_j^v(v_{\sigma j})}{\partial v_{\sigma j}}, j = 1, \ldots, r\right] \tag{13}$$

$$\widetilde{D}(v_\sigma) = D(q_\rho)|_{q_\rho = f^v(v_\sigma)} \tag{14}$$

$$\bar{D}(i_\rho, v_\sigma) = \begin{bmatrix} a_{11}(i_\rho, v_\sigma) & \cdots & a_{1r}(i_\rho, v_\sigma) \\ \vdots & \ddots & \vdots \\ a_{r1}(i_\rho, v_\sigma) & \cdots & a_{rr}(i_\rho, v_\sigma) \end{bmatrix} \tag{15}$$

with $a_{ij}(i_\rho, v_\sigma) = {i_\rho}^T C^{-1}(v_\sigma) \nabla_{v_\sigma} \widetilde{D}_{ij}(v_\sigma)$ *for* $i, j \in \{1, r\}$.

Corollary 1. *As a consequence of Theorem 1, recalling the definitions of the pseudo-inductor and the capacitor adopted in* (9) *and* (11) *respectively, the BM equations can be then re-written in the following more compact form*

$$\widetilde{Q}(x)\dot{x} = \nabla P(x) \tag{16}$$

with

$$\widetilde{Q}(x) = \begin{bmatrix} -\widetilde{L}(v_\sigma) & -\widetilde{M}(i_\rho, v_\sigma) \\ 0 & C(v_\sigma) \end{bmatrix}$$

and where $\widetilde{L}(v_\sigma) = \widetilde{D}(v_\sigma)$, $\widetilde{M}(i_\rho, v_\sigma) = \left[\dot{\bar{D}}(v_\sigma) - \hat{D}(i_\rho, v_\sigma)\right] C(v_\sigma)$.

Remark 1. The former result can been interpreted in two ways. From one side we established under which conditions–A1 and A2–a mechanical systems described by EL equations, through derivation of an *energy-based* function called Lagrangian, has a clear electrical counterpart based on the classical states analogy force/voltage and speed/current. On the other side, we state that this class of mechanical systems that can be electrically interpretable yields a *power-based* description in the BM framework. Under this second perspective we will present, in the further section, our main result.

3.1 Power-Based Passivity Properties

This section is dedicated to the derivation of passivity *sufficient* conditions for that class of mechanical systems that admits the power-based description given in (16). For that we have to find a *storage function candidate* and a corresponding set of *port variables*. It is then instrumental for the derivation of the next theorem to re-define the mixed-potential function $P(x)$ extracting the voltage sources $v_s \in \mathbb{R}^l$ with $l \leq r$, from the content term $F(i_\rho)$ as follows

$$P(x) = \widetilde{P}(x) - x^T B v_s \tag{17}$$

with $B = (B_s, 0)^T$ and $B_s \in \mathbb{R}^{r \times l}$.

Remark 2. In equation (4) we restricted our analysis to circuits having only voltage sources in series with inductors. This choice seems to be sensible considering that the mechanical counterpart of a current source is a velocity source which have no clear sense from a physical view point.

Storage Function Candidate

Following the procedure of [3], we can pre-multiply (16) by $\dot{x}^T$ obtaining

$$\dot{x}^T \widetilde{Q}(x)\dot{x} = \dot{x}^T \nabla_x \widetilde{P}(x) - \dot{x}^T B v_s$$

that can be re-arranged as follows

$$\frac{d\widetilde{P}}{dt}(x) = \dot{x}^T B v_s + \dot{x}^T \widetilde{Q}(x)\dot{x} \tag{18}$$

and which consists of the sum of two terms. The first one represents the inner product of the source variables in the suited form $\dot{x}^T B v_s = v_s^T \dot{\widehat{i}}_s$, where we assume the vector $i_s \in \mathbb{R}^l$ indicating the correspondent current terms flowing from each inductor series-connected voltage source.

The second one is a quadratic term. In general $\widetilde{Q}(x)$ is not symmetric and its symmetric part is sign indefinite making difficult the derivation of the power-balance inequality we are looking for. In order to overcome this drawback we follow the same procedure exploited in [3],[4],[5] that basically provides a method to describe the system (16) by another admissible pair, say $\widetilde{Q}_a(x)$ and $P_a(x)$. For instance, if the new pair fulfills the following conditions:

C1. $\widetilde{Q}_a^T(x) + \widetilde{Q}_a(x) \leq 0$

C2. $\widetilde{P}_a(x) : \mathbb{R}^{s+r} \to \mathbb{R}$ is positive semi-definite scalar function

we may state that

$$\frac{d\widetilde{P}_a}{dt}(x) \leq \dot{x}^T B v_s \tag{19}$$

being $\widetilde{P}_a(x)$ the storage function candidate related to $P_a(x)$ by (17), the pair $(v_s, \dot{\widehat{i}}_s)$ is passive and can serve as port-variables.

Power-Balance Inequality and Passivity Requirements

In the next theorem we will provide some conditions for passivity that may be useful for control in the power-based framework. In particular, we refer to a previous work of the second author [3] where the storage function has the dimension of power and is defined as a re-shaped mixed potential function $\widetilde{P}_a(x)$. This new function is then related to a new matrix $\widetilde{Q}_a(x)$ and both, having common solutions for (16), are related to the original pair $\widetilde{Q}(x), \widetilde{P}(x)$ by the following relations[2]

$$\begin{aligned} \widetilde{Q}_a(x) &= \left\{\lambda I + \tfrac{1}{2}\nabla^2 \widetilde{P}(x)\Pi(x) + \tfrac{1}{2}\nabla[\nabla^T \widetilde{P}(x)\Pi(x)]\right\}\widetilde{Q}(x) \\ \widetilde{P}_a(x) &= \lambda \widetilde{P}(x) + \tfrac{1}{2}\nabla^T \widetilde{P}(x)\Pi(x)\nabla \widetilde{P}(x) \end{aligned}$$

with $\Pi(x) \in \mathbb{R}^{r\times r}$ a symmetric matrix and $\lambda \in \mathbb{R}$ any constant.

Theorem 2. *Consider an electrical system for which the dynamics is described by* (16) *and assume A1 and A2 hold. Moreover, Assume that*

A3. (positivity) *pseudo-inductors and capacitors matrices are positive definite*
A4. (linearity in the content) $F(i_\rho) = -(1/2){i_\rho}^T R_\rho i_\rho$ *with the current-controlled resistor matrix* R_ρ *being constant and positive definite*
A5. (damping condition)

$$\left\| -2R_\rho^{-1}\widetilde{M}(x)C^{-1}(v_\sigma) + \widetilde{M}^T(x)\widetilde{L}^{-1}(v_\sigma)\widetilde{M}(x)C^{-1}(v_\sigma) + \beta(x)\right\| \leq 1$$

with

$$\beta(x) = \frac{\partial}{\partial v_\sigma}\left[{i_\rho}^T \widetilde{L}(v_\sigma)R_\rho^{-1}C^{-1}(v_\sigma)\right].$$

A6. (technical assumption)

$$\widetilde{L}(v_\sigma)R_\rho^{-1}C^{-1}(v_\sigma) \geq 0$$

then

$$\int_0^t v_s^T(t')\frac{di_s}{dt'}dt' \geq \widetilde{P}_a(x(t)) - \widetilde{P}_a(x(0)). \tag{20}$$

Proof. First, we set the matrix $\Pi(x)$ and the scalar λ in order to guarantee the semi-definite positivity of the storage function $\widetilde{P}_a(x)$ and to satisfy the following requirement[3]

$$\widetilde{Q}_a(x)^T + \widetilde{Q}_a(x) \leq 0. \tag{21}$$

[2] See [5] for a detailed proof of this statement.

[3] If these two conditions are matched the overall system, for which the dynamics can be written as $\widetilde{Q}^{-1}(x)\nabla P(x) = -\widetilde{Q}_a^{-1}(x)\nabla P_a(x) = (di_\rho/dt, dv_\sigma/dt)^T$, is then asymptotically stable.

Define

$$\begin{aligned}\lambda &= -1,\\ \Pi(x) &= diag[2R_\rho^{-1}, 2\widetilde{L}(v_\sigma)R_\rho^{-1}C^{-1}(v_\sigma)].\end{aligned}$$

Considering a mixed potential function $P(x)$ fitting the Assumption A4 and reminding that Assumption A1 $\Rightarrow \Lambda = I$, we obtain

$$\widetilde{Q}_a(x) = \begin{bmatrix} -\widetilde{L}(v_\sigma) & -\widetilde{M}(x) + 2L(v_\sigma)R_\rho^{-1} \\ -2R_\rho^{-1}\widetilde{L}(v_\sigma) & -[I - \beta(x)]C(v_\sigma) - 2R_\rho^{-1}\widetilde{M}(x) \end{bmatrix}$$

that, under Assumptions A3 and A5, satisfies (21). We refer to the appendix for a detailed development of the former statement. Furthermore, the storage function candidate becomes

$$\begin{aligned}\widetilde{P}_a(x) &= \frac{1}{2}(R_\rho i_\rho + v_\sigma)^T R_\rho^{-1}(R_\rho i_\rho + v_\sigma) + \\ &+ \frac{1}{2}{v_\sigma}^T R_\rho^{-1} v_\sigma + {i_\rho}^T \widetilde{L}(v_\sigma)R_\rho^{-1}C^{-1}(v_\sigma)i_\rho \end{aligned} \tag{22}$$

that, under Assumption A6, is clearly positive definite.

Remark 3. Assumption A5 is an important condition that can be satisfied for *small* values of the matrix R_ρ^{-1}—which represent the LTI resistors placed in series to each inductor— and/or with a *weak* mutual-coupling action provided by the presence of the matrix $\widetilde{M}(x)$. Since $\widetilde{M}(x)$ depends linearly on the current vector i_ρ—see $\widetilde{M}(x)$ definition provided in Theorem 1—, we can state that for *slow motion* or *well-damped* dynamics, A5 holds.

4 The Inverted Pendulum on a Cart

An interesting example of mechanical system to study is the inverted pendulum with rigid massless rod (of length l) placed on a cart as shown if Fig. 1. It is often used to test the performance of controllers that stabilize the pendulum mass m_2 to its natural unstable equilibrium point through a force F acting just on the cart of mass m_1. The equations describing the dynamics of the to masses could be computed considering as state variables the angular position of the row with the vertical axis θ and the cart distance $z - z_0$ to a fixed reference ($z_0 = 0$). The motion dynamic of each mass can be determined via the Euler-Lagrange equations

$$\begin{aligned}(m_1 + m_2)\ddot{z} + m_2 l\cos\theta\ddot{\theta} - m_2 l\sin\theta\dot{\theta}^2 &= F - R_1\dot{z}\\ m_2 l^2\ddot{\theta} + m_2 l\cos\theta\ddot{z} - m_2 gl\sin\theta &= -R_2\dot{\theta}\end{aligned} \tag{23}$$

where the generalized coordinates related to the position of each mass and its derivative representing the corresponding velocities are respectively

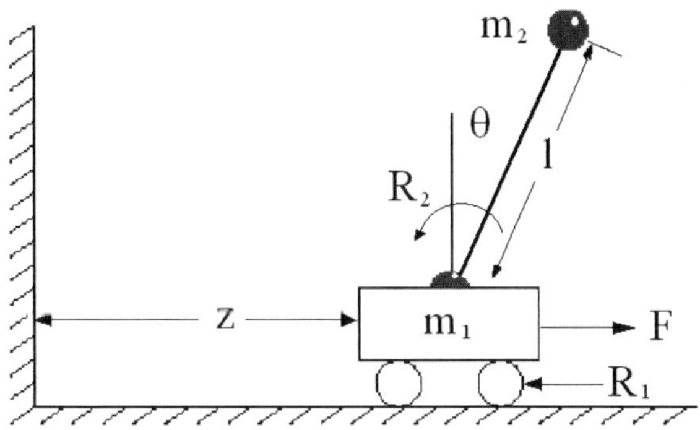

Fig. 1. Inverted pendulum on a cart

$$q = (z, \theta)^T \ , \quad \dot{q} = (\dot{z}, \dot{\theta})^T .$$

Applying the following coordinates transformation[4]

$$\begin{bmatrix} z \\ \theta \end{bmatrix} = \begin{bmatrix} C_1 v_{\sigma 1} \\ f_2^v(v_{\sigma 2}) = \arcsin\left(\frac{v_{\sigma 2}}{K}\right) \end{bmatrix} , \quad \begin{bmatrix} \dot{z} \\ \dot{\theta} \end{bmatrix} = \begin{bmatrix} i_{\rho 1} \\ i_{\rho 2} \end{bmatrix} \tag{24}$$

with $K = -m_2 g l$ and considering that Assumptions A1 and A2 are clearly satisfied we can express the motion equations (23) via the Brayton-Moser framework

$$\widetilde{Q}(x)\dot{x} = \nabla P(x) \tag{25}$$

with

$$\widetilde{Q}(x) = \begin{bmatrix} -(m_1+m_2) & -m_2 l \cos f_2^v(v_{\sigma 2}) & 0 & m_2 l \sin f_2^v(v_{\sigma 2}) i_{\rho_2} C_2(v_{\sigma 2}) \\ -m_2 l \cos f_2^v(v_{\sigma 2}) & -m_2 l^2 & 0 & 0 \\ 0 & 0 & C_1 & 0 \\ 0 & 0 & 0 & C_2(v_{\sigma 2}) \end{bmatrix}$$

and

$$P(x) = -v_s^T i_\rho + \frac{1}{2} {i_\rho}^T R_\rho i_\rho + {i_\rho}^T v_\sigma ,$$

being

$$v_s = (F, 0)^T \ , \quad R_\rho = diag(R_1, R_2) \ , \quad C_2(v_\sigma) = \frac{\partial f_2^v(v_{\sigma 2})}{\partial v_{\sigma 2}}$$

and $C_1 \in \mathbb{R}_+$ an arbitrary constant. Now that we have expressed the mechanical system model by (25) we can use the Theorem 2 in order to get the explicit passivity condition. By choosing

[4] The relation$(\dot{q}, q) \Leftrightarrow (i_\rho, v_\sigma)$ is one-to-one only when θ belongs to the open interval $(-\frac{\pi}{2}, \frac{\pi}{2})$.

$$\lambda = -1 \quad , \quad \Pi(x) = diag[2R_\rho^{-1}, 2\widetilde{L}(v_\sigma)R_\rho^{-1}C^{-1}(v_\sigma)],$$

after some algebraic computations we get the final local condition

$$\left\| \begin{bmatrix} 0 & 2m_2 l \sin f_2^v(v_{\sigma 2}) i_{\rho 2} R_1^{-1} \\ 0 & \frac{[m_2 l \sin f_2^v(v_{\sigma 2})]^2}{m_1+m_2-m_2\cos^2 f_2^v(v_{\sigma 2})} + \frac{i_{\rho 1}}{i_{\rho 2}} m_2 l \sin f_2^v(v_{\sigma 2}) C_2(v_{\sigma 2}) R_2^{-1} \end{bmatrix} \right\| \leq 1 \quad (26)$$

achieved for $C_1 \to \infty$. The former suitable choice of C_1 parameter is arbitrary because it depends on the coordinates transformation we arbitrary fixed. Of course, in order to apply Theorem 2 we have to verify, together with condition (26), that Assumption 6 holds, that means

$$\left\| \begin{bmatrix} 0 & m_2 l \cos f_2^v(v_{\sigma 2}) R_2^{-1} C_2^{-1} \\ 0 & m_2 l^2 R_2^{-1} C_2^{-1}(v_{\sigma 2}) \end{bmatrix} \right\| \geq 0. \tag{27}$$

From the overlap of (26) and (27), we deduce that

$$\int_0^t F\hat{\dot{i}}_{\rho 1}(\tau)d\tau \geq \widetilde{P}_a(t) - \widetilde{P}_a(0)$$

with $\widetilde{P}_a(x)$ given by (22), holds.

5 Conclusion and Outlooks

Our main purpose in this document was to present an alternative way to describe the dynamics of a large class of (possibly non-)linear mechanical systems within a framework–the Bryton-Moser equations–that relates the power to the trajectories of the system instead of energy, and derive from it sufficient conditions for passivity. This should be consider as a preliminary step towards stabilization of mechanical and electromechanical systems using passivity arguments–as already suggested in [5])–.

Acknowledgements

The work was partially supported through a European Community Marie Curie Fellowship, in the framework of the CTS (contract number: HPMT-CT-2001-00278) and partially financed by the european exchange program Van Gogh.

References

1. R. Ortega, A. Loria, P. J. Nicklasson, and H. Sira-Ramirez, *Passivity-based Control of Euler-Lagrange Systems.* Springer, London, 1998.

2. A. de Rinaldis and J. M. A. Scherpen, "An electrical interpretation of mechanical systems via the pseudo-inductor in the brayton-moser equations," *IEEE, proceedings of CDC-ECC, Seville, Spain*, pp. 5983–5988, december 2005.
3. D. Jeltsema, R. Ortega, and J. M. A. Scherpen, "On passivity and power-balance inequalities of nonlinear *RLC* circuits," *IEEE Trans. Circ. Syst.*, vol. 50, no. 9, pp. 1174–1179, September 2003.
4. R. K. Brayton and J. K. Moser, "A theory of nonlinear networks (part 1)," *Quarterly Of Applied Mathematics*, vol. XXII, no. 1, pp. 1–33, April 1964.
5. R. Ortega, D. Jeltsema, and J. M. A. Scherpen, "Power shaping: A new paradigm for stabilization of nonlinear rlc circuits," *IEEE Trans. Aut. Cont.*, vol. 48, no. 10, pp. 1762–1767, October 2003.
6. R. Horn and C. Johnson, *Matrix Analysis*. Cambridge University Press, UK, 1985.

Appendix

Here, we show that given Assumptions A3 and A5 of Theorem 2, the positivity of $Q_a(x)$ is established, i.e., (21) holds. Indeed, computing the symmetric part of $Q_a(x)$ we obtain

$$H_a(x) = \begin{bmatrix} \widetilde{L}(v_\sigma) & \widetilde{M}(x) \\ \widetilde{M}^T(x) & [I - \beta(x)]C(v_\sigma) + 2R_\rho^{-1}\widetilde{M}(x) \end{bmatrix}.$$

Then, provided the positivity of $\widetilde{L}(v_\sigma)$ by A3, we compute the Schur's complement of $H_a(x)$ and imposing its positivity we obtain

$$[I - \beta(x)]C(v_\sigma) + 2R_\rho^{-1}\widetilde{M}(x) \geq \widetilde{M}^T(x)\widetilde{L}^{-1}(v_\sigma)\widetilde{M}(x)$$

Let's re-write the above inequality as follows

$$I \geq -2R_\rho^{-1}\widetilde{M}(x)C^{-1}(v_\sigma) + \widetilde{M}^T(x)\widetilde{L}^{-1}(v_\sigma)\widetilde{M}(x)C^{-1}(v_\sigma) + \beta(x),$$

as a consequence of Perron's theorem[5] and reminding that the spectral norm applying on any squared matrix $A \in \mathbb{R}^{r\times r}$ is defined as

$$||A|| = \sqrt{\rho(A^T A)}$$

we have

$$1 \geq \left\| -2R_\rho^{-1}\widetilde{M}(x)C^{-1}(v_\sigma) + \widetilde{M}^T(x)\widetilde{L}^{-1}(v_\sigma)\widetilde{M}(x)C^{-1}(v_\sigma) + \beta(x) \right\|$$

which is true by Assumption A5.

[5] See lemma 8.4.2 of [6].

Power Shaping Control of Nonlinear Systems: A Benchmark Example

Eloísa García–Canseco[1], Romeo Ortega[1], Jacquelien M.A. Scherpen[2], and Dimitri Jeltsema[3]

[1] Laboratoire des Signaux et Systèmes, SUPELEC, Plateau de Moulon, 91192, Gif–sur–Yvette, France
garcia(ortega)@lss.supelec.fr
[2] Faculty of Mathematics and Natural Sciences, ITM, University of Groningen, Nijenborgh 4, 9747 AG Groningen, The Netherlands
j.m.a.scherpen@rug.nl
[3] Delft Center for Systems and Control, Delft University of Technology, Mekelweg 2, 2628 CD Delft, The Netherlands
d.jeltsema@tudelft.nl

Summary. It is well known that energy balancing control is stymied by the presence of pervasive dissipation. To overcome this problem in electrical circuits, the authors recently proposed the alternative paradigm of *power shaping*—where, as suggested by its name, stabilization is achieved shaping a function akin to power instead of the energy function. In this paper we extend this technique to general nonlinear systems and apply it for the stabilization of the benchmark tunnel diode circuit. It is shown that, in contrast with other techniques recently reported in the literature, e.g. piece–wise approximation of nonlinearities, power shaping yields a simple linear static state feedback that ensures (robust) global asymptotic stability of the desired equilibrium.

Keywords: Passivity–based control, nonlinear control, stability, nonlinear systems.

1 Introduction and Background

Passive systems constitute a very important class of dynamical systems for which the stored energy cannot exceed the energy supplied to them by the external environment—the difference being the dissipated energy. In view of this energy–balancing feature, it is clear that passivity is intimately related with the property of stability, a *sine qua non* condition for any controller design. Furthermore, invoking the universal principle of energy conservation, it may be argued that all physical systems are passive with respect to some suitably defined port variables that couple the system with the environment. It is not surprising then that, since the introduction of the first passivity–based controller (PBC) more than two decades ago [1, 2], we have witnessed an ever increasing popularity of passivity as a building block for controller design for all classes of physical systems.

F. Bullo et al. (Eds.): Lag. & Hamil. Methods for Nonlin. Ctrl., LNCIS 366, pp. 135–146, 2007.
springerlink.com

PBC can be used to stabilize a given equilibrium point. In this case we must modify the energy function—that will qualify as a Lyapunov function—to assign a minimum at this point, a step called energy shaping; which, combined with damping injection, constitute the two main stages of PBC [3, 4]. There are several ways to achieve energy shaping, the most physically appealing being the so–called *energy–balancing* PBC (or control by interconnection) method [4, 5]. With this procedure the energy function assigned to the closed–loop system is the difference between the total energy of the system and the energy supplied by the controller, hence the name energy balancing. Unfortunately, energy balancing PBC is stymied by the presence of pervasive dissipation, that is, the existence of dissipative elements, e.g. resistors, whose power does not vanish at the desired equilibrium point.

To put our contribution in perspective let us briefly recall the principles of energy–balancing control [6]. Consider a system whose state space representation is given by[1]

$$\begin{aligned}\dot{\boldsymbol{x}} &= \boldsymbol{f}(\boldsymbol{x}) + \boldsymbol{g}(\boldsymbol{x})\boldsymbol{u},\\ \boldsymbol{y} &= \boldsymbol{h}(\boldsymbol{x})\end{aligned} \tag{1}$$

where $\boldsymbol{x} \in \mathbb{R}^n$, and $\boldsymbol{u}, \boldsymbol{y} \in \mathbb{R}^m$ are the input and output vectors, respectively. We assume that the system (1) satisfies the energy–balance inequality, that is, along all trajectories compatible with $u : [0, t] \to \mathbb{R}^m$,

$$H(\boldsymbol{x}(t)) - H(\boldsymbol{x}(0)) \leq \int_0^t \boldsymbol{u}^\top(\tau)\boldsymbol{y}(\tau)d\tau \tag{2}$$

where $H : \mathbb{R}^n \to \mathbb{R}$ is the stored *energy* function.[2] In energy–balancing control, we look for a control such that the energy supplied by the controller can be expressed as a function of the state. Indeed, from (2) we see that for any function $\widehat{\boldsymbol{u}} : \mathbb{R}^n \to \mathbb{R}^m$ such that

$$-\int_0^t \widehat{\boldsymbol{u}}^\top(\boldsymbol{x}(\tau))\boldsymbol{h}(\boldsymbol{x}(\tau))d\tau = H_a(\boldsymbol{x}(t)) - H_a(\boldsymbol{x}(0)) \tag{3}$$

for some function $H_a : \mathbb{R}^n \to \mathbb{R}$, the control $\boldsymbol{u} = \widehat{\boldsymbol{u}}(\boldsymbol{x}) + \boldsymbol{v}$ will ensure that the closed–loop system satisfies

$$H_d(\boldsymbol{x}(t)) - H_d(\boldsymbol{x}(0)) \leq \int_0^t \boldsymbol{v}^\top(\tau)\boldsymbol{y}(\tau)d\tau$$

where $H_d(\boldsymbol{x}) = H(\boldsymbol{x}) + H_a(\boldsymbol{x})$ is the new total energy function. If, furthermore, $\boldsymbol{x}^\star = \arg\min H_d(\boldsymbol{x})$ then $\boldsymbol{x}^\star$ will be a stable equilibrium of the closed–loop system (with Lyapunov function the difference between the stored and the supplied energies $H_d(\boldsymbol{x})$).

[1] All vectors defined in the paper are *column* vectors, even the gradient of a scalar function that we denote with the operator $\nabla = \partial/\partial\boldsymbol{x}$. Also, we use $(\cdot)'$ to denote differentiation for functions of scalar arguments. To simplify the expressions, the arguments of all functions will be omitted, and will be explicitly written only the first time that the function is defined.

[2] Notice that no assumption of non–negativity on $H(\boldsymbol{x})$ is imposed. If $H(\boldsymbol{x}) \geq 0$, the system (1) is said to be *passive* with conjugated port variables $(\boldsymbol{u}, \boldsymbol{y})$.

Unfortunately, as shown in [6], energy–balancing stabilization is stymied by the existence of pervasive dissipation—term which refers to the existence of dissipative elements whose power does not vanish at the desired equilibrium point. More precisely, since solving (3) is equivalent to the solution of the PDE

$$(\dot{H}_a =) \quad \nabla H_a^\top [\boldsymbol{f} + \boldsymbol{g}\widehat{\boldsymbol{u}}^\top] = -\widehat{\boldsymbol{u}}^\top \boldsymbol{h}, \tag{4}$$

and the left hand side is equal to zero at $\boldsymbol{x}^\star$, it is clear that the method is applicable only to systems verifying $\widehat{\boldsymbol{u}}^\top(\boldsymbol{x}^\star)\boldsymbol{h}(\boldsymbol{x}^\star) = 0$. To overcome this obstacle in nonlinear RLC circuits the paradigm of *power shaping* was introduced in [7]—where, as suggested by its name, stabilization is achieved shaping the power instead of the energy function. The starting point for the method is a description of the circuit using Brayton–Moser equations [8]

$$\mathbf{Q}(\boldsymbol{x})\dot{\boldsymbol{x}} = \nabla P + \mathbf{G}(\boldsymbol{x})\boldsymbol{u}, \tag{5}$$

where $\mathbf{Q} : \mathbb{R}^n \to \mathbb{R}^{n\times n}$ is a *full rank* matrix containing the generalized inductance and capacitance matrices and $P : \mathbb{R}^n \to \mathbb{R}$ is the circuits mixed potential which has units of *power*, see [7, 9] for further details. We make the observation that if $\mathbf{Q} + \mathbf{Q}^\top \leq 0$ then the system satisfies the *power balance inequality*

$$P(\boldsymbol{x}(t)) - P(\boldsymbol{x}(0)) \leq \int_0^t \boldsymbol{u}^\top(\tau)\widetilde{\boldsymbol{y}}(\tau)d\tau$$

with $\widetilde{\boldsymbol{y}} = \widetilde{\boldsymbol{h}}(\boldsymbol{x}, \boldsymbol{u})$ and

$$\widetilde{\boldsymbol{h}}(\boldsymbol{x}, \boldsymbol{u}) := -\mathbf{G}^\top(\boldsymbol{x})\mathbf{Q}^{-1}(\boldsymbol{x})[\nabla P + \mathbf{G}(\boldsymbol{x})\boldsymbol{u}]. \tag{6}$$

This property follows immediately pre-multiplying (5) by $\dot{\boldsymbol{x}}^\top$ and then integrating. The mixed potential function is shaped with the control $\boldsymbol{u} = \widehat{\boldsymbol{u}}$ where

$$\mathbf{G}\widehat{\boldsymbol{u}} = \nabla P_a \tag{7}$$

for some $P_a : \mathbb{R}^n \to \mathbb{R}$. This yields the closed–loop system $\mathbf{Q}\dot{\boldsymbol{x}} = \nabla P_d$, with total power function

$$P_d(\boldsymbol{x}) := P(\boldsymbol{x}) + P_a(\boldsymbol{x}),$$

and the equilibrium will be stable if $\boldsymbol{x}^\star = \arg\min P_d(\boldsymbol{x})$.

Two key observations are, first, that the resulting controller is *power–balancing*, in the sense that the power function assigned to the closed–loop system is the difference between the total power of the system and the power supplied by the controller. Indeed, from (6) and (7) we have that

$$\dot{P}_a = -\widehat{\boldsymbol{u}}^\top(\boldsymbol{x})\widetilde{\boldsymbol{h}}(\boldsymbol{x}, \widehat{\boldsymbol{u}}(\boldsymbol{x})) \tag{8}$$

which, upon integration, establishes the claimed property. Second, in contrast with energy–balancing control, power–balancing is applicable to systems with pervasive dissipation. Indeed, in contrast with (4), the right hand side of (8) is *always* zero at the equilibrium, therefore, this equation may be solvable even if $\widehat{\boldsymbol{u}}^\top(\boldsymbol{x}^\star)\boldsymbol{h}(\boldsymbol{x}^\star) \neq 0$.

As indicated above, instrumental for the application of power shaping is the description of the system in the form (5). To make the procedure applicable to nonlinear systems described by (1) we apply in the paper Poincare's Lemma to derive necessary and sufficient conditions to achieve this transformation. We prove in this way that the power–shaping problem boils down to the solution of two linear homogeneous PDEs.

We illustrate the methodology using the textbook example of the tunnel diode [10]. In contrast with the existing techniques for this problem, e.g., approximating the nonlinearities using piecewise–affine functions and convex optimization techniques [11], or designs based on linear approximations [10], we show that power shaping yields a simple (partial static state feedback) linear controller that (robustly) globally asymptotically stabilizes the circuit.

2 Power–Shaping Control

The main contribution of this paper is contained in the following.

Proposition 1. *Consider the general nonlinear system (1). Assume*

A.1 There exist a matrix $\mathbf{Q} : \mathbb{R}^n \to \mathbb{R}^{n\times n}$, $|\mathbf{Q}| \neq 0$, *that*

i) solves the partial differential equation

$$\nabla(\mathbf{Q}\boldsymbol{f}) = [\nabla(\mathbf{Q}\boldsymbol{f})]^\top, \tag{9}$$

ii) and verifies $\mathbf{Q} + \mathbf{Q}^\top \leq 0$.

A.2 There exist a scalar function $P_a : \mathbb{R}^n \to \mathbb{R}$ *verifying*

iii)

$$\boldsymbol{g}^\perp \mathbf{Q}^{-1} \nabla P_a = 0,$$

where $\boldsymbol{g}^\perp(\boldsymbol{x})$ *is a full–rank left annihilator of* $\boldsymbol{g}$,[3] *and*

iv) $\boldsymbol{x}^\star = \arg\min P_d(\boldsymbol{x})$, *where*

$$P_d(\boldsymbol{x}) := \int [\mathbf{Q}(\boldsymbol{x})\boldsymbol{f}(\boldsymbol{x})]^\top d\boldsymbol{x} + P_a(\boldsymbol{x}) \tag{10}$$

Under these conditions, the control law

$$\boldsymbol{u} = \left(\boldsymbol{g}^\top \mathbf{Q}^\top \mathbf{Q} \boldsymbol{g}\right)^{-1} \boldsymbol{g}^\top \mathbf{Q}^\top \nabla P_a \tag{11}$$

ensures $\boldsymbol{x}^\star$ *is a (locally) stable equilibrium with Lyapunov function* P_d. *Assume, in addition,*

A.3 $\boldsymbol{x}^\star$ *is an isolated minimum of* P_d *and the largest invariant set contained in the set*

$$\{\boldsymbol{x} \in \mathbb{R}^n | \nabla^\top P_d (\mathbf{Q}^{-1} + \mathbf{Q}^{-\top}) \nabla P_d = 0\}$$

equals $\{\boldsymbol{x}^\star\}$.

[3] That is, $\boldsymbol{g}^\perp \boldsymbol{g} = 0$, and $\text{rank}(\boldsymbol{g}^\perp) = n - m$.

Then, the equilibrium is asymptotically stable and an estimate of its domain of attraction is given by the largest bounded level set $\{\boldsymbol{x} \in \mathbb{R}^n \mid P_d(\boldsymbol{x}) \leq c\}$.

Proof. The first part of the proof consist of showing that, under Assumption A.1, system (1) can be written in the form (5). To this end, invoking Poincare's lemma we have that (9) is *equivalent* to the existence of $P : \mathbb{R}^n \to \mathbb{R}$ such that

$$\mathbf{Q}\boldsymbol{f} = \nabla P. \tag{12}$$

Substituting (1) in the above equation and taking into account the full–rank property of $\mathbf{Q}$ in A.1, we get (5) with $\mathbf{G} := \mathbf{Q}\boldsymbol{g}$.

To prove the stability claim, we proceed as follows. Define $\mathbf{G}^{\perp}(\boldsymbol{x}) := \boldsymbol{g}^{\perp}(\boldsymbol{x})$ $\mathbf{Q}^{-1}(\boldsymbol{x})$, that is a full–rank left annihilator of $\mathbf{G}$, and the full–rank matrix

$$\begin{bmatrix} \mathbf{G}^{\perp} \\ \mathbf{G}^{\top} \end{bmatrix}. \tag{13}$$

Left–multiplying equation (5) by (13) yields

$$\begin{bmatrix} \mathbf{G}^{\perp} \\ \mathbf{G}^{\top} \end{bmatrix} \mathbf{Q}\dot{\boldsymbol{x}} = \begin{bmatrix} \mathbf{G}^{\perp}\nabla P \\ \mathbf{G}^{\top}\nabla P + \mathbf{G}^{\top}\mathbf{G}\boldsymbol{u} \end{bmatrix}. \tag{14}$$

Noticing from (10) and (12) that $P = P_d - P_a$, equation (14) becomes

$$\begin{bmatrix} \mathbf{G}^{\perp} \\ \mathbf{G}^{\top} \end{bmatrix} \mathbf{Q}\dot{\boldsymbol{x}} = \begin{bmatrix} \mathbf{G}^{\perp}(\nabla P_d - \nabla P_a) \\ \mathbf{G}^{\top}(\nabla P_d - \nabla P_a) + \mathbf{G}^{\top}\mathbf{G}\boldsymbol{u} \end{bmatrix}$$

Now, substituting the control action (11) and iii) of A.2., we finally get the closed–loop dynamics

$$\mathbf{Q}\dot{\boldsymbol{x}} = \nabla P_d.$$

Taking the time derivative of P_d along the closed–loop dynamics we have

$$\dot{P}_d = \frac{1}{2}\dot{\boldsymbol{x}}^{\top}(\mathbf{Q} + \mathbf{Q}^{\top})\dot{\boldsymbol{x}}.$$

Because of ii) of Assumption A.1, $\dot{P}_d \leq 0$ and P_d qualifies as a Lyapunov function. From the closed–loop equation we have $\dot{\boldsymbol{x}} = \mathbf{Q}^{-1}(\boldsymbol{x})\nabla P_d$, hence $\dot{P}_d$ can be rewritten as

$$\dot{P}_d = \frac{1}{2}\nabla^{\top} P_d \mathbf{Q}^{-\top}(\mathbf{Q} + \mathbf{Q}^{\top})\mathbf{Q}^{-1}\nabla P_d$$

Asymptotic stability follows immediately, with Assumption A.3, invoking La Salle's invariance principle. This completes the proof.

Assumption A.1 of Proposition 1 involves the solution of the PDE (9) subject to the sign constraint ii)—which may be difficult to satisfy. In [7] we have proposed a procedure to, starting from a pair $\{\mathbf{Q}, P\}$ describing the dynamics (5), explicitly generate alternative pairs $\{\widetilde{\mathbf{Q}}, \tilde{P}\}$ that also describe the dynamics. That is, that satisfy

$$\widetilde{\mathbf{Q}}\dot{\boldsymbol{x}} = \nabla\tilde{P} + \widetilde{\mathbf{G}}\boldsymbol{u}, \tag{15}$$

where $\widetilde{\mathbf{G}} = \widetilde{\mathbf{Q}}\boldsymbol{g}$. For ease of reference in the sequel, we repeat here this result adapting the notation to the present context.

Proposition 2. *[7] Let* $\mathbf{Q}$ *be an invertible matrix solution of (9) and define the full–rank matrix*

$$\widetilde{\mathbf{Q}}(\boldsymbol{x}) := \left[\frac{1}{2}\nabla\left(\mathbf{Q}\boldsymbol{f}\right)\mathbf{M} + \frac{1}{2}\nabla^{\top}(\mathbf{M}\mathbf{Q}\boldsymbol{f}) + \lambda\mathbf{I}\right]\mathbf{Q},$$

where $\lambda \in \mathbb{R}$ *and* $\mathbf{M} = \mathbf{M}^{\top} : \mathbb{R}^n \to \mathbb{R}^{n\times n}$, *are arbitrary. Then, system (5) is equivalently described by (15), where*

$$\tilde{P} := \lambda \int (\mathbf{Q}\boldsymbol{f})^{\top} d\boldsymbol{x} + \frac{1}{2}\boldsymbol{f}^{\top}\mathbf{Q}^{\top}\mathbf{M}\mathbf{Q}\boldsymbol{f}$$

Remark 1. Clearly, the power–shaping stage of the procedure—after transformation of the system (1) into the form (5)—is the same as the one proposed in [5] for energy shaping using interconnection and damping assignment passivity–based control (IDA–PBC). Additional remarks on the relation between these techniques may be found in [9, 12] and in the recent work [13] where a control by interconnection perspective is given to power shaping. Indeed, for port–controlled Hamiltonian (PCH) systems[4]

$$\begin{aligned}\dot{\boldsymbol{x}} &= [\mathbf{J}(\boldsymbol{x}) - \mathbf{R}(\boldsymbol{x})]\nabla H + \boldsymbol{g}(\boldsymbol{x})\boldsymbol{u} \\ \boldsymbol{y} &= \boldsymbol{g}^{\top}(\boldsymbol{x})\nabla H\end{aligned}$$

with *full–rank* matrix $\mathbf{J} - \mathbf{R}$, a trivial solution of (9) is obtained by setting $\mathbf{Q} = (\mathbf{J} - \mathbf{R})^{-1}$. This case has been studied in [14], where it is shown that PCH systems satisfy

$$H(\boldsymbol{x}(t)) - H(\boldsymbol{x}(0)) \leq \int_0^t \widetilde{\boldsymbol{y}}^{\top}(\tau)\boldsymbol{u}(\tau)d\tau$$

with the output (6), i.e. $\widetilde{\boldsymbol{y}} = -\boldsymbol{g}^{\top}[\mathbf{J} - \mathbf{R}]^{-\top}\{[\mathbf{J} - \mathbf{R}]\nabla H + \boldsymbol{g}\boldsymbol{u}\}$, and an energy–balancing interpretation is given to IDA–PBC. Equivalence between IDA–PBC and power shaping has been proved in [13], by viewing power–shaping as control by interconnection with port variables $(\boldsymbol{u}, \widetilde{\boldsymbol{y}})$, instead of the standard $(\boldsymbol{u}, \boldsymbol{y})$ [4].

3 The Tunnel Diode

In this section we illustrate the power shaping methodology of Proposition 1 with the benchmark example of the tunnel diode circuit. We show that this technique yields a simple linear controller that ensures robust global asymptotic stability of the desired equilibrium.

3.1 Model Description

Consider the nonlinear circuit of Figure 1 which represents the approximate behavior of a tunnel diode [10]. The dynamics of the circuit is given by

---

[4] We refer the reader to [4] for a complete treatment on PCH systems.

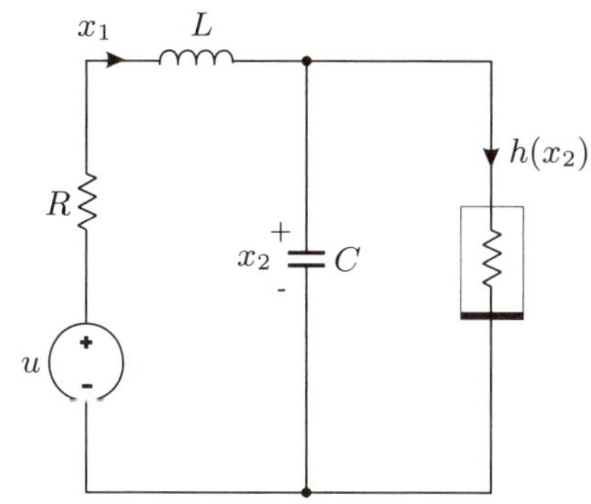

Fig. 1. Tunnel diode circuit

$$\begin{aligned} \dot{x}_1 &= -\tfrac{R}{L}x_1 - \tfrac{1}{L}x_2 + \tfrac{u}{L} \\ \dot{x}_2 &= \tfrac{1}{C}x_1 - \tfrac{1}{C}h(x_2) \end{aligned} \tag{16}$$

where x_1 is the current through the inductor L and x_2 the voltage across the capacitor C. The function $h : \mathbb{R} \to \mathbb{R}$ represents the characteristic curve of the tunnel diode depicted in Figure 2. The assignable equilibrium points of the circuit are determined by $x_1^\star = h(x_2^\star)$, with the corresponding constant control $u^\star = Rh(x_2^\star) + x_2^\star$. It is easy to see that, for all (non–zero) equilibrium states, the steady–state power extracted from the controller $(u^\star x_1^\star)$ is nonzero. Consequently, it is not possible to stabilize the circuit via *energy–balancing.* To state our main result we assume the following:

A.4 $\min_{x_2} h'(x_2) > -\frac{RC}{L}$.

As indicated in Remark 2, this assumption is made for simplicity and can be easily replaced by the knowledge of a lower bound on h'.[5]

3.2 Control Design

Proposition 3. *Consider the dynamic equations of the tunnel diode circuit (16), which verifies Assumption A.4. The power–shaping procedure of Proposition 1 yields a linear (partial) state feedback control*

$$u = -k(x_2 - x_2^\star) + u^\star. \tag{17}$$

If the tuning parameter $k > 0$ satisfies

$$k > -[1 + Rh'(x_2^\star)], \tag{18}$$

$\boldsymbol{x}^\star$ *is a* globally asymptotically stable *equilibrium of the closed loop with Lyapunov function*

[5] We notice that Assumption A.4 coincides with the constraint given in [15] to exclude the appearance of limit cycles in this kind of circuits.

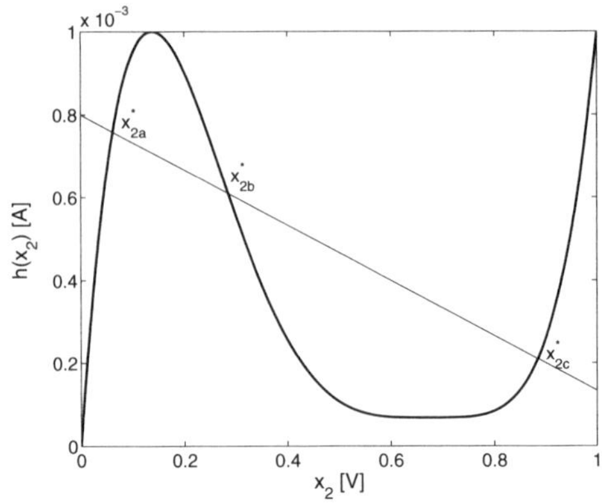

Fig. 2. Tunnel diode characteristic $h(x_2)$ and equilibrium points

$$P_d(\boldsymbol{x}) = \frac{R}{L}\int_0^{x_2} h(\tau)d\tau + \frac{1}{2C}(x_1 - h(x_2))^2 + \frac{k}{2L}(x_2 - x_2^\star)^2 + \frac{1}{2L}(x_2 - u^\star)^2. \quad (19)$$

Proof. We look for a matrix $\mathbf{Q}(\boldsymbol{x})$ such that Assumption A.1 of Proposition 1 is satisfied. Let us propose the simplest form

$$\mathbf{Q}(\boldsymbol{x}) = \begin{bmatrix} 0 & 1 \\ -1 & q(\boldsymbol{x}) \end{bmatrix}$$

where $q : \mathbb{R}^2 \to \mathbb{R}_{\leq 0}$ is a function to be defined. Computing $\mathbf{Q}\boldsymbol{f}$, and making q *function only* of x_2, we see that the integrability condition (9) reduces to $-\frac{1}{C}h' = \frac{R}{L} + \frac{q}{C}$. Hence, a suitable matrix is given by

$$\mathbf{Q}(x_2) = \begin{bmatrix} 0 & 1 \\ -1 & -\frac{RC}{L} - h'(x_2) \end{bmatrix}, \quad (20)$$

which is invertible for all x_2 and, under Assumption A.4, verifies $\mathbf{Q} + \mathbf{Q}^\top \leq 0$.

Condition iii) of Proposition 1 becomes $\frac{\partial P_a}{\partial x_1} = 0$, indicating that P_a cannot be a function of x_1. Hence, we fix $P_a = \Psi(x_2)$, where $\Psi(\cdot)$ is an arbitrary differentiable function that must be chosen so that $P_d = P + P_a$ has minimum at $\boldsymbol{x}^\star$. Computing P from (10) we get

$$P_d = \frac{R}{L}\int_0^{x_2} h(\tau)d\tau + \frac{1}{2C}(x_1 - h(x_2))^2 + \frac{1}{2L}x_2^2 + \Psi(x_2),$$

that should satisfy

$$\nabla P_d|_{\boldsymbol{x}=\boldsymbol{x}^\star} = \begin{bmatrix} 0 \\ \frac{R}{L}h(x_2^\star) + \frac{x_2^\star}{L} + \Psi'(x_2^\star) \end{bmatrix} \equiv \begin{bmatrix} 0 \\ 0 \end{bmatrix}$$

and

$$\nabla^2 P_d\big|_{\boldsymbol{x}=\boldsymbol{x}^\star} = \begin{bmatrix} \frac{1}{C} & -\frac{h'(x_2^\star)}{C} \\ -\frac{h'(x_2^\star)}{C} & \frac{1}{L} + \frac{R}{L}h'(x_2^\star) + \frac{(h'(x_2^\star))^2}{C} + \Psi''(x_2^\star) \end{bmatrix} > 0$$

An obvious selection is then to "complete the squares"

$$\Psi(x_2) = \frac{k}{2L}(x_2 - x_2^\star)^2 - \frac{x_2 u^\star}{L},$$

for which the conditions above are satisfied provided (18) holds.

Thus, the resulting Lyapunov function P_d is given by (19), which has a unique global minimum at $\boldsymbol{x}^\star$. From (11) we obtain the simple linear state feedback (17). This completes the proof.

To illustrate the general power–shaping procedure we have decided to start from a description of the circuit in the form (1) and explicitly solve the PDE (9). This step can be avoided writing the circuit in Brayton–Moser form (5) and invoking Proposition 2. In this case we have

$$\mathbf{Q} = \begin{bmatrix} -L & 0 \\ 0 & C \end{bmatrix}, \ \mathbf{G} = \begin{bmatrix} 1 \\ 0 \end{bmatrix}, \ P = -\int_0^{x_2} h(\tau)d\tau + \frac{R}{2}x_1^2 + x_1 x_2.$$

We note that the mixed potential P has, indeed, units of power and that $\mathbf{Q} + \mathbf{Q}^\top \nleq 0$, hence a new pair $\{\widetilde{\mathbf{Q}}, \tilde{P}\}$ must be generated. Selecting

$$\mathbf{M} = \begin{bmatrix} \frac{1}{L} & 0 \\ 0 & \frac{1}{C} \end{bmatrix}, \quad \lambda = -\frac{R}{L},$$

and invoking Proposition 2, the dynamic equations of the tunnel diode (16) can be equivalently written as (15), with

$$\widetilde{\mathbf{Q}}(x_2) = \begin{bmatrix} 0 & 1 \\ -1 & -\frac{RC}{L} - h'(x_2) \end{bmatrix},$$

which correspond to the matrix (20) obtained solving the PDE (9), the new mixed potential

$$\tilde{P}(\boldsymbol{x}) = \frac{R}{L}\int_0^{x_2} h(\tau)d\tau + \frac{1}{2C}(x_1 - h(x_2))^2 + \frac{1}{2L}x_2^2,$$

and $\widetilde{\mathbf{G}} = [0 \ \ -\frac{1}{L}]^\top$. Power shaping is completed along the lines of Proposition 1, but now considering the system in the form (15)

Remark 2. Assumption A.4 is satisfied if the resistance R is large enough. If it is not the case, a possible solution is to apply a preliminary feedback, $-R_a x_1$, with

$$R_a > -\left(\frac{L}{C}\min_{x_2} h'(x_2) + R\right).$$

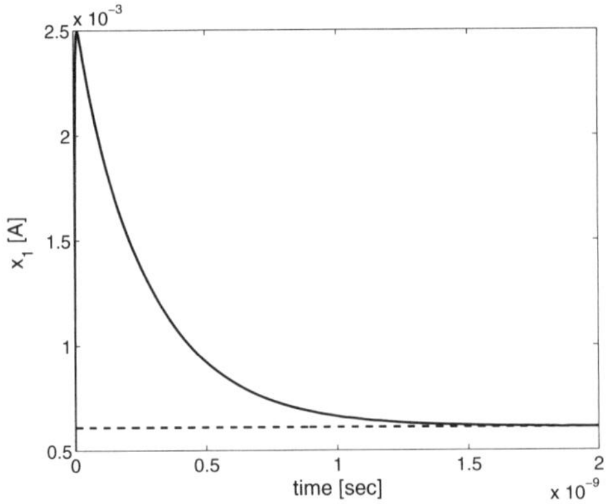

Fig. 3. Inductor current x_1

Remark 3. Knowledge of the system parameters in the power–shaping controller (17) is required only for the computation of $u^\star$. Doing a perturbation analysis it can be shown that an error on this constant induces a steady–state error, that can be reduced increasing the gain k—justifying the claimed "robust" qualifier. On the other hand, for all practical purposes, the term $u^\star$ can be replaced by an integral action on the voltage error. Current research is under way to study the stability properties of this PI control.

3.3 Simulation Results

The element values of the circuit, taken from [11, 10], are $R = 1.5$ kΩ , $L = 5$ nH, $C = 2$ pF, $u^\star = 1.2$ with the currents measured in mA, voltage in Volts and time in nanoseconds. The equation for the characteristic curve of the nonlinear resistor is

$$h(x_2) = 17.76x_2 - 103.79x_2^2 + 229.62x_2^3 - 226.31x_2^4 + 83.72x_2^5$$

It can be proved that Assumption A.4 is satisfied. In open loop the circuit has two stable equilibrium points, corresponding to $x_{2a}^\star$ and $x_{2c}^\star$ of Fig. 2.

The equilibrium to be stabilized is $x_{2b}^\star = 0.2853$ V, also indicated in Fig. 2. According to Proposition 3, the gain k of the controller should satisfy $k > 4.46$. The results of simulation are depicted in Figures 3 and 4 with the dashed line representing the desired equilibrium. The initial conditions were set as $x_1(0) = 0.0005$, $x_2(0) = 0.1$ and the gain $k = 5$.

If the equilibrium to be stabilized is $x_{2a}^\star$ (or $x_{2c}^\star$), then we have $h'(x_{2a}^\star) > 0$ (resp. $h'(x_{2c}^\star) > 0$) and the gain condition (18) is satisfied for any $k > 0$ (even some negative values of the gain k do not destabilize these equilibrium points)—this is, of course, consistent with the fact that $x_{2b}^\star$ is an (open–loop) unstable equilibrium, while the other two are stable.

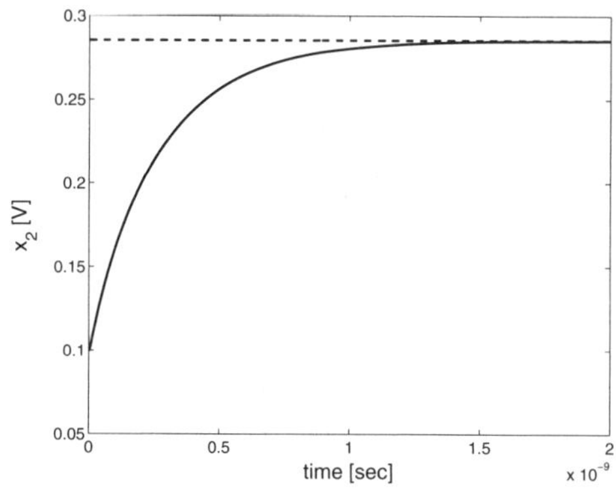

Fig. 4. Capacitor voltage x_2

4 Concluding Remarks

In this paper we have extended the *power–shaping* methodology, proposed in [7] for RLC circuits, to general nonlinear systems. We have illustrated this technique with the dynamic model of the tunnel diode. The resulting control law is a simple linear (partial) state feedback controller that ensures (robust) global asymptotic stability of the desired equilibrium point. The simplicity of this controller, which results from the effective exploitation of the physical structure of the system, should be contrasted with the daunting complexity of the "solution" proposed in [11]. This example, and many other that have been reported in the literature where PBC yields simple sensible solutions, see e.g. [3, 16] and the references therein, casts serious doubts on the pertinence of piece–wise approximation of nonlinearities to control physical systems.

Among the issues that remain open and are currently being explored are the solvability of the PDE (9) for different kind of systems and other applications of power shaping, for instance, to mechanical and electro-mechanical systems.

References

1. M. Takegaki and S. Arimoto. A new feedback for dynamic control of manipulators. *Trans. of the ASME: Journal of Dynamic Systems, Measurement and Control*, 102:119–125, 1981.
2. R. Ortega and M. Spong. Adaptive motion control of rigid robots: a tutorial. *Automatica*, 25(6):877–888, 1989.
3. R. Ortega, A. Loria, P.J. Nicklasson, and H. Sira-Ramirez. *Passivity–based control of Euler-Lagrange systems.* Communications and Control Engineering. Springer-Verlag, Berlin, 1998.
4. A.J. van der Schaft. *$\mathcal{L}_2$ –Gain and passivity techniques in nonlinear control.* Springer–Verlag, 2000.
5. R. Ortega, A.J. van der Schaft, B.M. Maschke, and G. Escobar. Interconnection and damping assignment passivity–based control of port–controlled Hamiltonian systems. *Automatica*, 38(4):585–596, 2002.

6. R. Ortega, A.J. van der Schaft, I. Mareels, and B. Maschke. Putting energy back in control. *IEEE Control Systems Magazine*, 21(2):18–33, 2001.
7. R. Ortega, D. Jeltsema, and J.M.A. Scherpen. Power shaping: a new paradigm for stabilization of nonlinear RLC circuits. *IEEE. Trans. on Automatic Control*, 48(10):1762– 1767, October 2003.
8. R.K. Brayton and J.K. Moser. A theory of nonlinear networks-I. *Quart. of App. Math*, 22:1–33, April 1964.
9. D. Jeltsema. *Modeling and control of nonlinear networks: a power–based perspective.* PhD thesis, Delft University of Technology, The Netherlands, May 2005.
10. H. Khalil. *Nonlinear Systems.* Prentice-Hall, 2nd. edition, 1996.
11. L. Rodriguez and S. Boyd. Piece-wise affine state feedback for piece-wise affine slab systems using convex optimization. *Systems & Control Letters*, 54:835–853, 2005.
12. G. Blankenstein. Power balancing for a new class of nonlinear systems and stabilization of RLC circuits. *International Journal of Control*, 78(3):159–171, 2005.
13. R. Ortega, A. van der Schaft, F. Castaños, and A. Astolfi. Control by state–modulated interconnection of port–hamiltonian systems. In *7th IFAC Symposium on Nonlinear Control Systems (NOLCOS'07)*, Pretoria, South Africa, August 22-24, 2007.
14. D. Jeltsema, R. Ortega, and J.M.A. Scherpen. An energy–balancing perspective of interconnection and damping assignment control of nonlinear systems. *Automatica*, 40(9):1643–1646, 2004.
15. J.K. Moser. Bistable systems of differential equations with applications to tunnel diode circuits. *IBM Journal of Res. Develop.*, 5:226–240, 1960.
16. R. Ortega and E. Garcia-Canseco. Interconnection and damping assignment passivity–based control: a survey. *European Journal of Control: Special Issue on Lagrangian and Hamiltonian Systems*, 10:432–450, December 2004.

Total Energy Shaping Control of Mechanical Systems: Simplifying the Matching Equations Via Coordinate Changes

Giuseppe Viola[1,⋆], Romeo Ortega[2,⋆⋆], Ravi Banavar[3], José Ángel Acosta[4,⋆⋆⋆], and Alessandro Astolfi[1,5]

[1] Dipartimento di Informatica, Sistemi e Produzione, Università di Roma "Tor Vergata", Via del Politecnico, 1, 00133 Roma, Italy
`{astolfi,viola}@disp.uniroma2.it`

[2] Laboratoire des Signaux et Systèmes, Supélec, Plateau du Moulon, 91192 Gif-sur Yvette, France
`ortega@lss.supelec.fr`

[3] Systems and Control Engineering, I. I. T. - Bombay, 101 A, ACRE Building Powai, Mumbai, 400 076, India
`banavar@iitb.ac.in`

[4] Depto. de Ingeniería de Sistemas y Automática, Escuela Superior de Ingenieros, Camino de los Descubrimientos s/n., 41092 Sevilla, Spain
`jaar@esi.us.es`

[5] Electrical Engineering Department, Imperial College, Exhibition Road, London, SW7 2AZ, UK
`a.astolfi@ic.ac.uk`

Summary. Total Energy Shaping is a controller design methodology that achieves (asymptotic) stabilization of mechanical systems endowing the closed–loop system with a Lagrangian or Hamiltonian structure with a desired energy function—that qualifies as Lyapunov function for the desired equilibrium. The success of the method relies on the possibility of solving two PDEs which identify the kinetic and potential energy functions that can be assigned to the closed–loop. Particularly troublesome is the PDE associated to the kinetic energy which is nonlinear and *non–homogeneous* and the solution, that defines the desired inertia matrix, must be positive definite. In this paper we prove that we can eliminate or simplify the forcing term in this PDE modifying the target dynamics and introducing a change of coordinates in the original system. Furthermore, it is shown that, in the particular case of transformation to the Lagrangian coordinates, the possibility of simplifying the PDEs is determined by the interaction between the Coriolis and centrifugal forces and the actuation structure. The example of a pendulum on a cart is used to illustrate the results.

⋆ The work of Giuseppe Viola was partially sponsored by the Control Training Site programme of the European Commission, contract number: HPMT-CT-2001-00278.
⋆⋆ Corresponding author.
⋆⋆⋆ The work of J. A. Acosta was supported by the Spanish Ministry of Science and Technology under grants DPI2003-00429 and DPI2004-06419.

F. Bullo et al. (Eds.): Lag. & Hamil. Methods for Nonlin. Ctrl., LNCIS 366, pp. 147–156, 2007.
springerlink.com

1 Introduction

Stabilization of underactuated mechanical systems shaping their energy function, but preserving the systems structure, has attracted the attention of control researchers for several years. While fully actuated mechanical systems admit an arbitrary shaping of the potential energy by means of feedback, and therefore stabilization to any desired equilibrium, this is in general not possible for underactuated systems. In certain cases this problem can be overcome by also modifying the kinetic energy of the system. The idea of total energy shaping was first introduced in [2] with the two main approaches being now: the method of controlled Lagrangians [6] and interconnection and damping assignment passivity–based control (IDA–PBC) [13], see also the closely related work [8]. In both cases stabilization (of a desired equilibrium) is achieved identifying the class of systems—Lagrangian for the first method and Hamiltonian for IDA–PBC—that can possibly be obtained via feedback. The conditions under which such a feedback law exists are called *matching conditions*, and consist of a set of nonlinear partial differential equations (PDEs). In case these PDEs can be solved the original control system and the target dynamic system are said to *match*.

A lot of research effort has been devoted to the solution of the matching equations. In [6] the authors give a series of conditions on the system and the assignable inertia matrices such that the PDEs can be solved. Also, techniques to solve the PDEs have been reported in [5],[4] and some geometric aspects of the equations are investigated in [10]. The case of underactuation degree one systems has been studied in detail in [3] and [1]. In the latter we proved that, if the inertia matrix and the force induced by the potential energy (on the unactuated coordinate) are independent of the unactuated coordinate, then the PDEs can be *explicitly solved*. In [11] explicit solutions are also given for a class of two degrees–of–freedom systems, that includes the interesting Acrobot example.

In this paper we pursue the investigation aimed at providing constructive solutions to the PDEs. We concentrate our attention on the PDE associated to the kinetic energy which is nonlinear and non–homogeneous and whose solution, that defines the desired inertia matrix, must be positive definite. We study the possibility of eliminating the forcing term in this PDE. Our main contribution is the proof that it is possible to achieve this objective *re–parametrizing the target dynamics* and introducing a *change of coordinates* in the original system. The class of coordinate changes that yields an homogenous PDE is a solution of another PDE—similar to the kinetic energy PDE—but this time without the requirement of positive definiteness. Furthermore, it is shown that, in the particular case of transformation to the Lagrangian coordinates, the possibility of simplifying the PDEs is determined by the interaction between the Coriolis forces and the actuation structure. We illustrate the result with the example of the pendulum on a cart.

Notation: Unless indicated otherwise, all vectors in the paper are *column* vectors, even the gradient of a scalar function: $\nabla_{(\cdot)} = \frac{\partial}{\partial(\cdot)}$—when clear from the context the subindex in ∇ will be omitted. To simplify the expressions, the arguments of all functions will be omitted, and will be explicitly written only the first time that the function is defined.

Caveat: This is an abridged version of the paper, where, for space reasons, all proofs have been omitted. The full version of the paper is available upon request to the authors.

2 Background on IDA–PBC and Problem Formulation

In [5] and [7], it has been shown that the PDEs of the controlled Lagrangian method and IDA–PBC are the same, therefore, in the sequel we will restrict our attention to IDA–PBC—for which a brief review is now presented. IDA–PBC was introduced in [13] to regulate the position of underactuated mechanical systems of the form

$$\Sigma : \begin{bmatrix} \dot{q} \\ \dot{p} \end{bmatrix} = \begin{bmatrix} 0 & I_n \\ -I_n & 0 \end{bmatrix} \begin{bmatrix} \nabla_q H \\ \nabla_p H \end{bmatrix} + \begin{bmatrix} 0 \\ G(q) \end{bmatrix} u, \tag{1}$$

where $q \in \mathbb{R}^n$, $p \in \mathbb{R}^n$ are the generalized position and momenta, respectively, $u \in \mathbb{R}^m$ and $G \in \mathbb{R}^{n\times m}$ with rank $G = m < n$,

$$H(q,p) = \frac{1}{2} p^\top M^{-1}(q) p + V(q) \tag{2}$$

is the total energy with $M = M^\top > 0$ the inertia matrix, and V the potential energy. The main result of [13] is the proof that for all matrices $M_d(q) = M_d^\top(q) \in \mathbb{R}^{n\times n}$ and functions $V_d(q)$ that satisfy the PDEs

$$G^\perp \left\{ M_d M^{-1} \nabla_q (p^\top M_d^{-1} p) - 2 J_2 M_d^{-1} p \right\} = G^\perp \nabla_q (p^\top M^{-1} p) \tag{3}$$

$$G^\perp M_d M^{-1} \nabla V_d = G^\perp \nabla V, \tag{4}$$

for some $J_2(q,p) = -J_2^\top(q,p) \in \mathbb{R}^{n\times n}$ and a full rank left annihilator $G^\perp(q) \in \mathbb{R}^{(n-m)\times n}$ of G, i.e., $G^\perp G = 0$ and $\mathrm{rank}(G^\perp) = n - m$, the system (1) in closed–loop with the IDA–PBC $u = \hat{u}(q,p)$, where

$$\hat{u}(q,p) = (G^\top G)^{-1} G^\top (\nabla_q H - M_d M^{-1} \nabla_q H_d + J_2 M_d^{-1} p), \tag{5}$$

takes the Hamiltonian form

$$\Sigma_d : \begin{bmatrix} \dot{q} \\ \dot{p} \end{bmatrix} = \begin{bmatrix} 0 & M^{-1} M_d \\ -M_d M^{-1} & J_2 \end{bmatrix} \begin{bmatrix} \nabla_q H_d \\ \nabla_p H_d \end{bmatrix}, \tag{6}$$

where the new total energy function is $H_d(q,p) = \frac{1}{2} p^\top M_d^{-1}(q) p + V_d(q)$. Further, if M_d is positive definite in a neighborhood of $q^\star \in \mathbb{R}^n$ and $q^\star = \arg\min V_d(q)$, then $(q^\star, 0)$ is a stable equilibrium point of (6) with Lyapunov function H_d.

Clearly, the success of IDA–PBC relies on the possibility of solving the PDEs (3) and (4). Particularly troublesome is the kinetic energy (KE) PDE (3), which is nonlinear and non–homogeneous, and whose solution must be positive definite. In this brief note we investigate the possibility of *eliminating the forcing term* $G^{\perp}\nabla_q(p^\top M^{-1}p)$ in this PDE—via coordinate changes and re-parametrization of the target dynamics.

The presence of the forcing term introduces a *quadratic term* in M_d in the KE PDE that renders very difficult its solution—even with the help of the free skew–symmetric matrix J_2. To reveal this deleterious effect let us eliminate the dependence on p and re-write the PDE in the equivalent form, see [13] and [1] for further details,

$$\begin{aligned}\sum_{i=1}^{n}\left[(G_k^{\perp}M_dM^{-1}e_i)\frac{\partial M_d}{\partial q_i}-(G_k^{\perp}e_i)M_d\frac{\partial M^{-1}}{\partial q_i}M_d\right]\\ =-[\mathcal{J}(q)\mathcal{A}_k^\top(q)+\mathcal{A}_k(q)\mathcal{J}^\top(q)],\end{aligned}\tag{7}$$

for $k=1,\ldots,n-m$, where $e_i\in\mathbb{R}^n$ is the i–th vector of the n–dimensional Euclidean basis,

$$\mathcal{J}(q)\triangleq\left[\alpha_1(q)\ \vdots\ \alpha_2(q)\ \vdots\ \cdots\ \vdots\ \alpha_{n_o}(q)\right]\in\mathbb{R}^{n\times n_o},$$

is a *free* matrix, $\alpha_i\in\mathbb{R}^n$, $i=1,\ldots,n_o\triangleq\frac{n}{2}(n-1)$ and we have defined the row vectors $G_k^{\perp}\in\mathbb{R}^{1\times n}$,

$$\begin{aligned}G^{\perp}&\triangleq\left[(G_1^{\perp})^\top\ldots(G_{n-m}^{\perp})^\top\right]^\top,\\ \mathcal{A}_k&\triangleq-\left[W_1\left(G_k^{\perp}\right)^\top,\ldots,W_{n_o}\left(G_k^{\perp}\right)^\top\right]\in\mathbb{R}^{n\times n_o},\end{aligned}$$

with the $W_i\in\mathbb{R}^{n\times n}$ skew–symmetric matrices with elements 1s and 0s. For instance, for $n=3$ we get

$$\begin{aligned}W_1&\triangleq\begin{bmatrix}0&1&0\\-1&0&0\\0&0&0\end{bmatrix},\quad W_2\triangleq\begin{bmatrix}0&0&1\\0&0&0\\-1&0&0\end{bmatrix},\\ W_3&\triangleq\begin{bmatrix}0&0&0\\0&0&1\\0&-1&0\end{bmatrix}.\end{aligned}\tag{8}$$

Remark 1. We bring to the readers attention the fact that the target dynamics Σ_d is *parameterized* by the triple $\{M_d,V_d,J_2\}$. Re–parameterization of the target dynamics is a key step introduced in this paper.

Remark 2. As shown in [1], with the definitions above, the free matrix J_2 can be written as

$$J_2=\sum_{i=1}^{n_o}p^\top M_d^{-1}\alpha_iW_i,\tag{9}$$

and the terms $G_k^{\perp}J_2$, that appear in (3), become $G_k^{\perp}J_2=p^\top M_d^{-1}\mathcal{J}\mathcal{A}_k^\top$. Equation (7) is obtained factoring (3) in the form $p^\top M_d^{-1}[\,\cdot\,]M_d^{-1}p$, taking the symmetric part of the matrices $\mathcal{J}\mathcal{A}_k^\top$ and setting the expression in brackets, which is independent of p, equal to zero.

Remark 3. In [1] we proved that, if $n - m = 1$ and the inertia matrix and the force induced by the potential energy (on the unactuated coordinate) are independent of the unactuated coordinate, then the PDEs can be *explicitly solved.* The first assumption on the inertia matrix implies, precisely, that the forcing term $G^\perp \nabla_q (p^\top M^{-1} p) = 0$, which is essential for the construction of the solutions (see Proposition 3).

3 Generating an Homogeneous Kinetic Energy PDE

Our strategy to eliminate the forcing term in the KE PDE consists of two steps. First, we express system (1) in the new coordinates $(q, \tilde{p})$, with $p = T(q)\tilde{p}$, where $T \in \mathbb{R}^{n \times n}$ is *full rank*, yielding:

$$\tilde{\Sigma} : \quad \begin{bmatrix} \dot{q} \\ \dot{\tilde{p}} \end{bmatrix} = \begin{bmatrix} 0 & T^{-\top} \\ T^{-1} & F_{22} \end{bmatrix} \begin{bmatrix} \nabla_q \tilde{H} \\ \nabla_{\tilde{p}} \tilde{H} \end{bmatrix} + \begin{bmatrix} 0 \\ T^{-1} G \end{bmatrix} u, \tag{10}$$

where

$$F_{22} = -T^{-1} \left[S(q, \tilde{p}) - S^\top (q, \tilde{p}) \right] T^{-\top},$$

$$\tilde{H}(q, \tilde{p}) = \frac{1}{2} \tilde{p}^\top T^\top (q) M^{-1}(q) T(q) \tilde{p} + V(q)$$

and

$$S(q, \tilde{p}) = \nabla_q (T(q) \tilde{p}). \tag{11}$$

It's worth noting that these equations can be seen as a particular form of the Boltzmann-Hamel equations (see [14]), in which $\tilde{p}$ is the vector of quasi-velocities, related to the velocity vector $\dot{q}$ by means of the relation $\tilde{p} = T^{-1}(q) M(q) \dot{q}$.

Second, we define *new target dynamics*, in the coordinates $(q, \tilde{p})$, as

$$\tilde{\Sigma}_d : \quad \begin{bmatrix} \dot{q} \\ \dot{\tilde{p}} \end{bmatrix} = \begin{bmatrix} 0 & F_{12} \\ -F_{12}^\top & \tilde{J}_2(q, \tilde{p}) \end{bmatrix} \begin{bmatrix} \nabla_q \tilde{H}_d \\ \nabla_{\tilde{p}} \tilde{H}_d \end{bmatrix}, \tag{12}$$

where

$$\tilde{H}_d(q, \tilde{p}) = \frac{1}{2} \tilde{p}^\top \tilde{M}_d^{-1}(q) \tilde{p} + \tilde{V}_d(q), \tag{13}$$

$F_{12} = M^{-1}(q) T(q) \tilde{M}_d(q)$, $\tilde{M}_d \in \mathbb{R}^{n \times n}$ and $\tilde{J}_2 = -\tilde{J}_2^\top$ is free. The proposed target dynamics are clearly "compatible" with the new system representation—in the sense that the first n equations are already "matched". In Section 4 we establish the connection between (12) and the target dynamic system (6) expressed in the new coordinates—see also Remark 7.

To state our main result we need the following assumption.

Assumption A. The full rank matrix T is such that, for $k = 1, \ldots, n - m$,

$$\sum_{i=1}^{n} \left[T^\top M^{-1} e_i G_k^\perp \frac{\partial T}{\partial q_i} + \frac{\partial T^\top}{\partial q_i} (e_i G_k^\perp)^\top M^{-1} T + G_k^\perp e_i T^\top \frac{\partial M^{-1}}{\partial q_i} T \right] = 0. \tag{14}$$

Proposition 1. *Consider the system (1) and the partial change of coordinates $p = T(q)\tilde{p}$ where T satisfies Assumption A. For all matrices $\tilde{M}_d(q) = \tilde{M}_d^\top(q) \in \mathbb{R}^{n\times n}$ and functions $\tilde{V}_d(q)$ that satisfy the PDEs*

$$G^\perp T\left[\tilde{M}_d T^\top M^{-1}\nabla_q(\tilde{p}^\top \tilde{M}_d^{-1}\tilde{p}) - 2\tilde{J}_2\tilde{M}_d^{-1}\tilde{p}\right] = 0 \tag{15}$$

$$G^\perp T\tilde{M}_d T^\top M^{-1}\nabla \tilde{V}_d = G^\perp \nabla V, \tag{16}$$

for some $\tilde{J}_2(q,\tilde{p}) = -\tilde{J}_2^\top(q,\tilde{p}) \in \mathbb{R}^{n\times n}$, the system (1) in closed–loop with the IDA–PBC $u = \hat{\tilde{u}}(q,\tilde{p})$, where

$$\hat{\tilde{u}}(q,\tilde{p}) = (G^\top G)^{-1}G^\top\left(\nabla_q H + \dot{T}\tilde{p} - T\tilde{M}_d T^\top M^{-1}\nabla_q \tilde{H}_d + T\tilde{J}_2\tilde{M}_d^{-1}\tilde{p}\right) \tag{17}$$

takes, in the coordinates $(q,\tilde{p})$, the Hamiltonian form (12), (13).

Proposition 2. *Assumption A holds with $T = M$ if and only if*

$$G^\perp(q)C(q,\dot{q})\dot{q} = 0, \tag{18}$$

where $C \in \mathbb{R}^{n\times n}$ is the matrix of Coriolis and centrifugal forces of the mechanical system (1).

Now we state a slightly modified version of a result reported in [1], which gives a constructive solution of the homogeneous KE PDE (15) for systems with underactuation degree one.

Proposition 3. *Consider equation (15). Suppose that $n - m = 1$, M does not depend on the unactuated coordinate, and the matrices G and T are function of a single element of q, say q_r, $r \in \{1,\ldots,n\}$. Then, for all desired locally positive definite inertia matrices of the form*

$$\tilde{M}_d(q_r) = \int_{q_r^\star}^{q_r} T^{-1}(\mu)G(\mu)\Psi(\mu)G^\top(\mu)T^{-T}(\mu)d\mu + \tilde{M}_d^0 \tag{19}$$

where the matrix function $\Psi = \Psi^\top \in \mathbb{R}^{(n-1)\times(n-1)}$ and the constant matrix $\tilde{M}_d^0 = (\tilde{M}_d^0)^\top > 0 \in \mathbb{R}^{n\times n}$, may be arbitrarily chosen, there exists a matrix $\tilde{J}_2$ such that the KE PDE (15) holds in a neighborhood of $q_r^\star$.

Remark 4. Comparing (3) with (15) we notice the absence of the forcing term in the latter—therefore, the PDE that needs to be solved is (in principle) simpler. As it will be shown in Proposition 4, this simplification has been achieved without modifying the potential energy PDE (4), but it is subject to the condition of finding a matrix T satisfying Assumption A.

Remark 5. It is interesting to compare the original KE PDE (7) and the additional PDE that needs to be solved (14). At first glance, it may be argued that (14) is as complicated as, if not more complicated than, (7). Notice, however, that the terms $e_i G_k^\perp$ in (14) are matrices with only one non–zero row, while

$G_k^\perp M_d M^{-1} e_i$ in (7) is a scalar that "mixes" all the terms in M_d. Also, we stress the fact that, contrary to M_d that must be symmetric and positive definite, the *only condition* on T is invertibility. For instance, in the example of the pendulum on a cart system, a solution to (14) is trivially obtained while no obvious solution for (7) is available.

Remark 6. To establish Proposition 1 we didn't need to use the transformed system dynamics (10) in Hamiltonian form. This is consistent with the fact that IDA–PBC is applicable to systems in the general form $\dot{x} = f(x) + g(x)u$; see e.g., [12].

Remark 7. Clearly, changing coordinates of the system *and* the target dynamics does not affect the matching conditions and, consequently, the PDEs will be equivalent. The subtlety here is that, as indicated in Remark 1, the target dynamic system Σ_d is parameterized by the triple $\{M_d, V_d, J_2\}$, while the system $\tilde{\Sigma}_d$ is parameterized by the triple $\{\tilde{M}_d, \tilde{V}_d, \tilde{J}_2\}$. Along the same lines, feedback actions of the form $u = \alpha(q,p) + \beta(q,p)v$, with $\beta \in \mathbb{R}^{m \times m}$ full rank, will not affect the PDEs—that "live in Ker G". Indeed, for a system of the form $\dot{x} = f(x) + g(x)u$ and target dynamics $\dot{x} = F(x)\nabla H_d$ the matching equations are $g^\perp f = g^\perp F \nabla H_d$, with $g^\perp g = 0$, independently of the feedback action.

4 Solving the Original PDEs

Proposition 1 establishes that, solving the new PDEs (15), (16), the system Σ described by equations (1) in closed–loop with the IDA–PBC (17) takes, in the coordinates $(q, \tilde{p})$, the Hamiltonian form $\tilde{\Sigma}_d$ described by equations (13). Three natural questions arise.

- What are the dynamics of the closed–loop system in the original coordinates (q,p)?
- What is the relationship between the solutions of the new matching problem $\{\tilde{M}_d, \tilde{V}_d, \tilde{J}_2\}$ and the solutions of the original matching problem $\{M_d, V_d, J_2\}$?
- What is the relationship between the original matching controller $\hat{u}(q,p)$ and the new one $\hat{\tilde{u}}(q,\tilde{p})$?

The answers to these questions are given in the proposition below. The rationale of the proposition is best explained referring to Fig. 1. The connections between the nodes Σ, $\tilde{\Sigma}$ and $\tilde{\Sigma}_d$ are given by Proposition 1. It remains to establish the connection with the original target dynamics node Σ_d. Towards this end, we write Σ_d, described by equations (6), in the new coordinates and prove the existence of a *bijective mapping* $\Psi : \{\tilde{M}_d, \tilde{V}_d, \tilde{J}_2\} \to \{M_d, V_d, J_2\}$, that makes the transformed system *equal* to $\tilde{\Sigma}_d$—that is, with the same structure matrix and the same Hamiltonian function.[1] This proves that Σ and Σ_d match and, consequently, the corresponding parameters $\{M_d, V_d, J_2\}$ solve the PDEs and define the control $\hat{u}(q,p)$.

[1] The mapping Ψ can also be derived computing the Poisson brackets of the coordinates $(q, \tilde{p})$, as done in [5] to establish the equivalence between controlled Lagrangians and IDA–PBC.

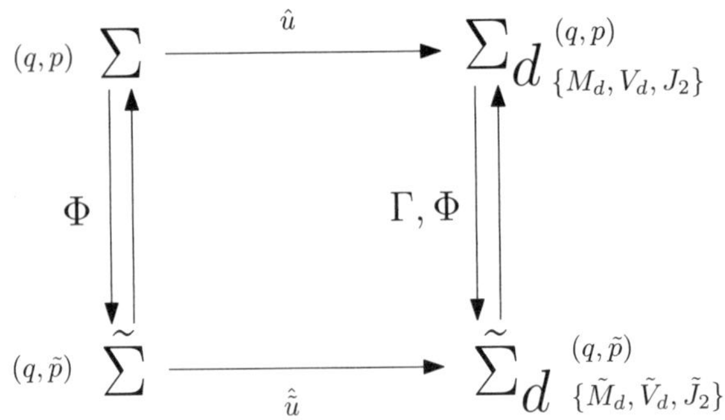

Fig. 1. Diagrammatic description of the systems transformations

Proposition 4. *The triple* $\{\tilde{M}_d, \tilde{V}_d, \tilde{J}_2\}$ *solves the new matching equations (15), (16) if and only if the triple* $\{M_d, V_d, J_2\}$ *solves the original matching equations (3),(4), where*[2]

$$\begin{aligned} M_d &= T\tilde{M}_d T^\top \\ V_d &= \tilde{V}_d \\ J_2(q,p) &= T\tilde{J}_2(q,T^{-1}p)T^\top + S(q,T^{-1}p)M^{-1}T\tilde{M}_d T^\top \\ &\quad - T\tilde{M}_d T^\top M^{-1} S^\top(q,T^{-1}p) \end{aligned} \tag{20}$$

and $S(q,\tilde{p})$ *is given in (11). Furthermore, the control (5) that matches* Σ *to* Σ_d *is obtained as* $\hat{u}(q,p) = \hat{\tilde{u}}(q, T^{-1}(q)p)$, *with* $\hat{\tilde{u}}$ *defined in (17).*

5 The Pendulum on a Cart Example

The dynamic equations of the pendulum on a cart are given by (1) with $n = 2$, $m = 1$, and

$$M(q_1) = \begin{bmatrix} 1 & b\cos q_1 \\ b\cos q_1 & c \end{bmatrix}, \; V(q_1) = a\cos q_1,$$

$$G = e_2, \quad a = \frac{g}{\ell}, \; b = \frac{1}{\ell}, \; c = \frac{M+m}{m\ell^2}$$

where q_1 denotes the pendulum angle with the upright vertical, q_2 the cart position, m and ℓ are, respectively, the mass and the length of the pendulum, M is the mass of the cart and g is the gravity acceleration. The equilibrium to be stabilized is the upward position of the pendulum with the cart placed in *any desired location*, which corresponds to $q_{1\star} = 0$ and an arbitrary $q_{2\star}$.

Noting that $G^\perp = e_1^\top$, the KE PDE (7) takes the form

$$\sum_{i=1}^{2} (e_1^\top M_d M^{-1} e_i) \frac{\partial M_d}{\partial q_i} - M_d \frac{\partial M^{-1}}{\partial q_1} M_d = - \begin{bmatrix} 0 & \alpha_1(q) \\ \alpha_1(q) & 2\alpha_2(q) \end{bmatrix}, \tag{21}$$

[2] For simplicity, we have omitted the argument q in the functions that depend only on q.

with α_i being free functions. Using these functions we can "solve" two of the three equations above, so it remains only one PDE to be solved. To simplify the expression of this equation we make M_d function only of q_1 leading to the ODE

$$\begin{aligned}(cm_{11} - bm_{12}\cos q_1)&\frac{dm_{11}}{dq_1}\\ &= \frac{2b\sin q_1}{c - b^2\cos^2 q_1}[b\cos q_1(cm_{11}^2 + m_{12}^2) - (c + b^2\cos^2 q_1)m_{11}m_{12}],\end{aligned}$$

where $m_{ij}(q_1)$ is the ij–element of the matrix M_d. Even using m_{12} as a degree of freedom finding a solution to this ODE is a daunting task.

To simplify the PDE we proceed, then, to apply the technique proposed in the paper. By computing the Coriolis and centrifugal forces matrix using the Christoffel symbols of the second kind [9], it is easy to see that the condition $G^{\perp}C\dot{q} = 0$ is satisfied. Therefore, we propose to take $T = M$ and search for a solution of the homogeneous PDE (15), which becomes:

$$G^{\perp}M\left[\tilde{M}_d\nabla_q(\tilde{p}^{\top}\tilde{M}_d^{-1}\tilde{p}) - 2\tilde{J}_2\tilde{M}_d^{-1}\tilde{p}\right] = 0. \tag{22}$$

It is easy to see that the hypotheses of Proposition 3 are satisfied with $r = 1$. By selecting

$$\Psi(\mu) = \frac{-k\sin\mu}{m_3 - b^2\cos^2\mu}, \quad \tilde{M}_d^0 = \begin{bmatrix} \frac{kb^2}{3}\cos^3 q_{1\star} & -\frac{kb}{2}\cos^2 q_{1\star} \\ -\frac{kb}{2}\cos^2 q_{1\star} & k\cos q_{1\star} + m_{22}^0 \end{bmatrix}$$

with $k > 0$ and $m_{22}^0 \geq 0$ free parameters, it follows that a solution is provided by

$$\tilde{M}_d = \begin{bmatrix} \frac{kb^2}{3}\cos^3 q_1 & -\frac{kb}{2}\cos^2 q_1 \\ -\frac{kb}{2}\cos^2 q_1 & k\cos q_1 + m_{22}^0 \end{bmatrix}, \quad \tilde{J}_2 = \tilde{p}^{\top}\tilde{M}_d^{-1}\begin{bmatrix}\alpha_1 \\ \alpha_2\end{bmatrix}\begin{bmatrix} 0 & 1 \\ -1 & 0\end{bmatrix}.$$

Such a solution is positive definite and bounded for all $q_1 \in (\frac{-\pi}{2}, \frac{\pi}{2})$.

Regarding the potential energy, it can be seen that a solution of the PDE (16) is given by

$$\tilde{V}_d = \frac{3a}{kb^2\cos^2 q_1} + \frac{P}{2}\left[q_2 - q_{2\star} + \frac{3}{b}\ln(\sec q_1 + \tan q_1) + \frac{6m_{22}^0}{kb}\tan q_1\right]^2,$$

with $P > 0$ arbitrary and $q_{2\star}$ the cart position to be stabilized. The asymptotic analysis of the closed–loop system and some simulations may be found in [1], where the problem has been solved by using Lagrangian coordinates and carrying out a partial feedback linearization.

6 Conclusions

In this paper we have investigated a way to simplify the solution of the matching equations of IDA-PBC for a class of underactuated mechanical systems. We have shown that it is possible to transform the KE PDE into an homogeneous one by using coordinate transformations and a re-parametrization of the target dynamics. This can be achieved provided that a solution of another PDE, involving the coordinates transformation matrix, can be found. This new PDE is similar to the kinetic energy PDE, but without the requirement of positive definiteness of its solutions. Moreover, it has been shown that in the particular case of transformation to the Lagrangian coordinates, the possibility of simplifying the PDEs is determined by the interaction between the Coriolis and centrifugal forces and the actuation structure. The proposed technique has been successfully applied to the pendulum on a cart.

References

1. J. A. Acosta, R. Ortega, A. Astolfi, and A. M. Mahindrakar. Interconnection and damping assignment passivity–based control of mechanical systems with underactuation degree one. *IEEE Trans. Automat. Contr.*, 50, 2005.
2. A. Ailon and R. Ortega. An observer-based controller for robot manipulators with flexible joints. *Syst.& Cont. Letters*, 21:329–335, 1993.
3. D. Auckly and L. Kapitanski. On the λ–equations for matching control laws. *SIAM J. Control and Optimization*, 41:1372–1388, 2002.
4. D. Auckly, L. Kapitanski, and W. White. Control of nonlinear underactuated systems. *Comm. Pure Appl. Math.*, 3:354–369, 2000.
5. G. Blankenstein, R. Ortega, and A.J. van der Schaft. The matching conditions of controlled lagrangians and interconnection assigment passivity based control. *Int J of Control*, 75:645–665, 2002.
6. A. Bloch, N. Leonard, and J. Marsden. Controlled lagrangians and the stabilization of mechanical systems. *IEEE Trans. Automat. Contr.*, 45, 2000.
7. D.E. Chang, A.M. Bloch, N.E. Leonard, J.E. Marsden, and C.A. Woolsey. The equivalence of controlled lagrangian and controlled hamiltonian systems for simple mechanical systems. *ESAIM: Control, Optimisation, and Calculus of Variations*, 45:393–422, 2002.
8. K. Fujimoto and T. Sugie. Canonical transformations and stabilization of generalized hamiltonian systems. *Systems and Control Letters*, 42:217–227, 2001.
9. R. Kelly, V. Santibanez, and A. Loria. *Control of robot manipulators in joint space.* Springer–Verlag, London, 2005.
10. A. Lewis. Notes on energy shaping. *43rd IEEE Conf Decision and Control, Dec 14–17, 2004, Paradise Island, Bahamas.*, 2004.
11. A. D. Mahindrakar, A. Astolfi, R. Ortega, and G. Viola. Further constructive results on interconnection and damping assignment control of mechanical systems: The acrobot example. *American Control Conference, Minneapolis, USA, 14-16 June 2006*, 2006.
12. R. Ortega and E. Garcia-Canseco. Interconnection and damping assignment passivity–based control: A survey. *European J of Control*, 10:432–450, 2004.
13. R. Ortega, M. Spong, F. Gomez, and G. Blankenstein. Stabilization of underactuated mechanical systems via interconnection and damping assignment. *IEEE Trans. Automat. Contr.*, AC–47:1218–1233, 2002.
14. E.T. Whittaker. *A treatise on the analytical dynamics of particles and rigid bodies.* Cambridge University Press, 1988.

Simultaneous Interconnection and Damping Assignment Passivity–Based Control: Two Practical Examples

Carles Batlle[1], Arnau Dòria-Cerezo[1], Gerardo Espinosa-Pérez[2], and Romeo Ortega[3]

[1] MA4, DEE and IOC, EPSEVG, UPC, Av. V. Balaguer s/n, Vilanova i la Geltrú, 08800, Spain
carles.batlle@upc.edu, arnau.doria@upc.edu

[2] DEPFI–UNAM, Apartado Postal 70-256, 04510 México D.F., Mexico
gerardoe@servidor.unam.mx

[3] Laboratoire des Signaux et Systémes, SUPELEC, Plateau du Moulon, Gif-sur-Yvette 91192, France
Romeo.Ortega@lss.supelec.fr

1 Introduction

Passivity–based control (PBC) is a generic name given to a family of controller design techniques that achieves system stabilization via the route of passivation, that is, rendering the closed–loop system passive with a desired storage function (that usually qualifies as a Lyapunov function for the stability analysis.) If the passivity property turns out to be output strict, with an output signal with respect to which the system is detectable, then asymptotic stability is ensured. See the monographs [5, 12], and [6] for a recent survey.

As is well–known, [12], a passive system can be rendered strictly passive simply adding a negative feedback loop around the passive output—an action sometimes called L_gV control, [10]. For this reason, it has been found convenient in some applications, in particular for mechanical systems, [11], [8], to split the control action into the sum of two terms, an energy–shaping term which, as indicated by its name, is responsible of assigning the desired energy/storage function to the passive map, and a second L_gV term that injects damping for asymptotic stability. The purpose of this paper is to bring to the readers attention the fact that splitting the control action in this way is not without loss of generality, and effectively reduces the set of problems that can be solved via PBC. This assertion is, of course, not surprising since it is clear that, to achieve strict passivity, the procedure described above is just one of many other possible ways. Our point is illustrated with the IDA–PBC design methodology proposed in [4]. To enlarge the set of systems that can be stabilized via IDA–PBC we suggest to carry out *simultaneously* the energy shaping and the damping injection stages and refer to this variation of the method as SIDA–PBC.

F. Bullo et al. (Eds.): Lag. & Hamil. Methods for Nonlin. Ctrl., LNCIS 366, pp. 157–169, 2007.
springerlink.com

We illustrate the application of SIDA–PBC with two practically important examples. First, we show that the fundamental problem of induction motor torque and rotor flux regulation *cannot be solved* with two stage IDA–PBC. It is, however, solvable with SIDA–PBC. Second, we prove that with SIDA–PBC we can shape the total energy of the *full (electrical and mechanical) dynamics* of a doubly–fed induction generator used in power flow regulation tasks while, as reported in [1], with two stage IDA–PBC only the electrical energy could be shaped. Simulation results of these examples are presented to illustrate the performance improvement obtained with SIDA–PBC.

2 PBC with Simultaneous Energy Shaping and Damping Injection

We consider the problem of stabilization of an equilibrium point for nonlinear systems of the form

$$\dot{x} = f(x,t) + g(x)u \tag{1}$$

where $x \in \mathbb{R}^n$ is the state vector, $u \in \mathbb{R}^m, m < n$ is the control action and $g(x)$ is assumed full rank. In *two–stage IDA–PBC* this objective is achieved as follows, see [4, 13] for further details. First, decompose the control signal in two terms

$$u = u_{es} + u_{di} \tag{2}$$

where u_{es} is responsible of the energy–shaping stage and u_{di} injects the damping. Second, solve the key *matching equation*[1]

$$g^{\perp}(x)f(x,t) = g^{\perp}(x)J_d(x,t)\nabla H_d \tag{3}$$

for some functions

$$J_d : \mathbb{R}^n \times \mathbb{R} \to \mathbb{R}^{n\times n}, \quad H_d : \mathbb{R}^n \to \mathbb{R},$$

satisfying the skew–symmetry condition for the interconnection matrix

$$J_d(x,t) + J_d^{\top}(x,t) = 0, \tag{4}$$

and the equilibrium assignment condition for the desired total stored energy

$$x_{\star} = \arg\min H_d(x) \tag{5}$$

with $x_{\star} \in \mathbb{R}^n$ the equilibrium to be stabilized[2] and $g^{\perp}(x) \in \mathbb{R}^{(n-m)\times n}$ a full–rank left–annihilator of $g(x)$, that is, $g^{\perp}(x)g(x) = 0$ and rank $g^{\perp}(x) = n - m$.

As shown in [4] system (1) in closed-loop with the control (2), with

$$u_{es} = [g^{\top}(x)g(x)]^{-1}g^{\top}(x)\{J_d(x,t)\nabla H_d - f(x,t)\}. \tag{6}$$

[1] All vectors in the paper are *column* vectors, even the gradient of a scalar function denoted $\nabla_{(\cdot)} = \frac{\partial}{\partial(\cdot)}$. When clear from the context the subindex will be omitted.

[2] That is, $x_{\star}$ is a member of the set $\{\bar{x} \in \mathbb{R}^n \mid g^{\perp}(\bar{x})f(\bar{x},t) = 0,\ \forall t \in \mathbb{R}\}$.

yields a port–controlled Hamiltonian (PCH) system of the form

$$\begin{aligned} \dot{x} &= J_d(x,t)\nabla H_d + g(x)u_{di} \\ y &= g^\top(x)\nabla H_d. \end{aligned} \tag{7}$$

The system (7) without damping injection term is conservative, i.e., $\dot{H}_d = 0$, with $x_\star$ a stable equilibrium (with Lyapunov function $H_d(x)$). To *add dissipation* we feedback the passive output y, for instance, with

$$u_{di} = -K_{di}y, \; K_{di} = K_{di}^\top > 0,$$

to finally obtain the PCH system with dissipation

$$\begin{aligned} \dot{x} &= [J_d(x,t) - R_d(x)]\nabla H_d + g(x)v \\ y &= g^\top(x)\nabla H_d. \end{aligned} \tag{8}$$

where the damping matrix $R_d(x) = R_d^\top(x) \geq 0$ is defined by

$$R_d(x) = g(x)K_{di}g^\top(x),$$

and we have added a signal v to (2) to define the port variables. Since the new closed–loop system (with $v = 0$) satisfies $\dot{H}_d = -y^\top K_{di}y$, it can be proved (see for example Lemma 3.2.8 of [12] for the autonomous systems case) that the equilibrium $x_\star$ will now be asymptotically stable if it is detectable from y, i.e., if the implication $(y(t) \equiv 0 \; \Rightarrow \lim_{t\to\infty} x(t) = x_\star)$ is true.

Obviously, the key for the success of IDA–PBC is the solution of the matching equation (3). With the motivation of enlarging the class of systems for which this equation is solvable we propose in this paper to avoid the decomposition of the control into energy–shaping and damping injection terms. Instead, we suggest to carry out *simultaneously* both stages and replace (3), with the SIDA–PBC matching equations

$$g^\perp(x)f(x,t) = g^\perp(x)F_d(x,t)\nabla H_d, \tag{9}$$

to replace the constraint (4) by the *strictly weaker* condition

$$F_d(x,t) + F_d^\top(x,t) \leq 0, \tag{10}$$

and define the control as

$$u = [g^\top(x)g(x)]^{-1}g^\top(x)\{F_d(x,t)\nabla H_d - f(x,t)\}.$$

Since the set of skew–symmetric matrices is strictly contained in the set of matrices with negative semi–definite symmetric part, it is clear that the set of functions $\{f(x,t), g(x)\}$ for which (3)—subject to the constraint (4)—is solvable is strictly smaller than the set for which (9), subject to (10), is solvable.

Remark 1. There exists several techniques to solve the matching equations (3) (resp., (9)), with two extremes cases being the purely algebraic approach of [2] and the PDE approach of [4]. In the former $H_d(x)$ is *a priori fixed*, which makes (3) (resp., (9)) an algebraic equation that is solved for $J_d(x,t)$ (resp., $F_d(x,t)$)—subject to the constraint (4) (resp., (10)). On the other hand, in the latter $J_d(x,t)$ (resp., $F_d(x,t)$) is fixed making (3) (resp., (9)) a PDE that is solved for $H_d(x)$. We refer the interested reader to [6] for a detailed discussion on these, as well as other, methods of solution of the matching equations. In this paper we will adopt the algebraic approach.

Remark 2. Similarly to IDA–PBC, application of SIDA–PBC also yields a closed–loop PCH system of the form (8) with

$$J_d(x,t) = \tfrac{1}{2}[F_d(x,t) - F_d^\top(x,t)],$$

$$R_d(x,t) = \tfrac{1}{2}[F_d(x,t) + F_d^\top(x,t)].$$

Remark 3. To make IDA–PBC applicable to non–autonomous systems, which will be required in the induction motor application, we have presented above a slight variation of the method. Notice that the matrices J_d and R_d may depend explicitly on time. Clearly, their skew–symmetry and non–negativity properties must now hold uniformly in time as well.

3 Induction Motor Control Via SIDA–PBC

In this section we will show that the problem of output feedback torque control of induction motors is *not solvable* via two–stage IDA–PBC but it is solvable with SIDA–PBC. An interesting feature of our SIDA–PBC is that we establish here (Lyapunov) stability of a *given equilibrium* that generates the desired torque and rotor flux amplitude.

The standard two-phase model, which rotates at an arbitrary speed $\omega_s \in \mathbb{R}$, is given by [13]

$$\begin{aligned} \dot{x}_{12} &= -\left[\gamma I_2 + (n_p\omega + u_3)\mathcal{J}\right] x_{12} \\ &\quad + \alpha_1 \left(I_2 - T_r n_p \omega \mathcal{J}\right) x_{34} + \alpha_2 u_{12} \end{aligned} \tag{11}$$

$$\dot{x}_{34} = -(\frac{1}{T_r} I_2 + \mathcal{J} u_3) x_{34} + \frac{L_{sr}}{T_r} x_{12} \tag{12}$$

$$\dot{\omega} = \alpha_3 x_{12}^\top \mathcal{J} x_{34} - \frac{\tau_L}{J_m} \tag{13}$$

in which $I_2 \in \mathbb{R}^{2\times 2}$ is the identity matrix, $\mathcal{J} = -\mathcal{J}^T \in \mathbb{R}^{2\times 2}$, $x_{12} \in \mathbb{R}^2$ are the stator currents, $x_{34} \in \mathbb{R}^2$ the rotor fluxes, $\omega \in \mathbb{R}$ the rotor speed, $u_{12} \in \mathbb{R}^2$ are the stator voltages, $\tau_L \in \mathbb{R}$ is the load torque and $u_3 := \omega_s - n_p\omega$. All the parameters are positive and defined in the usual way. Notice that, as first pointed out in the control literature in [7], the signal u_3 acts as an additional control input. We will select u_3 to transform the periodic orbits of the system into constant equilibria.

We are interested in this paper in the problem of regulation of the motor torque and the rotor flux amplitude

$$y = \begin{bmatrix} y_1 \\ y_2 \end{bmatrix} = \begin{bmatrix} J_m \alpha_3 x_{12}^\top \mathcal{J} x_{34} \\ |x_{34}| \end{bmatrix}, \tag{14}$$

to some constant desired values $y_\star = \text{col}(y_{1\star}, y_{2\star})$, where $|\cdot|$ is the Euclidean norm, assuming that the only signals *available for measurement* are x_{12} and ω.

To solve this problem using (S)IDA–PBC it is necessary to express the control objective in terms of a *desired equilibrium*. In this sense, it can be shown that for the induction motor model (11)–(13) with output functions (14) and

$$u_3 = u_{3\star} := \frac{R_r}{n_p} \frac{y_{1\star}}{y_{2\star}^2}. \tag{15}$$

the set of assignable equilibrium points, denoted $\text{col}(\bar{x}_{12}, \bar{x}_{34}, \bar{\omega}) \in \mathbb{R}^5$, which are compatible with the desired outputs $y_\star$ is defined by

$$\begin{aligned} \bar{x}_{12} &= \frac{1}{L_{sr}} \begin{bmatrix} 1 & -\frac{L_r}{n_p} \frac{y_{1\star}}{y_{2\star}^2} \\ \frac{L_r}{n_p} \frac{y_{1\star}}{y_{2\star}^2} & 1 \end{bmatrix} \bar{x}_{34} \\ |\bar{x}_{34}| &= y_{2\star} \end{aligned} \tag{16}$$

with $\bar{\omega}$ arbitrary.

Remark 4. From (13) and (14) we see that to operate the system in equilibrium, $y_{1\star} = \tau_L$—hence, to define the desired equilibrium the load torque *needs to be known.* In practical applications, an outer loop PI control around the velocity error is usually added. The output of the integrator, on one hand, provides an estimate of τ_L while, on the other hand, ensures that speed also converges to the desired value as shown via simulations below. A scheme that removes this assumption has recently been proposed in [3].

As indicated in Remark 1, in this paper we will adopt the *algebraic approach* to solve the matching equations. To this end, we will consider a quadratic in errors energy function of the form

$$H_d(x) = \frac{1}{2}(x - x_\star)^\top P (x - x_\star), \tag{17}$$

with $P = P^\top > 0$ a matrix to be determined. Moreover, the problem formulation is simplified by using the generic symbol $F(x,t)$ to denote either $J_d(x,t)$ or $F_d(x,t)$, and identifying the IDA and SIDA approaches by imposing either $F(x,t) + F^\top(x,t) = 0$ or $F(x,t) + F^\top(x,t) \leq 0$, respectively.

Since we are interested here in torque control, and this is only defined by the stator currents and the rotor fluxes, its regulation can be achieved applying IDA–PBC *to the electrical subsystem only.* Boundedness of ω will be established in a subsequent analysis.

If the electrical subsystem (11), (12), with $u_3 = u_{3\star}$ and $u = u_{12}$, is written in the form (1), then, selecting $g^{\perp} = \left[0_{2\times 2}\ I_2 \right]$, it is possible to notice that the matching equations (3) and (9) concern only the third and fourth rows of $f(x,t)$ and they take the form

$$-(\frac{1}{T_r}I_2 + \mathcal{J}u_{3\star})x_{34} + \frac{L_{sr}}{T_r}x_{12} = [F_3(x,t) \quad F_4(x,t)]\, P(x - x_\star), \tag{18}$$

where, to simplify the notation, we partition $F(x,t)$ into $F_i \in \mathbb{R}^{2\times 2}$, $i = 1,\ldots,4$, sub-matrices.

On the other hand, considering the constraint imposed by the possibility of measuring only x_{12} and ω, from (6) it is possible to see that the control can be written, factoring the components that depend on the unmeasurable quantity x_{34}, as

$$u_{12} = \hat{u}_{12}(x_{12},\omega) + \frac{1}{\alpha_2}S(x,t)x_{34}$$

with $\hat{u}_{12}(x_{12},\omega)$ given in (21). Hence, it is clear that to verify the output feedback condition it must be satisfied that

$$S(x,t) := \alpha_1\left[T_r n_p \omega(t)\mathcal{J} - I_2\right] + F_1(x,t)P_2 + F_2(x,t)P_3 = 0, \tag{19}$$

where we have partitioned the symmetric matrix P into $P_i \in \mathbb{R}^{2\times 2}$, $i = 1,2,3$, sub-matrices.

In order to show that the energy–shaping problem is *not solvable*, it is possible to establish that condition $F(x,t) + F^{\top}(x,t) = 0$ is equivalent to

$$G = P_3^{-1}(-G^{\top})P_3.$$

where

$$G := \frac{1}{\lambda}\left[\frac{L_{sr}}{T_r}P_3 + P_1\left(\alpha_1 I_2 - \lambda\beta_2\mathcal{J}\right)\right]$$

with $\lambda \in \mathbb{R}$, $\lambda \neq 0$, $\beta_2 \in \mathbb{R}$ and P_3 a full rank matrix.

Consequently, G must be similar to $-G^{\top}$, and both necessarily have the same eigenvalues. A necessary condition for the latter is that $\mathrm{trace}(G) = 0$, that clearly is not satisfied.

On the other hand, it can be shown that $F_2(t) = \alpha_1\left[I_2 - T_r n_p \omega(t)\mathcal{J}\right]P_3^{-1}$, $F_3 = \frac{L_{sr}}{T_r}P_1^{-1}$, $F_4 = -\left(\frac{1}{T_r}I_2 + u_{3\star}\mathcal{J}\right)P_3^{-1}$ and $P_2 = 0$, with $F_1(x,t), P_1, P_3$ free, provide a solution to (18) and (19). Moreover, if $P_1 = \frac{L_{sr}}{T_r}I_2$, $P_3 = \alpha_1 I_2$ and $F_1(t) = -\mathcal{K}(\omega(t))$, with

$$\mathcal{K}(\omega(t)) > \frac{L_{sr}}{L_s L_r - L_{sr}^2}\left[1 + \frac{1}{4}(T_r n_p \omega(t))^2\right] I_2 \tag{20}$$

then $F(t) + F^{\top}(t) < 0$.

The final part of the design is the explicit definition of the resulting controller, given by $u_3 = \frac{R_r}{n_p}\frac{y_{1\star}}{y_{2\star}^2}$ and $u_{12} = \hat{u}_{12}(x_{12}, \omega)$ with

$$\begin{aligned}\hat{u}_{12}(x_{12}, \omega) =& \frac{1}{\alpha_2}\left[\gamma I_2 + (n_p\omega + u_{3\star})\mathcal{J}\right] x_{12} - \frac{\alpha_1}{\alpha_2}\left(I_2 - T_r n_p \omega \mathcal{J}\right) x_{34\star} \\ & - \frac{L_{sr}}{\alpha_2 T_r}\mathcal{K}(\omega)(x_{12} - x_{12\star}) \end{aligned} \tag{21}$$

and $\mathcal{K}(\omega)$ satisfying (20), which guarantees that the equilibrium $x_\star$ is *globally exponentially stable* while ω remains bounded.

The performance of the proposed SIDA–PBC was investigated by simulations using the motor parameters reported in [7]. The rotor flux equilibrium value was set to $x_{34\star} = \mathrm{col}\,(\beta, 0)$ with $\beta = 2$, while $x_{12\star}$ were computed according to (16). In the experiment, with the motor at standstill and a startup zero load torque, a profile for this latter variable was considered going first to $\tau_L = 20Nm$ and later on to $\tau_L - 40Nm$. In Figure 1 it is shown how the generated torque regulation objective is achieved.

The second experiment was aimed to illustrate the claim stated in Remark 4 regarding the estimation of the load torque with speed control purposes. In this sense, the control input u_3 was set to

$$u_3 = \hat{u}_{3\star} = R_r \frac{\hat{y}_{1\star}}{y_{2\star}^2}$$

where the estimate of the load torque is obtained as the output of a PI controller, defined over the speed error between the actual and the desired velocities, of the form

$$\hat{y}_{1\star}(t) = k_p\left(\omega(t) - \omega_\star\right) + k_i \int_0^t \left(\omega(s) - \omega_\star\right) ds$$

Figure 2 shows the rotor speed behavior when the desired velocity is initially $\omega_\star = 100rpm$ and at $t = 50sec$ it is changed to $\omega_\star = 150rpm$. In this simulation it was considered $\tau_L = 10Nm$, $k_i = -.1$ and $k_p = -1$. All the other parameters were the same than in the first experiment.

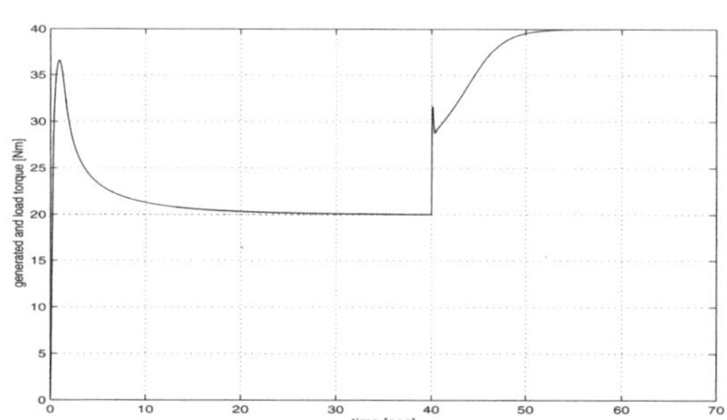

Fig. 1. Generated and load torque of the IM

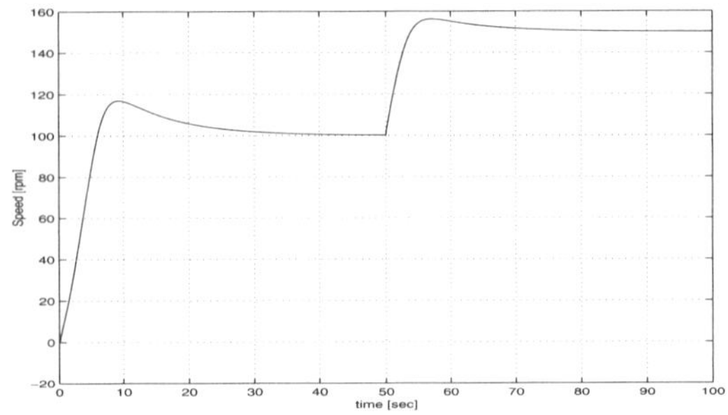

Fig. 2. Speed behavior with estimated load torque

4 Total Energy–Shaping of a Doubly–Fed Induction Generator

The second considered example in this paper is related with the control of a doubly-fed induction machine (DFIM) [9]. In this case the device acts as an energy–switching device between a local energy storing element (a flywheel) and the electrical power network. The control objective is to change the direction of the power flow (towards or from the flywheel) depending on the load demand. In [1] the equilibria associated to these regimes is stabilized with an IDA–PBC that shapes the *electrical energy*, treating the mechanical dynamics as a cascaded subsystem. The purpose of this section is to show that using SIDA–PBC it is possible to shape the energy function of the *complete* system dynamics, resulting in a controller with improved power–flow regulation performance due to the possibility of considering a fast response of the mechanical speed. To the best of our knowledge, this is the first control algorithm for this class of systems that provides for this additional degree of freedom.

We consider the configuration for the DFIM studied in [1] where a representation in the dq framework rotating at the (constant) angular speed of the AC source (ω_s) is assumed. The energy function of the overall system is $H(z) = \frac{1}{2}z^\top \mathcal{L}^{-1} z$, and the model is given by

$$\dot{z} = [J(i_s, \omega) - R]\nabla H + \begin{bmatrix} v_s \\ O_{2\times 1} \\ \tau_L \end{bmatrix} + \begin{bmatrix} O_{2\times 2} \\ I_2 \\ O_{1\times 2} \end{bmatrix} u \tag{22}$$

where

$$\mathcal{L} = \begin{bmatrix} L_s I_2 & L_{sr} I_2 & O_{2\times 1} \\ L_{sr} I_2 & L_r I_2 & O_{2\times 1} \\ O_{1\times 2} & O_{1\times 2} & J_m \end{bmatrix} = \mathcal{L}^\top > 0$$

is the generalized inductance matrix, $v_s \in \mathbb{R}^2, \tau_L \in \mathbb{R}$ are, respectively, the stator voltage and the external mechanical torque, which are *constant*, $u \in \mathbb{R}^2$ are the rotor *control* voltages, $z = \mathrm{col}(\lambda_s, \lambda_r, J_m\omega) \in \mathbb{R}^5$, where $\lambda_s, \lambda_r \in \mathbb{R}^2$ are the stator and rotor fluxes, respectively, and $\omega \in \mathbb{R}$ is the mechanical speed. The (skew–symmetric) structure and damping matrices are

$$J(i_s, \omega) := \begin{bmatrix} -\omega_s L_s \mathcal{J} & -\omega_s L_{sr} \mathcal{J} & O_{2\times 1} \\ -\omega_s L_{sr} \mathcal{J} & -(\omega_s - \omega) L_r \mathcal{J} & L_{sr} \mathcal{J} i_s \\ O_{1\times 2} & L_{sr} i_s^\top \mathcal{J} & 0 \end{bmatrix},$$

$$R := \begin{bmatrix} R_s I_2 & O_{2\times 2} & O_{2\times 1} \\ O_{2\times 2} & R_r I_2 & O_{2\times 1} \\ O_{1\times 2} & O_{1\times 2} & B_r \end{bmatrix} > 0,$$

All machine parameters are defined in the usual way. Also note that the vector $\chi = \mathrm{col}(i_s, i_r, \omega) \in \mathbb{R}^{5\times 5}$, where $i_s, i_r \in \mathbb{R}^2$ are the stator and rotor currents, respectively, satisfies $z = \mathcal{L}\chi$.

Clearly, the fixed point equations for (22) are given by $z_\star = \mathcal{L}\chi_\star$, with $\chi_\star := \mathrm{col}(i_{s\star}, i_{r\star}, \omega_\star)$ the solutions of

$$\begin{aligned} -(\omega_s L_s \mathcal{J} + R_s I_2) i_{s\star} - \omega_s L_{sr} \mathcal{J} i_{r\star} + v_s = 0 \\ L_{sr} i_{s\star}^\top \mathcal{J} i_{r\star} - B_r \omega_\star + \tau_L = 0, \end{aligned} \tag{23}$$

As discussed in [1], the direction of the power flow can be regulated commuting between two controllers that stabilize two different equilibrium points.

In order to obtain the controller that stabilizes the desired equilibrium, as done in Section 3 we will design our SIDA–PBC adopting the algebraic approach. For, we fix the desired energy function as

$$H_d(z) = \frac{1}{2}(z - z_\star)^\top P (z - z_\star),\ P = P^\top > 0. \tag{24}$$

Thus SIDA–PBC design reduces to finding a matrix $F_d(z)$ such that the right–hand term of (22) equals $F_d(z)P(z - z_\star)$ and verifying $F_d(z) + F_d^\top(z) \leq 0$. To simplify the solution we restrict P to be diagonal while

$$F_d(z) = \begin{bmatrix} F_{11}(z) & F_{12}(z) & O_{2\times 1} \\ F_{21}(z) & F_{22}(z) & F_{23}(z) \\ F_{31}^\top(z) & F_{32}^\top(z) & F_{33}(z) \end{bmatrix},$$

It can be shown that if $F_{11} = -\frac{1}{p_s}\left(\omega_s \mathcal{J} + \frac{L_r}{\mu} R_s I_2\right)$ and $F_{12} = \frac{L_{sr}}{p_r \mu} R_s I_2$ with $\mu := L_s L_r - L_{sr}^2 > 0$, with $F_{31} = \frac{L_{sr}}{p_s \mu} \mathcal{J} \lambda_{r\star}$, $F_{32}(z) = -\frac{L_{sr}}{p_r \mu} \mathcal{J} \lambda_s$, $F_{33} = -\frac{B_r}{p_\omega J_m}$, while

$$F_{21} = -F_{12},\ F_{23}(z) = -F_{32}(z),\ F_{22} = -\frac{k_r}{2p_r} I_2 < 0$$

yields

$$F_d(z) + F_d^\top(z) = \begin{bmatrix} -\frac{2L_r R_s}{p_s \mu} I_2 & O_{2\times 2} & \frac{L_{sr}}{p_s \mu} \mathcal{J} \lambda_{r\star} \\ O_{2\times 2} & -\frac{k_r}{p_r} I_2 & O_{2\times 1} \\ -\frac{L_{sr}}{p_s \mu} \lambda_{r\star}^\top \mathcal{J} & O_{1\times 2} & -\frac{2B_r}{p_\omega J_m} \end{bmatrix}. \tag{25}$$

A simple Schur's complement analysis establishes that $F_d(z) + F_d^\top(z) < 0$ if and only if the *free parameters* p_s and p_ω satisfy

$$p_s > \left(\frac{J_m L_{sr}^2}{4 B_r L_r R_s \mu} |\lambda_{r\star}|^2 \right) p_\omega . \tag{26}$$

Once we have solved the SIDA–PBC matching equations, the design is completed computing the controller which in this case is given as the static feedback control

$$\begin{aligned} u =& R_r i_r + (\omega_s - \omega)\mathcal{J}(L_{sr} i_s + L_r i_r) \\ & -k_s(L_s \tilde{i}_s + L_{sr}\tilde{i}_r) - k_r(L_{sr}\tilde{i}_s + L_r \tilde{i}_r) + k_\omega \mathcal{J} \lambda_s \tilde{\omega} \end{aligned} \tag{27}$$

where $k_r > 0$, $k_\omega > 0$ and $k_s > \frac{L_{sr}^2}{4 B_r L_r \mu} |\lambda_{r\star}|^2 k_\omega$. Considering this control law, the equilibrium $z_\star$ is *globally exponentially stable.*

In spite of the remarkable stability properties of the proposed scheme, it can be seen that (27) has an strong dependence on R_r, which is in general an uncertain parameter. With the aim of robustifying this control law it is possible to develope an adaptive version in the following way.

Replacing (27) by

$$\begin{aligned} u =& (\omega_s - \omega)\mathcal{J}(L_{sr} i_s + L_r i_r) \\ & -k_s(L_s \tilde{i}_s + L_{sr}\tilde{i}_r) \quad k_r(L_{sr}\tilde{i}_s + L_r \tilde{i}_r) + k_\omega \mathcal{J} \lambda_s \tilde{\omega} + \hat{R}_r i_r , \end{aligned} \tag{28}$$

where $\hat{R}_r$ is an estimate of R_r that we have to generate on-line, we define the parameter error $\tilde{R}_r = \hat{R}_r - R_r$. The closed-loop system has the form

$$\dot{z} = F \nabla H_d + \tilde{R}_r B i_r ,$$

where $B = [O_{2\times 2},\, I_2,\, O_{1\times 2}]^T$.

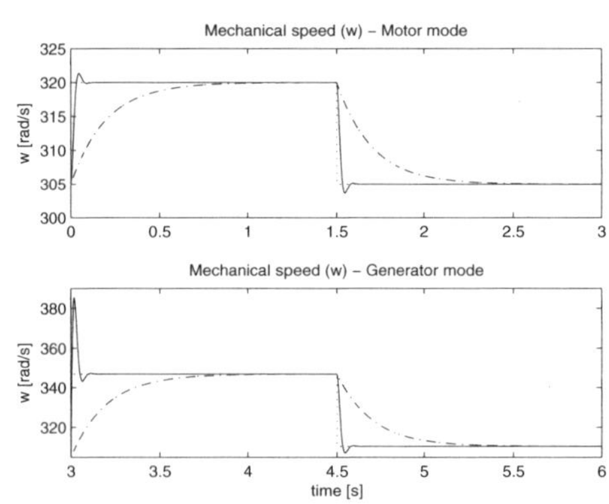

Fig. 3. Mechanical speed, ω, for SIDA–PBC (continuous line) and IDA–PBC (dashed line)

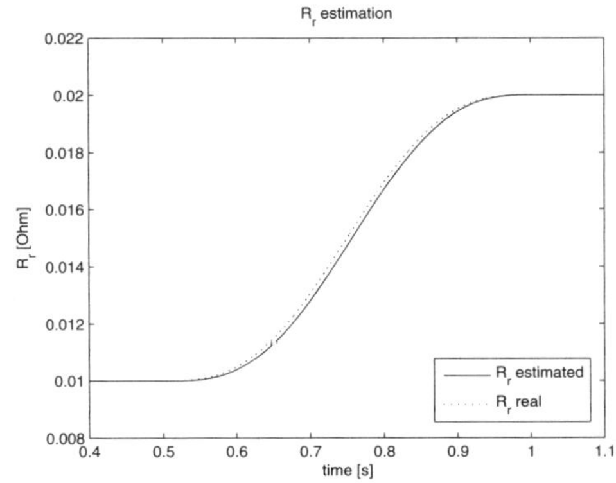

Fig. 4. Rotor resistance estimation, $\hat{R}_r$

To complete the design we propose a Lyapunov function

$$W(z, \tilde{R}_r) = H_d + \frac{1}{2k_a}\tilde{R}_r^2$$

where $k_a > 0$ is the adaptation gain. Its derivative yields, selecting $\dot{\tilde{R}}_r = -k_a(\nabla H_d)^\top B i_r$, that

$$\dot{W} = -(\nabla H_d)^\top F \nabla H_d \leq 0,$$

which proves the stability of the adaptive system. The resulting adaptation law is

$$\dot{\hat{R}}_r = -k_a(L_{sr}\tilde{i}_s^{\top} - L_r\tilde{i}_r^{\top})p_r i_r.$$

The usefulness of the controller (27) was illustrated by some simulations using the DFIM parameters of [1]. The controller parameters were selected as $k_s = 1000$, $k_r = 100$ and $k_\omega = 0.01$. In Figure 3 we compare the speed behavior of the new SIDA-PBC with the IDA–PBC reported in [1]—that shapes only the

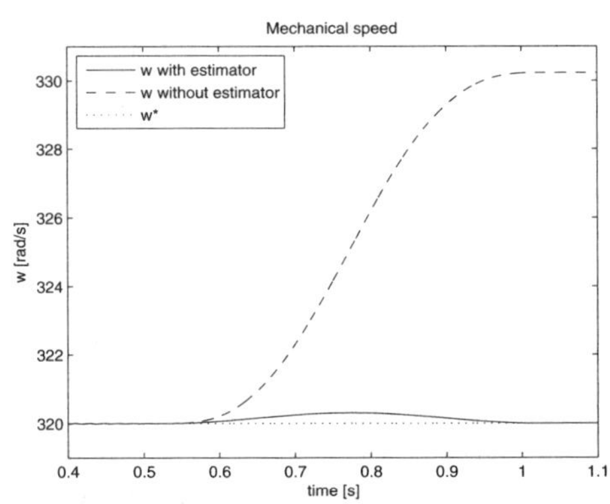

Fig. 5. Mechanical speed, with and without estimator

electrical energy. As can be noticed, SIDA–PBC achieves a much faster speed response.

To evaluate the adaptive controller (28) the same experiment as before was repeated but now varying the rotor resistance parameter R_r. At $t = 0.5s$ the value of R_r of the model is smoothly increased to $R_r = 0.02$, simulating temperature effects. Figures 4 and 5 show the estimation behavior of $\hat{R}_r$ and the dynamics of the mechanical speed, respectively.The convergence of the estimated value of $\hat{R}_r$ to the real value (Figure 4) ensures that the performance of the ideal system is recovered (Figure 5).

5 Conclusions

We have presented an extension of the highly successful IDA-PBC methodology, called SIDA–PBC, where the energy–shaping and damping injection tasks are not performed sequentially, but simultaneously. In this way we enlarge the class of systems that can be stabilized using PBC and, furthermore, through the consideration of a broader set of desired damping matrices, we provide the designer with more tuning knobs to improve performance.

This new idea has been applied to solve the long standing problem of IDA–PBC of induction motors, that turns out to be unsolvable with a two stage design. Also, by avoiding the classical nested–loop control configuration prevalent in electromechanical systems, we have been able to improve the mechanical response of a DFIM, working both as a motor and a generator. Experimental validation of the two control algorithms is currently being terminated and will be reported in the near future.

Acknowledgments

This work has been done in the context of the European sponsored project Geoplex with reference code IST-2001-34166. Further information is available at http://www.geoplex.cc. The work of Carles Batlle, Arnau Dòria-Cerezo and Gerardo Espinosa has been (partially) supported by the spanish project DPI2004-06871-CO2-02, the European Community Marie Curie Fellowship (in the framework of the European Control Training Site) and CONACyT (51050Y) and DGAPA-UNAM (IN103306), respectively.

References

1. Batlle, C., A. Dòria-Cerezo and R. Ortega (2005) Power flow control of a doubly-fed induction machine coupled to a flywheel European Journal of Control 11:3 209–221
2. Fujimoto, K. and T. Sugie (2001) Canonical transformations and stabilization of generalized hamiltonian systems Systems and Control Letters 42:3 217–227
3. Karagiannis, D., A. Astolfi, R. Ortega and M. Hilairet (2005) A nonlinear tracking controller for voltage-fed induction motors with uncertain load torque. Internal report, LSS Supelec, France

4. Ortega, R., A. van der Schaft, B. Maschke and G. Escobar (2002) Interconnection and Damping Assignment Passivity–based Control of Port–controlled Hamiltonian Systems AUTOMATICA 38:4 585–596
5. Ortega, R., A. Loria, P.J. Nicklasson and H. Sira-Ramírez (1998) Passivity–based Control of Euler–Lagrange Systems. Communications and Control Engineering. Springer-Verlag, Berlin
6. Ortega, R. and E. Garcia-Canseco (2004) Interconnection and Damping Assignment Passivity–Based Control: A Survey European Journal of Control 10:432–450
7. Ortega, R. and G. Espinosa (1993) Torque regulation of induction motors AUTOMATICA 47:8 621–633
8. Ortega, R., M. Spong, F. Gomez and G. Blankenstein (2002*a*) Stabilization of underactuated mechanical systems via interconnection and damping assignment IEEE Trans. Automatic Control 47:8 1218–1233
9. Peresada, S., A. Tilli and A. Tonelli (2004) Power control of a doubly fed induction machine via output feedback Control Engineering Practice 12:41–57
10. Sepulchre, R., M. Janković and P. Kokotović (1997) Constructive Nonlinear Control. Springer-Verlag, London
11. Takegaki, M. and S. Arimoto (1981) A new feedback for dynamic control of manipulators Trans. of the ASME: Journal of Dynamic Systems, Measurement and Control 102:119–125
12. van der Schaft, A. (2000) L_2-Gain and Passivity Techniques in Nonlinear Control. 2nd edition. Springer-Verlag, London
13. Krause, P.C., O. Wasynczuk and S.D. Sudhoff (1995) Analysis of Electric Machinery. IEEE Press, USA.

An Internal Model Approach to Implicit Fault Tolerant Control for Port-Hamiltonian Systems

Luca Gentili, Andrea Paoli, and Claudio Bonivento

CASY - DEIS University of Bologna, viale Risorgimento 2, 40136, Bologna, Italy
lgentili,apaoli,cbonivento@deis.unibo.it

In this paper an internal model based approach to implicit Fault Tolerant Control for port-Hamiltonian systems is presented: the main idea is to cast the problem into a regulation problem in presence of input disturbances representing exogenous effects of possible faults; this can be solved following an adaptive internal model based approach. The theoretical machinery exploited is specialized for the energy-based port-Hamiltonian formalism in order to prove the global asymptotical stability of the solution. Finally an application example is presented in order to deeply point out the effectiveness of the design procedure presented: a Fault Tolerant Control problem is solved for a magnetic levitation system affected by periodic voltage disturbances.

1 Introduction

In large systems, every component provides a certain function and the overall system works satisfactorily only if all components provide the service they are designed for. Therefore, a fault in a single component usually changes the performances of the overall system. For this reason faults have to be detected as quickly as possible and decisions that stop the propagation of their effects have to be made. A weak element in this framework are control loops. In fact automated systems are vulnerable to faults such as defects in sensors, in actuators and in controllers, which can cause undesired reactions and consequences as damage to technical parts of the plant, to personnel or to the environment. In this framework, the design of a Fault Tolerant Control (FTC) architecture is of crucial importance. FTC systems aims at *adapting* the control strategy to the presence of the fault in order to achieve prescribed performances also for the faulty system. See [2] for an exhaustive overview about FTC systems.

The most common approach in dealing with such a problem is to split the overall design in two distinct phases. The first phase addresses the so-called "Fault Detection and Isolation" (FDI) problem, which consists in designing a filter which, by processing input/output data, is able to detect the presence of an incipient fault and to isolate it from other faults and/or disturbances (see [7], [15] and the reference therein). Once the FDI filter has been designed, the

F. Bullo et al. (Eds.): Lag. & Hamil. Methods for Nonlin. Ctrl., LNCIS 366, pp. 171–182, 2007.
springerlink.com

second phase usually consists of designing a supervisory unit which, on the basis of the information provided by the FDI filter, reconfigures the control so as to compensate for the effect of the fault and to fulfill performances constraint.

In [4] authors introduce a new approach to FTC: the so-called *implicit FTC*. In this approach the control reconfiguration does not pass through an explicit FDI design but, indeed, is achieved by a proper design of a dynamic controller which is *implicitly* fault tolerant to all the possible faults whose model is embedded in the regulator through an *internal model*. In [3] and [18] this approach has been specialized for a robot manipulator which is forced to track a desired trajectory even in presence of malfunctioning causing periodic torque disturbances acting on joints.

In this paper a framework to deal with implicit FTC problems for port-Hamiltonian systems is presented. The main idea is to cast the FTC problem into a regulation and input disturbance suppression problem and to solve this problem with an adaptive internal model based regulator: this framework is justified by the observation that in many case the resulting effect of the occurrence of a fault is the arising of constant or periodical spurious disturbances superimposed to the control input variables (see [20], [6], [19], [25], [21]).

The theoretical machinery exploited in order to prove the global asymptotical stability of the solution exposed is the nonlinear regulation theory, specialized for the energy-based port-Hamiltonian formalism in order to take advantage of its peculiar properties. In [12] port-Hamiltonian systems were introduced as a generalization of Hamiltonian systems, described by Hamilton's canonical equations, which may represent general physical systems (e.g. mechanical, electric and electro-mechanical systems, nonholomic systems and their combinations)(see [17] for further references). Hence this formalism appears to perfectly suit in order to deal with general FTC design problems.

In Section 2, an implicit FTC problem is considered and cast into a problem of exogenous input disturbance suppression for a generic port-Hamiltonian system. The regulation problem is stated and an adaptive internal model based controller able to globally asymptotically solve this problem is introduced under certain assumption regarding the system into account.

It is worth to remark that the control algorithm presented in this section is able to deal with faults affecting the plant whose effects can be modeled as functions (of time) within a finitely-parametrized family of such functions: i.e. exogenous constant and sinusoidal disturbances characterized by unknown amplitude, phase and frequency; the hypothesis of not perfect knowledge of the characteristic frequencies introduces a complex issue to deal with: in the last years this problem has been pointed out and addressed using different design techniques (see [16], [11], [22], [13], [23] and references therein). In this paper a solution, relying on simple Lyapunov based consideration is presented.

In order to enlighten the practical effectiveness of the implicit FTC solution presented, in Section 3, the problem is solved for a magnetic levitation system affected by periodic voltage disturbances.

2 Input Disturbance Suppression Problem and Its Adaptive Solution

In this section we present a generic control algorithm able to solve an implicit FTC problem for port-Hamiltonian systems. The class of faults considered consists in all possible faults whose effects reflect into the arise of additive actuators disturbances modeled as exogenous input signals belonging to the class of constant and sinusoidal disturbances characterized by unknown amplitude, phase and frequency. The design procedure will define an implicit Fault Tolerant controller based on the internal model principle able to reject perfectly the disturbances due to the occurrence of a fault: the reconfiguration phase is hence implicitly attained while, testing the internal state of the controller, is it possible to perform the FDI phase.

Consider a generic port-Hamiltonian system with an exogenous disturbance $\delta(t)$ acting through the input channel:

$$\dot{x} = (J(x) - R(x))\frac{\partial H}{\partial x} + gu - g\delta \tag{1}$$

where $x \in \mathbb{R}^n$, $u \in \mathbb{R}^m$, $H : \mathbb{R}^n \to \mathbb{R}$ is the energy function (Hamiltonian function), $J(x)$ is a skew symmetric matrix ($J(x) = -J^{\mathrm{T}}(x)$), $R(x)$ is a symmetric semi-positive definite matrix ($R(x) = R^{\mathrm{T}}(x)$) and $g \in \mathbb{R}^{n\times m}$.

The disturbance Γw is generated by an exosystem defined by

$$\begin{cases} \dot{w} = Sw \\ \delta = \Gamma w \end{cases} \tag{2}$$

with $s = 2k + 1$, $w \in \mathbb{R}^s$; $\Gamma \in \mathbb{R}^{s\times m}$ is a known matrix and S is defined by

$$S = \mathrm{diag}\{S_0, S_1, \ldots, S_k\} \tag{3}$$

where $S_0 = 0$,

$$S_i = \begin{bmatrix} 0 & \omega_i \\ -\omega_i & 0 \end{bmatrix} \qquad \omega_i > 0 \qquad i = 1, \ldots, k\,. \tag{4}$$

The initial condition of the exosystem is $w(0) \in \mathcal{W}$, with $\mathcal{W} \subseteq \mathbb{R}^s$ bounded compact set.

In this discussion the dimension s of matrix S is known but all characteristic frequencies ω_i will be unknown but ranging within known compact sets, i.e. $\omega_i^{\min} \leq \omega_i \leq \omega_i^{\max}$.

In this set up the lack of knowledge of the exogenous disturbance reflects into the lack of knowledge of the initial state $w(0)$ of the exosystem and of the characteristic frequencies. For instance, any disturbance obtained by linear combination of a constant term and sinusoidal signals with unknown frequencies, amplitudes and phases are considered.

The problem to address is to regulate the system to the origin in despite of the presence of exogenous disturbances due to faults, hence obtaining a fault tolerant behavior for the system into account; it will be remarked later that the solution presented is even able to supply estimates of the disturbances occurred after the arise of a fault, solving the connected FDI problem.

The problem can be easily cast as a regulation problem (see [10], [5], [9], [1]) complicated by the lack of knowledge of the matrix S: the controller to be designed embeds an *internal model* of the exogenous disturbances and will be augmented by an adaptive mechanism in order to estimate the characteristic frequencies of the disturbances.

The internal model unit is designed according to the procedure proposed in [13] (*canonical internal model*). Given a symmetric, negative definite Hurwitz matrix F and any matrix G such that the couple (F, G) is controllable, denote by Y the unique nonsingular matrix solution of the Sylvester equation[1]

$$YS - FY = G\Gamma \tag{5}$$

and define the matrix $\Psi := \Gamma Y^{-1}$.

As matrix S is unknown, the solution Y of the Sylvester equation (5), and hence matrix Ψ are not known; for this reason an adaptation law is designed. The estimated term is $\hat{\Psi}$ and the adaptation error is easily defined as

$$\tilde{\Psi}_{ij} = \hat{\Psi}_{ij} - \Psi_{ij}\,, \quad \begin{array}{c} i = (1, \cdots, m) \\ j = (1, \cdots, s) \end{array} .$$

Considering now two vectors containing every element of matrix Ψ and $\hat{\Psi}$

$$\begin{aligned} \Phi &= \left(\Psi_{11} \cdots \Psi_{1s} \cdots \Psi_{m1} \cdots \Psi_{ms}\right)^{\mathrm{T}} \\ \hat{\Phi} &= \left(\hat{\Psi}_{11} \cdots \hat{\Psi}_{1s} \cdots \hat{\Psi}_{m1} \cdots \hat{\Psi}_{ms}\right)^{\mathrm{T}} \end{aligned}$$

and defining $\tilde{\Phi} = \hat{\Phi} - \Phi$ it is possible to state the main result in the following proposition.

Proposition 1. *Consider the port-Hamiltonian system (1), affected by exogenous signals generated by the autonomous system (2), (3), (4).*
Suppose that the following hypotheses hold:

H1: there exists two numbers $\eta_x \in \mathbb{R}^-$ and $\eta_\Psi \in \mathbb{R}$ and a matrix $Q \in \mathbb{R}^{n\times n}$ such that for all $\chi \in \mathbb{R}^s$ the following holds

$$-\frac{\partial^{\mathrm{T}} H}{\partial x} R(x) \frac{\partial H}{\partial x} + \frac{\partial^{\mathrm{T}} H}{\partial x} g \Psi \chi \leq \eta_x \|Qx\|^2 + \eta_\Psi \|Qx\| \|\chi\|\,; \tag{6}$$

H2: the origin of the auxiliary system

$$\begin{aligned} \dot{x} &= (J(x) - R(x)) \frac{\partial H}{\partial x} + g\sigma \\ \dot{\sigma} &= -g^{\mathrm{T}} \frac{\partial H}{\partial x} \end{aligned}$$

[1] Existence and uniqueness of the matrix Y follow from the fact that S and F have disjoint spectrum. The fact that Y is nonsingular can be easily proved using observability of the pairs (S, Γ) and controllability of the pair (F, G).

is the largest invariant set characterized by

$$\frac{\partial^{\mathrm{T}} H}{\partial x} R(x) \frac{\partial H}{\partial x} = 0 \,.$$

Then the controller (adaptive internal model unit):

$$\begin{cases} \dot{\xi} = (F + G\hat{\Psi})\xi - FAx + A(J(x) - R(x))\dfrac{\partial H}{\partial x} - Ag\hat{\Psi}Ax \\ \dot{\hat{\Phi}} = -\Pi(x,\xi)^{\mathrm{T}} \dfrac{\partial H}{\partial x} \\ u = \hat{\Psi}\xi - \hat{\Psi}Ax \,, \end{cases} \tag{7}$$

where A is chosen according to $Ag = G$, $\Pi(x,\xi)$ is a suitably defined updating term designed such that

$$\Pi(x,\xi)\tilde{\Phi} = g\tilde{\Psi}(\xi - Ax) \,,$$

F is a suitably defined symmetric Hurwitz matrix and G is a suitably defined matrix such that the couple (F, G) is controllable, is able to asymptotically stabilize the origin of system (1), zeroing the effect of the exogenous disturbances.

Sketch of the proof. Considering the Hamiltonian function $H(x)$ as Lyapunov function, it is possible to use hypothesis H1 and simple Young's inequality based consideration to prove that $\dot{H}(x) \leq 0$. LaSalle invariance principle and hypothesis H2 make then possible to state the asymptotic convergence of the state x to the origin. In next section this proof will be presented in extended form for the particular system considered.

Remark 1. It is worth to remark that, though the main hypotheses H1 and H2 could appear a bit conservative, they refer to the a system of the form (1) that could be not the original plant but the port-Hamiltonian formulation of the original system already controlled to attain specific tasks; this a priori control action could be suitably designed such that the resulting system satisfies conditions H1 and H2. In particular, it could be easily shown that hypothesis H1 is always verified if the system into account is characterized by a quadratic Hamiltonian function. To enlighten the effectiveness of the property remarked here, in section 3 a magnetic levitation system is taken into account: the original system does not satisfy both conditions but, with a suitably defined control action able to perform even a regulation objective and a change of coordinates, the resulting system is in the form (1) satisfying H1 and H2; hence the control algorithm introduced can be used to solve the FTC problem. Moreover a control procedure able to impose particular shape and properties to the controlled system is the well known IDA-PBC (see [24], [14] and reference therein for a survey about this control strategy): this control strategy is able to design a suitable port-Hamiltonian controller such that the interconnected system (original plant and IDA-PBC controller) results in the form of a desired objective port-Hamiltonian system. It is easy to realize that one of the characteristic step of the IDA-PBC control strategy is the definition of a target system, usually described by the classical port-Hamiltonian structure, characterizing the resulting dynamic after that the controller is designed and connected: this target system could be assumed to be of the form (1), imposing moreover that all the assumption imposed by Proposition 1 are satisfied.

An interesting research topic, that is still in development, regards the condition to impose to the original system, and to the original problem, such that, for example, an IDA-PBC control strategy makes it possible to cast the problem in the framework presented.

Remark 2. It is interesting to see that, thanks to Proposition 1, the FDI phase can be carried out by testing the state of the internal model unit which automatically activates to offset the presence of the fault. Hence this phase, which is usually the starting point for the design of the FTC system, is postponed to the reconfiguration phase. More in detail it is possible to show that the control action u is composed by a stabilizing term asymptotically vanishing ($\hat{\Psi} Ax$) and by an internal model term ($\hat{\Psi}\xi$) that asymptotically rejects the disturbance: hence testing the internal model state ξ and the adaptation term $\hat{\Psi}$ it is possible to reconstruct asymptotically the fault effect Γw acting on the system.

In the following section, an interesting example is presented to point out the main properties of the FTC algorithm presented above.

3 An Example: Magnetic Levitation System

In this section we consider a typical example of electro-mechanical system such as the magnetic levitation plant shown in fig. 1. A typical problem, in applications where magnetic levitation principle is used for low-friction rotating machinery, is the arising of sinusoidal disturbances superimposed to the electrical variables when some faults such as any radial imbalance in the rotating components occur (see [25], [19], [20], [21], [6]). Moreover, magnitude and frequency of such disturbances are directly related to severity of the imbalance and rotational speed

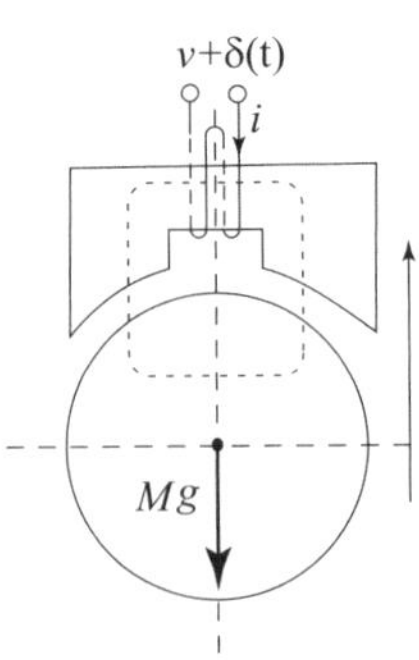

Fig. 1. A sphere in a vertical magnetic field

and therefore unknown. Hence it is very interesting to show how it is possible to apply the general framework illustrated in section 2 to a maglev system subject to an exogenous disturbance $\delta(t)$ superimposing to the control voltage.

In this section we are going to introduce the port-Hamiltonian model of the system (see [8]) and to design a suitable control action in order to solve a regulation objective while satisfying the requirements of Proposition 1; this make it possible to design an adaptive internal model based controller to avoid the effect of the disturbances.

Calling x_1 the vertical position of the sphere (whose mass is defined by M), x_2 its velocity, $x_3 = L(x_1)i$, where $L(x_1)$ is the system inductance and i is the current flowing into the electrical circuit, v the voltage control input, r the resistance of the coil and g the gravitational constant, it is possible to introduce the model of the system as

$$\dot{x} = (J - R)\frac{\partial H}{\partial x} + g(v + \delta) \tag{8}$$

where $x = \begin{pmatrix} x_1 \; x_2 \; x_3 \end{pmatrix}^{\mathrm{T}}$,

$$J = \begin{bmatrix} 0 & 1 & 0 \\ -1 & 0 & 0 \\ 0 & 0 & 0 \end{bmatrix}, \quad R = \begin{bmatrix} 0 & 0 & 0 \\ 0 & 0 & 0 \\ 0 & 0 & r \end{bmatrix}, \quad g = \begin{bmatrix} 0 \\ 0 \\ 1 \end{bmatrix},$$

the Hamiltonian $H(x)$ is defined by

$$H(x) = \frac{x_3^2}{2L(x_1)} + \frac{x_2^2}{2M} + Mgx_1 ,$$

and

$$L(x_1) = \frac{c_1}{c_2 - x_1} + L_0 ,$$

with c_1, c_2 and L_0 suitably defined physical parameters.

3.1 Nominal Regulation Control Design

In this section we design a state feedback controller able to solve a regulation problem for system (8) satisfying, in the meantime, all condition imposed by Proposition 1.

Denoting with x_1^{des} the desired position of the sphere, it is possible to introduce a change of coordinates and a control action to solve the regulation problem and satisfy the requirements of Proposition 1: define error coordinates as

$$\begin{aligned} \tilde{x}_1 &:= x_1 - x_1^{des} \\ \tilde{x}_2 &:= x_2 - x_2^{des} = x_2 - M\dot{x}_1^{des} = x_2 \\ \tilde{x}_3 &:= x_3 - x_3^{des}(\tilde{x}_1, \tilde{x}_2, x_1^{des}) = x_3 - \sqrt{2L^2\left(\frac{\partial L}{\partial x_1}\right)^{-1}\left(Mg - \tilde{x}_1 - \frac{\tilde{x}_2}{M}\right)} . \end{aligned} \tag{9}$$

Simple computations make it possible to write time derivatives of these new error coordinates as

$$\begin{aligned}\dot{\tilde{x}}_1 &= \frac{\tilde{x}_2}{M} \\ \dot{\tilde{x}}_2 &= \frac{(\tilde{x}_3 + 2x_3^{des})}{2L^2}\frac{\partial L}{\partial x_1}\tilde{x}_3 - \tilde{x}_1 - \frac{\tilde{x}_2}{M} \\ \dot{\tilde{x}}_3 &= -K_1\tilde{x}_3 - \frac{(\tilde{x}_3 + 2x_3^{des})}{2L^2}\frac{\partial L}{\partial x_1}\frac{\tilde{x}_2}{M} + v' - \delta(t)\end{aligned} \tag{10}$$

where the control action has been set to

$$v = v' + \frac{r}{L}x_3^{des} + \dot{x}_3^{des} - \frac{(\tilde{x}_3 + 2x_3^{des})}{2L^2}\frac{\partial L}{\partial x_1}\frac{\tilde{x}_2}{M} - K\tilde{x}_3 \tag{11}$$

with $K = K_1 - r/L$ and K_1 positive design gain.

Defining a new Hamiltonian function as

$$\tilde{H}(\tilde{x}) := \frac{1}{2}\tilde{x}_1^2 + \frac{\tilde{x}_2^2}{2M} + \frac{1}{2}\tilde{x}_3^2$$

is possible to write (10) as a pHs

$$\dot{\tilde{x}} = [\tilde{J}(\tilde{x}) - \tilde{R}]\frac{\partial \tilde{H}}{\partial \tilde{x}} + g(v' + \delta(t)) \tag{12}$$

with

$$\tilde{J}(\tilde{x}) = \begin{bmatrix} 0 & 1 & 0 \\ -1 & 0 & \tilde{J}_{23}(\tilde{x}) \\ 0 & -\tilde{J}_{23}(\tilde{x}) & 0 \end{bmatrix}, \quad \tilde{J}_{23}(\tilde{x}) = \frac{(\tilde{x}_3 + 2x_3^{des})}{2L^2}\frac{\partial L}{\partial x_1}$$

and

$$\tilde{R} = \begin{bmatrix} 0 & 0 & 0 \\ 0 & 1 & 0 \\ 0 & 0 & K_1 \end{bmatrix}.$$

It is easy to realize that, in absence of external disturbances ($\delta(t) \equiv 0$), the controller (11) is able to solve the regulation problem as, asymptotically, error system (12) converges to the origin.

Moreover, it is worth to remark the instrumental role played by the change of coordinates (9): the origin of the auxiliary system

$$\begin{aligned}\dot{\tilde{x}} &= [\tilde{J}(\tilde{x}) - \tilde{R}]\frac{\partial \tilde{H}}{\partial \tilde{x}} + g\sigma \\ \dot{\sigma} &= -g^{\mathrm{T}}\frac{\partial \tilde{H}}{\partial \tilde{x}}\end{aligned} \tag{13}$$

is now the unique invariant set characterized by

$$\frac{\partial^{\mathrm{T}} \tilde{H}}{\partial \tilde{x}}\tilde{R}\frac{\partial \tilde{H}}{\partial \tilde{x}} = 0,$$

and the Hamiltonian function $\tilde{H}(\tilde{x})$ is now quadratic; hypotheses H1 and H2 of Proposition 1 are therefore satisfied.

3.2 Adaptive Internal Model Design

State again the input disturbance suppression problem considering the voltage disturbance $\delta(t) = \Gamma w$ as generated by a neutrally stable autonomous exosystem like the one defined by (2), (3) and (4): the problem fits in the framework presented in section 2 and it is possible to design an adaptive internal model controller following Proposition 1.

The resulting control law generated by the adaptive internal model unit is defined by

$$\begin{cases} \dot{\xi} &= (F + G\hat{\Psi})\xi - FG\tilde{x}_3 - GK_1\tilde{x}_3 - G\dfrac{(\tilde{x}_3 + 2x_3^{des})}{2L^2}\dfrac{\partial L}{\partial x_1}\dfrac{\tilde{x}_2}{M} - G\hat{\Psi}G\tilde{x}_3 \\ \dot{\hat{\Phi}} &= -(\xi - G\tilde{x}_3)\tilde{x}_3 \\ v' &= \hat{\Psi}\xi - \hat{\Psi}G\tilde{x}_3 \end{cases} \tag{14}$$

The remaining part of the section is devoted to prove the asymptotic convergence of the error state $\tilde{x}$ to the origin in presence of disturbance δ.

Defining the changes of coordinate

$$\chi = \xi - Yw - G\tilde{x}_3 \tag{15}$$

system (12) with controller (14) identifies an interconnection described by:

$$\dot{\bar{x}} = [\bar{J}(\bar{x}) - \bar{R}]\frac{\partial H_x(\bar{x})}{\partial \bar{x}} + \Lambda \tag{16}$$

with

$$\bar{x} = \begin{pmatrix} \tilde{x} \; \chi \; \tilde{\Phi} \; w \end{pmatrix}^{\mathrm{T}},$$

the Hamiltonian function $H_x(\bar{x})$ defined by

$$H_x(\bar{x}) = \tilde{H}(\tilde{x}) + \frac{1}{2}\chi^{\mathrm{T}}\chi + \frac{1}{2}\tilde{\Phi}^{\mathrm{T}}\tilde{\Phi} + \frac{1}{2}w^{\mathrm{T}}w$$

the skew-symmetric interconnection matrix $\bar{J}(\bar{x})$ defined by:

$$\bar{J}(\bar{x}) = \begin{pmatrix} \tilde{J}(\tilde{x}) & 0 & g(\chi + Yw)^{\mathrm{T}} & 0 \\ 0 & 0 & 0 & 0 \\ -(\chi + Yw)g^{\mathrm{T}} & 0 & 0 & 0 \\ 0 & 0 & 0 & S \end{pmatrix}$$

the positive semi-definite damping matrix $\bar{R}$ defined as:

$$\bar{R} = \begin{pmatrix} \tilde{R} & 0 & 0 & 0 \\ 0 & -F & 0 & 0 \\ 0 & 0 & 0 & 0 \\ 0 & 0 & 0 & 0 \end{pmatrix}$$

and Λ defined by:

$$\Lambda = \begin{pmatrix} g\Psi\chi & 0 & 0 & 0 \end{pmatrix}^{\mathrm{T}} .$$

Consider the Lyapunov function $V = H_x(\bar{x})$: simple computations show that its time derivative is

$$\dot{V} = -\frac{\partial^{\mathrm{T}} \tilde{H}}{\partial \tilde{x}} \tilde{R} \frac{\partial \tilde{H}}{\partial \tilde{x}} + \frac{\partial^{\mathrm{T}} \tilde{H}}{\partial \tilde{x}} g\Psi\chi + \chi^{\mathrm{T}} F \chi .$$

As (6) holds then there exist real numbers $\eta_{K_1} \in \mathbb{R}^-$, $\eta_F \in \mathbb{R}^-$ and $\eta_\Psi \in \mathbb{R}$, such that

$$\dot{V} \leq -\|\tilde{x}_2\|^2 + \eta_{K_1}\|\tilde{x}_3\|^2 + \eta_F\|\chi\|^2 + \eta_\Psi\|\tilde{x}_3\|\|\chi\| .$$

Using a Young's inequality argumentation is is possible to state that

$$\dot{V} \leq -\|\tilde{x}_2\|^2 + \eta_{K_1}\|\tilde{x}_3\|^2 + \eta_F\|\chi\|^2 + \frac{\eta_\Psi}{2}\varepsilon\|\tilde{x}_3\|^2 + \frac{\eta_\Psi}{2\varepsilon}\|\chi\|^2 ,$$

for a certain value of ε. Choosing $\varepsilon = -\eta_{K_1}/\eta_\Psi$, it comes out that

$$\dot{V} \leq -\|\tilde{x}_2\|^2 + \frac{\eta_{K_1}}{2}\|\tilde{x}_3\|^2 + \left(\eta_F - \frac{\eta_\Psi^2}{2\eta_x}\right)\|\chi\|^2 , \qquad (17)$$

hence choosing matrix F such that

$$\eta_F < \frac{\eta_\Psi^2}{2\eta_{K_1}} ,$$

it turns out that $\dot{V} \leq 0$ and, for LaSalle invariance principle, system's trajectory are asymptotically captured by the largest invariant set characterized by $\dot{V} = 0$. This set is the same considered for the auxiliary system (13) augmented with the autonomous dynamics $\dot{w} = Sw$; hence the system asymptotically converge to

$$\lim_{t\to\infty} (\tilde{x}, \chi) = (0, 0)$$

solving the input disturbance suppression problem.

4 Conclusions

In this paper an internal model based approach to implicit FTC for port-Hamiltonian systems is presented. The main idea is to cast the original problem into a regulation problem complicated by the presence of input disturbances and to solve this new problem with an adaptive internal model based approach. This adaptive internal model design procedure is presented exploiting the energy-based characteristic properties of this formalism in order to prove the global asymptotical stability of the solution.

To deeply point out the effectiveness of the design procedure presented, an implicit fault tolerant regulation control problem is solved for a magnetic levitation system.

References

1. Astolfi A, Isidori A, Marconi L (2003) A note on disturbance suppression for hamiltonian systems by state feedback. 2nd IFAC Workshop LHMNLC, Seville, Spain
2. Blanke M, Kinnaert M, Lunze J, Staroswiecki M (2003) Diagnosis and fault-tolerant control. Springer-Verlag
3. Bonivento C, Gentili L, Paoli A (2004a) Internal model based fault tolerant control of a robot manipulator. 43rd Conference on Decision and Control, Paradise Island, Bahamas
4. Bonivento C, Isidori A, Marconi L, Paoli A (2004b) Implicit fault tolerant control: Application to induction motors. Automatica 40(3):355–371
5. Byrnes C, Delli Priscoli F, Isidori A (1997a) Output regulation of uncertain non-linear systems. Birkhäuser, Boston
6. Canudas de Wit C, Praly L (2000) Adaptive eccentricity compensation. IEEE Transactions on Control Systems Technology 8(5):757–766
7. Frank P (1990) Fault diagnosis in dynamic systems using analitical and knowledge based redundancy: a survey and some new results. Automatica 26(3):459–474
8. Fujimoto K, Sakurama K, Sugie T (2003) Trajectory tracking control of port-controlled hamiltonian systems via generalized canonical transformations. Automatica 39(12):2059–2069
9. Gentili L, van der Schaft A (2003) Regulation and input disturbance suppression for port-controlled Hamiltonian systems. 2nd IFAC Workshop LHMNLC, Seville, Spain
10. Isidori A (1995) Nonlinear Control Systems. Springer-Verlag, London
11. Isidori A, Marconi L, Serrani A (2003) Robust Autonomous Guidance: An Internal Model-based Approach. Limited series Advances in Industrial Control, Springer Verlag, London
12. Maschke B, van der Schaft A (1992) Port-controlled hamiltonian system: modelling origins and system theoretic approach. 2nd IFAC NOLCOS, Bordeaux, France
13. Nikiforov V (1998) Adaptive non-linear tracking with complete compensation of unknown disturbances. European Journal of Control 4:132–139
14. Ortega R (2003) Some applications and recent results on passivity based control. 2nd IFAC Workshop on Lagrangian and Hamiltonian Methods for Nonlinear Control, Seville, Spain
15. Patton R, Frank P, Clark R (2000) Issues of fault diagnosis for dynamical systems. Springer-Verlag, London
16. Serrani A, Isidori A, Marconi L (2001) Semiglobal output regulation with adaptive internal model. IEEE Transaction On Automatic Control 46(8):1178–1194
17. van der Schaft A (1999) L_2-gain and Passivity Techniques in Nonlinear Control. Springer-Verlag, London
18. Bonivento C, Gentili L, Paoli A (2005) Fault tolerant tracking of a robot manipulator: an internal model based approach. Current Trends in Nolinear Systems and Control, Birkhäuser, Boston:271–287
19. Bonivento C, Gentili L, Marconi L (2005) Balanced Robust Regulation of a Magnetic Levitation System. IEEE Tran. Control System Technology 13(6):1036–1044
20. Alleyne A (2000) Control of a class of nonlinear systems subject to periodic exogenous signals. IEEE Trans. on Control Systems Technology 8(2):279–287
21. Charara A, De Miras J, Caron B (1996) Nonlinear control of a magnetic levitation system without premagnetization. IEEE Transaction on Control Systems Technology 4(5):513–523

22. Marino R, Santosuosso G L, Tomei P (2003) Robust adaptive compensation of biased sinusoidal disturbances with unknown frequency. Automatica 19(10):1755–1761
23. Bodson M, Douglas S C (1997) Adaptive algorithms for the rejection of periodic disturbances with unknown frequencies. Automatica 33(12):2213-2221
24. Ortega R, van der Schaft A, Maschke B, Escobar G (1999) Interconnection and damping assignment passivity-based control of port-controlled Hamiltonian systems. Automatica 38(4):585–596
25. Gentili L, Marconi L (2003) Robust Nonlinear Disturbance Suppression of a Magnetic Levitation System. Automatica 39(4):735–742

On the Geometric Reduction of Controlled Three-Dimensional Bipedal Robotic Walkers

Aaron D. Ames[1], Robert D. Gregg[2], Eric D.B. Wendel[3], and Shankar Sastry[4]

[1] Control and Dynamical Systems, California Institute of Technology, Pasadena, CA 91125
ames@cds.caltech.edu

[2] Department of Electrical and Computer Engineering, University of Illinois at Urbana-Champaign, Urbana, IL 61801
rgregg@uiuc.edu

[3] Sensis Corporation, Campbell, CA 95008
eric.wendel@sensis.com

[4] Department of Electrical Engineering and Computer Sciences, University of California, Berkeley, CA 94720
sastry@eecs.berkeley.edu

Summary. The purpose of this paper is to apply methods from geometric mechanics to the analysis and control of bipedal robotic walkers. We begin by introducing a generalization of Routhian reduction, *functional* Routhian Reduction, which allows for the conserved quantities to be functions of the cyclic variables rather than constants. Since bipedal robotic walkers are naturally modeled as hybrid systems, which are inherently nonsmooth, in order to apply this framework to these systems it is necessary to first extend functional Routhian reduction to a hybrid setting. We apply this extension, along with potential shaping and controlled symmetries, to derive a feedback control law that provably results in walking gaits on flat ground for a three-dimensional bipedal walker given walking gaits in two dimensions.

1 Introduction

Geometric reduction plays an essential role in understanding physical systems modeled by Lagrangians or Hamiltonians; the simplest being Routhian reduction first discovered in the 1860's (cf. [6]). In the case of Routhian reduction, symmetries in the system are characterized by *cyclic* variables, which are coordinates of the configuration space that do not appear in the Lagrangian. Using these symmetries, one can reduce the dimensionality of the phase space (by "dividing" out by the symmetries) and define a corresponding Lagrangian on this reduced phase space. The main result of geometric reduction is that we can understand the behavior of the full-order system in terms of the behavior of the reduced system and vice versa.

In classical geometric reduction the conserved quantities used to reduce and reconstruct systems are constants; this indicates that the "cyclic" variables

F. Bullo et al. (Eds.): Lag. & Hamil. Methods for Nonlin. Ctrl., LNCIS 366, pp. 183–196, 2007.
springerlink.com

eliminated when passing to the reduced phase space are typically uncontrolled. Yet it is often the case that these variables are the ones of interest—it may be desirable to *control* the cyclic variables while not affecting the reduced order system. This motivates an extension of Routhian reduction to the case when the conserved quantities are functions of the cyclic variables instead of constants.

These concepts motivate our main goal:

> *Develop a feedback control law that results in walking gaits on flat ground for a three-dimensional bipedal robotic walker given walking gaits for a two-dimensional bipedal robotic walker.*

In order to achieve this goal, we begin by considering Lagrangians that are cyclic except for an additional non-cyclic term in the potential energy, i.e., *almost-cyclic* Lagrangians. When Routhian reduction is performed with a function (of a cyclic variable) the result is a Lagrangian on the reduced phase-space: the *functional Routhian*. We are able to show that the dynamics of an almost-cyclic Lagrangian satisfying certain initial conditions project to dynamics of the corresponding functional Routhian, and dynamics of the functional Routhian can be used to reconstruct dynamics of the full-order system. In order to use this result to develop control strategies for bipedal walkers, it first must be generalized to a hybrid setting. That is, after discussing how to explicitly obtain a hybrid system model of a bipedal walker (Section 2), we generalize functional Routhian reduction to a hybrid setting (Section 3), demonstrating that hybrid flows of the reduced and full order system are related in a way analogous to the continuous result.

We then proceed to consider two-dimensional (2D) bipedal walkers. It is well-known that 2D bipedal walkers can walk down shallow slopes without actuation (cf. [7], [3]). [10] used this observation to develop a positional feedback control strategy that allows for walking on flat ground. In Section 4, we use these results to obtain a hybrid system, $\mathscr{H}_{2D}^{s}$, modeling a 2D bipedal robot that walks on flat ground.

In Section 5 we consider three-dimensional (3D) bipedal walkers. Our main result is a positional feedback control law that produces walking gaits in three dimensions. To obtain this controller we shape the potential energy of the Lagrangian describing the dynamics of the 3D bipedal walker so that it becomes an almost-cyclic Lagrangian, where the cyclic variable is the roll (the unstable component) of the walker. We are able to control the roll through our choice of a non-cyclic term in the potential energy. Since the *functional Routhian* hybrid system obtained by reducing this system is $\mathscr{H}_{2D}^{s}$, by picking the "correct" function of the roll, we can force the roll to go to zero for certain initial conditions. That is, we obtain a non-trivial set of initial conditions that provably result in three-dimensional walking.

Space constraints prevent us from including proofs for the results stated in this paper; these can be found in [1].

2 Lagrangian Hybrid Systems

We begin this section by defining (simple) hybrid systems and hybrid flows (as introduced in [2]); for more on general hybrid systems, see [5] and the references therein. We then turn our attention to introducing a special class of hybrid systems that will be important when discussing bipedal robots: *unilaterally constrained Lagrangian hybrid systems.* It will be seen that bipedal robotic walkers are naturally modeled by systems of this form.

Definition 1. *A simple hybrid system*[1] *is a tuple:*

$$\mathscr{H} = (D, G, R, f),$$

where

- *D is a smooth manifold, called the domain,*
- *G is an embedded submanifold of D called the guard,*
- *$R : G \to D$ is a smooth map called the reset map (or impact equations),*
- *f is a vector field or control system (in which case we call $\mathscr{H}$ a controlled hybrid system) on D, i.e., $\dot{x} = f(x)$ or $\dot{x} = f(x, u)$, respectively.*

Hybrid flows. A *hybrid flow* (or *execution*) is a tuple

$$\chi^{\mathscr{H}} = (\Lambda, \mathcal{I}, \mathcal{C}),$$

where

- $\Lambda = \{0, 1, 2, \ldots\} \subseteq \mathbb{N}$ is a finite or infinite indexing set.
- $\mathcal{I} = \{I_i\}_{i \in \Lambda}$ is a *hybrid interval* where $I_i = [\tau_i, \tau_{i+1}]$ if $i, i+1 \in \Lambda$ and $I_{N-1} = [\tau_{N-1}, \tau_N]$ or $[\tau_{N-1}, \tau_N)$ or $[\tau_{N-1}, \infty)$ if $|\Lambda| = N$, N finite. Here, $\tau_i, \tau_{i+1}, \tau_N \in \mathbb{R}$ and $\tau_i \leq \tau_{i+1}$.
- $\mathcal{C} = \{c_i\}_{i \in \Lambda}$ is a collection of integral curves of f, i.e., $\dot{c}_i(t) = f(c_i(t))$ for all $i \in \Lambda$.

We require that the following conditions hold for every $i, i+1 \in \Lambda$,

$$\begin{array}{ll} \text{(i)} & c_i(\tau_{i+1}) \in G, \\ \text{(ii)} & R(c_i(\tau_{i+1})) = c_{i+1}(\tau_{i+1}). \end{array}$$

The initial condition for the hybrid flow is $c_0(\tau_0)$.

Lagrangians. Let Q be a *configuration space*, assumed to be a smooth manifold, and TQ the tangent bundle of Q. In this paper, we will consider Lagrangians $L : TQ \to \mathbb{R}$ describing mechanical, or robotic, systems; that is, Lagrangians given in coordinates by:

$$L(q, \dot{q}) = \frac{1}{2}\dot{q}^T M(q)\dot{q} - V(q), \tag{1}$$

[1] So named because of their simple discrete structure, i.e., a simple hybrid system has a single domain, guard and reset map.

where $M(q)$ is the inertial matrix, $\frac{1}{2}\dot{q}^T M(q)\dot{q}$ is the kinetic energy and $V(q)$ is the potential energy. In this case, the Euler-Lagrange equations yield the equations of motion for the system:

$$M(q)\ddot{q} + C(q,\dot{q})\dot{q} + N(q) = 0,$$

where $C(q,\dot{q})$ is the *Coriolis matrix* (cf. [8]) and $N(q) = \frac{\partial V}{\partial q}(q)$. The Lagrangian vector field, f_L, associated to L takes the familiar form:

$$(\dot{q},\ddot{q}) = f_L(q,\dot{q}) = \left(\dot{q}, M(q)^{-1}(-C(q,\dot{q})\dot{q} - N(q))\right).$$

Controlled Lagrangians. We will also be interested in controlled Lagrangians. In this case, the equations of motion for the system have the form:

$$M(q)\ddot{q} + C(q,\dot{q})\dot{q} + N(q) = Bu,$$

where we assume that B is an invertible matrix. The result is a control system of the form:

$$(\dot{q},\ddot{q}) = f_L(q,\dot{q},u) = \left(\dot{q}, M(q)^{-1}(-C(q,\dot{q})\dot{q} - N(q) + Bu)\right).$$

In the future, it will be clear from context whether for a Lagrangian L we are dealing with a corresponding vector field $(\dot{q},\ddot{q}) = f_L(q,\dot{q})$ or a control system $(\dot{q},\ddot{q}) = f_L(q,\dot{q},u)$.

Unilateral constraints. It is often the case that the set of admissible configurations for a mechanical system is determined by a *unilateral constraint function*, which is a smooth function $h : Q \to \mathbb{R}$ such that $h^{-1}(0)$ is a manifold, i.e., 0 is a regular value of h. For bipedal walkers this function is the height of the non-stance (or swing) foot above the ground. In this case, we can explicitly construct the domain and the guard of a hybrid system:

$$\begin{aligned} D_h &= \{(q,\dot{q}) \in TQ : h(q) \geq 0\}, \\ G_h &= \{(q,\dot{q}) \in TQ : h(q) = 0 \text{ and } dh_q\dot{q} < 0\}, \end{aligned}$$

where in coordinates $dh_q = \left(\frac{\partial h}{\partial q_1}(q) \cdots \frac{\partial h}{\partial q_n}(q) \right)$.

Definition 2. *We say that $\mathscr{H} = (D,G,R,f)$ is a unilaterally constrained Lagrangian hybrid system w.r.t. a Lagrangian $L : TQ \to \mathbb{R}$ and a unilateral constraint function $h : Q \to \mathbb{R}$ if $D = D_h$, $G = G_h$ and $f = f_L$.*

Impact Equations. In order to determine the impact equations (or reset map) for the hybrid system $\mathscr{H}$, we typically will utilize an additional constraint function. A *kinematic* constraint function is a smooth function $\Upsilon : Q \to \mathbb{R}^{\upsilon}$ ($\upsilon \geq 1$); this function usually describes the position of the end-effector of a kinematic chain, e.g., in the case of bipedal robots, this is the position of the swing foot. Using this kinematic constraint function one obtains a reset map $R(q,\dot{q}) = (q, P_q(\dot{q}))$, where $P_q : T_qQ \to T_qQ$ with

$$P_q(\dot{q}) = \dot{q} - M(q)^{-1} d\Upsilon_q^T (d\Upsilon_q M(q)^{-1} d\Upsilon_q^T)^{-1} d\Upsilon_q \dot{q}.$$

This reset map models a perfectly plastic impact without slipping and was derived using the set-up in [4] together with block-diagonal matrix inversion.

3 Functional Routhian Reduction

In this section, we introduce a variation of classical Routhian reduction termed *functional Routhian reduction.* Conditions are given on when this type of reduction can be performed on continuous and hybrid systems.

Shape space. We begin by considering an abelian Lie group, $\mathbb{G}$, given by:

$$\mathbb{G} = \underbrace{(\mathbb{S}^1 \times \mathbb{S}^1 \times \cdots \times \mathbb{S}^1)}_{m-\text{times}} \times \mathbb{R}^p,$$

with $k = m + p = \dim(\mathbb{G})$; here $\mathbb{S}^1$ is the circle. The starting point for classical Routhian reduction is a configuration space of the form $Q = S \times \mathbb{G}$, where S is called the *shape space*; we denote an element $q \in Q$ by $q = (\theta, \varphi)$ where $\theta \in S$ and $\varphi \in \mathbb{G}$. Note that we have a projection map $\pi : TS \times T\mathbb{G} \to TS$ where $(\theta, \dot{\theta}, \varphi, \dot{\varphi}) \mapsto (\theta, \dot{\theta})$.

Almost-Cyclic Lagrangians. We will be interested (in the context of bipedal walking) in Lagrangians of a very special form. We say that a Lagrangian $L_\lambda : TS \times T\mathbb{G} \to \mathbb{R}$ is *almost-cyclic* if, in coordinates, it has the form:

$$L_\lambda(\theta, \dot{\theta}, \varphi, \dot{\varphi}) = \frac{1}{2} \begin{pmatrix} \dot{\theta} \\ \dot{\varphi} \end{pmatrix}^T \begin{pmatrix} M_\theta(\theta) & 0 \\ 0 & M_\varphi(\theta) \end{pmatrix} \begin{pmatrix} \dot{\theta} \\ \dot{\varphi} \end{pmatrix} - V_\lambda(\theta, \varphi), \tag{2}$$

where

$$V_\lambda(\theta, \varphi) = \widetilde{V}(\theta) - \frac{1}{2}\lambda(\varphi)^T M_\varphi^{-1}(\theta)\lambda(\varphi)$$

for some function $\lambda : \mathbb{G} \to \mathbb{R}^k$ with a symmetric Jacobian, i.e.,

$$\left(\frac{\partial \lambda}{\partial \varphi}\right)^T = \frac{\partial \lambda}{\partial \varphi}.$$

Here $M_\theta(\theta) \in \mathbb{R}^{n \times n}$ and $M_\varphi(\theta) \in \mathbb{R}^{k \times k}$ are both symmetric positive definite matrices with $n = \dim(S)$.

Momentum maps. Fundamental to reduction is the notion of a momentum map $J : TQ \to \mathbb{R}^k$, which makes explicit the conserved quantities in the system. In the framework we are considering here,

$$J(\theta, \dot{\theta}, \varphi, \dot{\varphi}) = \frac{\partial L_\lambda}{\partial \dot{\varphi}}(\theta, \dot{\theta}, \varphi, \dot{\varphi}) = M_\varphi(\theta)\dot{\varphi}.$$

Typically, one sets the momentum map equal to a constant $\mu \in \mathbb{R}^k$; this defines the conserved quantities of the system. In our framework, we will breach this convention and set J equal to a function: this motivates the name *functional Routhian reduction.*

Functional Routhians. For an almost-cyclic Lagrangian L_λ as given in (2), define the corresponding *functional Routhian* $\widetilde{L} : TS \to \mathbb{R}$ by:

$$\widetilde{L}(\theta, \dot{\theta}) = \left[L_\lambda(\theta, \dot{\theta}, \varphi, \dot{\varphi}) - \lambda(\varphi)^T \dot{\varphi} \right] \Big|_{J(\theta,\dot{\theta},\varphi,\dot{\varphi})=\lambda(\varphi)}$$

Because $J(\theta, \dot{\theta}, \varphi, \dot{\varphi}) = \lambda(\varphi)$ implies that $\dot{\varphi} = M_\varphi^{-1}(\theta)\lambda(\varphi)$, by direct calculation the functional Routhian is given by:

$$\widetilde{L}(\theta, \dot{\theta}) = \frac{1}{2}\dot{\theta}^T M_\theta(\theta)\dot{\theta} - \widetilde{V}(\theta).$$

That is, any Lagrangian of the form given in (1) is the functional Routhian of an almost-cyclic Lagrangian.

We can relate solutions of the Lagrangian vector field $f_{\widetilde{L}}$ to solutions of the Lagrangian vector field f_{L_λ} and vice versa (in a way analogous to the classical Routhian reduction result, see [6]).

Theorem 1. *Let L_λ be an almost-cyclic Lagrangian, and $\widetilde{L}$ the corresponding functional Routhian. Then $(\theta(t), \dot{\theta}(t), \varphi(t), \dot{\varphi}(t))$ is a solution to the vector field f_{L_λ} on $[t_0, t_F]$ with*

$$\dot{\varphi}(t_0) = M_\varphi^{-1}(\theta(t_0))\lambda(\varphi(t_0)),$$

if and only if $(\theta(t), \dot{\theta}(t))$ is a solution to the vector field $f_{\widetilde{L}}$ and $(\varphi(t), \dot{\varphi}(t))$ satisfies:

$$\dot{\varphi}(t) = M_\varphi^{-1}(\theta(t))\lambda(\varphi(t)).$$

We now have the necessary material needed to introduce our framework for hybrid functional Routhian reduction.

Definition 3. *If $\mathscr{H}_\lambda = (D_h, G_h, R, f_{L_\lambda})$ is a unilaterally constrained Lagrangian hybrid system, $\mathscr{H}_\lambda$ is almost-cyclic if the following conditions hold:*

- $Q = S \times \mathbb{G}$,
- $h : Q = S \times \mathbb{G} \to \mathbb{R}$ *is cyclic,* $\frac{\partial h}{\partial \varphi} = 0$, *and so can be viewed as a function* $\tilde{h} : S \to \mathbb{R}$,
- $L_\lambda : TS \times T\mathbb{G} \to \mathbb{R}$ *is almost-cyclic,*
- $\pi_\varphi(R(\theta, \dot{\theta}, \varphi, \dot{\varphi})) = \varphi$, *where π_φ is the projection onto the φ-component,*
- *The following diagram commutes:*

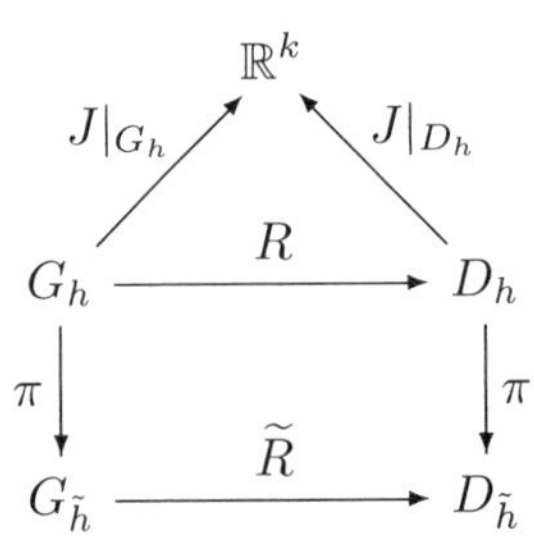

for some map $\widetilde{R} : G_{\tilde{h}} \to D_{\tilde{h}}$.

Hybrid functional Routhian. If $\mathscr{H}_\lambda = (D_h, G_h, R, f_{L_\lambda})$ is an almost-cyclic unilaterally constrained Lagrangian hybrid system, we can associate to this hybrid system a reduced hybrid system, termed a *functional Routhian* hybrid system, denoted by $\widetilde{\mathscr{H}}$ and defined by:

$$\widetilde{\mathscr{H}} := (D_{\tilde{h}}, G_{\tilde{h}}, \widetilde{R}, f_{\widetilde{L}}).$$

The following theorem quantifies the relationship between $\mathscr{H}_\lambda$ and $\widetilde{\mathscr{H}}$.

Theorem 2. *Let $\mathscr{H}_\lambda$ be a cyclic unilaterally constrained Lagrangian hybrid system, and $\widetilde{\mathscr{H}}$ the associated functional Routhian hybrid system. Then $\chi^{\mathscr{H}_\lambda} = (\Lambda, \mathcal{I}, \{(\theta_i, \dot{\theta}_i, \varphi_i, \dot{\varphi}_i)\}_{i\in\Lambda})$ is a hybrid flow of $\mathscr{H}_\lambda$ with*

$$\dot{\varphi}_0(\tau_0) = M_\varphi^{-1}(\theta_0(\tau_0))\lambda(\varphi_0(\tau_0)),$$

if and only if $\chi^{\widetilde{\mathscr{H}}} = (\Lambda, \mathcal{I}, \{\theta_i, \dot{\theta}_i\}_{i\in\Lambda})$ is a hybrid flow of $\widetilde{\mathscr{H}}$ and $\{(\varphi_i, \dot{\varphi}_i)\}_{i\in\Lambda}$ satisfies:

$$\dot{\varphi}_i(t) = M_\varphi^{-1}(\theta_i(t))\lambda(\varphi_i(t)), \qquad \varphi_{i+1}(\tau_{i+1}) = \varphi_i(\tau_{i+1}).$$

4 Controlled Symmetries Applied to 2D Bipedal Walkers

In this section, we begin by studying the standard model of a two-dimensional bipedal robotic walker walking down a slope (walkers of this form have been well-studied by [7] and [3], to name a few). We then use controlled symmetries to shape the potential energy of the Lagrangian describing this model so that it can walk stably on flat ground.

2D biped model. We begin by introducing a model describing a controlled bipedal robot walking in two dimensions down a slope of γ degrees. That is, we explicitly construct the controlled hybrid system

$$\mathscr{H}_{2\mathrm{D}}^\gamma = (D_{2\mathrm{D}}^\gamma, G_{2\mathrm{D}}^\gamma, R_{2\mathrm{D}}, f_{2\mathrm{D}}).$$

describing this system.

The configuration space for the 2D biped is $Q_{2\mathrm{D}} = \mathbb{T}^2$, the two-dimensional torus, and the Lagrangian describing this system is:

$$L_{2\mathrm{D}}(\theta, \dot{\theta}) = \frac{1}{2}\dot{\theta}^T M_{2\mathrm{D}}(\theta)\dot{\theta} - V_{2\mathrm{D}}(\theta),$$

where $\theta = (\theta_{\mathrm{ns}}, \theta_{\mathrm{s}})^T$. Table 1 gives $M_{2\mathrm{D}}(\theta)$ and $V_{2\mathrm{D}}(\theta)$.

Using the controlled Euler-Lagrange equations, the dynamics for the walker are given by:

$$M_{2\mathrm{D}}(\theta)\ddot{\theta} + C_{2\mathrm{D}}(\theta, \dot{\theta})\dot{\theta} + N_{2\mathrm{D}}(\theta) = B_{2\mathrm{D}}u.$$

These equations yield the control system: $(\dot{\theta}, \ddot{\theta}) = f_{2\mathrm{D}}(\theta, \dot{\theta}, u) := f_{L_{2\mathrm{D}}}(\theta, \dot{\theta}, u)$.

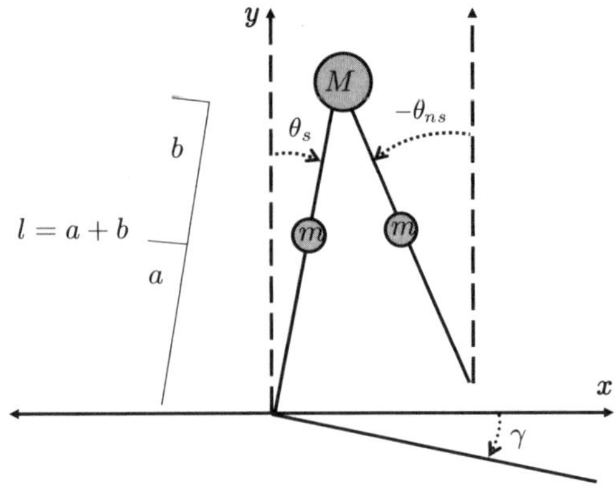

Fig. 1. Two-dimensional bipedal robot

We construct D^γ_{2D} and G^γ_{2D} by applying the methods outlined in Section 2 to the unilateral constraint function: $h^\gamma_{2D}(\theta) = \cos(\theta_s) - \cos(\theta_{ns}) + (\sin(\theta_s) - \sin(\theta_{ns}))\tan(\gamma)$, which gives the height of the foot of the walker above the slope with normalized unit leg length.

Finally, the reset map R_{2D} is given by:

$$R_{2D}(\theta, \dot{\theta}) = \left(S_{2D}\theta, P_{2D}(\theta)\dot{\theta}\right),$$

where S_{2D} and $P_{2D}(\theta)$ are given in Table 1. Note that this reset map was computed using the methods outlined in Section 2 coupled with the condition that the stance foot is fixed (see [4] for more details).

Setting the control $u = 0$ yields the standard model of a 2D passive bipedal robot walking down a slope. For such a model, it has been well-established (for example, in [3]) that for certain γ, $\mathscr{H}^\gamma_{2D}$ has a walking gait. For the rest of the paper we pick, once and for all, such a γ.

Controlled Symmetries. Controlled symmetries were introduced in [9] and later in [10] in order to shape the potential of bipedal robotic walkers to allow for stable walking on flat ground based on stable walking down a slope. We will briefly apply the results of this work to derive a feedback control law that yields a hybrid system, $\mathscr{H}^s_{2D}$, with stable walking gaits on flat ground.

The main idea of [10] is that inherent symmetries in $\mathscr{H}^\gamma_{2D}$ can be used to "rotate the world" (via a group action) to allow for walking on flat ground. Specifically, we have a group action $\Phi : \mathbb{S}^1 \times Q_{2D} \to Q_{2D}$ denoted by:

$$\Phi(\gamma, \theta) := (\theta_{ns} - \gamma, \theta_s - \gamma)^T,$$

for $\gamma \in \mathbb{S}^1$. Using this, define the following feedback control law:

$$u = K^\gamma_{2D}(\theta) := B^{-1}_{2D} \frac{\partial}{\partial \theta} \left(V_{2D}(\theta) - V_{2D}(\Phi(\gamma, \theta))\right).$$

Table 1. Additional equations for $\mathscr{H}_{2\mathrm{D}}$ and $\mathscr{H}_{3\mathrm{D}}$

Additional equations for $\mathscr{H}_{2\mathrm{D}}$:

$$M_{2\mathrm{D}}(\theta) = \begin{pmatrix} \frac{l^2 m}{4} & -\frac{l^2 m\cos(\theta_{\mathrm{s}}-\theta_{\mathrm{ns}})}{2} \\ -\frac{l^2 m\cos(\theta_{\mathrm{s}}-\theta_{\mathrm{ns}})}{2} & \frac{l^2 m}{4} + l^2(m+M) \end{pmatrix} \qquad S_{2\mathrm{D}} = \begin{pmatrix} 0 & 1 \\ 1 & 0 \end{pmatrix}$$

$$V_{2\mathrm{D}}(\theta) = \frac{1}{2} gl((3m+2M)\cos(\theta_{\mathrm{s}}) - m\cos(\theta_{\mathrm{ns}})) \qquad B_{2\mathrm{D}} = \begin{pmatrix} -1 & 0 \\ 1 & 1 \end{pmatrix}$$

$$P_{2\mathrm{D}}(\theta) = \frac{1}{-3m-4M+2m\cos(2(\theta_{\mathrm{s}}-\theta_{\mathrm{ns}}))} \begin{pmatrix} 2m\cos(\theta_{\mathrm{ns}}-\theta_{\mathrm{s}}) & m-4(m+M)\cos(2(\theta_{\mathrm{ns}}-\theta_{\mathrm{s}})) \\ m & -2(m+2M)\cos(\theta_{\mathrm{ns}}-\theta_{\mathrm{s}}) \end{pmatrix}$$

Additional equations for $\mathscr{H}_{3\mathrm{D}}$:

$$m_{3\mathrm{D}}(\theta) = \frac{1}{8}(l^2(6m+4M) + l^2(m\cos(2\theta_{\mathrm{ns}}) - 8m\cos(\theta_{\mathrm{ns}})\cos(\theta_{\mathrm{s}}) + (5m+4M)\cos(2\theta_{\mathrm{s}}))$$

$$V_{3\mathrm{D}}(\theta,\varphi) = V_{2\mathrm{D}}(\theta)\cos(\varphi)$$

$$p_{3\mathrm{D}}(\theta) = \frac{-m\cos(2\theta_{\mathrm{ns}}) + 8(m+M)\cos(\theta_{\mathrm{ns}})\cos(\theta_{\mathrm{s}}) - m(2+\cos(2\theta_{\mathrm{s}}))}{6m+4M+(5m+4M)\cos(2\theta_{\mathrm{ns}}) - 8m\cos(\theta_{\mathrm{ns}})\cos(\theta_{\mathrm{s}}) + m\cos(2\theta_{\mathrm{s}})}$$

Applying this control law to the control system $(\dot{q},\ddot{q}) = f_{2\mathrm{D}}(\theta,\dot{\theta},u)$ yields the dynamical system:

$$(\dot{\theta},\ddot{\theta}) = f^{\gamma}_{2\mathrm{D}}(\theta,\dot{\theta}) := f_{2\mathrm{D}}(\theta,\dot{\theta},K^{\gamma}_{2\mathrm{D}}(\theta))$$

which is just the vector field associated to the Lagrangian

$$L^{\gamma}_{2\mathrm{D}}(\theta,\dot{\theta}) = \frac{1}{2}\dot{\theta}^T M_{2\mathrm{D}}(\theta)\dot{\theta} - V^{\gamma}_{2\mathrm{D}}(\theta),$$

where $V^{\gamma}_{2\mathrm{D}}(\theta) := V_{2\mathrm{D}}(\Phi(\gamma,\theta))$. That is, $f^{\gamma}_{2\mathrm{D}} = f_{L^{\gamma}_{2\mathrm{D}}}$.

Now define, for some γ that results in stable passive walking for $\mathscr{H}^{\gamma}_{2\mathrm{D}}$,

$$\mathscr{H}^{\mathrm{s}}_{2\mathrm{D}} := (D^0_{2\mathrm{D}}, G^0_{2\mathrm{D}}, R_{2\mathrm{D}}, f^{\gamma}_{2\mathrm{D}}),$$

which is a unilaterally constrained Lagrangian hybrid system. In particular, it is related to $\mathscr{H}^{\gamma}_{2\mathrm{D}}$ as follows:

Theorem 3 ([10]). *$\chi^{\mathscr{H}^{\mathrm{s}}_{2\mathrm{D}}} = (\Lambda,\mathcal{I},\{(\Phi(\gamma,\theta_i),\dot{\theta}_i)\}_{i\in\Lambda})$ is a hybrid flow of $\mathscr{H}^{\mathrm{s}}_{2\mathrm{D}}$ if $\chi^{\mathscr{H}^{\gamma}_{2\mathrm{D}}} = (\Lambda,\mathcal{I},\{(\theta_i,\dot{\theta}_i)\}_{i\in\Lambda})$ is a hybrid flow of $\mathscr{H}^{\gamma}_{2\mathrm{D}}$.*

The importance of this theorem lies in the fact that it implies that if $\mathscr{H}^{\gamma}_{2\mathrm{D}}$ walks (stably) on a slope, then $\mathscr{H}^{\mathrm{s}}_{2\mathrm{D}}$ walks (stably) on flat ground.

5 Functional Routhian Reduction Applied to 3D Bipedal Walkers

In this section we construct a control law that results in stable walking for a simple model of a 3D bipedal robotic walker. In order to achieve this goal, we shape the potential energy of this model via feedback control so that when hybrid functional Routhian reduction is carried out, the result is the 2D walker $\mathscr{H}^{\mathrm{s}}_{\mathrm{2D}}$ introduced in the previous section. We utilize Theorem 2 to demonstrate that this implies that the 3D walker has a walking gait on flat ground (in three dimensions). This is the main contribution of this work.

3D biped model. We now introduce the model describing a controlled bipedal robot walking in three dimensions on flat ground, i.e., we will explicitly construct the controlled hybrid system describing this system:

$$\mathscr{H}_{\mathrm{3D}} = (D_{\mathrm{3D}}, G_{\mathrm{3D}}, R_{\mathrm{3D}}, f_{\mathrm{3D}}).$$

The configuration space for the 3D biped is $Q_{\mathrm{3D}} = \mathbb{T}^2 \times \mathbb{S}$ and the Lagrangian describing this system is given by:

$$L_{\mathrm{3D}}(\theta, \dot{\theta}, \varphi, \dot{\varphi}) = \frac{1}{2} \begin{pmatrix} \dot{\theta} \\ \dot{\varphi} \end{pmatrix}^T \begin{pmatrix} M_{\mathrm{2D}}(\theta) & 0 \\ 0 & m_{\mathrm{3D}}(\theta) \end{pmatrix} \begin{pmatrix} \dot{\theta} \\ \dot{\varphi} \end{pmatrix} - V_{\mathrm{3D}}(\theta, \varphi),$$

where $m_{\mathrm{3D}}(\theta)$ is given in the Table 1. Note that, referring to the notation introduced in Section 3, $M_\theta(\theta) = M_{\mathrm{2D}}(\theta)$ and $M_\varphi(\theta) = m_{\mathrm{3D}}(\theta)$. Also note that L_{3D} is nearly cyclic; it is only the potential energy that prevents its cyclicity. This will motivate the use of a control law that shapes this potential energy.

Using the controlled Euler-Lagrange equations, the dynamics for the walker are given by:

$$M_{\mathrm{3D}}(q)\ddot{q} + C_{\mathrm{3D}}(q, \dot{q})\dot{q} + N_{\mathrm{3D}}(q) = B_{\mathrm{3D}} u,$$

with $q = (\theta, \varphi)$ and

$$B_{\mathrm{3D}} = \begin{pmatrix} B_{\mathrm{2D}} & 0 \\ 0 & 1 \end{pmatrix}.$$

These equations yield the control system: $(\dot{q}, \ddot{q}) = f_{\mathrm{3D}}(q, \dot{q}, u) := f_{L_{\mathrm{3D}}}(q, \dot{q}, u)$.

We construct D_{3D} and G_{3D} by applying the methods outlined in Section 2 to the unilateral constraint function

$$h_{\mathrm{3D}}(\theta, \varphi) = h^0_{\mathrm{2D}}(\theta) = \cos(\theta_{\mathrm{s}}) - \cos(\theta_{\mathrm{ns}}).$$

This function gives the normalized height of the foot of the walker above flat ground with the implicit assumption that $\varphi \in (-\pi/2, \pi/2)$ (which allows us to disregard the scaling factor $\cos(\varphi)$ that would have been present). The result is that h_{3D} is cyclic.

Finally, the reset map R_{3D} is given by:

$$R_{\mathrm{3D}}(\theta, \dot{\theta}, \varphi, \dot{\varphi}) = \left(S_{\mathrm{2D}}\theta, P_{\mathrm{2D}}(\theta)\dot{\theta}, \varphi, p_{\mathrm{3D}}(\theta)\dot{\varphi} \right)$$

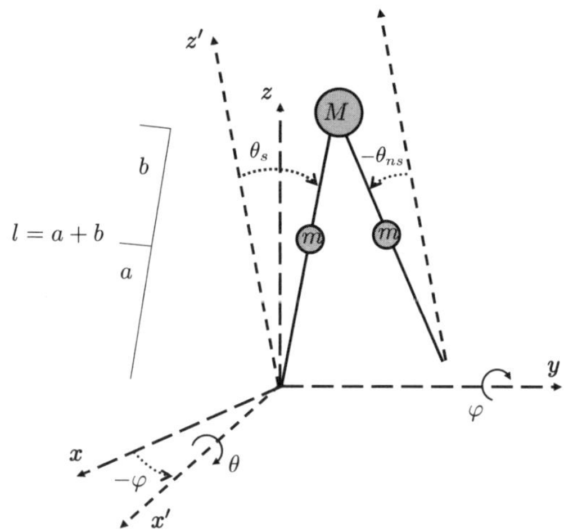

Fig. 2. Three-dimensional bipedal robot

where $p_{3\mathrm{D}}(\theta)$ is given in Table 1. Note that this map was again computed using the methods outlined in Section 2 coupled with the condition that the stance foot is fixed.

Control law construction. We now proceed to construct a feedback control law for $\mathscr{H}_{3\mathrm{D}}$ that makes this hybrid system an almost-cyclic unilaterally constrained Lagrangian hybrid system, $\mathscr{H}^{\alpha}_{3\mathrm{D}}$. We will then demonstrate, using Theorem 2, that $\mathscr{H}^{\alpha}_{3\mathrm{D}}$ has a walking gait by relating it to $\mathscr{H}^{\mathrm{s}}_{2\mathrm{D}}$.

Define the feedback control law parameterized by $\alpha \in \mathbb{R}$:

$$\begin{aligned} u &= K^{\alpha}_{3\mathrm{D}}(q) \\ &:= B^{-1}_{3\mathrm{D}} \frac{\partial}{\partial q}\left(V_{3\mathrm{D}}(q) - V^{\gamma}_{2\mathrm{D}}(\theta) + \frac{1}{2}\frac{\alpha^2 \varphi^2}{m_{3\mathrm{D}}(\theta)} \right) \end{aligned}$$

Applying this control law to the control system $(\dot{q}, \ddot{q}) = f_{3\mathrm{D}}(q, \dot{q}, u)$ yields the dynamical system:

$$(\dot{q}, \ddot{q}) = f^{\alpha}_{3\mathrm{D}}(q, \dot{q}) := f_{3\mathrm{D}}(q, \dot{q}, K^{\alpha}_{3\mathrm{D}}(q)),$$

which is just the vector field associated to the almost-cyclic Lagrangian

$$L^{\alpha}_{3\mathrm{D}}(\theta, \dot{\theta}, \varphi, \dot{\varphi}) = \frac{1}{2}\begin{pmatrix} \dot{\theta} \\ \dot{\varphi} \end{pmatrix}^{T} \begin{pmatrix} M_{2\mathrm{D}}(\theta) & 0 \\ 0 & m_{3\mathrm{D}}(\theta) \end{pmatrix} \begin{pmatrix} \dot{\theta} \\ \dot{\varphi} \end{pmatrix} - V^{\alpha}_{3\mathrm{D}}(\theta, \varphi),$$

where

$$V^{\alpha}_{3\mathrm{D}}(\theta, \varphi) = V^{\gamma}_{2\mathrm{D}}(\theta) - \frac{1}{2}\frac{\alpha^2 \varphi^2}{m_{3\mathrm{D}}(\theta)}.$$

That is, $f^{\alpha}_{3\mathrm{D}} = f_{L^{\alpha}_{3\mathrm{D}}}$.

Let

$$\mathscr{H}_{3D}^{\alpha} := (D_{3D}, G_{3D}, R_{3D}, f_{3D}^{\alpha}),$$

which is a unilaterally constrained Lagrangian hybrid system.

Applying hybrid functional Routhian reduction. Using the methods outlined in Section 3, there is a momentum map $J_{3D} : TQ_{3D} \to \mathbb{R}$ given by:

$$J_{3D}(\theta, \dot{\theta}, \varphi, \dot{\varphi}) = m_{3D}(\theta)\dot{\varphi}.$$

Setting $J_{3D}(\theta, \dot{\theta}, \varphi, \dot{\varphi}) = \lambda(\varphi) = -\alpha\varphi$ implies that

$$\dot{\varphi} = -\frac{\alpha\varphi}{m_{3D}(\theta)}.$$

The importance of $\mathscr{H}_{3D}^{\alpha}$ is illustrated by:

Proposition 1. *$\mathscr{H}_{3D}^{\alpha}$ is an almost-cyclic unilaterally constrained Lagrangian hybrid system. Moreover, the following diagram commutes:*

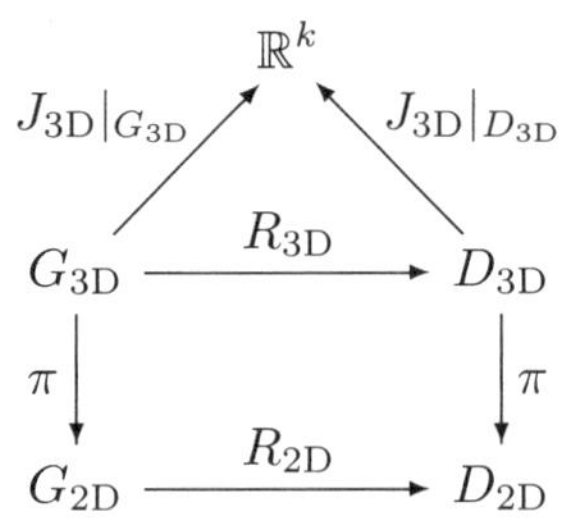

Therefore, $\mathscr{H}_{2D}^{s}$ is the functional Routhian hybrid system associated with $\mathscr{H}_{3D}^{\alpha}$.

This result allows us to prove—using Theorem 2—that the control law used to construct $\mathscr{H}_{3D}^{\alpha}$ in fact results in walking in three dimensions.

Theorem 4. *$\chi^{\mathscr{H}_{3D}^{\alpha}} = (\Lambda, \mathcal{I}, \{(\theta_i, \dot{\theta}_i, \varphi_i, \dot{\varphi}_i)\}_{i\in\Lambda})$ is a hybrid flow of $\mathscr{H}_{3D}^{\alpha}$ with*

$$\dot{\varphi}_0(\tau_0) = -\frac{\alpha\varphi_0(\tau_0)}{m_{3D}(\theta_0(\tau_0))}, \tag{3}$$

if and only if $\chi^{\mathscr{H}_{2D}^{s}} = (\Lambda, \mathcal{I}, \{\theta_i, \dot{\theta}_i\}_{i\in\Lambda})$ is a hybrid flow of $\mathscr{H}_{2D}^{s}$ and $\{(\varphi_i, \dot{\varphi}_i)\}_{i\in\Lambda}$ satisfies:

$$\dot{\varphi}_i(t) = -\frac{\alpha\varphi_i(t)}{m_{3D}(\theta_i(t))}, \qquad \varphi_{i+1}(\tau_{i+1}) = \varphi_i(\tau_{i+1}). \tag{4}$$

To better understand the implications of Theorem 4, suppose that $\chi^{\mathscr{H}_{3D}^{\alpha}} = (\Lambda, \mathcal{I}, \{(\theta_i, \dot{\theta}_i, \varphi_i, \dot{\varphi}_i)\}_{i\in\Lambda})$ is a hybrid flow of $\mathscr{H}_{3D}^{\alpha}$. If this hybrid flow has an initial condition satisfying (3) with $\alpha > 0$ and the corresponding hybrid flow, $\chi^{\mathscr{H}_{2D}^{s}} = (\Lambda, \mathcal{I}, \{\theta_i, \dot{\theta}_i\}_{i\in\Lambda})$, of $\mathscr{H}_{2D}^{s}$ is a walking gait in 2D:

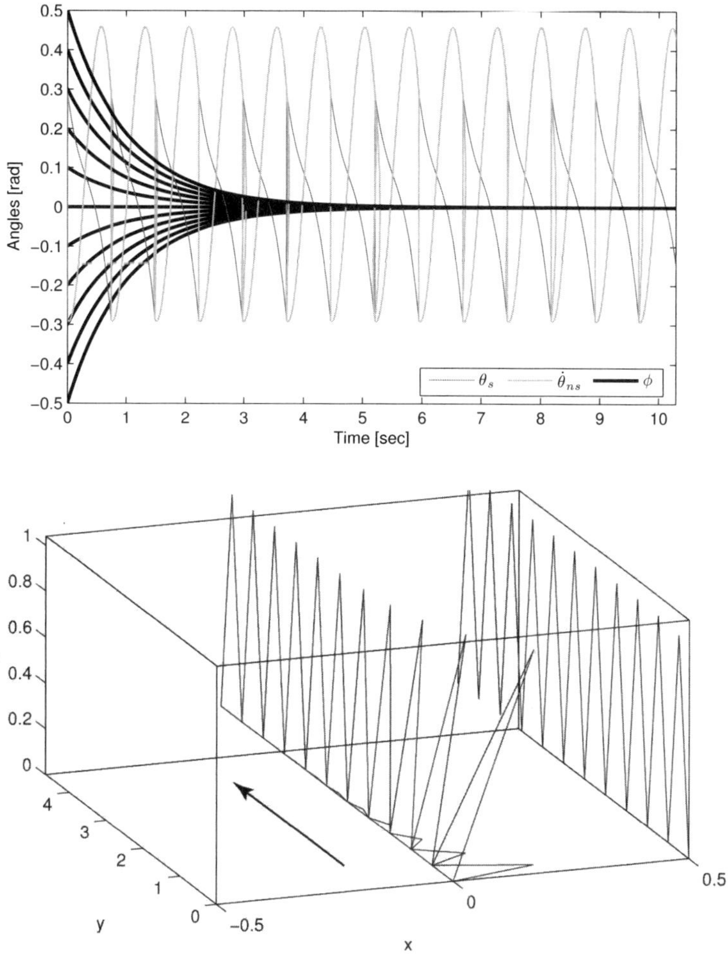

Fig. 3. $\theta_{\rm ns}$, $\theta_{\rm s}$ and φ over time for different initial values of φ (top). A walking gait for the three-dimensional bipedal robot (bottom).

$$\Lambda = \mathbb{N}, \quad \lim_{i\to\infty} \tau_i = \infty, \quad \theta_i(\tau_i) = \theta_{i+1}(\tau_{i+1}),$$

then the result is walking in three dimensions. This follows from the fact that θ and $\dot{\theta}$ will have the same behavior over time for the full-order system—the bipedal robot will walk. Moreover, since Theorem 4 implies that (4) holds, the walker stabilizes to the "upright" position. That is, the roll φ will tend to zero as time goes to infinity since (4) essentially defines a stable linear system $\dot{\varphi} = -\alpha\varphi$ (because $m_{\rm 3D}(\theta_i(t)) > 0$ and $\alpha > 0$), which controls the behavior of φ when (3) is satisfied. This convergence can be seen in Fig. 3 along with a walking gait of the 3D walker.

References

1. A. D. Ames. *A Categorical Theory of Hybrid Systems.* PhD thesis, University of California, Berkeley, 2006.
2. A. D. Ames and S. Sastry. Hybrid Routhian reduction of hybrid Lagrangians and Lagrangian hybrid systems. In *American Control Conference*, Minneapolis, MN, 2006.
3. A. Goswami, B. Thuilot, and B. Espiau. Compass-like biped robot part I : Stability and bifurcation of passive gaits. Rapport de recherche de l'INRIA, 1996.
4. J.W. Grizzle, G. Abba, and F. Plestan. Asymptotically stable walking for biped robots: Analysis via systems with impulse effects. *IEEE Transactions on Automatic Control*, 46(1):51–64, 2001.
5. J. Lygeros, K. H. Johansson, S. Simic, J. Zhang, and S. Sastry. Dynamical properties of hybrid automata. *IEEE Transactions on Automatic Control*, 48:2– 17, 2003.
6. J. E. Marsden and T. S. Ratiu. *Introduction to Mechanics and Symmetry*, volume 17 of *Texts in Applied Mathematics.* Springer, 1999.
7. T. McGeer. Passive dynamic walking. *International Journal of Robotics Research*, 9(2):62–82, 1990.
8. R. M. Murray, Z. Li, and S. Sastry. *A Mathematical Introduction to Robotic Manipulation.* CRC Press, 1993.
9. M. W. Spong and F. Bullo. Controlled symmetries and passive walking. In *IFAC World Congress*, Barcelona, Spain, 2002.
10. M. W. Spong and F. Bullo. Controlled symmetries and passive walking. *IEEE Transactions on Automatic Control*, 50(7):1025– 1031, 2005.

Gait Generation for a Hopping Robot Via Iterative Learning Control Based on Variational Symmetry

Satoshi Satoh[1], Kenji Fujimoto[2], and Sang-Ho Hyon[3,4]

[1] Department of Mechanical Science and Engineering, Graduate School of Engineering, Nagoya University, Furo-cho, Chikusa-ku, Nagoya, 464-8603, Japan
s_satou@nuem.nagoya-u.ac.jp
[2] Department of Mechanical Science and Engineering, Graduate School of Engineering, Nagoya University, Furo-cho, Chikusa-ku, Nagoya, 464-8603, Japan
k.fujimoto@ieee.org
[3] Computational Brain Project, ICORP, JST
[4] ATR, Computational Neuroscience Laboratories 2-2-2 Hikaridai, Kyoto, 619-0288, Japan
sangho@atr.jp

Summary. This paper proposes a novel framework to generate optimal gait trajectories for a one-legged hopping robot via iterative learning control. This method generates gait trajectories which are solutions of a class of optimal control problems without using precise knowledge of the plant model. It is expected to produce natural gait movements such as that of a passive walker. Some numerical examples demonstrate the effectiveness of the proposed method.

1 Introduction

Passive dynamic walking originally studied by McGeer [1] inspires many researchers to work on a gait generation problem for walking robots. They try to design more natural and less energy consuming gait trajectories than those produced by conventional walking control such as ZMP based control. Behavior analysis of passive walkers were investigated, e.g, in [2, 3]. There are some results on gait generation based on passive dynamic walking [4, 5, 6] by designing appropriate feedback control systems such that the closed loop systems behave like passive walkers. In particular, a hopping robot modelled in [7] has a hopping gait for which the input signal coincides with zero, that is, this robot can be regarded as a passive walker walking on a horizontal plane. Furthermore, an adaptive control system for this robot to achieve a walking gait with zero input was proposed in the authors' former result [8].

The objective of this paper is to generate optimal walking gait trajectories for a hopping robot via iterative learning control without using precise knowledge of the plant model. To this end, we formulate an optimal control type cost function and try to find a control input minimizing it by iterative learning technique based

F. Bullo et al. (Eds.): Lag. & Hamil. Methods for Nonlin. Ctrl., LNCIS 366, pp. 197–208, 2007.
springerlink.com

on *variational symmetry* of Hamiltonian control systems [9], which can solve a class of optimal control problems by iteration of experiments. For this purpose, two novel techniques with respect to iterative learning control are proposed: One is a technique to take the time derivatives of the output signal into account in the iterative learning control by employing a pseudo adjoint of the time derivative operator. The other is a cost function to achieve time symmetric gait trajectories to guarantee stable walking without a fall. Furthermore, the proposed learning scheme is applied to the hopping robot in [7] and the corresponding numerical simulations demonstrate its advantage.

2 Iterative Learning Control Based on Variational Symmetry

This section refers to the iterative learning control (ILC) method based on variational symmetry in [9] briefly.

2.1 Variational Symmetry of Hamiltonian Systems

Consider a Hamiltonian system with dissipation and a controlled Hamiltonian $H(x,u,t)$ described by

$$\Sigma : \begin{cases} \dot{x} = (J-R)\dfrac{\partial H(x,u,t)}{\partial x}^{\mathrm{T}}, & x(t^0) = x^0 \\ y = -\dfrac{\partial H(x,u,t)}{\partial u}^{\mathrm{T}} \end{cases} . \tag{1}$$

Here $x(t) \in \mathbb{R}^n$, $u(t) \in \mathbb{R}^m$ and $y(t) \in \mathbb{R}^r$ describe the state, the input and the output, respectively. The structure matrix $J \in \mathbb{R}^{n\times n}$ and the dissipation matrix $R \in \mathbb{R}^{n\times n}$ are skew-symmetric and symmetric positive semi-definite, respectively. The matrix R represents dissipative elements. In this paper the Hamiltonian system in (1) is written as $y = \Sigma(u)$. For this system, the following theorem holds. This property is called *variational symmetry* of Hamiltonian control systems.

Theorem 1. *[9] Consider the Hamiltonian system in (1). Suppose that J and R are constant and that there exists a nonsingular matrix $T \in \mathbb{R}^{n\times n}$ satisfying*

$$J = -TJ\,T^{-1}$$
$$R = TR\,T^{-1}$$
$$\frac{\partial^2 H(x,u,t)}{\partial (x,u)^2} = \begin{pmatrix} T & 0 \\ 0 & I \end{pmatrix} \frac{\partial^2 H(x,u,t)}{\partial (x,u)^2} \begin{pmatrix} T^{-1} & 0 \\ 0 & I \end{pmatrix} .$$

Suppose moreover that $J-R$ is nonsingular. Then the variational system $\mathrm{d}\Sigma$ of Σ and its adjoint $(\mathrm{d}\Sigma)^*$ of Σ have almost the same state-space realizations.

Remark 1. Suppose the Hessian of the Hamiltonian with respect to (x, u) satisfies

$$\frac{\partial^2 H(x,u,t)}{\partial(x,u)^2}(t-t^0) = \frac{\partial^2 H(x,u,t)}{\partial(x,u)^2}(t^1-t), \quad {}^\forall t \in [t^0, t^1].$$

Then under appropriate initial conditions of Σ,

$$(\mathrm{d}\Sigma(u))^*(v) \approx \mathcal{R} \circ (\Sigma(u + \mathcal{R}(v)) - \Sigma(u)) \tag{2}$$

holds when v is small where $\mathcal{R}$ is a time-reversal operator defined by $\mathcal{R}(u)(t - t^0) = u(t^1 - t)$ for $\forall t \in [t^0, t^1]$.

Equation (2) implies that we can calculate the input-output mapping of the variational adjoint by only using the input-output data of the original system.

2.2 Optimal Control Via Iterative Learning

Let us consider the system Σ in (1) and a cost function $\Gamma : L_2^m[t^0, t^1] \times L_2^r[t^0, t^1] \to \mathbb{R}$ as follows

$$\Gamma(u,y) = \frac{1}{2}\int_{t^0}^{t^1} \Big(u(t)^{\mathrm{T}} \Lambda_u u(t) + (y(t) - y^d(t))^{\mathrm{T}} \Lambda_y (y(t) - y^d(t)) \Big) \mathrm{d}t,$$

where $y^d \in L_2^r[t^0, t^1]$ represents a desired output and $\Lambda_u \in \mathbb{R}^{m\times m}$ and $\Lambda_y \in \mathbb{R}^{r\times r}$ are positive definite matrices. The objective is to find the optimal input minimizing the cost function $\Gamma(u, y)$. Note that the Fréchet derivative of Γ is $\mathrm{d}\Gamma(u, y)$. It follows from well-known Riesz's representation theorem that there exists an operator $\Gamma'(u, y)$ such that

$$\begin{aligned} \mathrm{d}(\Gamma(u,y)) &= \mathrm{d}\Gamma(u,y)(\mathrm{d}u, \mathrm{d}y) \\ &= \langle \Gamma'(u,y), (\mathrm{d}u, \mathrm{d}y)\rangle_{L_2}. \end{aligned}$$

Here we can calculate

$$\begin{aligned} \mathrm{d}(\Gamma(u,y)) &= \langle (\Lambda_u u, \Lambda_y(y - y^d)), (\mathrm{d}u, \mathrm{d}y)\rangle_{L_2} \\ &= \langle \Lambda_u u, \mathrm{d}u\rangle_{L_2} + \langle \Lambda_y(y-y^d), \mathrm{d}\Sigma(u)(\mathrm{d}u)\rangle_{L_2} \\ &= \langle \Lambda_u u + (\mathrm{d}\Sigma(u))^* \Lambda_y (y - y^d), \mathrm{d}u\rangle_{L_2}. \end{aligned}$$

Therefore the steepest descent method implies that we should change the input u such that

$$\mathrm{d}u = -K\big(\Lambda_u u + (\mathrm{d}\Sigma(u))^* \Lambda_y(y - y^d)\big),$$

where K is an appropriate positive gain. Hence the iteration law should be taken as

$$u_{(i+1)} = u_{(i)} - K_{(i)}\big(\Lambda_u u_{(i)} + (\mathrm{d}\Sigma(u))^* \Lambda_y (y_{(i)} - y^d)\big). \tag{3}$$

Here i denotes the i-th iteration in laboratory experiments. Suppose Equation (2) holds, then ILC law based on variational symmetry is given by

$$u_{(2i-1)} = u_{(2i-2)} + \mathcal{R}(\Lambda_y(y_{(2i-2)} - y^d)) \tag{4}$$

$$u_{(2i)} = u_{(2i-2)} - K_{(2i-2)}\big(\Lambda_u u_{(2i-2)} + \mathcal{R}(y_{(2i-1)} - y_{(2i-2)})\big) \tag{5}$$

provided that the initial input $u_{(0)}$ is equivalent to zero.

This pair of iteration laws Equations (4) and (5) implies that this learning procedure needs two steps laboratory experiments. In the ($2i$-1)-th iteration, we can calculate the input and output signals of $(\mathrm{d}\Sigma(u))^*$ by Equation (2) with the output signal of $\Sigma(u + \mathcal{R}(v))$. Input for the $2i$-th iteration is generated by (3) with these signals.

3 Extension of ILC for Time Derivatives

Let us recall that there is a constraint with respect to cost functions in the iterative learning control method in [9]. For the system Σ in (1), a possible choice of an optimal control type cost function used in the iterative learning control is a functional of u and y, and it is not possible to choose a functional of $\dot{y}$ the time derivative of the output. However, the signal $\dot{y}$ often plays an important role in control of mechanical systems. In particular, it is important to check the behavior of $\dot{y}$ for the gait trajectory generation problem. In this section, we extend the iterative learning control method referred in the previous section to take the time derivative $\dot{y}$ into account.

Let us consider the Hamiltonian system in (1) and suppose that the following assumption holds.

assumptionps 1. *The conditions* $\mathrm{d}y(t^0) = 0$ *and* $\mathrm{d}y(t^1) = 0$ *hold.*

In the iterative learning control, it is assumed that all the initial conditions are the same in each laboratory experiment in general. Therefore the condition $\mathrm{d}y(t^0) = 0$ always holds. But the other one $\mathrm{d}y(t^1) = 0$ does not hold in general. In order to let the latter condition $\mathrm{d}y(t^1) = 0$ hold approximately, we can employ an optimal control type cost function such as $\int_{t^1-\epsilon}^{t^1} \|y(t) - y^d(t)\|^2 \mathrm{d}t$ with a small constant $\epsilon > 0$ as in [10].

3.1 Pseudo Adjoint of the Time Derivative Operator

Here we investigate a pseudo adjoint of the time derivative operator to take account of the time derivative of the output signal $\dot{y}$ in the iterative learning control procedure.

Consider a differentiable signal $\xi \in L_2[t^0, t^1]$ and a time derivative operator $D(\cdot)$ which maps the signal ξ into its time derivative is defined by

$$D(\xi)(t) := \frac{\mathrm{d}\xi(t)}{\mathrm{d}t}. \tag{6}$$

Let us provide the following lemma to define the pseudo adjoint of the time derivative operator.

Lemma 1. *Consider differentiable signals ξ and $\eta \in L_2[t^0, t^1]$. Suppose that the signal ξ satisfies the condition*

$$\xi(t^0) = \xi(t^1) = 0. \tag{7}$$

Then the following equation holds

$$\langle \eta, D(\xi) \rangle_{L_2} = \langle -D(\eta), \xi \rangle_{L_2}. \tag{8}$$

Proof. Consider the inner product of η and $D(\xi)$. Let us calculate that

$$\langle \eta, D(\xi) \rangle_{L_2} = \int_{t^0}^{t^1} \eta(t)^{\mathrm{T}} \frac{\mathrm{d}\xi(t)}{\mathrm{d}t} \mathrm{d}t = \Big[\eta(t)^{\mathrm{T}} \xi(t)\Big]_{t^0}^{t^1} - \int_{t^0}^{t^1} \frac{\mathrm{d}\eta(t)}{\mathrm{d}t}^{\mathrm{T}} \xi(t) \mathrm{d}t.$$

Since $\xi(t)$ satisfies the condition (7), $\Big[\eta(t)^{\mathrm{T}} \xi(t)\Big]_{t^0}^{t^1} = 0$ holds and we can calculate that

$$\langle \eta, D(\xi) \rangle_{L_2} = -\int_{t^0}^{t^1} \frac{\mathrm{d}\eta(t)}{\mathrm{d}t}^{\mathrm{T}} \xi(t) \mathrm{d}t = \langle -D(\eta), \xi \rangle_{L_2}. \tag{9}$$

Then (9) implies (8). □

This lemma implies

$$D^* = -D$$

for a certain class of input signals.

3.2 Application to Iterative Learning Control

Here we take the following cost function to illustrate the proposed method

$$\Gamma(\dot{y}) = \frac{1}{2} \int_{t^0}^{t^1} \Big((\dot{y}(t) - \dot{y}^d(t))^{\mathrm{T}} \Lambda_{\dot{y}} (\dot{y}(t) - \dot{y}^d(t)) \Big) \mathrm{d}t. \tag{10}$$

Here $\dot{y}^d$ is a differentiable desired velocity which satisfies $\dot{y}^d \in L_2[t^0, t^1]$. Suppose that the output y satisfies Assumption 1. Then we have

$$\mathrm{d}(\Gamma(\dot{y})) = \langle \Lambda_{\dot{y}} (\dot{y} - \dot{y}^d), \mathrm{d}\dot{y} \rangle_{L_2}.$$

The authors' former result [9] can not directly apply to this cost function (10) because it contains $\dot{y}$. Here let us rewrite $\dot{y}$ as $\dot{y} = D(y)$ with the time derivative operator $D(\cdot)$ defined in (6). Then we have

$$\mathrm{d}\dot{y} = \mathrm{d}D(y)(\mathrm{d}y).$$

Note that the time derivative operator is linear, we obtain $\mathrm{d}\dot{y} = D(\mathrm{d}y)$. Using Assumption 1 and (8), we can obtain

$$\begin{aligned} \mathrm{d}(\Gamma(\dot{y})) &= \langle \Lambda_{\dot{y}} (\dot{y} - \dot{y}^d), D(\mathrm{d}y) \rangle_{L_2} \\ &= \langle -D\Big(\Lambda_{\dot{y}} (\dot{y} - \dot{y}^d)\Big), \mathrm{d}y \rangle_{L_2} \\ &= \langle (\mathrm{d}\Sigma(u))^* \Big(-D(\Lambda_{\dot{y}} (\dot{y} - \dot{y}^d))\Big), \mathrm{d}u \rangle_{L_2}. \end{aligned}$$

As we mentioned above, the proposed method allows one to apply the ILC method to cost functions of state variables which do not appear in the output y.

Table 1. Notations

Notation	Meaning	Unit
r_0	natural leg length	m
m	total mass	kg
g	gravity acceleration	m/s^2
K_l	leg spring stiffness	kgm^2
K_h	hip spring stiffness	kgm^2
T_s	stance time	s
T_f	flight time	s

4 Optimal Gait Generation

In this section, a cost function to generate a symmetric gait is proposed, to which the proposed method in the previous section is applied.

4.1 Description of the Plant

Let us consider a passive hopping robot in [7, 8] depicted in Fig. 1. The body and the leg have mass m_b and m_l and moment of inertia J_b and J_u respectively. The distance between the hip joint and the center of mass of the leg is d. Let us define the equivalent leg inertia $J_l := J_u + m_b m_l d^2/(m_b + m_l)$. Let us also define the control force of the leg ρ and the control torque of the hip joint τ. Table 1 shows other notations. Furthermore, some assumptions are assumed on this robot. An important one is as follows. See [7, 8] for the rest of them.

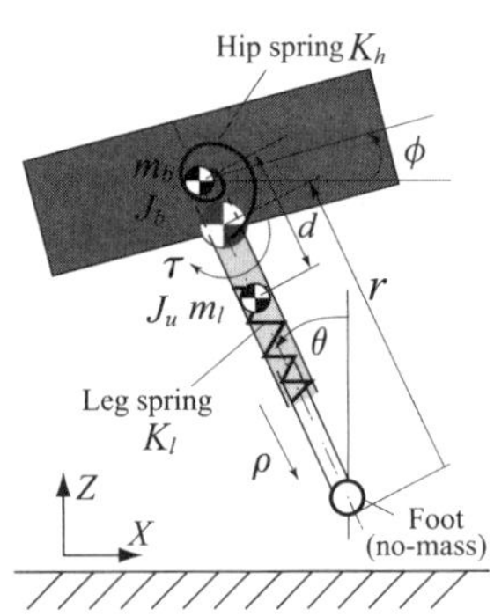

Fig. 1. Description of the plant

assumptionps 2. *The foot does not bounce back nor slip on the ground (inelastic impulsive impact).*

A notion of phases is introduced: the stance phase and the flight phase. When the leg touches the ground the robot is said to be in the stance phase, and when

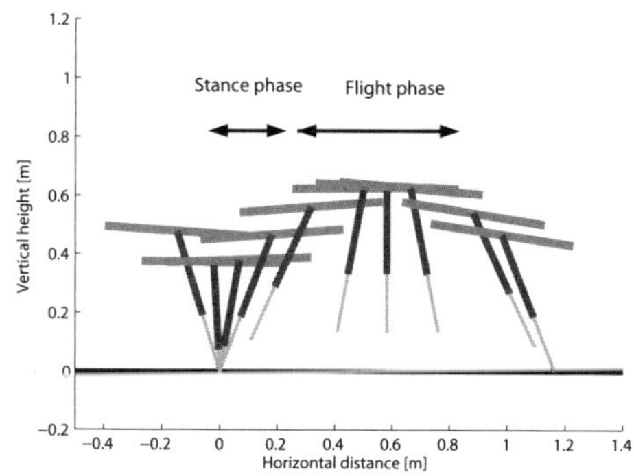

Fig. 2. Subsequent one step of hopping. (The robot moves from left to right.)

the leg is above the ground it is said to be in the flight phase. The stance time represents the time interval during the stance phase and the flight time is defined in a similar way. The robot moves between these two phases alternately. See Fig. 2.

In the stance phase, let us define the generalized coordinate q as $q := (r, \theta, \phi)^{\mathrm{T}} \in \mathbb{R} \times \mathbb{S}^1 \times \mathbb{S}^1$, the generalized momentum p as $p := (p_r, p_\theta, p_\phi)^{\mathrm{T}} \in \mathbb{R}^3$, input u as $u := (\rho, \tau)^{\mathrm{T}} \in \mathbb{R}^2$ and the inertia matrix $M(q)$ as

$$M(q) := \begin{pmatrix} m & 0 & 0 \\ 0 & J_l + mr^2 & 0 \\ 0 & 0 & J_b \end{pmatrix} \in \mathbb{R}^{3\times 3}.$$

Then, the dynamics is described as a Hamiltonian system in (1) with the Hamiltonian $H(q, p, u)$ represented as

$$\begin{aligned} H(q,p,u) &= \frac{1}{2} p^{\mathrm{T}} M(q)^{-1} p - mgr(1-\cos\theta) \\ &\quad + \frac{1}{2} K_l (r - r_0)^2 + \frac{1}{2} K_h (\theta - \phi)^2 - u^{\mathrm{T}} \begin{pmatrix} 1 & 0 & 0 \\ 0 & 1 & -1 \end{pmatrix} q. \end{aligned} \qquad (11)$$

Let us consider the dynamics in the flight phase as below

$$\begin{cases} \ddot{x} = 0 \\ \ddot{z} = -g \\ J_l \ddot{\theta} + J_b \ddot{\phi} = 0 \\ J_b \ddot{\phi} = K_h(\theta - \phi) + \tau_f \end{cases} .$$

Here the variables x and z represent the horizontal and vertical positions of the center of mass and τ_f is the control torque.

4.2 Problem Setting

This section sets the control problem to get the periodic gait based on [11]. Consider the behavior in a flight phase and let $\mathcal{J}$ denote the map from the initial state to the terminal state in this phase. This map can be regarded as a discrete map which connects two adjacent stance phases. Then the desired map of $\mathcal{J}$ is given by

$$\mathcal{J}_s := (r, \theta, \phi, p_r, p_\theta, p_\phi) \mapsto (r, -\theta, -\phi, -p_r, p_\theta, p_\phi). \tag{12}$$

Let us explain why such a map $\mathcal{J}_s$ in (12) is desired. Let us define the flow Φ_t in the stance phase with no inputs, i.e., $\rho \equiv \tau \equiv 0$ by $\Phi_t := (q(t^0), p(t^0)) \mapsto (q(t^0 + t), p(t^0 + t))$. For Equations (11) and (12), the Hamiltonian $H(q, p, 0)$ is invariant with respect to $\mathcal{J}_s$. Therefore $\mathcal{J}_s$ and Φ_t satisfy $\mathcal{J}_s \circ \Phi_{T_s} = \mathrm{id}$ where id represents the identity mapping. If this equation holds, then a periodic gait is generated.

The leg angle θ is the most important state variable, because it has a direct effect to avoid falling. However, it is difficult to control the variable θ in the stance phase, since this robot has no foot. As in [11], we apply no input in the stance phase, and try to control the variable θ to let $\mathcal{J} = \mathcal{J}_s$ hold in the flight phase.

In [8], dead-beat control is used. Although it works well, it requires the precise knowledge of the plant system. Here we try to use iterative learning control based on variational symmetry with a special cost function given in the following section which will generate an optimal flow in the flight phase without the precise knowledge of the system.

4.3 Application of Iterative Learning Control

Let us define the desired values of θ and $\dot{\theta}$ as follows (we let $t^0 = 0$ for simplicity in what follows)

$$\theta^d := \theta\big|_{t=T_s+T_f} = -\theta\big|_{t=T_s} \tag{13}$$

$$\dot{\theta}^d := \dot{\theta}\big|_{t=T_s+T_f} = \dot{\theta}\big|_{t=T_s}. \tag{14}$$

As for the model mentioned above, energy dissipation occurs at the touchdown. Let E_- and E_+ represent the energies just before the touchdown and just after it. Then ΔE the variation of the energy between them can be calculated as

$$\Delta E := E_- - E_+ = \frac{mJ_l}{2(J_l + mr_0^2)}\mu_-^2 \tag{15}$$

by [8], where μ_- is defined by

$$\mu_- := v_{x\_} \cos\theta_- + v_{z\_} \sin\theta_- + \frac{r_0}{J_l + mr_0^2} p_{\theta_-}.$$

Here $v_{x\_}$ and $v_{z\_}$ represent the horizontal and vertical velocity of the center of mass. If the total mechanical energy is completely preserved, that is, a condition

$$\mu_- = 0 \tag{16}$$

holds at the touchdown, then it is expected that periodic gait trajectories are autonomously generated. In fact, [8] implies that the condition (16) is satisfied if the control objects (13) and (14) are achieved, and if the initial condition is appropriately chosen (the way how to choose is described in [8]).

Now we propose a novel cost function as

$$\Gamma(\theta,\dot{\theta},u) := \frac{K_\theta}{2}\|\theta-(-\mathcal{R}(\theta))\|_{L_2}^2 + \frac{K_{\dot{\theta}}}{2}\|\dot{\theta}-\mathcal{R}(\dot{\theta})\|_{L_2}^2 + \frac{K_u}{2}\|u\|_{L_2}^2 \tag{17}$$

where K_θ, $K_{\dot{\theta}}$ and K_u represent appropriate positive constants. $\mathcal{R}$ is the time-reversal operator as defined in Sect. 2. The first and the second terms in the right hand side of Equation (17) are expected to make Assumption 1 approximately hold. It is expected that we can generate an optimal trajectory satisfying Equations (13) and (14) which minimizes the L_2 norm of the control input. Furthermore, there is no energy transfer except for the control input.

Let us recall the fact that gait trajectories are essentially periodic. However, ILC can not generate periodic trajectories. Let us connect the stance flow Φ_t and that generated by the cost function (17). Take the connected trajectory as a single period of a periodic gait trajectory. Therefore if Equation (16) holds, then we can generate the optimal periodic trajectory.

Now, let us define the input $u=\tau_f$ and the output $y=\theta$. We can calculate the iteration law as in (3) and in (11) (Derivation of the law is omitted due to limitations of space)

$$u_{(2i-1)} = u_{(2i-2)} + \mathcal{R}\Big(2K_\theta(\mathrm{id}+\mathcal{R})(y_{(2i-2)}) - 2K_{\dot{\theta}}(\mathrm{id}-\mathcal{R})(\ddot{y}_{(2i-2)})\Big) \tag{18}$$

$$u_{(2i)} = u_{(2i-2)} - K_{(2i-2)}\Big(K_u u_{(2i-2)} + \mathcal{R}\big(y_{(2i-1)} - y_{(2i-2)}\big)\Big). \tag{19}$$

5 Simulation

We apply the proposed iteration law in (18) and (19) to the hopping robot introduced in the previous section. We proceed 50 steps of the learning algorithm, that is, we execute 100 simulations.

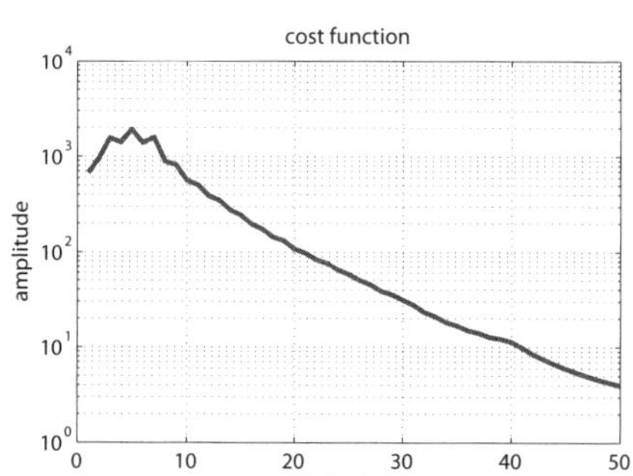

Fig. 3. Cost function

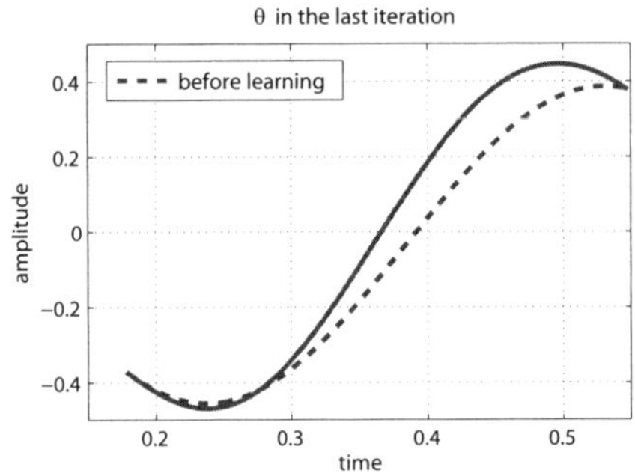

Fig. 4. Response of θ

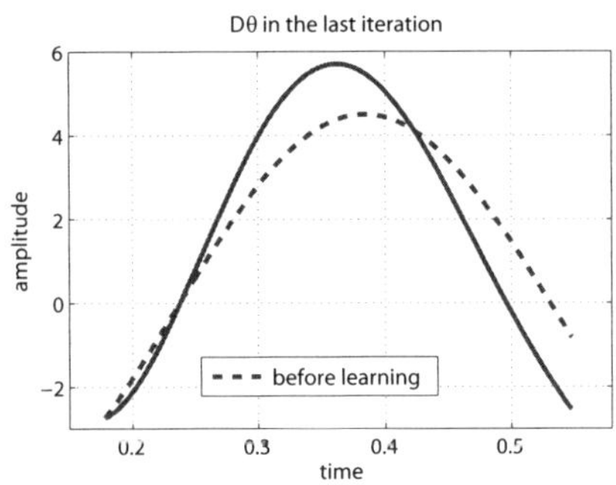

Fig. 5. Response of $\dot{\theta}$

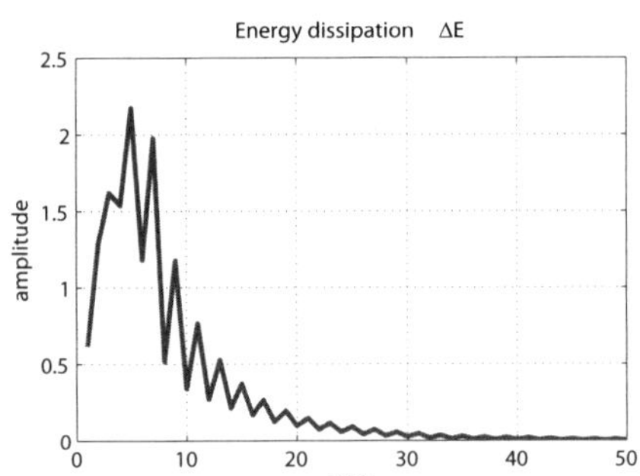

Fig. 6. Variation of the Energy ΔE

Fig. 3 shows that the cost function (17) almost decreases at each experiment. This implies that the output trajectory converges to the optimal one smoothly. Since we choose the initial input $u_{(0)}(t) \equiv 0$, Assumption 1 does not hold in the beginning. This is probably the reason why the cost function increases in the beginning of the learning procedure. In Figs. 4 and 5, the solid lines show the responses of θ and $\dot{\theta}$ at the last step and the dotted lines depict the initial trajectories corresponding to 0 input. These figures show that both θ and $\dot{\theta}$ converge to the trajectories satisfying Equations (13) and (14). Furthermore, Fig. 6 exhibits that the variation of the energy at the touchdown ΔE in (15) converges to zero automatically.

6 Conclusion

In this paper, we have proposed an extension of the iterative learning control based on variational symmetry employing a pseudo adjoint of the time derivative operator. This allows one to execute iterative learning with optimal control type cost function including time derivatives of the output signals. Application of this method to gait generation problem for a hopping robot derives an optimal gait trajectories without using precise knowledge of the plant. Finally, numerical simulations have demonstrated the effectiveness of the proposed method.

Although we succeed in generating optimal gait trajectories minimizing the L_2 norm of the control input, they are *not optimal* in a sense that the hopping gaits with zero input, *passive gaits*, are not obtained. This is because the algorithm proposed in this paper can not take into account the variation of the initial conditions. However passive gaits depend on their initial conditions. It means that only the input iteration can not generate such gaits. In our recent result in [12], we propose a novel algorithm to generate *passive gaits* by employing an update law for the initial conditions as well as that for the feedforward input.

The proposed cost function can not deal with asymmetric gaits. Our another result in [13], a certain class of non-symmetric periodic gaits are considered.

References

1. T. McGeer, "Passive dynamic walking," *Int. J. Robotics Research*, vol. 9, no. 2, pp. 62–82, 1990.
2. K. Osuka and K. Kirihara, "Motion analysis and experiments of passive walking robot quartet ii," in *Proc. the 2000 IEEE Int. Conf. on Robotics and Automation*, 2000, pp. 3052–3056.
3. A. Sano, Y. Ikemata, and H. Fujimoto, "Analysis of dynamics of passive walking from storage energy and supply rate," in *Proc. the 2003 IEEE Int. Conf. on Robotics and Automation*, 2003, pp. 2478–2483.
4. A. Goswami, B. Espiau, and A. Keramane, "Limit cycles in a passive compass gait biped and passivity-mimicking control laws," *Autonomous Robots*, vol. 4, no. 3, pp. 273–286, 1997.
5. M. W. Spong, "Passivity-based control of the compass gait biped," in *Proc. of IFAC World Congress*, 1999, pp. 19–23.
6. F. Asano, M. Yamakita, N. Kamamichi, and Z. W. Luo, "A novel gait generation for biped walking robots based on mechanical energy constraint," *IEEE Trans. Robotics and Automation*, vol. 20, no. 3, pp. 565–573, 2004.
7. M. Ahmadi and M. Buehler, "Stable control of a simulated one-legged running robot with hip and leg compliance," *IEEE Trans. Robotics and Automation*, vol. 13, no. 1, pp. 96–104, 1997.

8. S. Hyon and T. Emura, "Energy-preserving control of passive one-legged running robot," *Advanced Robotics*, vol. 18, no. 4, pp. 357–381, 2004.
9. K. Fujimoto and T. Sugie, "Iterative learning control of Hamiltonian systems: I/O based optimal control approach," *IEEE Trans. Autom. Contr.*, vol. 48, no. 10, pp. 1756–1761, 2003.
10. K. Fujimoto, T. Horiuchi, and T. Sugie, "Optimal control of Hamiltonian systems with input constraints via iterative learning," in *Proc. 42nd IEEE Conf. on Decision and Control*, 2003, pp. 4387–4392.
11. S. Hyon, "Hamiltonian-based running control of dynamic legged robots," *Systems, Control and Information*, vol. 49, no. 7, pp. 260–265, 2005, (in Japanese).
12. S. Satoh, K. Fujimoto, and S. Hyon, "Gait generation for passive running via iterative learning control," in *Proc. IEEE/RSJ International Conference on Intelligent Robots and Systems*, 2006, pp. 5907–5912.
13. "Biped gait generation via iterative learning control including discrete state transitions," 2007, submitted.

On the Interconnection Structures of Irreversible Physical Systems

Damien Eberard[1], Bernhard Maschke[2], and Arjan J. van der Schaft[1]

[1] Institute for Mathematics and Computer Science, University of Groningen, the Netherlands
eberard@math.rug.nl

[2] Laboratoire d'Automatique et de Génie des Procédés UMR CNRS 5007, Université Lyon 1, France
maschke@lagep.univ-lyon1.fr

Summary. An energy balance equation with respect to a control contact system provides port outputs which are conjugated to inputs. These conjugate variables are used to define the composition of port contact systems in the framework of contact geometry. We then propose a power-conserving interconnection structure, which generalizes the interconnection by Dirac structures in the Hamiltonian formalism. Furthermore, the composed system is again a port contact system, as illustrated on the example of a gas-piston system undergoing some irreversible transformation.

1 Introduction

The properties of the control systems arising from models of physical systems reveal to be extremely useful for the design of control laws of nonlinear systems not only for stabilizing purposes but also for the design of the closed-loop behavior [19]. For electro-mechanical systems, the Lagrangian and Hamiltonian framework revealed to be best suited to represent their physical properties. Their state space is naturally endowed with symplectic, Poisson or Dirac structures arising from the variational formulation and eventual symmetries [1] [3] [11] or directly from the interconnection structure of complex physical systems like electrical or hydraulic networks and spatial mechanisms [12] and constraints [20] [21]. However the Hamiltonian and Lagrangian framework represent only the dynamics of reversible physical systems (in the sense of Thermodynamics). If the dissipation can no more be neglected, usually the Hamiltonian systems are augmented with a dissipative output feedback term [4] and lead to define dissipative Hamiltonian systems (defined on a so-called Leibniz bracket [15]) augmented with input and outputs maps [19]. From a thermodynamic perspective this means that the Hamiltonian function represent the free energy of the physical system which is not conserved. However there is a way of representing simultaneously (internal) energy conservation and irreversibility which uses model structures arising from Irreversible Thermodynamics and which has been developed in the context of Chemical Engineering. These systems are defined on the differentiable manifolds endowed with a contact structure which is canonically associated with the phase

F. Bullo et al. (Eds.): Lag. & Hamil. Methods for Nonlin. Ctrl., LNCIS 366, pp. 209–220, 2007.
springerlink.com

space in Reversible Thermodynamics [9] [3] [16]. On these contact manifolds, a class of control systems has been defined, called *conservative control contact systems*, which encompasses both reversible and irreversible systems and allows to express in both cases the conservation of the total energy [5] [6].

In this paper we shall consider port contact systems as defined in [7] and define their composition by power continuous interconnection structure which strictly generalize Dirac structures used for the composition of port Hamiltonian systems [21].

The sketch of the paper is the following. In the second section we shall briefly recall the definition and motivation of conservative contact systems. In the third section, we shall recall the definition of the port conjugated variables and then define the composition (or interconnection) of conservative contact systems. This is then illustrated in details on the example of a gas in a cylinder submitted to irreversible transformations.

2 Conservative Contact Systems

In this section we shall recall the motivation and definition of conservative contact systems. The reader is referred to [11] [3] and [1] for a detailed definition of the objects of contact geometry and will find a very brief summary in the appendix of this paper.

2.1 Definition and Motivation

The properties of thermodynamical systems are mathematically defined as the space of 1-jets of real functions of its extensive variables corresponding to their fundamental equation [8] [9] [13]. If one denotes the manifold of extensive variables (excluding the internal energy) by $\mathcal{N}$, it is known that the space of 1-jets of functions on $\mathcal{N}$ may be identified with $\mathbb{R} \times T^*\mathcal{N}$ [11]. This space is called the Thermodynamic Phase Space and has a canonical geometric structure, called contact form, which plays an analogous role as the Liouville form for cotangent bundle of differentiable manifolds. It may then be shown that the Thermodynamic Properties of physical systems (including thermodynamical systems but also mechanical and electro-magnetic systems) is defined by a Legendre submanifold of the associated Thermodynamic Phase Space $\mathcal{T}$ [9] [3] [13] [14].

It has been shown in previous work that the dynamics of *open* physical systems undergoing reversible or irreversible transformations may also be formulated in terms of contact geometry [5] [7] [6]. This has lead to the definition of a class of control contact systems that we have called conservative control contact systems [5] [6].

Definition 1. *A* conservative control contact system *is defined by the* $m+5$-*tuple* $(\mathcal{M}, \mathcal{E}, \mathcal{L}, K_0, \ldots, K_m, \mathcal{U})$, *where* $(\mathcal{M}, \mathcal{E})$ *is a strict contact manifold with* $\mathcal{L}$ *a given Legendre submanifold, the* K_i*'s are smooth real-valued functions on* $\mathcal{M}$, *called* contact Hamiltonians, *which are identically zero on* $\mathcal{L}$, *and* $\mathcal{U}$ *denotes* *the space of inputs functions u_j. The dynamics is then given by the differential equation:*

$$\frac{d}{dt}(x^0, x, p) = X_{K_0} + \sum_{j=1}^{m} u_j X_{K_j} . \tag{1}$$

This system may be interpreted in the context of physical system' modelling as follows. Firstly the differential equation (1) defines *a control contact system* in a very similar way to control Hamiltonian systems [17] [18]. It is defined by an internal contact Hamiltonian K_0 generating the drift dynamics X_{K_0} and by interaction contact Hamiltonians K_j defining the external action on the system by the control vector fields X_{K_j}. It is interesting to note that, for physical systems, *the contact Hamiltonians have the dimension of power* and that they are defined by the *law of fluxes* of the system (heat conduction, chemical reaction kinetics or diffusion).

However the system is also defined by a second objet: the Legendre submanifold $\mathcal{L}$ which represents the Thermodynamic properties of the system. Practically this Legendre submanifold is generated by a potential energy function (for instance, the internal energy or free energy of a thermodynamic system, the kinetic and potential energy of a mechanical system). The contact Hamiltonians have to satisfy the *compatibility conditions, i.e.* $K_{i|\mathcal{L}} \equiv 0$, which are essential as they express the first principle of Thermodynamics. In other words the conservative control contact system leaves invariant the Legendre submanifold. More precisely these conditions follow from the following result [16].

Theorem 1. *Let $(\mathcal{M}, \mathcal{E})$ be a strictly contact manifold and denote θ its contact form. Let $\mathcal{L}$ denote a Legendre submanifold. Then X_f is tangent to $\mathcal{L}$ if and only if f is identically zero on $\mathcal{L}$.*

Finally note that actually only the restriction of the conservative contact system to the Legendre submanifold (where the first principle is satisfied) is relevant for the description of the dynamics of the system.

Example 1. (Lift of a dissipative Hamiltonian system) This example has been treated in details in [6]. Consider an autonomous *dissipative* Hamiltonian system defined on $\mathcal{N}$ by the equation

$$\dot{x} = (J(x) - D(x)) \frac{\partial H_0}{\partial x}(x) , \tag{2}$$

where D is the symmetric positive definite matrix of friction. Notice that the tensor $J - D$ defines a Leibniz bracket [15]. It has been shown that it may be embedded into a contact vector field considering:

- the extended base manifold $\mathcal{N}_e = \mathbb{R} \times \mathcal{N}$, and its associated *extended* thermodynamic phase space $\mathcal{T}_e = \mathbb{R} \times T^*\mathcal{N}_e \ni (x^0, x, S, p, p_S)$
- the Legendre submanifold generated by H_e: $H_e(x, S) = H_0(x) + T_0 S$
- the contact Hamiltonian function

$$K_e = -\langle p, \dot{x} \rangle + \frac{p_S}{T_0} \frac{\partial H_0}{\partial x}^t D(x) \frac{\partial H_0}{\partial x} . \tag{3}$$

3 Interconnection of Port Contact Systems

In this section, the definition of port outputs conjugated to the control inputs is recalled [7] [6]. The composition of port contact systems is then defined using a power continuous interconnection structure which is not necessarily a Dirac structure, generalizing hence the result in [7]. This latter result is illustrated on the example of a gas under in a cylinder under a piston undergoing some irreversible processes.

3.1 Port Contact Systems and Losslessness

Consider a differentiable real-valued function f on $\mathcal{M}$ and its variation with respect to a conservative control contact system (of definition 1). A straightforward calculation, given below in canonical coordinates[1], leads to the following balance equation :

$$\frac{\mathrm{d}f}{\mathrm{d}t} = \sum_{j=1}^{m} y_f^j + s_f \,, \tag{4}$$

where y_f^j denotes the *f-conjugated output variable* associated with the input u_j:

$$y_f^j = \{K_j, f\} + f \frac{\partial K_j}{\partial x^0} \,, \tag{5}$$

and s_f denotes the *source term* defined by:

$$s_f = \{K_0, f\} + f \frac{\partial K_0}{\partial x^0} \,. \tag{6}$$

For a conserved quantity, the source term is expected to be zero. However, as has been shown in [7] there is no reason to require it on the entire state space but rather only on the Legendre submanifold. This lead to the following definition of a conserved quantity.

Definition 2. *A* conserved quantity *of a conservative control contact system is a real-valued function f defined on $\mathcal{M}$ such that*

$$s_{f|\mathcal{L}} = 0 \,. \tag{7}$$

Definition 3 ([7]). *A* port contact system *is a control contact system with the additional condition that there exists a generating function U of a Legendre submanifold that is a conserved quantity, completed with the U-conjugated output y_U^j defined as in (5).*

Example 2. [5] Consider a port Hamiltonian system defined on a manifold $\mathcal{N}$ endowed with the pseudo-Poisson tensor Λ, the Hamiltonian $H_0(x) \in C^\infty(\mathcal{N})$, an

[1] Recall that the *Jacobi bracket* $\{\cdot,\cdot\}$ of functions on $\mathcal{M}$ is defined as $\{f, g\} = i([X_f, X_g])\theta$, where θ is the contact form defining $\mathcal{E}$, and $[\cdot,\cdot]$ denotes the Lie bracket.

input vector $u(t) = (u_1, \ldots, u_m)^T$ function of t, m input vector fields $g_1, \ldots, g_m$ on $\mathcal{N}$, and the equations :

$$\begin{cases} \dot{x} = \Lambda^{\#}(\mathrm{d}_x H_0(x)) + \sum_{i=1}^m u_i(t)\, g_i(x) \\ {y_p}^j = L_{g_j}.H_0(x) \end{cases} \tag{8}$$

Its lift on the thermodynamic phase space $\mathcal{T} = \mathbb{R} \times T^*\mathcal{N}$ with canonical coordinates (ϵ, x, p), is a port contact system with internal contact Hamiltonian $K_0 = \Lambda(p, dH_0)$, the internal contact interaction contact Hamiltonian are $K_j = \langle \mathrm{d}H_0 - p, g_j \rangle$. The port conjugated output variables defined in (5) becomes

$$y^j_{H_0} = \frac{\partial H_0}{\partial x}^T g_j = L_{g_j} H_0 \,. \tag{9}$$

It is remarkable that they correspond precisely to the outputs called *port outputs* defined in (8).

3.2 Interconnection of Two Port Contact Systems

In this section we consider the interconnection or composition of port contact systems.

Consider now two differential manifolds $\mathcal{N}_i$ of dimension n_i with coordinates $x_i = (x_i^1, \ldots, x_i^{n_i})$, for $i = 1, 2$. Each 1-jet space $\mathcal{T}^i$ over $\mathcal{N}_i$ is endowed with a canonical contact structure whose contact form is denoted by θ_i. We now construct the composed state space in the same way. Denote by $\mathcal{N}$ the whole product base space $\mathcal{N}_1 \times \mathcal{N}_2$. Then, the 1-jet bundle $\mathcal{T}$ over $\mathcal{N}$ is also endowed with a canonical contact form θ whose local expression is

$$\theta = \mathrm{d}x^0 - \sum_{j=1}^{n_1+n_2} p_j\, \mathrm{d}x^j, \tag{10}$$

where $x^j = x_1^j$ and $p_j = p_j^1$ if $1 \leq j \leq n_1$, else $x^j = x_2^{j-n_1}$ and $p_j = p_{j-n_1}^2$ if $n_1 + 1 \leq j \leq n_1 + n_2$.

According to definition 3, consider two port contact systems $(\mathcal{N}_i, U_i, K_j^i)$ on $\mathcal{T}^i$ with contact Hamiltonian K_i defined as

$$K_i = K_0^i + \sum_{j=1}^{n_i} u_j^i\, K_j^i\,, \tag{11}$$

satisfying the invariance condition with respect to the conserved quantity U_i, $i = 1, 2$. We define the new (conserved) generating function U of the Legendre submanifold of the composed state space as $U_1 + U_2$.

Denote by m the number of input variables involved in the interconnection ($m \leq min(m_1, m_2)$). Without loss of generality we may suppose that the first m variables are involved in the interconnection. Denote $u = (u_j^1, u_j^2)$ and $y = (y_1^j, y_2^j)$, $j = 1, \ldots, m$.

Proposition 1. *The composition of two port contact systems* $(\mathcal{N}_i, U_i, K_j^i)_{i=1,2}$ *with respect to a* power continuous interconnection relation

$$u = \Phi(y) \quad \textit{together with} \quad \Phi(y)^T y = 0\,, \tag{12}$$

is the port contact system on the Thermodynamical Phase Space $\mathbb{R}\times T^*(\mathcal{N}_1 \times \mathcal{N}_2)$ *defined with respect to the Legendre* $\mathcal{L}_{U_1+U_2}$ *and the contact Hamiltonian* $K(x,p,\Phi(y))$, *where* $K(x,p,u^1,u^2) = K_1 + K_2$.

It is obvious to see that the invariance condition is satisfied by K on $\mathcal{L}_U$. We now show that U is a conserved quantity of the interconnected system thus obtained, when restricted to the Legendre submanifold $\mathcal{L}_U$. Indeed, let us compute its time-derivative

$$\frac{\mathrm{d}U}{\mathrm{d}t}_{|\mathcal{L}_U} = \sum_{j=1}^{m} \left[u_1^j \{K_j^1, U_1\} + u_2^j \{K_j^2, U_2\} \right]_{|\mathcal{L}_U}, \tag{13}$$

which is zero by (12) and (5).

The interconnection defined in the proposition 1, strictly generalize the interconnections defined by Dirac structures [20] [21]. Indeed the map Φ in equation (12) defines a power continuous relation but is not necessarily linear hence does not define a vector bundle. As an illustration of this feature we shall present in the next paragraph an example of such a non-linear power continuous interconnection.

3.3 A Gas in a Cylinder Under a Piston

In this paragraph will shall consider a system composed of a gas in a cylinder closed by a piston subject to the gravity. In a first instance, we shall consider that this system undergoes *reversible* transformations. In this case the interconnection structure defining the interaction between the gas and the piston is defined by a Dirac structure (as has been shown in [10]). In a second instance, we shall consider that the system undergoes some *irreversible* transformations due to mechanical friction or viscosity of the gas. In this case the interconnection structure is defined by some map Φ which is non-linear.

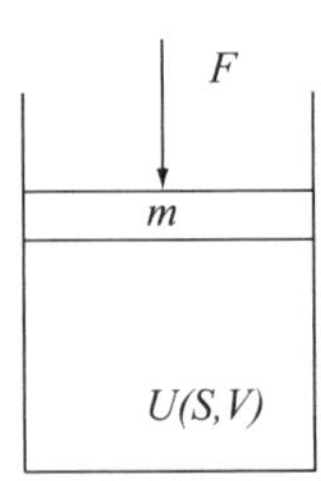

Fig. 1. A gas under a piston

Reversible Transformations Gas Under a Piston

This system may be decomposed in two elementary subsystems, namely the ideal gas undergoing some mechanical work and a mass (the piston) submitted to external forces.

The dynamics of the ideal gas undergoing some mechanical work may be defined as a conservative contact system defined on the Thermodynamic Phase Space $\mathcal{T}_{gas} = \mathbb{R} \times \mathbb{R}^6 \ni \{x^0, x^j, p_j\}$ where x^i denotes the extensive variables and p_i the conjugated intensive variables. Its thermodynamic properties are given by the Legendre submanifold $\mathcal{L}_U$ generated by the internal energy. As the gas is considered to be in equilibrium in the control volume, the drift dynamics is of course zero. And the external mechanical work provided by an external pressure P^e and variation of volume f_V^e leads to the interaction contact Hamiltonian:

$$K_{gas}^i = (p_2 - P)\ f_V^e\,, \tag{14}$$

with the U-conjugated port output $y_U = -P$.

The dynamics of the piston is given as the lift of a standard Hamiltonian systems on the associated Thermodynamic Phase Space following [5]:

$$\mathcal{T}_{mec} = \mathbb{R} \times \mathbb{R}^4 \ni \left\{x^0, x^{pot}, x^{kin}, p_{pot}, p_{kin}\right\}\,, \tag{15}$$

where x^{pot} denotes the altitude of the piston and x^{kin} its kinetic momentum. The Legendre submanifold $\mathcal{L}_{mec}$ is generated by the total mechanical energy: $H_0 = \frac{1}{2m}\, {x^{kin}}^2 + mgx^{pot}$, which defines the two intensive variables, the velocity v of the piston and the gravity force F:

$$\left.\begin{array}{l} p_{pot}|_{\mathcal{L}_{mec}} = F = mg \\ p_{kin}|_{\mathcal{L}_{mec}} = v = \frac{x^{kin}}{m} \end{array}\right\}. \tag{16}$$

The drift dynamics of the piston may be represented by a conservative contact system with internal contact Hamiltonian

$$K_{mec}^0 = -\left(p_{pot},\, p_{kin}\right) \begin{pmatrix} 0 & 1 \\ -1 & 0 \end{pmatrix} \begin{pmatrix} F \\ v \end{pmatrix}. \tag{17}$$

Associated with the external force F^e exerted on the piston is the interaction contact Hamiltonian:

$$K_{mec}^i = (p_{kin} - v)\ F^e\,, \tag{18}$$

and the H_0-conjugated output: $y_{H_0} = v$, *i.e.* the velocity of the piston. In this first case, following [10], we consider that the composed system is *reversible* and the interconnection is given by the linear relation :

$$\begin{pmatrix} f_V^e \\ F^e \end{pmatrix} = \begin{pmatrix} 0 & A \\ -A & 0 \end{pmatrix} \begin{pmatrix} (-P) \\ v \end{pmatrix}, \tag{19}$$

where A denotes the area of the piston and defining a Dirac structure on the port variables $f_V^e, F^e, (-P), v$.

The dynamics of the gas with piston is then given as the composed conservative contact system on the composed Thermodynamic Phase Space:

$$\begin{aligned} \mathcal{T}_{GP} &= \mathbb{R} \times \mathbb{R}^{10} \\ &\ni \left\{x^0, x^j, p_j, x^{pot}, x^{kin}, p_{pot}, p_{kin}\right\}_{j=1,..,3}, \end{aligned} \tag{20}$$

with Legendre submanifold generated by the sum of the internal energy and the mechanical energy $U(S, V, N) + H_0(x^{pot}, x^{kin})$ with contact Hamiltonian:

$$K_{tot} = K^0_{mec} + (p_2 - P)\ Av + (p_{kin} - v)\ AP \tag{21}$$

It is interesting to notice that the x-component of the contact field $X_{K_{tot}}$ restricted to $\mathcal{L}_{U+H_0}$ is precisely the port Hamiltonian system proposed in [10].

Irreversible Transformations

In this second case we shall assume that there is some mechanical friction and that the *lost mechanical energy is converted entirely into a heat flow in the gas.*

Firstly the thermodynamical model of the gas, $\mathcal{L}_U$, is remained unchanged. But its dynamics has to be changed as follows, as now there might be a heat flow induced by the mechanical losses. Denoting by $f^e_S = \frac{Q_e}{T}$ the flow of entropy associated with the external heat flow Q_e, there has to be *an additional interaction contact Hamiltonian* to consider:

$$K_S = (p_1 - T)\ f^e_S\,, \tag{22}$$

with its conjugate port output $y^s_U = T$.

Consider that the mechanical losses come from a viscous friction with coefficient ν. The friction losses will be generated by an additional interaction Hamiltonian associated with the dissipative force F^d:

$$K^d_{mec} = (p_{kin} - v)\ F^d\,, \tag{23}$$

with conjugated output v.

The dissipation losses will be taken into account as *an additional interconnection relation* associated with the transformation of part of the mechanical energy in heat. Firstly the friction force is defined and secondly the power continuity of the interconnection defined as follows :

$$(F^d, f^e_S) = \Phi(v, T) = (\nu v, (-v/T)\nu v) \tag{24}$$

An essential feature of this power continuous interconnection is, that it is not a Dirac structure on the port variables (F^d, f^e_S, v, T) as the map Φ in (24) is nonlinear.

The *composed system piston and gas* is defined on $\mathcal{T}_{GP}$, defined in (20), with the canonical contact structure. The properties of the systems are defined precisely as for the reversible case by the Legendre submanifold $\mathcal{L}_{U+H_0}$. The contact Hamiltonian is obtained according to the proposition 1

$$K_{irr} = K_{tot} + \big(p_{kin} - (p_1/T)v\big)\nu v\,. \tag{25}$$

It is immediately seen that the contact Hamiltonian satisfies the invariance condition $K_{irr}|_{\mathcal{L}_{U+H_0}} = 0$. Note that an analogous expression hold if one consider that the gas undergoes some irreversible transformation due to its viscosity. The dynamics restricted to the the Legendre submanifold and projected on the extensive coordinates is:

$$\left.\begin{array}{l}
\left.\frac{dx^S}{dt}\right|_{\mathcal{L}} = \frac{dS}{dt} = -\frac{\partial K_{tot}}{\partial p_S} = \frac{1}{T}\nu v^2 \\
\left.\frac{dx^V}{dt}\right|_{\mathcal{L}} = \frac{dV}{dt} = -\frac{\partial K_{tot}}{\partial p_V} = A\,v \\
\left.\frac{dx^N}{dt}\right|_{\mathcal{L}} = \frac{dN}{dt} = -\frac{\partial K_{tot}}{\partial p_N} = 0 \\
\left.\frac{dx^{pot}}{dt}\right|_{\mathcal{L}} = \frac{dz}{dt} = -\frac{\partial K_{tot}}{\partial p_{pot}} \quad v \\
\left.\frac{dx^{kin}}{dt}\right|_{\mathcal{L}} = \frac{d\pi}{dt} = -\frac{\partial K_{tot}}{\partial p_{kin}} = -F + A\,P = -mg + AP
\end{array}\right\} \tag{26}$$

It may be noted that this formulation of an irreversible transformation of a gas-cylinder system encompasses (by setting $\nu = 0$) the formulation of reversible transformation using a port Hamiltonian system defined on a Dirac structures proposed [10].

4 Conclusion

In this paper we have suggested a definition of power continuous interconnection of conservative contact systems which strictly generalizes the Dirac structures which define the interconnection of port Hamiltonian systems [21]. However the power continuity properties of the interconnection allow to compose conservative contact systems. We have illustrated this in details on the example of a gas in a cylinder and submitted to mechanical work by a piston. On this example we have considered firstly the reversible case, according to the example in [10], and secondly the irreversible by considering some mechanical friction and the entropy balance associated with it.

Acknowledgments

This work was partially supported by the EU-project GeoPleX (www.geoplex.cc) EU-IST-2001-34166.

References

1. Ralph Abraham and Jerrold E. Marsden. *Foundations of Mechanics.* Addison, ii edition, March 1994. ISBN 0-8053-0102-X.
2. V.I. Arnold. *Equations Diffrentielles Ordinaires.* Editions Mir, Moscou, 3-ime dition edition, 1978.

3. V.I. Arnold. *Mathematical Methods of Classical Mechanics.* Springer, ii edition, 1989. ISBN 0-387-96890-3.
4. M. Dalsmo and A.J. van der Schaft. On representations and integrability of mathematical structures in energy-conserving physical systems. *SIAM Journal of Control and Optimization*, 37(1):54–91, 1999.
5. D. Eberard, B.M. Maschke, and A.J. van der Schaft. Conservative systems with ports on contact manifolds. In *Proc. IFAC World Congress*, Prague, July 2005.
6. D. Eberard, B.M. Maschke, and A.J. van der Schaft. Port contact systems for irreversible thermodynamical systems. In *Proc. 44th IEEE Conference on Decision and Control and European Control Conference ECC 2005*, Seville, Spain, December, 12th-15th 2005.
7. D. Eberard, B.M. Maschke, and A.J. van der Schaft. Systems theory of interconnected port contact systems. In *Proc. 2005 International Symposium on Nonlinear Theory and its Applications*, Bruges, Belgium, October 18 - 21 2005.
8. J.W. Gibbs. *Collected Works*, volume I: Thermodynamics. Longmans, New York, 1928.
9. R. Herman. *Geometry, Physics and Systems.* Dekker, New-York, 1973.
10. R. Jongschaap and H.C. ttinger. The mathematical representation of driven thermodynamical systems. *J. Non-Newtonian Fluid Mech.*, 120:3–9, 2004.
11. Paulette Libermann and Charles-Michel Marle. *Symplectic Geometry and Analytical Mechanics.* D. Reidel Publishing Company, Dordrecht, Holland, 1987. ISBN 90-277-2438-5.
12. B.M. Maschke, A.J. van der Schaft, and P.C. Breedveld. An intrinsic Hamiltonian formulation of network dynamics: Non-standard Poisson structures and gyrators. *Journal of the Franklin institute*, 329(5):923 966, 1992. Printed in Great Britain.
13. R. Mrugala. Geometrical formulation of equilibrium phenomenological thermodynamics. *Reports on Mathematical Physics*, 14(3):419–427, 1978.
14. R. Mrugala. A new representation of Thermodynamic Phase Space. *Bull. of the Polish Academy of Sciences*, 28(1):13–18, 1980. series on Pysical Sci. and Astronomy.
15. J.-P. Ortega and V. Planas-Bielsa. Dynamics on Leibniz manifolds. *J. of Geometry and Physics*, 52:pp. 1–27, 2004.
16. J.C. Schon R. Mrugala, J.D. Nulton and P. Salamon. Contact structure in thermodynamic theory. *Reports in Mathematical Physics*, 29(1):109–121, 1991.
17. R.W.Brockett. *Geometric Control Theory*, volume 7 of *Lie groups: History, Frontiers and Applications*, chapter Control theory and analytical mechanics, pages 1–46. Math.Sci.Press,, Brookline, 1977. C.Martin and R.Herman eds.
18. A. van der Schaft. *Three Decades of Mathematical System Theory*, volume 135 of *Lect. Notes Contr. Inf. Sci.*, chapter System Theory and Mechanics, page 426452. Springer, Berlin, 1989.
19. A.J. van der Schaft. *L_2-Gain and Passivity Techniques in Nonlinear Control.* Springer Communications and Control Engineering series. Springer-Verlag, London, 2nd revised and enlarged edition, 2000. first edition Lect. Notes in Control and Inf. Sciences, vol. 218, Springer-Verlag, Berlin, 1996.
20. A.J. van der Schaft and B.M. Maschke. The Hamiltonian formulation of energy conserving physical systems with external ports. *Archiv für Elektronik und Übertragungstechnik*, 49(5/6):362–371, 1995.
21. A.J. van der Schaft and B.M. Maschke. *Modelling and Control of Mechanical Systems*, chapter Interconnected Mechanical systems. Part1: Geometry of interconnection and implicit Hamiltonian systems, pages 1–16. Imperial College Press, London, 1997. ISBN 1-86094-058-7.

5 Appendix: Reminder on Contact Geometry

In this section we shall briefly recall the basic concepts of contact geometry (following [11] and [3]).

Let $\mathcal{M}$ be an $2n+1$-dimensional, connected, differentiable manifold of class C^∞.

Definition 4 ([11]). *A Pfaffian equation on $\mathcal{M}$ is a vector subbundle $\mathcal{E}$ of rank 1 of $T^*\mathcal{M}$. The pair $(\mathcal{M}, \mathcal{E})$ is a* strictly contact structure *if there exists a form θ of constant class $2n+1$, called* contact form, *that determines $\mathcal{E}$.*

Using Darboux's theorem, one shows the existence of canonical coordinates $(x^0, x^1, \ldots, x^n, p_1, \ldots, p_n)$ in a neighborhood V of any $x \in \mathcal{M}$ such that

$$\theta_{|V} = \mathrm{d}x^0 - p_i \mathrm{d}x^i. \tag{27}$$

Definition 5 ([11]). *A* Legendre submanifold *of a $(2n+1)$-dimensional contact manifold $(\mathcal{M}, \mathcal{E})$ is an n-dimensional submanifold $\mathcal{L}$ of $\mathcal{M}$ that is an integral manifold of $\mathcal{E}$.*

Legendre submanifolds are locally generated by some generating function.

Theorem 2 ([2]). *For a given set of canonical coordinates and any partition $I \cup J$ of the set of indices $\{1, \ldots, n\}$ and for any differentiable function $F(x^I, p_J)$ of n variables, $i \in I, j \in J$, the formulas*

$$x^0 = F - p_J \frac{\partial F}{\partial p_J}, \quad x^J = -\frac{\partial F}{\partial p_J}, \quad p_I = \frac{\partial F}{\partial x^I} \tag{28}$$

define a Legendre submanifold of $\mathbb{R}^{2n+1}$ denoted $\mathcal{L}_F$. Conversely, every Legendre submanifold of $\mathbb{R}^{2n+1}$ is defined in a neighborhood of every point by these formulas, for at least one of the 2^n possible choices of the subset I.

Finally we shall recall the definition of the class of vector fields, called contact vector fields, which preserve the contact structure and may be characterized using the following result.

Proposition 2 ([11]). *A vector field X on $(\mathcal{M}, \mathcal{E})$ is an* contact vector field *if and only if there exists a differentiable function ρ such that*

$$\mathcal{L}(X)\,\theta = \rho\,\theta, \tag{29}$$

where $\mathcal{L}(X)$ denotes the Lie derivative with respect to the vector field X. When ρ vanishes, X is an infinitesimal automorphism of the contact structure.

The set of contact vector fields forms a Lie subalgebra of the Lie algebra of vector fields on $\mathcal{M}$.

Analogously to the case of Hamiltonian vector fields, one may associate some generating function to the contact vector fields. Actually there exists an isomorphism Φ between contact vector fields and differentiable function on $\mathcal{M}$ which associate to a contact vector field X a function called *contact Hamiltonian* and defined by :

$$\Phi(X) = i(X)\theta, \tag{30}$$

where $i(X)$ denotes the contraction of a form by the vector field X. In the sequel we shall denote the contact vector field associated with a function f by :

$$X_f = \Phi^{-1}(f). \tag{31}$$

The contact vector field X_f may be expressed in canonical coordinates in terms of the generating function, as follows :

$$\begin{aligned} X_f = &\left(f - \sum_{k=1}^{n} p_k \frac{\partial f}{\partial p_k}\right) \frac{\partial}{\partial x^0} + \frac{\partial f}{\partial x^0} \left(\sum_{k=1}^{n} p_k \frac{\partial}{\partial p_k}\right) \\ &+ \sum_{k=1}^{n} \left(\frac{\partial f}{\partial x^k} \frac{\partial}{\partial p_k} - \frac{\partial f}{\partial p_k} \frac{\partial}{\partial x^k}\right). \end{aligned} \tag{32}$$

Furthermore the isomorphism Φ transports the Lie algebra structure on the differentiable function on $\mathcal{M}$ and defines the following bracket that we shall use in the sequel :

$$\{f, g\} = i([X_f, X_g])\theta, \tag{33}$$

whose expression in canonical coordinates is given by

$$\begin{aligned} \{f, g\} = &\sum_{k=1}^{n} \left(\frac{\partial f}{\partial x^k} \frac{\partial g}{\partial p_k} - \frac{\partial g}{\partial x^k} \frac{\partial f}{\partial p_k}\right) \\ &+ \left(f - \sum_{k=1}^{n} p_k \frac{\partial f}{\partial p_k}\right) \frac{\partial g}{\partial x^0} - \left(g - \sum_{k=1}^{n} p_k \frac{\partial g}{\partial p_k}\right) \frac{\partial f}{\partial x^0} . \end{aligned} \tag{34}$$

Rolling Problems on Spaces of Constant Curvature

Velimir Jurdjevic[1] and Jason Zimmerman[2]

[1] University of Toronto, Toronto, Canada
jurdj@math.utoronto.ca
[2] University of Toronto, Toronto, Canada
a.n.j@sympatico.ca

Summary. The rolling sphere problem on $\mathbb{E}^n$ consists of determining the path of minimal length traced by the point of contact of the oriented unit sphere $\mathbb{S}^n$ as it rolls without slipping between two boundary points of $\mathbb{E}^n \times SO_n$. This problem is extended to the following cases of rolling: $\mathbb{H}^n$ on $\mathbb{E}^n$, $\mathbb{S}^n_\rho$ on $\mathbb{S}^n_\sigma$, and $\mathbb{H}^n_\rho$ on $\mathbb{H}^n_\sigma$, where $\sigma \neq \rho$ are the radii of the spheres or hyperboloids. The term "rolling" is generalized to an isometric sense: the length of a curve is measured using the Riemannian metric of the stationary manifold while the orientation of the rolling object is described by a matrix from its isometry group. These problems constitute left-invariant optimal control problems on Lie groups, whose Hamiltonian equations reveal certain integrals of motion and show, on the level of Lie algebras, that all of the above problems are governed by a single set of equations.

Keywords: Hamiltonian systems, isometric, rolling, sub-Riemannian, sphere.

1 Introduction: Euclidean Cases

The rolling sphere problems arise from the following kinematic question posed by Hammersley [3] and later by Brockett and Dai [2]:

Suppose that an oriented ball is rolled on a plane by moving a plate that remains parallel to the plane while touching the ball. Assume that the ball *rolls without slipping*: it does not slide on either plate and the moving plate cannot rotate about its normal axis. How should the ball be rolled from one orientation to another and from one point of the stationary plate to another so that the path on the stationary plate traced by the point of contact has minimal length?

To formulate this kinematic problem as a mathematical problem, it is natural to identify the stationary plate with the plane $\mathbb{E}^2 = \{x \in R^3 : x_1 = 0\}$ in the ambient space $\mathbb{R}^3$ having canonical frame $\{e_1, e_2, e_3\}$. The ball is modelled by the unit sphere $\mathbb{S}^2$ whose center has its first coordinate equal to 1. The orientation of the sphere is defined by a matrix R in $SO_3(\mathbb{R})$ satisfying $f_i = R\, e_i$, where $\mathcal{F} = (f_1, f_2, f_3\}$ is an orthonormal frame affixed to the sphere at its center. In this setting, each point (x, R) of the configuration space $\mathbb{E}^2 \times SO_3$ corresponds to the sphere having orientation R whose point of contact with the stationary plane is given by x.

F. Bullo et al. (Eds.): Lag. & Hamil. Methods for Nonlin. Ctrl., LNCIS 366, pp. 221–231, 2007.
springerlink.com

The assumption of rolling without slipping means that the velocity of the point of contact relative to the stationary plate is equal to zero at all times during a roll. The fact that the motions of the sphere are controlled by the moving plate implies that the rolling along a particular curve $x(t)$ in the stationary plane determines the orientation $R(t)$ of the sphere during the roll: that is, curves $(x(t), R(t))$ and $(x(t), S(t))$ in $\mathbb{E}^2 \times SO_3(R)$ correspond to the actual motions of the sphere if and only if $R(t) = S(t)$ for all t. It was then shown in Jurdjevic [5] that the motions of the oriented sphere subject to the above constraints are given by the following system of differential equations in $G = \mathbb{E}^2 \times SO_3$:

$$\frac{\mathrm{d}x}{\mathrm{d}t} = \begin{pmatrix} 0 \\ u \end{pmatrix} \qquad \frac{\mathrm{d}R}{\mathrm{d}t} = R \begin{pmatrix} 0 & -u^{\mathrm{T}} \\ u & \mathbf{0} \end{pmatrix} \tag{1}$$

where $u \in L^\infty([0,T], \mathbb{R}^2)$ and $[0,T]$ is a finite interval; $u(t)$ will be called the *control*. The solutions $g(t) = (x(t), R(t))$ of (1) are then referred to as the *trajectories generated by the control* $u(t)$.

Since $\int_0^T \sqrt{u_1^2(t) + u_2^2(t)}\ \mathrm{d}t$ is the length of $x(t)$ in the interval $[0,T]$, the kinematic problem of Hammersley and Brockett is naturally phrased as an optimal two-point boundary value control problem on G. To arrive at the precise formulation, define $\int_0^T \sqrt{u_1^2(t) + u_2^2(t)}\ \mathrm{d}t$ to be the *length* of a trajectory $g(t) = (x(t), R(t))$ in the interval $[0,T]$. If g_0 and g_1 are any fixed points in G, let $Traj(g_0, g_1)$ denote the set of all trajectories $g(t) = (x(t), R(t))$ of (1) satisfying the boundary conditions $g(0) = g_0$, $g(T) = g_1$ (T depends on the control). The kinematic problem of Hammersley and Brockett can be stated as the problem of finding a trajectory in $Traj(g_0, g_1)$ whose length is minimal among all trajectories in $Traj(g_0, g_1)$. It is well known (for example, see Liu and Sussmann [9]) that the preceding problem can be recast as a time-optimal problem of finding a trajectory $g(t) = (x(t), R(t))$ of (1) with $g(0) = g_0$ that reaches g_1 in a minimum time by the trajectories generated by controls $u(t)$ constrained by $\|u(t)\| = 1$. In such a case the length of an optimal curve is equal to the minimal time of transfer from g_0 to g_1.

The extension of Hammersley-Brockett problem to arbitrary dimensions is immediate: simply replace equation (1) by an analogous equation in $G = \mathbb{E}^n \times SO_{n+1}$, where the control u is in $L^\infty([0,T], \mathbb{R}^n)$ with $[0,T]$ compact, and then paraphrase the planar optimal control problem in $G = \mathbb{E}^n \times SO_{n+1}$; this problem is studied in detail in Zimmerman [10]. However, it is less immediate that the n-dimensional Hammersley-Brockett problem has a natural formulation in the Lorentzian geometry of the ambient space $\mathbb{R}^{n+1}$ with the Euclidean inner product $\sum_{i=1}^{n+1} x_i y_i$ replaced by the Lorentzian inner product $x_1 y_1 - \sum_{i=2}^{n+1} x_i y_i$. In this geometry, the unit sphere $\mathbb{S}^n = \{x \in \mathbb{R}^{n+1} : \sum_{i=1}^{n+1} x_i^2 = 1\}$ is replaced by the unit hyperboloid $\mathbb{H}^n = \{x \in \mathbb{R}^{n+1} : x_1^2 - \sum_{i=2}^{n+1} x_i^2 = 1, x_1 < 0\}$ and the orientation of the hyperboloid is given by a matrix R in $SO_0(1,n)$, the connected component of the isometry group of $\mathbb{H}^n$ through the identity. The configuration space of the hyperboloid "rolling" on the Euclidean space $\mathbb{E}^n = \{x \in \mathbb{R}^{n+1} : x_1 = 0\}$ is equal to $G = \mathbb{E}^n \times SO_0(1,n)$.

The hyperbolic Hammersley-Brockett problem can be phrased together with its Euclidean antecedent as a sub-Riemannian optimal control problem on $G = \mathbb{E}^n \times SO_\varepsilon$ with $\varepsilon = \pm 1$, where $SO_1 = SO_{n+1}$ and $SO_{-1} = SO_0(1,n)$, in terms of the trajectories of

$$\frac{\mathrm{d}x}{\mathrm{d}t} = \begin{pmatrix} 0 \\ u \end{pmatrix} \qquad \frac{\mathrm{d}R}{\mathrm{d}t} = R \begin{pmatrix} 0 & -\varepsilon u^{\mathrm{T}} \\ u & \mathbf{0} \end{pmatrix} \tag{2}$$

where the control $u \in L^\infty([0,T], \mathbb{R}^n)$ and $[0,T]$ is a finite interval.

Definition 1. *Let $Traj_\varepsilon(g_0, g_1)$ denote the set of all trajectories of (2) that conform to the boundary condtions $g(0) = g_0$, $g(T) = g_1$. The Euclidean rolling sphere problem (**ERSP**) is the optimal control problem of finding a trajectory $g(t) = (x(t), R(t))$ in $Traj_\varepsilon(g_0, g_1)$ whose length $\int_0^T \sqrt{u_1^2(t) + \cdots + u_n^2(t)}\, dt$ is minimal relative to $Traj_\varepsilon(g_0, g_1)$.*

2 Non-Euclidean Rolling Problems

In contrast to the **ERSP**, where the stationary manifold is a Euclidean space $\mathbb{E}^n$, the non-Euclidean rolling sphere problem (**NRSP**) pertains to the two situations in which the manifold $\mathbb{S}^n_\rho$ or $\mathbb{H}^n_\rho$ rolls on the corresponding manifold $\mathbb{S}^n_\sigma$ or $\mathbb{H}^n_\sigma$ with $\sigma \neq \rho$. Additional definitions and notations are required to formulate these non-Euclidean cases. To begin with, amalgamate the Euclidean and the Lorentzian geometry of $\mathbb{R}^{n+1}$ in terms of a single parameter ε and write $\langle x, y\rangle_\varepsilon = x_1 y_1 + \varepsilon \sum_{i=2}^{n+1} x_i y_i$, for $\varepsilon = \pm 1$. The form corresponding to $\varepsilon = 1$ will be called *Euclidean* while the other will be called *hyperbolic*. These forms can be used to define the *Euclidean/hyperbolic length* $\|x\|_\varepsilon$ of any vector $x \in \mathbb{R}^{n+1}$ to be $\|x\|_\varepsilon = \sqrt{|\langle x, x\rangle|_\varepsilon}$; in the Euclidean case abbreviate $\|x\|_1$ to $\|x\|$. The group that leaves $\langle x, y\rangle_\varepsilon$ invariant will be denoted by O_ε and the connected components that contain the group identity will be denoted by SO_ε. It is well known that SO_ε is equal to SO_{n+1} or $SO_0(1,n)$ for ε equal to 1 or -1 respectively.

Now let $S_{\sigma,\varepsilon}(c) = \left\{ x \in \mathbb{R}^{n+1} \middle| \langle x - c, x - c\rangle_\varepsilon = \sigma^2 \right\}$ so that $S_{\sigma,1}(c)$ is the Euclidean sphere of radius σ centered at c, while $S_{\sigma,-1}(c)$ is the hyperbolic sphere of radius σ centered at c. A hyperbolic sphere is a manifold comprised of two disconnected sheets: $S^+_{\sigma,-1}(c)$ is the sheet lying in the region $x_1 > c_1$ and $S^-_{\sigma,-1}(c)$ is the sheet lying in the region $x_1 < c_1$. In general there are two kinds of rolling: rolling on the "outside" of the stationary manifold and rolling on the "inside" of the stationary manifold. Since conceptually there is little difference between the two kinds of rolling, only the outside rollings will be considered in this paper. To be consistent with the Euclidean situations, certain matching of sheets need to made in the hyperbolic case. For that reason, define $S^-_{\rho,-1}(c)$ to be the rolling sphere and $S^+_{\sigma,-1}(c)$ to be the stationary sphere. For convenience, let us refer to any $S^\pm_{\sigma,\varepsilon}(c)$ as a "sphere", and if $\sigma = 1$ then write $S_\varepsilon(c)$ in lieu of $S_{1,\varepsilon}(c)$.

An *oriented sphere* is a sphere $S_{\sigma,\varepsilon}(c)$ together with a frame of vectors $\mathcal{F} = \{f_1, f_2, ..., f_{n+1}\}$. Only frames $\mathcal{F}$ that satisfy $\mathcal{F} = R\mathcal{E}$ for some $R \in SO_\varepsilon$, where $\mathcal{E} = \{e_1, e_2, ..., e_{n+1}\}$ denotes the canonical frame in $\mathbb{R}^{n+1}$, will be considered.

(In this notation $\mathcal{F} = R\mathcal{E}$ means that $f_i = Re_i$ for $i = 1, ..., n+1$.) The matrix R will be referred to as the *orientation* of the sphere. Strictly speaking, an oriented sphere lives in the space $\mathbb{R}^{n+1} \times SO_\varepsilon$, but we will visualize the frame $\mathcal{F}$ as being attached to the centre of the sphere and view the oriented sphere as an object in $\mathbb{R}^{n+1}$.

Let us now turn our attention back to the **NRSP**, where the stationary manifold M_ε is either the Euclidean sphere $S_{\sigma,\varepsilon}(0)$ or the hyperbolic one-sheeted manifold $S^+_{\sigma,\varepsilon}(0)$ and the rolling manifold $M_{\rho,\varepsilon}$ is either the Euclidean sphere $S_{\rho,1}(c_0)$ or the hyperboloid $S^-_{\rho,-1}(c_0)$ with $\sigma \neq \rho$. It will be convenient to write $S_{\rho,\varepsilon}(c_0)$ for any of the preceding cases. "Rolling" motions of an oriented sphere $(S_{\rho,\varepsilon}(c_0), R_0)$ arise from the action of $SE_\varepsilon = \mathbb{R}^{n+1} \rtimes SO_\varepsilon$, a generalization of the group of motions, on $\mathbb{R}^{n+1} \times SO_\varepsilon$: the action of $(\bar{x}, \bar{R}) \in SE_\varepsilon$ on a point $(p, R) \in \mathbb{R}^{n+1} \times SO_\varepsilon$ is given by the formula

$$(\bar{x}, \bar{R}) \cdot (p, R) = (\bar{x} + c_0 + \bar{R}(p - c_0), \bar{R}R) \tag{3}$$

Any curve in SE_ε, that moves an oriented sphere $(S_{\rho,\varepsilon}(c_0), R_0)$ through $\mathbb{R}^{n+1}$ in such a way that the sphere always maintains tangential contact with a stationary manifold M, is said to induce an *isometric rolling* of $(S_{\rho,\varepsilon}(c_0), R_0)$ on M. See Jurdjevic and Zimmerman [8] for a formal definition of isometric rolling.

Proposition 1. *(Jurdjevic and Zimmerman [8]) Suppose that the sphere $S_{\rho,\varepsilon}(c_0)$, with initial orientation R_0, rolls isometrically without slipping on the outside of $M_\varepsilon = S_{\sigma,\varepsilon}(0)$ from an initial time 0 to some finite terminal time $T > 0$. Then the path $x(t)$ traced by the point of contact and the orientation $R(t)$ are given by the equations*

$$x = \sigma S e_0 \qquad R = \hat{S}^{-1} S$$

where $(S(t), \hat{S}(t))$ are the solution curves in $G_{\varepsilon,\varepsilon} = SO_\varepsilon \times SO_\varepsilon$ of the equations

$$\frac{dS}{dt} = \alpha S U_\varepsilon \qquad \frac{d\hat{S}}{dt} = \beta \hat{S} U_\varepsilon \tag{4}$$
$$\alpha = \frac{\rho}{\rho + \sigma} \qquad \beta = \alpha - 1.$$

The matrix $U_\varepsilon(t)$ is of the form $\begin{pmatrix} 0 & -\varepsilon u(t)^{\mathrm{T}} \\ u(t) & 0 \end{pmatrix}$ for some $u \in L^\infty([0, T], \mathbb{R}^n)$ and $S(0)$ is any matrix in SO_ε that satisfies $\sigma S(0) e_1 = \frac{\sigma}{\sigma+\rho} c_0$ with $\hat{S}(0) = R_0^{-1} S(0)$.

To complete the formulation of **NRSP**, the notion of length for the solutions of equations (4) must be introduced. If $g(t) = (S(t), \hat{S}(t))$ is any solution of (4) on a finite interval $[0, T]$ containing points t_0 and t_1, then the *length* $\left\| g_{[t_0,t_1]} \right\|_\varepsilon$ of $g(t)$ between $g(t_0)$ and $g(t_1)$ is defined to be $\int_{t_0}^{t_1} \|u(t)\| \, dt$. Suppose that a curve $(x(t), R(t))$ defined on a finite interval $[0, T]$ corresponds to an isometric rolling of the sphere $S_{\rho,\varepsilon}(c_0)$ on the stationary manifold M_ε. According to the proposition above, the curve traced by the point of contact with the stationary sphere is given by $x(t) = \sigma S(t)e_0$ where $(S(t), \hat{S}(t))$ is a solution curve of (4), and the orientation $R(t)$ is given by $R(t) = \hat{S}^{-1}(t)S(t)$. It is then shown in Jurdjevic and Zimmerman [8] that the length of $x(t)$ is minimal among all rollings of $S_{\rho,\varepsilon}(c_0)$ that connect the points $(x(0), R(0))$ and $(x(T), R(T))$ if and only if the length of $g(t) = (S(t), \hat{S}(t))$ is minimal among all solutions of (4) that connect the submanifolds $N_0 = \{(\check{S}, R(0)^{-1}\check{S}) \in G_{\varepsilon,\varepsilon} : \sigma\check{S}e_1 = x(0)\}$ and $N_1 = \{(\check{S}, R(T)^{-1}\check{S}) \in G_{\varepsilon,\varepsilon} : \sigma\check{S}e_1 = x(T)\}$. Thus the **NRSP** can be stated entirely in terms of control system (4):

Definition 2. *Let $Traj_{\varepsilon,\varepsilon}(N_0, N_1)$ denote the set of all trajectories of (4) that conform to the boundary condtions $g(0) \in N_0$, $g(T) \in N_1$. Then the* ***NRSP*** *is the optimal control problem of finding a trajectory $g(t) = (S(t), \hat{S}(t))$ in $Traj_{\varepsilon,\varepsilon}(N_0, N_1)$ whose length $\int_0^T \|u(t)\|\,dt$ is minimal relative to $Traj_{\varepsilon,\varepsilon}(N_0, N_1)$.*

3 Hamiltonians and Extremal Equations

The Pontryagin maximum principle of optimal control (**PMP**) states that each optimal trajectory is the projection of an extremal curve in the cotangent bundle of the underlying manifold on which the variational problem is defined. Since each rolling sphere problem is defined on a Lie group G, the associated extremal curves evolve on the cotangent bundle T^*G. To preserve the left-invariant symmetries in our variational problems, the cotangent bundle T^*G is realized as the product $G \times \mathfrak{g}^*$, where $\mathfrak{g}^*$ denotes the dual of the Lie algebra $\mathfrak{g}$ of G. In this identification an element ξ in the cotangent space T^*_gG is identified with $(g, \ell) \in G \times \mathfrak{g}^*$ via the relation $\xi(gA) = \ell(A)$ for each $A \in \mathfrak{g}$.

The extremal equations obtained through (**PMP**) will be absorbed in the following general framework, explained in full detail in Jurdjevic [5] or Jurdjevic [6]. Functions H on $G \times \mathfrak{g}^*$ are called Hamiltonians. They are called *left invariant Hamiltonians* if they are constant on the first factor. Stated differently, left-invariant Hamiltonians coincide with the functions on $\mathfrak{g}^*$. Any left-invariant vector field $X(g) = gA$ induces a left-invariant Hamiltonian $H(\ell) = \ell(A)$, where $\ell \in \mathfrak{g}^*$. Such a Hamiltonian is called the *Hamiltonian* of X.

In general, $\mathbf{H}$ denotes the *Hamiltonian vector field* associated with any Hamiltonian H. When H is a left invariant Hamiltonian then $\mathrm{d}H$ is a linear function on $\mathfrak{g}^*$ and hence is an element of $\mathfrak{g}$. In such a case, $\mathbf{H}(g, \ell) = (g\mathrm{d}H, -ad^*(\mathrm{d}H_\ell)(\ell))$ and the integral curves $(g(t), \ell(t))$ of $\mathbf{H}$ are the solution curves

$$\frac{\mathrm{d}g}{\mathrm{d}t} = g\mathrm{d}H_\ell \qquad \frac{\mathrm{d}\ell}{\mathrm{d}t} = -ad^*(\mathrm{d}H_\ell)(\ell) \tag{5}$$

where $ad^*(\mathrm{d}H_\ell(\ell))(A) = \ell([\mathrm{d}H_\ell, A])$ for all $A \in \mathfrak{g}$. The preceding differential equations will be referred to as *Hamiltonian equations* on a Lie group G.

For our purposes it will be convenient to consider the Hamiltonian equations on $G \times \mathfrak{g}$ rather than on $G \times \mathfrak{g}^*$. On semi-simple Lie algebras, any scalar multiple of the Killing form is non-degenerate and invariant (in the sense that

$\langle [A, B], C \rangle = \langle A, [B, C] \rangle)$, and hence provides a natural identification of $\mathfrak{g}^*$ with $\mathfrak{g}$. In this identification, each $\ell \in \mathfrak{g}^*$ is identified with $L \in \mathfrak{g}$ via the relation $\ell(A) = \langle L, A \rangle$ for all $A \in \mathfrak{g}$. Then equations (5) are identified with equations

$$\frac{\mathrm{d}g}{\mathrm{d}t} = g\mathrm{d}H \qquad \frac{\mathrm{d}L}{\mathrm{d}t} = [\mathrm{d}H, L] \tag{6}$$

With this setting in mind, let us now return to the problems at hand. The sub-Riemannian differential control systems in this paper are of the form

$$\frac{\mathrm{d}g}{\mathrm{d}t} = \sum_{i=1}^{n} u_i X_i(g) \tag{7}$$

where $X_1, ..., X_n$ are left invariant vector fields on a Lie group G. It is well-known that minimizing length is equivalent to minimizing energy on a fixed interval. Since it is more convenient to work with energy rather than length, each of our variational problems asks for the minimum of $\frac{1}{2}\int_0^T \|u(t)\|^2 \, \mathrm{d}t$ over the solution curves of system (7) subject to the appropriate boundary conditions with $T > 0$ fixed. For this class of left-invariant optimal control problems it is a simple matter, based on the maximality condition of **PMP,** to show that the normal extremals are the integral curves of a single Hamiltonian function on $\mathfrak{g}^*$ given by

$$\mathcal{H} = \frac{1}{2}\sum_{i=1}^{n} H_i^2 \tag{8}$$

where $H_1, ..., H_k$ are the Hamiltonians of the vector fields $X_1, ..., X_n$. The abnormal extremals will be ignored since the Goh condition (Agrachev and Sachkov [1]) can be used to show that every optimal solution is the projection of a normal extremal curve (Jurdjevic and Zimmerman [8]).

The **ERSP** is defined on $G_\varepsilon = \mathbb{R}^n \times SO_\varepsilon$ with Lie algebra $\mathfrak{g}_\varepsilon$. If so_ε denotes the Lie algebra of SO_ε, then $so_\varepsilon = \mathfrak{p}_\varepsilon \oplus \mathfrak{k}$, where $\mathfrak{p}_\varepsilon$ is the collection of matrices $\left\{ \begin{pmatrix} 0 & -\varepsilon p^{\mathrm{T}} \\ p & \mathbf{0} \end{pmatrix} \middle| \, p \in \mathbb{R}^n \right\}$, and $\mathfrak{k}$ is equal to the Lie algebra of the isotropy group K_{e_1} consisting of the antisymmetric matrices $\left\{ \begin{pmatrix} 0 & 0 \\ 0 & \tilde{K} \end{pmatrix} \middle| \, \tilde{K}^{\mathrm{T}} = -\tilde{K} \right\}$. The reader can readily verify the relations

$$[\mathfrak{k}, \mathfrak{k}] = \mathfrak{k} \qquad [\mathfrak{k}, \mathfrak{p}_\varepsilon] = \mathfrak{p}_\varepsilon \qquad [\mathfrak{p}_\varepsilon, \mathfrak{p}_\varepsilon] = \mathfrak{k} \tag{9}$$

Then points $\check{\ell}$ of $\mathfrak{g}_\varepsilon^*$ will be identified with elements $\check{L} = (a, L) = (a, P + K)$ of $\mathfrak{g}_\varepsilon$ where $a \in \mathbb{R}^n$, $P \in \mathfrak{p}_\varepsilon$ and $K \in \mathfrak{k}$.

Similar to Zimmerman [10], the normal extremal equations associated with the **ERSP** are the solutions of

$$\begin{aligned} &\frac{\mathrm{d}x}{\mathrm{d}t} = a + p \quad &&\frac{\mathrm{d}R}{\mathrm{d}t} = RU_\varepsilon \quad &&\frac{\mathrm{d}a}{\mathrm{d}t} = 0 \\ &\frac{\mathrm{d}K}{\mathrm{d}t} = [U_\varepsilon, P] \quad &&\frac{\mathrm{d}P}{\mathrm{d}t} = [U_\varepsilon, K] && \end{aligned} \tag{10}$$

where $U_\varepsilon = \begin{pmatrix} 0 & -\varepsilon(a+p)^{\mathrm{T}} \\ a+p & \mathbf{0} \end{pmatrix}$.

Let us now turn to the extremal equations for the **NRSP** on the cotangent bundle of $G_{\varepsilon,\varepsilon} = SO_\varepsilon \times SO_\varepsilon$. Letting $\mathfrak{g}_{\varepsilon,\varepsilon}$ denote the Lie algebra $so_\varepsilon \times so_\varepsilon$ of $G_{\varepsilon,\varepsilon}$, then each $\breve{L} = (L, \hat{L}) \in \mathfrak{g}_{\varepsilon,\varepsilon}$ will be written as $L = K + P$, and $\hat{L} = \hat{K} + \hat{P}$ according to the Cartan decompositions (9) in each factor. Now set $E_{i,\varepsilon} = \begin{pmatrix} 0 & -\varepsilon\tilde{e}_i^{\mathrm{T}} \\ \tilde{e}_i & \mathbf{0} \end{pmatrix}$ for $i = 1, \ldots, n$, where $\{\tilde{e}_1, \tilde{e}_2, ..., \tilde{e}_n\}$ denotes the canonical basis of $\mathbb{R}^n$. The Hamiltonians H_i, that correspond to the vector fields $X_i(S, \hat{S}) = \left(\alpha S E_{i,\varepsilon}, \beta \hat{S} E_{i,\varepsilon}\right)$ and that span the distribution associated with the **NRSP**, are given by:

$$H_i(\breve{\ell}) = \breve{\ell}(X_i) = \alpha p_i + \beta \hat{p}_i \tag{11}$$

Therefore, normal extremal curves are the integral curves of the Hamiltonian vector field $\mathbf{H}$ associated with Hamiltonian $\mathcal{H} = \frac{1}{2}\sum_{i=1}^{n} (\alpha p_i + \beta \hat{p}_i)^2$ and are given by the following equations:

$$\frac{\mathrm{d}}{\mathrm{d}t}(S, \hat{S}) = (S, \hat{S})(\mathrm{d}\mathcal{H}) \qquad \frac{\mathrm{d}\breve{L}}{\mathrm{d}t} = \left[\mathrm{d}\mathcal{H}, \breve{L}\right]$$

It follows that $\mathrm{d}\mathcal{H} = (\alpha U_\varepsilon, \beta U_\varepsilon)$ with $U_\varepsilon = \alpha P + \beta \hat{P}$, and the above equations can be written more explicitly as:

$$\begin{aligned} \frac{\mathrm{d}S}{\mathrm{d}t} &= \alpha S U_\varepsilon & \frac{\mathrm{d}\hat{S}}{\mathrm{d}t} &= \beta \hat{S} U_\varepsilon \\ \frac{\mathrm{d}K}{\mathrm{d}t} &= \alpha \left[U_\varepsilon, P\right] & \frac{\mathrm{d}\hat{K}}{\mathrm{d}t} &= \beta \left[U_\varepsilon, \hat{P}\right] \\ \frac{\mathrm{d}P}{\mathrm{d}t} &= \alpha \left[U_\varepsilon, K\right] & \frac{\mathrm{d}\hat{P}}{\mathrm{d}t} &= \beta \left[U_\varepsilon, \hat{K}\right] \end{aligned} \tag{12}$$

The extremal equations for the **NRSP** must also meet the transversality conditions specified by **PMP**. Recall that the boundary submanifolds for the lifted problem are

$$\begin{aligned} N_0 &= \left\{ \left(\breve{S}, R_0^{-1}\breve{S}\right) \in G_{\varepsilon,\varepsilon} \left| \sigma \breve{S} e_1 = x_0 \right. \right\} \\ N_1 &= \left\{ \left(\breve{S}, S_0^{-1}\breve{S}\right) \in G_{\varepsilon,\varepsilon} \left| \sigma \breve{S} e_1 = y_0 \right. \right\} \end{aligned}$$

The tangent space of N_0 at $(\breve{S}, R_0^{-1}\breve{S})$ is equal to $\left\{ \left(\breve{S}B, R_0^{-1}\breve{S}B\right) | B \in \mathfrak{k} \right\}$, and the tangent space of N_1 at $(\breve{S}, S_0^{-1}\breve{S})$ is equal to $\left\{ \left(\breve{S}B, S_0^{-1}\breve{S}B\right) | B \in \mathfrak{k} \right\}$. Therefore each extremal curve $\breve{\ell}(t) = (\ell(t), \hat{\ell}(t))$ satisfies $(\ell(0) + \hat{\ell}(0))(B) = 0$ and $(\ell(T) + \hat{\ell}(T))(B) = 0$ for all $B \in \mathfrak{k}$. After the identifications with matrices, $K(0) + \hat{K}(0) = 0$ and $K(T) + \hat{K}(T) = 0$. But then $K(t) + \hat{K}(t) = 0$ for all t because $K(t) + \hat{K}(t)$ is constant, as becomes evident upon a cursory inspection of equations (12).

The transversality condition $K(t) + \hat{K}(t) = 0$ should be added to the normal extremal equations (12), in which case it is easy to see that $\breve{A} = \beta P(t) + \alpha \hat{P}(t)$ is another constant of motion because $\beta \frac{\mathrm{d}P}{\mathrm{d}t}(t) + \alpha \frac{\mathrm{d}\hat{P}}{\mathrm{d}t}(t) = \alpha\beta[U_\varepsilon(t), K(t) + \hat{K}(t)] = 0$.

With the aid of this constant of motion equations, (12) can be brought to a form that reveals remarkable similarity to the equations (10) that correspond to the **ERSP**. This form is obtained as follows. Along extremal curves, $P(t) = \frac{\breve{A}}{\beta} - \frac{\alpha}{\beta}\hat{P}(t)$ which further implies that $U_\varepsilon(t) = \frac{\beta^2-\alpha^2}{\beta}\hat{P}(t) + \frac{\alpha}{\beta}\breve{A}$. Let

$$\begin{aligned} \bar{K} &= (\beta^2 - \alpha^2)\hat{K} & \bar{P} &= (\alpha^2 - \beta^2)\hat{P} \\ \bar{A} &= -\alpha\breve{A} & \bar{U}_\varepsilon &= -\beta U_\varepsilon = \bar{P} + \bar{A} \end{aligned}$$

and in addition, let $\bar{R} = I_{1,n}\hat{S}I_{1,n}$ where $I_{1,n} = \begin{pmatrix} 1 & 0 \\ 0 & -\tilde{I} \end{pmatrix}$. The rescaling factor $\beta^2 - \alpha^2$ is not equal to 0 since $\sigma \neq \rho$, and consequently the preceding transformations are invertible. In terms of the transformed variables, equations (12) become:

$$\begin{aligned} \frac{\mathrm{d}S}{\mathrm{d}t} &= \frac{\rho}{\sigma}S\bar{U}_\varepsilon & \frac{\mathrm{d}\bar{R}}{\mathrm{d}t} &= \bar{R}\bar{U}_\varepsilon \\ \frac{\mathrm{d}K}{\mathrm{d}t} &= \frac{\rho}{\sigma}\left[\bar{U}_\varepsilon, P\right] & \frac{\mathrm{d}\bar{K}}{\mathrm{d}t} &= \left[\bar{U}_\varepsilon, \bar{P}\right] \\ \frac{\mathrm{d}P}{\mathrm{d}t} &= \frac{\rho}{\sigma}\left[\bar{U}_\varepsilon, K\right] & \frac{\mathrm{d}\bar{P}}{\mathrm{d}t} &= \left[\bar{U}_\varepsilon, \bar{K}\right] \end{aligned} \tag{13}$$

The equations for the variables $\bar{R}$, $\bar{K}$, $\bar{P}$, $\bar{A}$, $\bar{U}_\varepsilon$ take exactly the same form as in the Euclidean case. Furthermore, the transformed variables have no bearing on the solution $S(t)$ nor on the length of the projected curve $x(t) = \sigma S(t)e_1$. If we agree to set $\rho = 1$, $\sigma = \infty$, $S = I$ for the **ERSP**, then the following equations with $\bar{U}_\varepsilon = \bar{A} + \bar{P}$ describe all our rolling sphere problems:

$$\begin{aligned} \frac{\mathrm{d}x}{\mathrm{d}t} &= \rho S\bar{U}_\varepsilon e_1 & \frac{\mathrm{d}S}{\mathrm{d}t} &= \frac{\rho}{\sigma}S\bar{U}_\varepsilon & \frac{\mathrm{d}\bar{R}}{\mathrm{d}t} &= \bar{R}\bar{U}_\varepsilon \\ \frac{\mathrm{d}\bar{K}}{\mathrm{d}t} &= \left[\bar{U}_\varepsilon, \bar{P}\right] & \frac{\mathrm{d}\bar{P}}{\mathrm{d}t} &= \left[\bar{U}_\varepsilon, \bar{K}\right] & \frac{\mathrm{d}\bar{A}}{\mathrm{d}t} &= 0 \end{aligned}$$

4 Integrability

Let us now restrict our attention to solutions of the extremal equations for low dimensions under the assumption that the constant $A = \begin{pmatrix} 0 & -\varepsilon a^{\mathrm{T}} \\ a & \mathbf{0} \end{pmatrix}$ is conveniently reduced such that $a = \mu\tilde{e}_1$ (integrability of the rolling sphere problems is studied in more detail in Jurdjevic and Zimmerman [8]).

$\mathbf{n = 2.}$ In this case write

$$P = \begin{pmatrix} 0 & -\varepsilon p_1 & -\varepsilon p_2 \\ p_1 & 0 & 0 \\ p_2 & 0 & 0 \end{pmatrix} \qquad K = \begin{pmatrix} 0 & 0 & 0 \\ 0 & 0 & -q \\ 0 & q & 0 \end{pmatrix}$$

so that the extremal equations take the form

$$\frac{\mathrm{d}p_1}{\mathrm{d}t} = -qp_2 \quad \frac{\mathrm{d}p_2}{\mathrm{d}t} = q(\mu + p_1) \quad \frac{\mathrm{d}q}{\mathrm{d}t} = -\varepsilon\mu p_2 \tag{14}$$

As in Jurdjevic [4], the solutions of (14) are most naturally expressed through the mathematical pendulum defined by the integrals of motion $\chi_1 = \|p\|^2 + \varepsilon q^2$, $\chi_2 = \|A + P\|$. These integrals define the energy E of the pendulum as

$$\begin{aligned} E &= \frac{1}{2}\|K\|^2 - \varepsilon\langle A, P\rangle = \frac{q^2}{2} - \varepsilon\mu p_1 \\ &= \frac{\varepsilon}{2}\left(\chi_1 - \chi_2^2 + \mu^2\right) \end{aligned}$$

The angle $\theta(t)$ of the pendulum is defined by

$$p_1(t) + \mu = \chi_2 \cos\theta(t) \qquad p_2(t) = \chi_2 \sin\theta(t) \tag{15}$$

Since $-\chi_2 \sin\theta(t)\frac{\mathrm{d}\theta}{\mathrm{d}t} = \frac{\mathrm{d}p_1}{\mathrm{d}t} = -q\chi_2 \sin\theta(t)$,

$$\frac{\mathrm{d}\theta}{\mathrm{d}t} = q = \pm\sqrt{2\left(E + \varepsilon\mu(\chi_2\cos\theta - \mu)\right)}.$$

Thus $\theta(t)$ can be expressed in terms of elliptic integrals, and the remaining variables p_1, p_2 are then obtained from equations (15). The remaining integrations down to the group level, and further down to the stationary manifold, can be carried out in a manner analogous to that used by Jurdjevic [6] in integrating the equations for the elastic problem.

n = 3. Here write

$$L = K + P = \begin{pmatrix} 0 & -\varepsilon p^{\mathrm{T}} \\ p & \tilde{K} \end{pmatrix} \quad \tilde{K} = \begin{pmatrix} 0 & -q_3 & q_2 \\ q_3 & 0 & -q_1 \\ -q_2 & q_1 & 0 \end{pmatrix}$$

As in the case with $n = 2$, the integrals of motion $\chi_1 = \|q\|^2 + \varepsilon\|q\|^2$ and $\chi_2 = \|A + P\|$ imply the existence of a related conserved quantity E given by

$$E = \frac{1}{2}\|K\|^2 - \varepsilon\langle A, P\rangle = \frac{1}{2}\|q\|^2 - \varepsilon\mu p_1$$

which can be identified with the energy function for the heavy top of Lagrange (**LHT**). In this analogy with the top $\tilde{K}$ is viewed as the moments matrix with all the principal moments of inertia equal to 1, while the term $\varepsilon\mu p_1$ corresponds to external torque due to gravity.

These mechanical analogues, the mathematical pendulum for $n = 2$ together with **LHT** for $n = 3$, as remarkable as they may seem at first, are a part of a larger pattern observed in Jurdjevic [5] that mechanical tops form invariant subsystems of elastic problems. This topic is further explored in Jurdjevic and Zimmerman [8], where it is shown that all solutions of the Euler-Griffiths elastic problems investigated in Jurdjevic and Monroy-Perez [7] are also solutions of the rolling sphere problems. To elaborate more amply, note first that q_1 is constant along the extremal curves and therefore, the "reduced" energy $E_0 = q_2^2+q_3^2-\varepsilon\mu p_1$ is also an integral of motion. It follows that the "reduced" curvature function $\xi = q_2^2 + q_3^2$ satisfies the differential equation $\frac{d\xi}{dt} = \mu(q_2p_3 - q_3p_2)$, and therefore

$$\begin{aligned}\frac{1}{4\mu^2}\left(\frac{d\xi}{dt}\right)^2 &= (q_2^2+q_3^2)(p_2^2+p_3^2) - (q_2p_2+q_3p_3)^2\\ &= \xi\left(\chi_2^2 - (\mu+p_1)^2\right) - (c - q_1p_1)^2\\ &= \begin{array}{c}\xi\left(\chi_2^2 - \left(\mu + \frac{\varepsilon}{2\mu}(\xi - E_0)\right)^2\right)\\ -\left(c - \frac{\varepsilon q_1}{2\mu}(\xi - E_0)\right)^2\end{array}\end{aligned}$$

where $c = p \cdot q$. Hence ξ is solvable in terms of elliptic integrals. The spherical variables, defined by the relations $\mu + p_1 = \chi_2\cos\theta$, $p_2 = \chi_2\sin\theta\sin\psi$, $p_3 = \chi_2\sin\theta\cos\psi$, conform to the differential equation $-\sin\theta\frac{d\theta}{dt} = \frac{dp_1}{dt} = \frac{\varepsilon}{2}\frac{d\xi}{dt}$ along the extremal curves. Hence,

$$\begin{aligned}\left(\frac{d\theta}{dt}\right)^2 &= 2\left(E_0 - \varepsilon(\chi_2\cos\theta + \mu)\right)\\ &\quad - \frac{1}{\chi_2^2\sin^2\theta}\left(c - q_1(\chi_2\cos\theta + \mu)\right)^2\end{aligned}$$

is also solvable in terms of elliptic integrals. The angle θ is known as the nutation angle. The reader is referred to Jurdjevic [5] for the relation of ψ to θ and to the remaining details of integration on so_ε.

References

1. Agrachev A, Sachkov Y (2004) Control Theory from the Geometric Viewpoint. Springer-Verlag, Berlin
2. Brockett RW, Dai L (1993) Non-holonomic kinematics and the role of elliptic functions in constructive controllability. In: Li Z, Canny JF (eds) Nonholonomic Motion Planning. Kluwer Academic, Boston
3. Hammersley J (1983) Oxford commemoration ball. In: Kingman JFC, Reuter GEH (eds) Probability, Statistics, and Analysis. Cambridge University Press, Cambridge
4. Jurdjevic V (1993) The geometry of the plate-ball problem. Arch Rat Mech Anal 124:305–328
5. Jurdjevic V (1997) Geometric Control Theory. Cambridge University Press, Cambridge
6. Jurdjevic V (2005) Hamiltonian systems on complex Lie groups and their homogeneous spaces. Mem Amer Math Soc vol 178, no 838

7. Jurdjevic V, Monroy-Perez F (2002) Variational problems on Lie groups and their homogeneous spaces: elastic curves, tops, and constrained geodesic problems. In: Anzaldo-Meneses A et al. (eds) Contemporary Trends in Nonlinear Geometric Control Theory and its Applications. World Scientific, Singapore
8. V. Jurdjevic, Zimmerman J (2006) Rolling Sphere Problems on Spaces of Constant Curvature. A preprint submitted for publication.
9. Liu W, Sussmann H (1995) Shortest paths for sub-Riemannian metrics on rank 2 distributions. Mem Amer Math Soc vol 118, no 564
10. Zimmerman J (2005) Optimal control of the sphere S^n rolling on E^n. Math Control Signals Systems 17:14–37

Dirac Structures and the Legendre Transformation for Implicit Lagrangian and Hamiltonian Systems

Hiroaki Yoshimura[1] and Jerrold E. Marsden[2]

[1] Department of Mechanical Engineering, Waseda University, Ohkubo, Shinjuku, Tokyo 169-8555, Japan
yoshimura@waseda.jp
[2] Control and Dynamical Systems, California Institute of Technology, Pasadena, CA 91125, USA
marsden@cds.caltech.edu

Summary. This paper begins by recalling how a constraint distribution on a configuration manifold induces a Dirac structure together with an implicit Lagrangian system, a construction that is valid even for degenerate Lagrangians. In such degenerate cases, it is shown in this paper that an implicit Hamiltonian system can be constructed by using a generalized Legendre transformation, where the primary constraints are incorporated into a generalized Hamiltonian on the Pontryagin bundle. Some examples of degenerate Lagrangians for L–C circuits, nonholonomic systems, and point vortices illustrate the theory.

1 Introduction

In recent years, the theory of implicit Hamiltonian systems has been developed along with associated formulations of physical systems, such as L-C circuits and nonholonomic systems. This is a useful analytical tool, in which Dirac structures are employed to help understand how interconnected system elements are energetically related and are systematically incorporated into the Hamiltonian formalism; see, for instance, [9, 2, 8, 1]. The notion of Dirac structures, which was first developed in [3], is also relevant to Dirac's theory of constraints for degenerate Lagrangian systems. However, research has only just begun on the theory of implicit Lagrangian systems, and, in addition there is a need to understand how they are related to implicit Hamiltonian systems as well as with Dirac's theory of constraints.

Recently, the theory of implicit Lagrangian systems, namely, a Lagrangian analogue of implicit Hamiltonian systems, has been developed by [10, 11]. This theory, which also makes use of Dirac structures, has similar examples that can be systematically treated from the Lagrangian viewpoint, namely nonholonomic mechanical systems and degenerate Lagrangian systems, such as L-C circuits.

In the present paper, we investigate systems with degenerate Lagrangians and, following [10], we first show how to construct a Dirac structure on the cotangent

F. Bullo et al. (Eds.): Lag. & Hamil. Methods for Nonlin. Ctrl., LNCIS 366, pp. 233–247, 2007.
springerlink.com

bundle T^*Q induced from a constraint distribution on a configuration manifold Q. Second, we demonstrate how an implicit Lagrangian system can be constructed from the induced Dirac structure. Using this framework, we show how to construct an implicit Hamiltonian system from a given, possibly degenerate, Lagrangian. To do this, we make use of a generalized Legendre transformation for degenerate Lagrangians to define a Hamiltonian on a constraint momentum space $P \subset T^*Q$ and also define a generalized Hamiltonian on the Pontryagin bundle $TQ \oplus T^*Q$ by combining primary constraints in the sense of Dirac with the Hamiltonian. Thus, we show how degenerate Lagrangian systems that are useful in L-C circuits as well as in nonholonomic systems, can be represented in the context of both implicit Lagrangian systems and Hamiltonian systems. Lastly, we illustrate an example of degenerate Lagrangians for point vortices and the KdV equations.

2 Induced Dirac Structures

Dirac Structures. We begin by reviewing the definition of a Dirac structure on a vector space, following [3].

Let V be an n-dimensional vector space, V^* be its dual space, and let $\langle \cdot\,, \cdot \rangle$ be the natural paring between V^* and V. Define the symmetric paring $\langle\!\langle \cdot, \cdot \rangle\!\rangle$ on $V \oplus V^*$ by

$$\langle\!\langle\, (v, \alpha), (\bar{v}, \bar{\alpha})\, \rangle\!\rangle = \langle \alpha, \bar{v} \rangle + \langle \bar{\alpha}, v \rangle,$$

for $(v, \alpha), (\bar{v}, \bar{\alpha}) \in V \oplus V^*$. A *Dirac structure* on V is a subspace $D \subset V \oplus V^*$ such that $D = D^{\perp}$, where $D^{\perp}$ is the orthogonal of D relative to the pairing $\langle\!\langle \cdot, \cdot \rangle\!\rangle$.

Let M be a smooth differentiable manifold whose tangent bundle is denoted as TM and whose cotangent bundle is denoted as T^*M. Let $TM \oplus T^*M$ denote the Whitney sum bundle over M; that is, it is the bundle over the base M and with fiber over the point $x \in M$ equal to $T_xM \times T_x^*M$. An *(almost) Dirac structure* on M is a subbundle $D \subset TM \oplus T^*M$ that is a Dirac structure in the sense of vector spaces at each point $x \in M$.

In geometric mechanics, (almost) Dirac structures provide a simultaneous generalization of both two-forms (not necessarily closed, and possibly degenerate) as well as almost Poisson structures (that is brackets that need not satisfy the Jacobi identity). An *integrable Dirac structure*, which corresponds in geometric mechanics to assuming the two-form is closed or to assuming Jacobi's identity for the Poisson tensor, is one that satisfies

$$\langle \pounds_{X_1} \alpha_2, X_3 \rangle + \langle \pounds_{X_2} \alpha_3, X_1 \rangle + \langle \pounds_{X_3} \alpha_1, X_2 \rangle = 0,$$

for all pairs of vector fields and one-forms (X_1, α_1), (X_2, α_2), (X_3, α_3) that take values in D and where $\pounds_X$ denotes the Lie derivative along the vector field X on M.

Induced Dirac Structures. We now construct induced Dirac structure, an essential ingredient in the setting of implicit Lagrangian systems; see [10].

Let Q be an n-dimensional configuration manifold, whose kinematic constraints are given by a constraint distribution $\Delta_Q \subset TQ$, which is defined, at each $q \in Q$, by

$$\Delta_Q(q) = \{v \in T_qQ \mid \langle \omega^a(q), v\rangle = 0,\ a = 1, ..., m\},$$

where ω^a are m one-forms on Q. Define the distribution Δ_{T^*Q} on T^*Q by

$$\Delta_{T^*Q} = (T\pi_Q)^{-1}(\Delta_Q) \subset TT^*Q,$$

where $T\pi_Q : TT^*Q \to TQ$ is the tangent map of $\pi_Q : T^*Q \to Q$, while the annihilator of Δ_{T^*Q} can be defined for each $z = (q,p) \in T^*Q$, by

$$\Delta^{\circ}_{T^*Q}(z) = \{\alpha_z \in T^*_zT^*Q \mid \langle \alpha_z,\ w_z\rangle = 0 \ \text{ for all}\ w_z \in \Delta_{T^*Q}(z)\}.$$

Let Ω be the canonical symplectic structure on T^*Q and $\Omega^\flat : TT^*Q \to T^*T^*Q$ be the associated bundle map. Then, a Dirac structure D_{Δ_Q} on T^*Q induced from the constraint distribution Δ_Q can be defined for each $z = (q,p) \in T^*Q$, by

$$\begin{aligned} D_{\Delta_Q}(z) = \{\, (w_z, \alpha_z) \in T_zT^*Q \times T^*_zT^*Q \mid\ & w_z \in \Delta_{T^*Q}(z), \\ & \text{and}\ \ \alpha_z - \Omega^\flat(z)\cdot w_z \in \Delta^{\circ}_{T^*Q}(z)\,\}. \end{aligned}$$

Local Representation. Let us choose local coordinates q^i on Q so that locally, Q is represented by an open set $U \subset \mathbb{R}^n$. The constraint set Δ_Q defines a subspace of TQ, which we denote by $\Delta(q) \subset \mathbb{R}^n$ at each point $q \in U$. If the dimension of the constraint space is $n-m$, then we can choose a basis $e_{m+1}(q), e_{m+2}(q), \ldots, e_n(q)$ of $\Delta(q)$.

The constraint sets can be also represented by the annihilator of $\Delta(q)$, which is denoted by $\Delta^\circ(q)$, spanned by such one-forms that we write as $\omega^1, \omega^2, \ldots, \omega^m$. Since the cotangent bundle projection $\pi_Q : T^*Q \to Q$ is locally denoted as $(q,p) \mapsto q$, its tangent map may be locally given by $T\pi_Q : (q,p,\dot q,\dot p) \mapsto (q,\dot q)$. So, we can locally represent Δ_{T^*Q} as

$$\Delta_{T^*Q} \cong \left\{v_{(q,p)} = (q,p,\dot q,\dot p) \mid q \in U, \dot q \in \Delta(q)\right\}.$$

Then, the annihilator of Δ_{T^*Q} is locally represented as

$$\Delta^{\circ}_{T^*Q} \cong \left\{\alpha_{(q,p)} = (q,p,\alpha,w) \mid q \in U,\ \alpha \in \Delta^\circ(q) \text{ and } w = 0\right\}.$$

Because of the local formula $\Omega^\flat(z)\cdot v_z = (q,p,-\dot p,\dot q)$, the condition $\alpha_z - \Omega^\flat(z)\cdot v_z \in \Delta^{\circ}_{T^*Q}$ reads

$$\alpha + \dot p \in \Delta^\circ(q), \quad \text{and} \quad w - \dot q = 0.$$

Thus, the induced Dirac structure is locally represented by

$$D_{\Delta_Q}(z) = \{((q,p,\dot q,\dot p),(q,p,\alpha,w)) \mid \dot q \in \Delta(q),\ w = \dot q,\ \alpha + \dot p \in \Delta^\circ(q)\}. \quad (1)$$

3 Implicit Lagrangian Systems

Dirac Differential Operator. Let $L : TQ \to \mathbb{R}$ be a Lagrangian (possibly degenerate). The differential of L is the map $\mathbf{d}L : TQ \to T^*TQ$, which is locally given, for each $(q, v) \in TQ$, by

$$\mathbf{d}L = \left(q, v, \frac{\partial L}{\partial q}, \frac{\partial L}{\partial v} \right).$$

Define the *Dirac differential* of a Lagrangian L, to be the map

$$\mathfrak{D}L : TQ \to T^*T^*Q$$

defined by

$$\mathfrak{D}L = \gamma_Q \circ \mathbf{d}L.$$

Here, the map $\gamma_Q : T^*TQ \to T^*T^*Q$ is the natural symplectomorphism (see [10]), which is defined by

$$\gamma_Q = \Omega^\flat \circ \kappa_Q^{-1},$$

where $\Omega^\flat : TT^*Q \to T^*T^*Q$ is the induced map from Ω and $\kappa_Q : TT^*Q \to T^*TQ$ is the natural symplectomorphism (see [7]). In coordinates, the symplectomorphism $\gamma_Q : T^*TQ \to T^*T^*Q$ is given by

$$(q, \delta q, \delta p, p) \mapsto (q, p, -\delta p, \delta q)$$

and hence the Dirac differential of L is locally given, at each $(q, v) \in TQ$, by

$$\mathfrak{D}L = \left(q, \frac{\partial L}{\partial v}, -\frac{\partial L}{\partial q}, v \right). \tag{2}$$

Implicit Lagrangian Systems. An *implicit Lagrangian system* is a triple (L, Δ_Q, X), which satisfies the condition

$$(X, \mathfrak{D}L) \in D_{\Delta_Q}, \tag{3}$$

where $X : \Delta_Q \oplus P \subset TQ \oplus T^*Q \to TT^*Q$ is a partial vector field defined at points $(v, p) \in \Delta_Q \times P$, where $P = \mathbb{F}L(\Delta_Q)$; that is, X assigns a vector in T_pT^*Q to each point $(q, v, p) \in \Delta_Q \oplus P$. We write $X(q, v, p) = (q, p, \dot{q}, \dot{p})$, so that $\dot{q}$ and $\dot{p}$ are functions of (q, v, p).

Equality of base points in (3) implies that p is given by the Legendre transformation, and so one can equivalently say that X depends only on (q, v) with p determined by the Legendre transform. That is, equation (3) means that for each $(q, v) \in \Delta_Q \subset TQ$, we have

$$(X(q, v, p), \mathfrak{D}L(q, v)) \in D_{\Delta_Q}(q, p), \tag{4}$$

where $(q, p) = \mathbb{F}L(q, v)$. It follows from equations (1), (2) and (4) that

$$p = \frac{\partial L}{\partial v}, \quad \dot{q} \in \Delta(q), \quad \dot{q} = v, \quad \text{and} \quad \dot{p} - \frac{\partial L}{\partial q} \in \Delta^\circ(q). \tag{5}$$

A *solution curve* of an implicit Lagrangian system (L, Δ_Q, X) is a curve $(q(t), v(t), p(t)) \in TQ \oplus T^*Q$, $t_1 \leq t \leq t_2$, such that it is an integral curve of X in the sense that the time derivative of $(q(t), p(t)) = \mathbb{F}L(q(t), v(t))$ coincides with the value of $X(q(t), v(t), p(t))$, which is a vector in T^*Q at the point $(q(t), p(t)) = \mathbb{F}L(q(t), v(t))$.

Note that for the case $\Delta_Q = TQ$ the condition of an implicit Lagrangian system is equivalent to the Euler–Lagrange equations $\dot{p} = \partial L/\partial q$ together with the second order condition $\dot{q} = v$.

Energy Conservation. We now show that energy is conserved for any implicit Lagrangian system (L, Δ_Q, X). Define the *generalized energy* E on $TQ \oplus T^*Q$ by

$$E(q, v, p) = \langle p, v \rangle - L(q, v).$$

Let $(q(t), v(t))$, $t_1 \leq t \leq t_2$, be the solution curve of implicit Lagrangian systems together with $(q(t), p(t)) = \mathbb{F}L(q(t), v(t))$; thus,

$$\begin{aligned}\frac{d}{dt}E(q, v, p) &= \langle \dot{p}, v \rangle + \langle p, \dot{v} \rangle - \frac{\partial L}{\partial q}\dot{q} - \frac{\partial L}{\partial v}\dot{v} \\ &= \left\langle \dot{p} - \frac{\partial L}{\partial q}, v \right\rangle\end{aligned}$$

which vanishes since $\dot{q} = v \in \Delta(q)$, $p = \partial L/\partial v$ and $\dot{p} - \partial L/\partial q \in \Delta^\circ(q)$.

Remark. Using the generalized energy E on $TQ \oplus T^*Q$, the condition for an implicit Lagrangian system (L, Δ_Q, X), namely, $(X, \mathfrak{D}L) \in D_{\Delta_Q}$, can be restated as $(X, \mathbf{d}E|_{TT^*Q}) \in D_{\Delta_Q}$ together with the Legendre transform $P = \mathbb{F}L(\Delta_Q)$. Namely, the following relation holds, for each $(q, v) \in \Delta_Q$,

$$\left(X(q, v, p), \mathbf{d}E(q, v, p)|_{T_{(q,p)}T^*Q}\right) \in D_{\Delta_Q}(q, p),$$

together with $(q, p) = \mathbb{F}L(q, v)$. The restriction $\mathbf{d}E(q, v, p)|_{T_{(q,p)}T^*Q}$ is understood in the sense that $T_{(q,p)}T^*Q$ is naturally included in $T_{(q,v,p)}(TQ \oplus T^*Q)$.

Coordinate Representation. In coordinates, since the one-forms $\omega^1, ..., \omega^m$ span a basis of the annihilator $\Delta^\circ(q)$ at each $q \in U \subset \mathbb{R}^n$, it follows that equation (5) can be represented in terms of Lagrange multipliers μ_a, $a = 1, ..., m$ as follows:

$$\begin{aligned}\begin{pmatrix} \dot{q}^i \\ \dot{p}_i \end{pmatrix} &= \begin{pmatrix} 0 & 1 \\ -1 & 0 \end{pmatrix}\begin{pmatrix} -\frac{\partial L}{\partial q^i} \\ v^i \end{pmatrix} + \begin{pmatrix} 0 \\ \mu_a\, \omega_i^a(q) \end{pmatrix}, \\ p_i &= \frac{\partial L}{\partial v^i}, \\ 0 &= \omega_i^a(q)\, v^i,\end{aligned} \tag{6}$$

where $\omega^a = \omega_i^a\, dq^i$.

Later, we shall see that L–C circuits, which are a typical degenerate Lagrangian system, can be represented by equation (6) in the context of implicit Lagrangian systems [10].

Example: Lagrangians Linear in the Velocity. Consider a system with the Lagrangian $L : TQ \to \mathbb{R}$ given by

$$L(q^i, v^i) = \langle \alpha_i(q^j),\, v^i \rangle - h(q^i),\ i, j = 1, ..., n,$$

where α is a one-form on Q and h is a function on Q. This form arises in various physical systems such as point vortices and the KdV equation (see, for instance, [5, 6]). It is obvious that the Lagrangian is degenerate.

Since there are no kinematic constraints, it follows from equation (6) that equations of motion are given by

$$\begin{aligned} \dot{q}^i &= v^i, \\ \dot{p}_i &= \frac{\partial L}{\partial q^i} = \frac{\partial \alpha_j(q)}{\partial q^i} v^j - \frac{\partial h(q)}{\partial q^i}, \\ p_i &= \frac{\partial L}{\partial v^i} = \alpha_i(q). \end{aligned}$$

4 Implicit Hamiltonian Systems

Degenerate Lagrangians. Let Q be a manifold, L be a Lagrangian on TQ and Δ_Q a given constraint distribution on Q. The *constraint momentum space* $P \subset T^*Q$ is defined to be the image of Δ_Q under the Legendre transform $\mathbb{F}L : TQ \to T^*Q$; namely, $P = \mathbb{F}L(\Delta_Q)$, which in coordinates, is represented by

$$(q^i, p_i) = \left(q^i, \frac{\partial L}{\partial v^i} \right), \quad i = 1, ..., n.$$

Now, suppose that L is degenerate; that is,

$$\det \left[\frac{\partial^2 L}{\partial v^i \partial v^j} \right] = 0; \ \ i, j = 1, ..., n,$$

and also that the dimension of P_q at each $q \in Q$ is a fixed integer k $(0 \leq k < n)$, and the submanifold P can be represented by, at each $q \in Q$,

$$P_q = \left\{ p \in T_q^*Q \mid \phi_A(q, p) = 0,\ A = k+1, ..., n \right\}, \tag{7}$$

where ϕ_A, $A = k+1, ..., n$, are functions on T_q^*Q. The functions $\phi_A(q,p) = 0$ in equation (7) are called *primary constraints* when $\Delta_Q = TQ$ (see, for instance, [4]), and we shall continue to call them primary constraints even in the case of $\Delta_Q \subset TQ$. Needless to say, if the Lagrangian is regular, then there are no primary constraints.

Thus, we can choose the local coordinates $(q^i,\, p_\lambda)$, $i = 1, ..., n;\ \lambda = 1, ..., k$ for $P \subset T^*Q$ together with the *partial Legendre transform*

$$p_\lambda = \frac{\partial L}{\partial v^\lambda}, \quad \lambda = 1, ..., k,$$

where

$$\det \left[\frac{\partial^2 L}{\partial v^\lambda \partial v^\mu} \right] \neq 0; \;\; \lambda,\, \mu = 1, ..., k,$$

and $v = (v^\lambda, v^A)$ are local coordinates for T_qQ and with the constraints $v \in \Delta(q)$.

Generalized Legendre Transform. Define a generalized energy E on the Pontryagin bundle $TQ \oplus T^*Q$ by

$$\begin{aligned} E(q^i, v^i, p_i) &= p_i\, v^i - L(q^i, v^i) \\ &= p_\lambda\, v^\lambda + p_A\, v^A - L(q^i, v^\lambda, v^A), \end{aligned}$$

where $p_i = (p_\lambda, p_A)$. Then, the Hamiltonian H_P on P can be defined by

$$H_P(q^i, p_\lambda) = \mathrm{stat}_{\,v^i}\, E(q^i, v^i, p_i)|P,$$

where $\mathrm{stat}_{\,v^i}$ is the stationarity operator (defining a critical point in the variable v). In view of the primary constraints in (7), we can define a *generalized Hamiltonian* H by

$$H(q^i, v^i, p_i) = H_P(q^i, p_\lambda) + \phi_A(q^i, p_i)\, v^A.$$

which has the property that $H|P = H_P$ (but it does depend on how we split the coordinates for p_i and v^i). In the above, v_A, $A = k+1, ..., n$, are local coordinates for an $(n-k)$-dimensional subspace of T_qQ, which can be regarded as Lagrange multipliers for the primary constraints $\phi_A(q^i, p_i) = 0$. The range of the index A varies according to the degeneracy of the Lagrangian, namely, $0 \leq k < n$. So, the generalized Hamiltonian H may be regarded as a function on $TQ \oplus T^*Q$.

Implicit Hamiltonian Systems. The differential of the generalized Hamiltonian $H : TQ \oplus T^*Q \to \mathbb{R}$ is in coordinates given by, for each $(q, v, p) \in TQ \oplus T^*Q$,

$$\mathbf{d}H(q, v, p) = \left(\frac{\partial H}{\partial q}, \frac{\partial H}{\partial v}, \frac{\partial H}{\partial p} \right),$$

where we can obtain the primary constraints by setting

$$\frac{\partial H}{\partial v} = \phi_A(q^i, p_i) = 0, \quad A = k+1, ..., n.$$

Meanwhile, since $\mathbf{d}H(q, v, p)$ takes its values in $T^*_{(q,v,p)}(TQ \oplus T^*Q)$, the restriction of the differential of H to $T_{(q,p)}T^*Q$ is

$$\mathbf{d}H(q, v, p)|_{T_{(q,p)}T^*Q} = \left(\frac{\partial H}{\partial q}, \frac{\partial H}{\partial p} \right).$$

Then, an *implicit Hamiltonian system* is a triple (H, Δ_Q, X), which satisfies the condition

$$(X(q,p), \mathbf{d}H(q,v,p)|_{T_{(q,p)}T^*Q}) \in D_{\Delta_Q}(q,p), \tag{8}$$

where $X = (q, p, \dot{q}, \dot{p})$ is a vector filed on T^*Q.

The local expression for implicit Hamiltonian systems in equation (8) is given by

$$\dot{q} = \frac{\partial H(q,v,p)}{\partial p} \in \Delta(q), \quad \dot{p} + \frac{\partial H(q,v,p)}{\partial q} \in \Delta^{\circ}(q) \tag{9}$$

and with the primary constraints

$$\frac{\partial H(q,v,p)}{\partial v} = \phi_A(q,p) = 0. \tag{10}$$

Coordinate Representation. In coordinates, recall the one-forms $\omega^1, ..., \omega^m$ span a basis of the annihilator $\Delta^{\circ}(q)$ at each $q \in U \in \mathbb{R}^n$, and it follows from equations (9) and (10) that

$$\begin{aligned} \begin{pmatrix} \dot{q}^i \\ \dot{p}_i \end{pmatrix} &= \begin{pmatrix} 0 & 1 \\ -1 & 0 \end{pmatrix} \begin{pmatrix} \dfrac{\partial H(q,v,p)}{\partial q^i} \\ \dfrac{\partial H(q,v,p)}{\partial p^i} \end{pmatrix} + \begin{pmatrix} 0 \\ \mu_a \, \omega_i^a(q) \end{pmatrix}, \\ 0 &= \omega_i^a(q) \, \frac{\partial H(q,v,p)}{\partial p^i}, \\ 0 &= \phi_A(q^i, p_i), \quad A = k+1, ..., n, \end{aligned} \tag{11}$$

where $\omega^a = \omega_i^a \, dq^i$ and we employed the Lagrange multipliers μ_a, $a = 1, ..., m$.

Example of a Lagrangian Linear in the Velocity. Again let us consider the example the Lagrangian $L : TQ \to \mathbb{R}$, which is given by

$$L(q^i, v^i) = \langle \alpha_i(q^j), \, v^i \rangle - h(q^i), \; i, j = 1, ..., n.$$

By a direct computation, we obtain the primary constraints as

$$\phi_i(q^j, p_j) = p_i - \frac{\partial L}{\partial v^i} = p_i - \alpha_i(q^j) = 0,$$

so that the submanifold P is the graph of α in T^*Q. Define a generalized energy E on $TQ \oplus T^*Q$ by

$$\begin{aligned} E(q^i, v^i, p_i) &= p_i \, v^i - L(q^i, v^i) \\ &= (p_i - \alpha_i(q^j)) \, v^i + h(q^i) \end{aligned}$$

and the Hamiltonian H_P on P can be defined by

$$H_P(q^i, p_i) = \operatorname{stat}_{v^i} E(q^i, v^i, p_i)|P = h(q^i),$$

where $p_i = \alpha_i(q^j)$. Hence, the generalized Hamiltonian H on $TQ \oplus T^*Q$ is given by

$$\begin{aligned} H(q^i, v^i, p_i) &= H_P(q^i, p_i) + \phi_i(q^i, p_i)\, v^i \\ &= h(q^i) + (p_i - \alpha_i(q^j))\, v^i, \end{aligned}$$

where we note $H|P = H_P$. Therefore, the equations of motion are given, in the context of implicit Hamiltonian systems, by

$$\begin{aligned} \dot{q}^i &= \frac{\partial H}{\partial p_i} = v^i, \\ \dot{p}^i &= -\frac{\partial H}{\partial q^i} = \frac{\partial \alpha_j(q)}{\partial q^i}\, v^j - \frac{\partial h(q)}{\partial q^i}, \\ \frac{\partial H}{\partial v^i} &= \phi_i(q^j, p_j) = p_i - \alpha_i(q^j) = 0. \end{aligned}$$

5 Examples of L–C Circuits

As an Implicit Lagrangian System. Consider the illustrative example of an L–C circuit shown in Fig. 1, which was also investigated in [8]. In the L–C circuit, the configuration space W is a 4-dimensional vector space, that is, $W = \mathbb{R}^4$. Then, we have $TW\,(\cong W \times W)$ and $T^*W\,(\cong W \times W^*)$. Let $q = (q_L, q_{C_1}, q_{C_2}, q_{C_3}) \in W$ denote charges and $f = (f_L, f_{C_1}, f_{C_2}, f_{C_3}) \in T_qW$ currents associated with the L–C circuit.

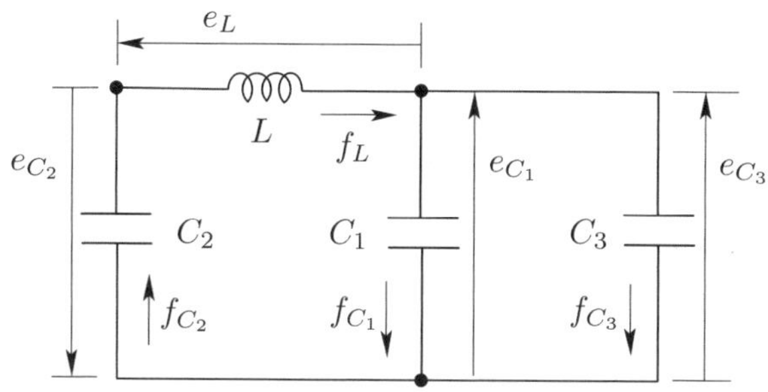

Fig. 1. L-C Circuit

The set of currents satisfying the KCL (Kirchhoff current law) constraints forms a constraint subspace $\Delta \subset TW$, which we shall call the *constraint KCL space* that is defined, for each $q \in W$, by

$$\Delta(q) = \{f \in T_qW \mid \langle \omega^a, f\rangle = 0,\ a = 1, 2\},$$

where $f = (f_1, f_2, f_3, f_4) = (f_L, f_{C_1}, f_{C_2}, f_{C_3})$ and ω^a denote 2-independent covectors (or one-forms) represented, in coordinates, by

$$\omega^a = \omega_k^a \, dq^k, \quad a = 1, 2; \; k = 1, ..., 4,$$

where $q = (q^1, q^2, q^3, q^4) = (q_L, q_{C_1}, q_{C_2}, q_{C_3})$. In this example, the coefficients ω_k^a are given in matrix representation by

$$\omega_k^a = \begin{pmatrix} -1 & 0 & 1 & 0 \\ 0 & -1 & 1 & -1 \end{pmatrix}.$$

Consistent with the general theory, the induced distribution Δ_{T^*W} on T^*W is defined by the KCL constraint distribution $\Delta \subset TW$ by

$$\Delta_{T^*W} = (T\pi_W)^{-1}(\Delta) \subset TT^*W,$$

where $\pi_W : T^*W \to W$ is the canonical projection and $T\pi_W : TT^*W \to TW$. Recall that the constraint set $\Delta \subset TW$ is represented as the simultaneous kernel of a number of constraint one-forms; that is, the annihilator of $\Delta(q)$, which is denoted by $\Delta^\circ(q)$, is spanned by such one-forms, that we write as $\omega^1, \omega^2, \ldots, \omega^m$. Now writing the projection map $\pi_W : T^*W \to W$ locally as $(q,p) \mapsto q$, its tangent map is locally given by $T\pi_W : (q, p, \dot{q}, \dot{p}) \mapsto (q, \dot{q})$. Then, we can locally represent Δ_{T^*W} as

$$\Delta_{T^*W} \cong \left\{ v_{(q,p)} = (q, p, \dot{q}, \dot{p}) \mid q \in U, \dot{q} \in \Delta(q) \right\}.$$

Let points in T^*T^*W be locally denoted by $\alpha_{(q,p)} = (q, p, \alpha, w)$, where α is a covector and w is a vector, and the annihilator of Δ_{T^*W} is

$$\Delta^\circ_{T^*W} \cong \left\{ \alpha_{(q,p)} = (q, p, \alpha, w) \mid q \in U, \; \alpha \in \Delta^\circ(q) \text{ and } w = 0 \right\}.$$

Recall also from equation (1) that the Dirac structure D_Δ on T^*W induced from the KCL constraint distribution Δ is locally given, for each $(q,p) \in T^*W$, by

$$D_\Delta(q,p) = \{((q, p, \dot{q}, \dot{p}), (q, p, \alpha, w)) \mid \dot{q} \in \Delta(q), \; w = \dot{q}, \; \alpha + \dot{p} \in \Delta^\circ(q)\}.$$

Let $T : TW \to \mathbb{R}$ be the magnetic energy of the L-C circuit, which is defined by the inductance L such that

$$T_q(f) = \frac{1}{2} L \, (f_L)^2,$$

and let $V : W \to \mathbb{R}$ be the electric potential energy of the L-C circuit, which is defined by capacitors C_1, C_2, and C_3 such that

$$V(q) = \frac{1}{2} \frac{(q_{C_1})^2}{C_1} + \frac{1}{2} \frac{(q_{C_2})^2}{C_2} + \frac{1}{2} \frac{(q_{C_3})^2}{C_3}.$$

Then, we can define the Lagrangian of the L–C circuit $\mathcal{L} : TW \to \mathbb{R}$ by

$$\begin{aligned} \mathcal{L}(q, f) &= T_q(f) - V(q) \\ &= \frac{1}{2} L \, (f_L)^2 - \frac{1}{2} \frac{(q_{C_1})^2}{C_1} - \frac{1}{2} \frac{(q_{C_2})^2}{C_2} - \frac{1}{2} \frac{(q_{C_3})^2}{C_3}. \end{aligned}$$

It is obvious that the Lagrangian $\mathcal{L} : TW \to \mathbb{R}$ of the L-C circuit is degenerate, since

$$\det\left[\frac{\partial^2 \mathcal{L}}{\partial f^i \partial f^j}\right] = 0; \quad i, j = 1, ..., 4.$$

The *constraint flux linkage subspace* is defined by the Legendre transform:

$$P = \mathbb{F}\mathcal{L}(\Delta) \subset T^*W.$$

In coordinates, $(q, p) = \mathbb{F}\mathcal{L}(q, f) \in T^*W$, and it follows

$$(p_L, p_{C_1}, p_{C_2}, p_{C_3}) = \left(\frac{\partial \mathcal{L}}{\partial f_L}, \frac{\partial \mathcal{L}}{\partial f_{C_1}}, \frac{\partial \mathcal{L}}{\partial f_{C_2}}, \frac{\partial \mathcal{L}}{\partial f_{C_3}}\right),$$

from which we obtain

$$p_L = L\, f_L$$

and with the constraints

$$p_{C_1} = 0,\ p_{C_2} = 0,\ p_{C_3} = 0,$$

which correspond to primary constraints in the sense of Dirac. Needless to say, the primary constraints form the constraint flux linkage subspace $P \subset T^*W$, which immediately reads

$$(q, p) = (q_L, q_{C_1}, q_{C_2}, q_{C_3},\ p_L, 0, 0, 0) \in P.$$

Let $X : TW \oplus T^*W \to TT^*W$ be a partial vector field on T^*W, defined at each point in P, with components denoted by

$$X(q, f, p) = (\dot{q}_L, \dot{q}_{C_1}, \dot{q}_{C_2}, \dot{q}_{C_3},\ \dot{p}_L, 0, 0, 0)\,.$$

Since the differential of the Lagrangian $\mathbf{d}\mathcal{L}(q, f) = (\partial\mathcal{L}/\partial q, \partial\mathcal{L}/\partial f)$ is given by

$$\mathbf{d}\mathcal{L}(q, f) = \left(0, -\frac{q_{C_1}}{C_1}, -\frac{q_{C_2}}{C_2}, -\frac{q_{C_3}}{C_3},\ L f_L, 0, 0, 0\right),$$

the Dirac differential of the Lagrangian $\mathfrak{D}\mathcal{L}(q, f) = (-\partial\mathcal{L}/\partial q, f)$ is given by

$$\mathfrak{D}\mathcal{L}\,(q, f) = \left(0, \frac{q_{C_1}}{C_1}, \frac{q_{C_2}}{C_2}, \frac{q_{C_3}}{C_3},\ f_L, f_{C_1}, f_{C_2}, f_{C_3}\right)$$

together with $p = \partial\mathcal{L}/\partial f$.

Thus, the L-C circuit can be represented in the context of implicit Lagrangian systems $(\mathcal{L}, \Delta, X)$ by requiring that, for each $(q, f) \in \Delta \subset TW$,

$$(X(q, f, p), \mathfrak{D}\mathcal{L}(q, f)) \in D_\Delta(q, p)$$

holds and with the Legendre transform $(q, p) = \mathbb{F}\mathcal{L}(q, f)$. Therefore, the implicit Lagrangian system for this L-C circuit may be locally described by

$$\begin{pmatrix} \dot{q}_L \\ \dot{q}_{C_1} \\ \dot{q}_{C_2} \\ \dot{q}_{C_3} \\ \hline \dot{p}_L \\ 0 \\ 0 \\ 0 \end{pmatrix} = \left(\begin{array}{cccc|cccc} 0 & 0 & 0 & 0 & 1 & 0 & 0 & 0 \\ 0 & 0 & 0 & 0 & 0 & 1 & 0 & 0 \\ 0 & 0 & 0 & 0 & 0 & 0 & 1 & 0 \\ 0 & 0 & 0 & 0 & 0 & 0 & 0 & 1 \\ \hline -1 & 0 & 0 & 0 & 0 & 0 & 0 & 0 \\ 0 & -1 & 0 & 0 & 0 & 0 & 0 & 0 \\ 0 & 0 & -1 & 0 & 0 & 0 & 0 & 0 \\ 0 & 0 & 0 & -1 & 0 & 0 & 0 & 0 \end{array}\right) \begin{pmatrix} 0 \\ \frac{q_{C_1}}{C_1} \\ \frac{q_{C_2}}{C_2} \\ \frac{q_{C_3}}{C_3} \\ \hline f_L \\ f_{C_1} \\ f_{C_2} \\ f_{C_3} \end{pmatrix} + \begin{pmatrix} 0 & 0 \\ 0 & 0 \\ 0 & 0 \\ 0 & 0 \\ \hline -1 & 0 \\ 0 & -1 \\ 1 & 1 \\ 0 & -1 \end{pmatrix} \begin{pmatrix} \mu_1 \\ \mu_2 \end{pmatrix}$$

together with the Legendre transformation

$$p_L = L\, f_L.$$

The above equations of motion are supplemented by the KCL constraints

$$\begin{pmatrix} 0 \\ 0 \end{pmatrix} = \begin{pmatrix} -1 & 0 & 1 & 0 \\ 0 & -1 & 1 & -1 \end{pmatrix} \begin{pmatrix} f_L \\ f_{C_1} \\ f_{C_2} \\ f_{C_3} \end{pmatrix}.$$

Finally, we can obtain the implicit Lagrangian system for this L-C circuit as

$$\begin{aligned} &\dot{q}_L = f_L,\ \dot{q}_{C_1} = f_{C_1},\ \dot{q}_{C_2} = f_{C_2},\ \dot{q}_{C_3} = f_{C_3}, \\ &\dot{p}_L = -\mu_1, \\ &\mu_2 = -\frac{q_{C_1}}{C_1},\ \mu_1 = -\mu_2 + \frac{q_{C_2}}{C_2},\ \mu_2 = -\frac{q_{C_3}}{C_3}, \\ &p_L = L\, f_L, \\ &f_L = f_{C_2},\ f_{C_1} = f_{C_2} - f_{C_3}. \end{aligned}$$

Representation as an Implicit Hamiltonian System. Next, let us illustrate this example of an L-C circuit in the context of implicit Hamiltonian systems via the generalized Legendre transformation.

First, define the generalized energy E on $TW \oplus T^*W$ by

$$\begin{aligned} E(q^i, f^i, p_i) &= p_i\, f^i - \mathcal{L}(q^i, f^i) \\ &= p_L\, f_L + p_{C_1}\, f_{C_1} + p_{C_2}\, f_{C_2} + p_{C_3}\, f_{C_3} \\ &\quad - \frac{1}{2} L\, (f_L)^2 + \frac{1}{2}\frac{(q_{C_1})^2}{C_1} + \frac{1}{2}\frac{(q_{C_2})^2}{C_2} + \frac{1}{2}\frac{(q_{C_3})^2}{C_3}. \end{aligned}$$

In the above, $(p_1, p_2, p_3, p_4) = (p_L, p_{C_1}, p_{C_2}, p_{C_3})$. Therefore, we can define the constrained Hamiltonian H_P on $P \subset T^*W$ by

$$\begin{aligned} H_P(q^i, p_\lambda) &= \operatorname{stat}_{f^i} E(q^i, f^i, p_i)|P \\ &= \frac{1}{2} L^{-1}\, (p_L)^2 + \frac{1}{2}\frac{(q_{C_1})^2}{C_1} + \frac{1}{2}\frac{(q_{C_2})^2}{C_2} + \frac{1}{2}\frac{(q_{C_3})^2}{C_3}, \end{aligned}$$

where $\lambda = 1$, that is, $p_1 = p_L$ and we employed the inverse partial Legendre transformation

$$f_L = L^{-1} p_L.$$

Since the primary constraints are given by

$$\phi_2 = p_{C_1} = 0,\ \phi_3 = p_{C_2} = 0,\ \phi_4 = p_{C_3} = 0,$$

we can define a generalized Hamiltonian H on $TW \oplus T^*W$ by

$$\begin{aligned} H(q^i, f^i, p_i) &= H_P(q^i, p_\lambda) + \phi_A(q^i, p_i)\, f^A \\ &= \frac{1}{2} L^{-1} (p_L)^2 + \frac{1}{2}\frac{(q_{C_1})^2}{C_1} + \frac{1}{2}\frac{(q_{C_2})^2}{C_2} + \frac{1}{2}\frac{(q_{C_3})^2}{C_3} \\ &\qquad + p_{C_1} f_{C_1} + p_{C_2} f_{C_2} + p_{C_3} f_{C_3}. \end{aligned}$$

The differential of H is given by

$$\mathbf{d}H = \left(q^i, f^i, p_i, \frac{\partial H}{\partial q^i}, \frac{\partial H}{\partial f^i}, \frac{\partial H}{\partial p_i} \right).$$

Considering the primary constraints, we can set

$$\frac{\partial H}{\partial f^A} = \phi_A(q^i, p_i) = p_A = 0, \qquad A = 2, 3, 4.$$

restriction $\mathbf{d}H(q,f,p) : T_{(q,f,p)}(TW \oplus T^*W) \to \mathbb{R}$ to $T_{(q,p)}T^*W$ is

$$\mathbf{d}H(q,f,p)|_{T_{(q,p)}T^*W} = \left(\frac{\partial H}{\partial q^i}, \frac{\partial H}{\partial p_i} \right),$$

which gives

$$\mathbf{d}H(q,f,p)|_{T_{(q,p)}T^*W} = \left(0, \frac{q_{C_1}}{C_1}, \frac{q_{C_2}}{C_2}, \frac{q_{C_3}}{C_3},\ L^{-1} p_L, f_{C_1}, f_{C_2}, f_{C_3} \right).$$

Hence, the L-C circuit can be represented as an implicit Hamiltonian system (H, Δ, X) that satisfies, for each $(q,p) \in T^*W$,

$$(X(q,p), \mathbf{d}H(q,f,p)|_{T_{(q,p)}T^*W}) \in D_\Delta(q,p)$$

together with the primary constraints

$$\frac{\partial H}{\partial f} = 0.$$

Recall that the vector filed X on T^*W is given in coordinates by

$$X(q,p) = (\dot{q}_L, \dot{q}_{C_1}, \dot{q}_{C_2}, \dot{q}_{C_3}, \dot{p}_L, 0, 0, 0).$$

Then, it follows from equation (11) that the implicit Hamiltonian system for the L-C circuit can be represented in coordinates as

$$
\begin{pmatrix} \dot{q}_L \\ \dot{q}_{C_1} \\ \dot{q}_{C_2} \\ \dot{q}_{C_3} \\ \hline \dot{p}_L \\ 0 \\ 0 \\ 0 \end{pmatrix} = \left(\begin{array}{cccc|cccc} 0 & 0 & 0 & 0 & 1 & 0 & 0 & 0 \\ 0 & 0 & 0 & 0 & 0 & 1 & 0 & 0 \\ 0 & 0 & 0 & 0 & 0 & 0 & 1 & 0 \\ 0 & 0 & 0 & 0 & 0 & 0 & 0 & 1 \\ \hline -1 & 0 & 0 & 0 & 0 & 0 & 0 & 0 \\ 0 & -1 & 0 & 0 & 0 & 0 & 0 & 0 \\ 0 & 0 & -1 & 0 & 0 & 0 & 0 & 0 \\ 0 & 0 & 0 & -1 & 0 & 0 & 0 & 0 \end{array}\right) \begin{pmatrix} 0 \\ \frac{q_{C_1}}{C_1} \\ \frac{q_{C_2}}{C_2} \\ \frac{q_{C_3}}{C_3} \\ \hline L^{-1} p_L \\ f_{C_1} \\ f_{C_2} \\ f_{C_3} \end{pmatrix} + \begin{pmatrix} 0 & 0 \\ 0 & 0 \\ 0 & 0 \\ 0 & 0 \\ \hline -1 & 0 \\ 0 & -1 \\ 1 & 1 \\ 0 & -1 \end{pmatrix} \begin{pmatrix} \mu_1 \\ \mu_2 \end{pmatrix},
$$

where the primary constraints

$$
p_{C_2} = p_{C_3} = p_4 = 0
$$

have been incorporated. The above equations of motion are accompanied with the KCL constraints

$$
\begin{pmatrix} 0 \\ 0 \end{pmatrix} = \begin{pmatrix} -1 & 0 & 1 & 0 \\ 0 & -1 & 1 & -1 \end{pmatrix} \begin{pmatrix} L^{-1} p_L \\ f_{C_1} \\ f_{C_2} \\ f_{C_3} \end{pmatrix}.
$$

Finally, the implicit Hamiltonian system for the L-C circuit can be locally given as follows:

$$
\begin{aligned}
&\dot{q}_L = L^{-1} p_L, \; \dot{q}_{C_1} = f_{C_1}, \; \dot{q}_{C_2} = f_{C_2}, \; \dot{q}_{C_3} = f_{C_3}, \\
&\dot{p}_L = -\mu_1, \\
&\mu_2 = -\frac{q_{C_1}}{C_1}, \; \mu_1 = -\mu_2 + \frac{q_{C_2}}{C_2}, \; \mu_2 = -\frac{q_{C_3}}{C_3}, \\
&L^{-1} p_L = f_{C_2}, \; f_{C_1} = f_{C_2} - f_{C_3}.
\end{aligned}
$$

It seems that this Hamiltonian view of this electric circuit is consistent with that presented in [8] and [2]. Note that the present approach derives the Hamiltonian structure in a systematic way from a degenerate Lagrangian, whereas the direct Hamiltonian approach requires some ingenuity to derive and its applicability to all cases is not clear.

6 Conclusions

The paper started by reviewing how a Dirac structure on a cotangent bundle is induced from a constraint distribution. In this context, implicit Lagrangian systems can be introduced in association with this induced Dirac structure, which is available for degenerate Lagrangians. It was shown how an implicit Hamiltonian system can be defined by a generalized Legendre transformation, starting with a generalized Hamiltonian on the Pontryagin bundle of a configuration manifold by incorporating the primary constraints that are present due to the possible degeneracy of the Lagrangian. The techniques were illustrated via some examples of degenerate Lagrangians with constraints, namely for L–C circuits and point vortices, as well as for nonholonomic systems, where the Lagrangian is typically nondegenerate, but constraints are present.

Acknowledgments. This research was partially supported by JSPS Grant 16560216 and NSF-ITR Grant ACI-0204932.

References

1. Blankenstein, G. and T. S. Ratiu [2004], Singular reduction of implicit Hamiltonian systems, *Rep. Math. Phys.* **53**, 211–260.
2. Bloch, A. M. and Crouch, P. E. [1997], Representations of Dirac structures on vector spaces and nonlinear L-C circuits. Differential Geometry and Control (Boulder, CO, 1997), 103–117, *Proc. Sympos. Pure Math.* **64**. Amer. Math. Soc., Providence, RI.
3. Courant, T. and A. Weinstein [1988], Beyond Poisson structures. In *Action hamiltoniennes de groupes. Troisième théorème de Lie (Lyon, 1986)*, volume 27 of *Travaux en Cours*, pages 39–49. Hermann, Paris.
4. Dirac, P. A. M. [1950], Generalized Hamiltonian dynamics, *Canadian J. Math.* **2**, 129–148.
5. Marsden, J. E. and T. S. Ratiu [1999], *Introduction to Mechanics and Symmetry*, volume 17 of *Texts in Applied Mathematics.* Springer-Verlag, second edition.
6. Rowley, C. W. and J. E. Marsden [2002], Variational integrators for degenerate Lagrangians, with application to point vortices. *Proceedings of the 41st IEEE Conference on Decision and Control, December 2002.*
7. Tulczyjew, W. M. [1977], The Legendre transformation, *Ann. Inst. H. Poincaré*, Sect. A, **27**(1), 101–114.
8. van der Schaft, A. J. [1998], Implicit Hamiltonian systems with symmetry, *Rep. Math. Phys.* **41**, 203–221.
9. van der Schaft, A. J. and B. M. Maschke [1995], The Hamiltonian formulation of energy conserving physical systems with external ports, *Archiv für Elektronik und Übertragungstechnik* **49**, 362–371.
10. Yoshimura, H. and Marsden, J. E.[2006], Dirac structures in mechanics. Part I: Implicit Lagrangian systems, *Journal of Geometry and Physics* **57-1**, 133–156.
11. Yoshimura, H. and Marsden, J. E.[2006], Dirac structures in mechanics. Part II: Variational structures. *Journal of Geometry and Physics*, **57-1**, 209–250.

Control of a Class of 1-Generator Nonholonomic System with Drift Through Input-Dependent Coordinate Transformation

Tsuyoshi Sagami, Mitsuji Sampei, and Shigeki Nakaura

Dept. of Mechanical and Control Engineering, Tokyo Institute of Technology
{sagami@sc., sampei@, nakaura@}ctrl.titech.ac.jp

1 Introduction

In this paper, a new control strategy for a class of 1-generator nonholonomic systems with drift is proposed:

$$\begin{aligned}
y_1^{(k_1)} &= u_1 \\
y_2^{(k_2)} &= u_2 \\
y_3^{(k_3)} &= \ell_{31}(y_3, \ldots, y_3^{(k_3)}) \\
&\quad + \ell_{32}(y_2, \ldots, y_2^{(k_2-1)}) \times \ell_g(u_1, y_1, \ldots, y_1^{(k_1-1)}) \\
&\ \ \vdots \\
y_n^{(k_n)} &= \ell_{n1}(y_n, \ldots, y_n^{(k_n)}) \\
&\quad + \ell_{n2}(y_{n-1}, \ldots, y_{n-1}^{(k_{n-1}-1)}) \times \ell_g(u_1, y_1, \ldots, y_1^{(k_1-1)}),
\end{aligned} \tag{1}$$

where each $\ell_*(\cdot)$ represents a homogeneous linear term. This system is an asymmetric affine nonholonomic system of the form $\dot{x} = f(x) + g(x)u$ with 1-generator controllability structure. including such as high-order chained form [5], underactuated surface vessel [4] and so on.

The proposed strategy is based on a coordinate transformation which is directly controlled by one of the input to the system. By this transformation, the original system is decomposed into two linear subsystems; one is a linear time-invariant subsystem which governs the *generator* dynamics, and the other is a linear time varying representing the remaining dynamics. Control of the original nonholonomic system is reduced to the linear control problem under a certain input constraint.

For a similar class of systems, Imura et. al. [1] have presented the control method for 3-variable 2nd-order chained form. Their method is based on a tracking to a pre-designed trajectory which satisfies the nonholonomic constraints. However, generalization to the n-variable and higher order case is not discussed therein.

Laiou and Astolfi [2] have introduced a controller design for the *generalized high-order chained system*, based on a state-dependent discontinuous coordinate

F. Bullo et al. (Eds.): Lag. & Hamil. Methods for Nonlin. Ctrl., LNCIS 366, pp. 249–258, 2007.
springerlink.com

transformation. The resultant controller is discontinuous and achieves an *almost exponential stability*, namely, the exponential stability except for the case in which the initial conditions lies in a sub-manifold of zero-measure. This method, however, cannot be combined with non-feedback method such as partial feedback with partial trajectory-planning.

Marchand and Alamir[3] also have presented an input-dependent coordinate transformation and designed a stabilizing controller for the 1st-order chained form. Nevertheless, the targeted system is the 1st-order chained form, and the high-order chained form system has not been considered in their work.

Our method can be applied to a class similar to the one presented in [2] and also is flexible in that the resultant control strategy is not necessarily be a pure feedback.

This paper is organized as follows. Section 2 is the main result, in which our proposed transformation, basic control strategy is introduced.

Section 3 contains a design example; state regulation of the nonholonomic underactuated surface vessel is solved by our strategy. Finally, section 4 includes some concluding remarks.

2 The Method

The essence of our method is the transformation of the original nonholonomic system (1) into the following form:

$$\frac{\mathrm{d}x_1}{\mathrm{d}t} = A_g x_1 + B_g v_1 \tag{2a}$$

$$\frac{\mathrm{d}\xi}{\mathrm{d}t} = \left(A_\xi + \frac{\dot{v}_1}{v_1} G_\xi \right) \xi + B_\xi u_2. \tag{2b}$$

In this form, the input $v_1 := \ell_g(u_1, y_1, \ldots, y_1^{(k_1-1)})$ represents the *generator* part of the original system.

Once the original nonholonomic system (1) is transformed into the proposed form (2), the problem is to control these two linear systems under the constraint $v_1 \neq 0$. The task can be performed utilizing various linear tools.

In this section, the main result is presented. Firstly, proposed coordinate transformation is presented in 2.1. Secondly, a basic control strategy of the transformed system is presented in 2.2 with an example to design a discontinuous controller.

2.1 Coordinate Transformation

Proposed transformation is summarized in the following theorem.

Theorem 1. *Let us define the state variable of the asymmetric affine nonholonomic system* (1) *as*

$$x_i = \begin{bmatrix} x_{i1} \\ x_{i2} \\ \vdots \\ x_{i\,k_i} \end{bmatrix} := \begin{bmatrix} y_i \\ \dot{y}_i \\ \vdots \\ y_i^{(k_i-1)} \end{bmatrix} \tag{3a}$$

$$\bar{x} := \begin{bmatrix} x_n \\ x_{n-1} \\ \vdots \\ x_2 \end{bmatrix} \tag{3b}$$

Then the system (1) is transformed into two linear subsystems (2) by the following input transformation:

$$v_1 := \ell_g(u_1, y_1, \ldots, y_1^{(k_1-1)}) = \ell_g(u_1, x_1), \tag{4}$$

and the input-dependent coordinate transformation:

$$\xi := \begin{bmatrix} \xi_n \\ \xi_{n-1} \\ \vdots \\ \xi_2 \end{bmatrix} = \operatorname{diag}\left(\frac{1}{{v_1}^{n-2}} \boldsymbol{I}_{k_n}, \ldots, \frac{1}{{v_1}^{i-2}} \boldsymbol{I}_{k_i}, \ldots, \boldsymbol{I}_{k_2} \right) \bar{x}, \tag{5}$$

in which $\boldsymbol{I}_*$'s denote the identity.

Proof. For the x_1-subsystem, one have

$$\begin{aligned} v_1 &= \ell_g(u_1, x_1) \\ &= a_u^g u_1 + \sum_{i=1}^{k_1} a_i^g x_{1i}, \quad \text{where } a_u^g \neq 0, a_i^g \neq 0, \end{aligned}$$

since $\ell_g(u_1, y_1, \ldots, y_1^{(k_1-1)}) = \ell_g(u_1, x_1)$ is a homogeneous linear term by the assumption. Thus, by letting

$$u_1 = \frac{1}{a_u^g} \left(v_1 - \sum_{i=1}^{k_1} a_i^g x_{1i} \right), \tag{6}$$

the subsystem of x_1 is the following linear time-invariant subsystem:

$$\begin{aligned} \frac{\mathrm{d}x_1}{\mathrm{d}t} &= \begin{bmatrix} 0 & 1 & 0 & \cdots & 0 \\ 0 & 0 & 1 & \cdots & 0 \\ \vdots & \vdots & \vdots & \ddots & \vdots \\ 0 & 0 & 0 & \cdots & 1 \\ 0 & 0 & 0 & \cdots & 0 \end{bmatrix} x_1 + \begin{bmatrix} 0 \\ 0 \\ \vdots \\ 0 \\ 1 \end{bmatrix} u_1 \\ &= \begin{bmatrix} 0 & 1 & 0 & \cdots & 0 \\ 0 & 0 & 1 & \cdots & 0 \\ \vdots & \vdots & \vdots & \ddots & \vdots \\ 0 & 0 & 0 & \cdots & 1 \\ -a_1^g & -a_2^g & -a_3^g & \cdots & -a_g^{k_1} \end{bmatrix} x_1 + \begin{bmatrix} 0 \\ 0 \\ \vdots \\ 0 \\ \frac{1}{a_u^g} \end{bmatrix} v_1 \\ &=: A_g x_1 + B_g v_1 \end{aligned}$$

and thus one have

$$\frac{\mathrm{d}x_1}{\mathrm{d}t} = A_g x_1 + B_g v_1. \tag{2a}$$

Next, for x_2, one have

$$\frac{\mathrm{d}\xi_2}{\mathrm{d}t} = A_2\xi_2 + B_2 u_2, \tag{7}$$

from the coordinate transformation $x_2 = \xi_2$. Similarly, for $x_i(i = 3, \dots, n)$, Letting

$$\ell_{i1}(y_i, \dots, {y_i}^{(k_i-1)}) =: \sum_{j=1}^{k_i} a_j^i x_{ij} \tag{8}$$

$$\ell_{i2}(y_{i-1}, \dots, {y_{i-1}}^{(k_{i-1}-1)}) =: \sum_{j=1}^{k_i} g_j^i x_{i-1j} \tag{9}$$

results in

$$\begin{aligned}
\dot{x}_{i1} &= x_{i2} \\
&\vdots \\
\dot{x}_{ik_i-1} &= x_{ik_i} \\
\dot{x}_{ik_i} &= \sum_{j=1}^{k_i} a_j^i x_{ij} + v_1 \sum_{j=1}^{k_{i-1}} g_j^i x_{i-1j}
\end{aligned}$$

which is equivalent to

$$\begin{aligned}
\frac{\mathrm{d}x_i}{\mathrm{d}t} &= \begin{bmatrix} 0 & 1 & \cdots & 0 \\ \vdots & \vdots & \ddots & \vdots \\ 0 & 0 & \cdots & 1 \\ a_1^i & a_2^i & \cdots & a_i^{k_i} \end{bmatrix} x_i + v_1 \begin{bmatrix} 0 & \cdots & 0 \\ \vdots & & \vdots \\ 0 & \cdots & 0 \\ g_1^i & \cdots & g_{k_{i-1}}^i \end{bmatrix} x_{i-1} \\
&=: \tilde{A}_i x_i + v_1 \hat{A}_i x_{i-1}.
\end{aligned}$$

Finally, since the time derivative of ξ is as follows:

$$\begin{aligned}
\frac{\mathrm{d}\xi_i}{\mathrm{d}t} &= \frac{\mathrm{d}}{\mathrm{d}t}\left(\frac{1}{{v_1}^{i-2}} x_i\right) \\
&= -(i-2)\frac{\dot{v}_1}{{v_1}^{i-1}} x_i + \frac{1}{{v_1}^{i-2}} \dot{x}_i \\
&= -(i-2)\frac{\dot{v}_1}{v_1} \cdot \frac{x_i}{{v_1}^{i-2}} + \frac{1}{{v_1}^{i-2}}\left(\tilde{A}_i x_i + v_1 \hat{A}_i x_{i-1}\right) \\
&= -(i-2)\frac{\dot{v}_1}{v_1} \cdot \frac{x_i}{{v_1}^{i-2}} + \tilde{A}_i \frac{x_i}{{v_1}^{i-2}} + \hat{A}_i \frac{x_{i-1}}{{v_1}^{i-3}} \\
&= \left[\tilde{A}_i - (i-2)\frac{\dot{v}_1}{v_1} I_{k_i}\right] \xi_i + \hat{A}_i \xi_{i-1},
\end{aligned}$$

by letting

$$A_\xi := \begin{bmatrix} \tilde{A}_{k_n} & \hat{A}_{k_n} & \boldsymbol{O} & \cdots & \boldsymbol{O} \\ \boldsymbol{O} & \tilde{A}_{k_{n-1}} & \hat{A}_{k_{n-1}} & \cdots & \boldsymbol{O} \\ \vdots & \vdots & \ddots & \ddots & \vdots \\ \boldsymbol{O} & \boldsymbol{O} & \cdots & \tilde{A}_{k_3} & \hat{A}_{k_3} \\ \boldsymbol{O} & \boldsymbol{O} & \cdots & \boldsymbol{O} & A_{k_2} \end{bmatrix}, \tag{10}$$

$$B_\xi := \begin{bmatrix} \boldsymbol{O} \\ \boldsymbol{O} \\ \vdots \\ \boldsymbol{O} \\ B_2 \end{bmatrix} = \begin{bmatrix} 0 \\ 0 \\ \vdots \\ 0 \\ 1 \end{bmatrix}, \tag{11}$$

and

$$G_\xi := \mathrm{diag}\left[(n-2)\boldsymbol{I_{k_n}}, \ldots, (i-2)\boldsymbol{I_{k_i}}, \ldots, \boldsymbol{O}_{k_2}\right], \tag{12}$$

one have the ξ-subsystem (2b). □

2.2 Control of the Transformed System

In this subsection, basic strategy is shown for designing the controller of the transformed system (2). The policy is as follows:

i) Control two subsystems (2a) and (2b) simultaneously.
ii) Keep the input v_1 non-vanishing to prevent the singularity of the transformation (5).
iii) Stabilize the ξ-subsystem (2b) regarding as if $\dot{v}_1/v_1$ is a *system disturbance*.
iv) Keep the input v_1 be bounded when $t \to \infty$. This assures that $\xi \to 0$ implies $\bar{x} \to 0$.

As far as the above conditions are satisfied, one can choose the various methods appropriate for the situations. Among them, discontinuous controller design is presented below in this paper.

Firstly, we introduce the following proposition.

Proposition 1. *Suppose that the subsystem* (2a) *is controllable or at least uncontrollable poles are real and stable. Then there exists a linear feedback for the subsystem* (2a) *such that:*

i) There exists a set of initial conditions of x_1 such that $v_1 \neq 0$, $\forall t > 0$, which implies that nonsingularity of the coordinate transformation (5).
ii) $\dot{v}_1/v_1$ is bounded $\forall t > 0$ and converges to a real value as $t \to \infty$. Furthermore, this limit is not depends on the initial condition $x_1(0)$.
iii) the ξ-subsystem becomes a so-called linearly bounded *nonlinear system of the form*

$$\frac{\mathrm{d}\xi}{\mathrm{d}t} = (A_\xi + \alpha G_\xi + \Delta(t))\xi + B_\xi u_2, \tag{13}$$

where $\Delta(t)$ is the matrix function which satisfies

$$\lim_{t\to\infty} \Delta(t) = 0, \text{ and } \int_0^\infty \|\Delta(t)\|\,\mathrm{d}t < \infty. \tag{14}$$

Proof. By the assumption, one can make the poles of the closed loop system of x_1 to be *stable real* poles: $-\alpha_1, \ldots, -\alpha_{k_1}$, where $0 < \alpha_1 < \cdots < \alpha_{k_1}$ by some linear feedback. This means that the input of the subsystem (2a) can be parametrized by a set of linear functions $\phi_i(\cdot)$, $i = 1, \ldots, k_1$ such as

$$v_1(t) = \sum_{i=0}^{k_1} \phi_i(x_1(0)) \exp(-\alpha_i t).$$

Since the coefficient $\phi_i(x_1(0))$ only depends on the initial condition $x_1(0)$, it is possible to choose a initial condition such that $v_1(t)$ converges to zero without changing its sign. For example, selecting such that all $\phi_i(x_1(0))$ have same sign is sufficient.

For such initial condition $x_1(0)$, $\dot{v}_1/v_1$ is bounded for all $t > 0$ and converges to the *slowest pole* $-\alpha_1$:

$$\frac{\dot{v}_1}{v_1} = \frac{\sum_{i=0}^{k_1} -\alpha_i \phi_i(x_1(0)) \exp(-\alpha_i t)}{\sum_{i=0}^{k_1} \phi_i(x_1(0)) \exp(-\alpha_i t)} \to -\alpha_1 \quad (t \to \infty).$$

Therefore, the subsystem of ξ can be represented as the form

$$\frac{\mathrm{d}\xi}{\mathrm{d}t} = (A_\xi + \alpha G_\xi + \Delta(t))\xi + B_\xi u_2,$$

where

$$\Delta(t) = \left(\alpha_1 + \frac{\dot{v}_1}{v_1}\right) G_\xi,$$

and $\Delta(t)$ exponentially converges to zero matrix.

Finally, exponential rate of convergence of $\dot{v}_1/v_1$ implies the boundedness of $\int_0^\infty \|\Delta(t)\|\,\mathrm{d}t$. □

By Proposition 1, the linear time-varying subsystem (2b) can be represented as a linear system with asymptotically converging disturbance (13). Such kind of systems can be stabilized by the linear control. For example Slotine and Li shows the following result.

Lemma 1. *Quoted from [6, p. 115] Consider a linear time-varying system of the form:*

$$\dot{x} = (A + \Delta(t))x \tag{15}$$

where the matrix A *is constant and Hurwitz, and the time-varying matrix* $\Delta(t)$ *is such that*

$$\lim_{t\to\infty} \Delta(t) = 0, \text{ and } \int_0^{\infty} \|\Delta(t)\| \, \mathrm{d}t < \infty.$$

Then the system (15) *is globally exponentially stable.*

As has been shown above, one can design a linear feedback controller which 1) converges the x_1 for some initial conditions and 2) globally exponentially stabilize ξ with Proposition 1 and Lemma 1. Furthermore, it implies the convergence of the state of the original system.

The reason is as follows. By the definition (5), one have the following inverse transformation (nonsingular if $v_1 = 0$):

$$\bar{x} = \operatorname{diag}\left({v_1}^{n-2} \boldsymbol{I}_{k_n}, \ldots, {v_1}^{i-2} \boldsymbol{I}_{k_i}, \ldots . \boldsymbol{I}_{k_2}\right) \xi.$$

Thus $\xi \to 0$ and $v_1 \to 0$ implies $\bar{x} \to 0$.

It is to be noted that the discontinuous controller design is one of the applications of our approach. Since the input v_1 of the x_1-subsystem (2a) is regarded as a *system disturbance* in the ξ-subsystem (2b), feedback for the x_1 is not essential. In other words, as far as the *system disturbance* $\dot{v}_1/v_1$ does not violate the stability of the ξ-subsystem, one can control the x_1-subsystem arbitrarily, with such as feedforward controller, discrete-valued controller and so on. This property is of importance for a practical purpose. Because such systems that does not allow any static-smooth feedback controller requires some auxiliary controller to escape the states from the singularity sub-manifold. Our proposed approach allows us to use the same feedback controller for the ξ-subsystem, which improve the partial regulation of the system state.

3 Numerical Example

In this section, control of an underactuated surface vessel is presented as a numerical example of the proposed method. It is known that there exists a coordinate and input transformations by which the vessel system is transformed into the equation of the form (1). Our method is applied to the transformed system, and a discontinuous controller is designed.

3.1 Underactuated Surface Vessel

Let us consider the equation of motion of an under-actuated surface vessel [4]:

$$\ddot{x} = u_1 \tag{16a}$$

$$\ddot{y} = u_2 \tag{16b}$$

$$\ddot{\psi} = u_1 \tan\psi + \frac{c_y}{m}\left(\dot{x}\tan\psi - \dot{y}\right), \tag{16c}$$

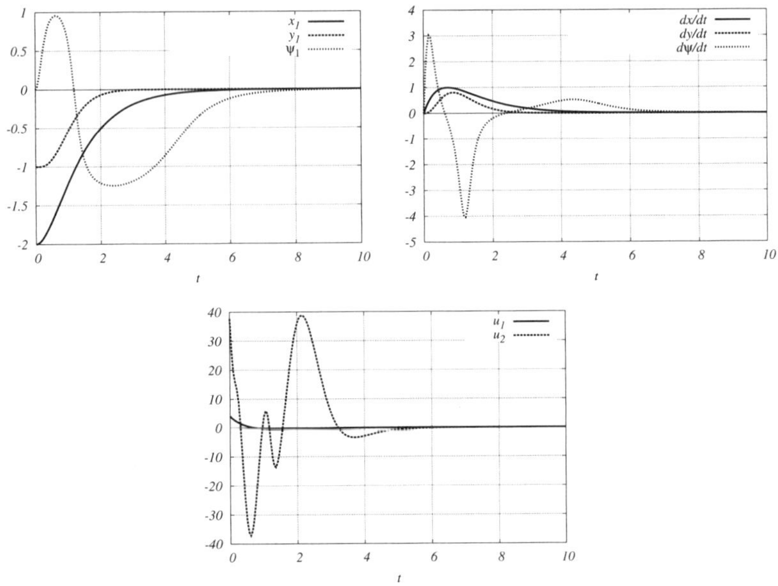

Fig. 1. Control of a Surface Vessel: state/input

in which x and y is a position, ψ the azimuth angle of the vessel, and the inputs u_1, u_2 are some function of its surge force and angular acceleration.

Laiou and Astolfi has shown that the system (16) can be transformed into the following form:

$$\ddot{z}_{11} = w_1, \tag{17a}$$

$$\ddot{z}_{21} = w_2, \tag{17b}$$

$$\ddot{z}_{31} = z_{21}w_1 + \dot{z}_{11}z_{21} - \dot{z}_{31}, \tag{17c}$$

by some coordinate and input transformation.

Since (17) is a subclass of the (1), it can be transformed into the proposed form (2) by the following input and coordinate transformation:

$$v_1 := w_1 + z_{12}$$

$$\xi := \mathrm{diag}\left(\frac{1}{v_1}, \frac{1}{v_1}, 1, 1\right) \begin{bmatrix} z_{31} \\ z_{32} \\ z_{21} \\ z_{22}. \end{bmatrix}$$

A discontinuous controller is designed with the result in Proposition 1. Feedback gain is chosen as follows

$$\begin{aligned} w_1 &= \left[-2.0 \ -2.0\right] z_1 \\ w_2 &= \left[-1.51 \times 10^2 \ -1.09 \times 10^2 \ -3.74 \times 10 \ -8.70\right] \xi. \end{aligned} \tag{18}$$

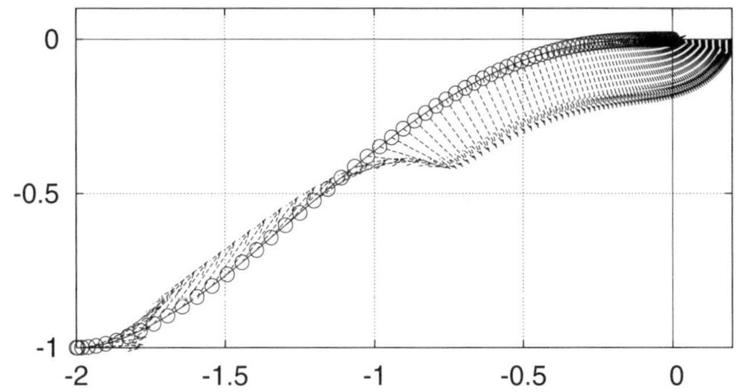

Fig. 2. Control of a Surface Vessel: trajectory

Figure 1 shows the (exponential) convergence of the states, time-derivative of the states, and inputs of the vessel.

Figure 2 is a plot of the trajectory (line and circles) and orientation (arrows) of the vessel system with the controller (18).

Note that the set of initial conditions of z_1 such that $v_1 = 0$ is

$$\{z_1 \mid z_{11}(0) + z_{12}(0) = 0\} \tag{19}$$

in this case. Unless the initial condition $z_1(0)$ lines in this set, the state and input converges to the origin exponentially.

4 Conclusion

In this paper, a new methodology for a class of nonholonomic system is presented. Proposed approach utilizes a special coordinate transformation which is dependent on one of the input to the system.

The essence of the method is to decompose the original nonholonomic system into two linear subsystems by the proposed transformation. Once the original system is transformed into the proposed form, control problem is reduced to the linear control problem with an input constraint($v_1 \neq 0$). Therefore, various methods can be used for controlling the transformed subsystems.

Among them, discontinuous feedback controller design is presented as an example. In this example, resultant controller drives the state to the origin, if the initial value of the one subsystem satisfies a certain condition.

References

1. J. Imura, K. Kobayashi, and T. Yoshikawa. Nonholonomic control of 3 link planar manipulator with a free joint. *Decision and Control, 1996., Proceedings of the 35th IEEE*, 2, 1996.

2. M.C. Laiou and A. Astolfi. Discontinuous control of high-order generalized chained systems. *Systems and Control Letters*, 37(5):309–322, 1999.
3. N. Marchand and M. Alamir. Discontinuous exponential stabilization of chained form systems. *Automatica*, 39(2):343–348, 2003.
4. M. Reyhanoglu. Exponential Stabilization of an Underactuated Autonomous Surface Vessel. *Automatica*, 33(12):2249–2254, 1997.
5. T Sagami, M Sampei, and S Nakaura. Input-dependent Coordinate Transformation for High-order Chained System. *Proc. 49th Annual Conference of the ISCIE*, pages 427–428, 2005. Japanese.
6. J.J.E. Slotine and W. Li. *Applied nonlinear control.* Prentice Hall Englewood Cliffs, NJ, 1991.

Principal Subspace Flows Via Mechanical Systems on Grassmann Manifolds

Uwe Helmke* and P.S. Krishnaprasad

Institute of Mathematics, University of Würzburg,
Am Hubland, 97074 Würzburg, Germany
Department of Electrical and Computer Engineering,
University of Maryland, College Park, MD 20742-3285, USA

Summary. We propose a class of second order mechanical systems on Grassmann manifolds that converge to the dominant eigenspace of a given symmetric matrix. Such second order flows for principal subspace analysis are derived from a modification of the familiar Euler–Lagrange equation by inserting a suitable damping term. The kinetic energy of the system is defined by the Riemannian metric on the Grassmannian while the potential energy is given by a trace function, that is defined by the symmetric matrix. Convergence of the algorithm to the dominant subspace is shown for generic initial conditions and a comparison with the Oja flow is made.

1 Introduction

In this paper we explore the computational abilities of second order mechanical systems as a tool for principal subspace analysis (PSA). This amounts to study certain optimization problems on Grassmann manifolds and complements previous approaches via gradient flows solving eigenvalue problems; see [11]. Specifically, we focus on trace minimization problems on Grassmann manifolds for computing the first few dominant eigenvectors of a positive semidefinite covariance matrix. Using the Hessian as a dissipation term, we derive an apparently new class of mechanical systems on Grassmann manifolds for which–under mild conditions–global convergence to the principal subspace of a given positive definite covariance matrix can be established.

Our results are obtained as a special case of a more general approach to construct second order mechanical systems on an arbitrary Riemannian manifold for cost function optimization. Explicitly, we consider a damped version

$$\nabla_{\dot{x}}\dot{x} = -\mathrm{grad}\ \mathrm{f(x)} - \mathrm{Hess_f(x)\dot{x}}$$

of the Euler–Lagrange equations on Riemannian manifolds, with the Hessian of the potential function entering as a friction term. We refer to these equations as the Newton Euler-Lagrange equations, as they coincide for $\nabla_{\dot{x}}\dot{x} = 0$ with

* Partially supported by the German Research Foundation under grant KONNEW HE 1858/10-1

F. Bullo et al. (Eds.): Lag. & Hamil. Methods for Nonlin. Ctrl., LNCIS 366, pp. 259–267, 2007.
springerlink.com

the continuous analog of the Newton method on a manifold. As is well known, the dynamics of dissipative mechanical systems may exhibit rather unintuitive behaviour, such as inherent instability properties. Thus the task of designing well-behaved second order dynamics on a manifold is a nontrivial one. For general results on dissipative mechanical systems on Lie groups and principal bundles we refer to [5, 6]. Here our purpose is to insert damping terms that preserve the well-known convergence properties of gradient flows. Although some specific dissipation terms were already introduced in [5, 6], such as e.g. a double bracket dissipation based on Brockett's work [7], the idea of systematically using Hessian friction terms for constrained optimization on manifolds has so far been explored only in a few papers. The only prior work we are aware of are the papers by [3, 1, 2], as well as that by [8]. In [3, 1], the authors consider the Euler–Lagrange equations on a Hilbert space, modified by a suitable Hessian damping term. Pointwise convergence to the critical points of a real analytic potential function is shown using Lyapunov techniques and the Lojasiewicz inequality. However their analysis is limited to the Hilbert space case and the more interesting problem of mechanical systems on Riemannian manifolds has not beeen considered. Since a constant scalar product on the ambient Hilbert space is used, one cannot easily extend their analysis via local coordinate charts to a manifold. In [2], a partial extension is given to gradient flows on Riemannian manifolds, but again this analysis does not immediately carry over to second order dynamics on manifolds. Brockett's work [8] is somewhat closer in spirit to what we are doing. His starting point is a Lagrangian mechanical system on a Riemannian manifold, coupled to a heath bath. As we will see, there is a straightforward geometric procedure to write down such equations in the deterministic case.

2 Newton–Euler–Lagrange Equations

We now introduce the equation of motion for the second order mechanical systems we are interested in. Let $(M, < \cdot, \cdot >)$ denote an n-dimensional Riemannian manifold and let $f : M \to \mathbb{R}$ denote a smooth function. In Lagrangian mechanics, the cost function $f : M \to \mathbb{R}$ defines the potential function, while the Riemannian metric defines the kinetic energy $K : TM \to \mathbb{R}$

$$K(x; v) := \frac{1}{2}\|v\|_x^2.$$

The associated *Lagrangian* and *Hamiltonian*, respectively, are the smooth functions on the tangent bundle $L, H : TM \to \mathbb{R}$, defined by

$$L(x, v) = K(x; v) - f(x), H(x, v) = K(x; v) + f(x).$$

The pair (M, L) is often called a simple mechanical system on M. In Lagrangian mechanics one considers the problem of finding the extrema of the variational problem for the energy functional

$$E(x) = \frac{1}{2}\int_a^b \|\dot{x}(t)\|^2 dt - \int_a^b f(x(t))dt.$$

Let grad f denote the Riemannian gradient vector field of f on M and $\mathrm{H_f(x)} : \mathrm{T_xM} \to \mathrm{T_xM}$ denote the Riemannian Hesse operator, i.e.

$$\mathrm{H_f(x)}\xi = \nabla_\xi \mathrm{gradf(x)} \quad \forall \xi \in \mathrm{T_xM}.$$

Thus $\mathrm{H_f(x)}$ denotes the Hesse operator of f that is associated with the Levi-Civita connection ∇ on M. Let $\nabla_{\dot{x}}\dot{x}$ denote the geodesic operator. With the terms at hand we can define the Newton Euler–Lagrange equation.

Definition 1. *The* **Newton Euler-Lagrange** *equation on the Riemannian manifold M is the second order differential equation*

$$\nabla_{\dot{x}}\dot{x} = -\mathrm{grad\ f(x)} - \mathrm{H_f(x)\dot{x}}. \tag{1}$$

Note that this equation differs from the usual Euler–Lagrange equation

$$\nabla_{\dot{x}}\dot{x} = -\mathrm{grad\ f(x)}$$

just by the presence of the dissipation term defined by the Hessian operator.

Our goal is to analyze (1) as a general second order optimization method, and explore its behavior for specific computational tasks, such as for principal and minor component analysis. Before entering into the discussion of principal component analysis we make some general comments on the convergence properties of (1). First note, that for any solution $x(t)$ of (1), the Hamiltonian function satisfies

$$\begin{aligned}\frac{d}{dt}H(x(t),\dot{x}(t)) &= <\dot{x}, \nabla_{\dot{x}}\dot{x}> + <\mathrm{grad\ f(x)}, \dot{\mathrm{x}}> \\ &= - <\mathrm{H_f(x)}\dot{\mathrm{x}}, \dot{\mathrm{x}}> .\end{aligned}$$

Thus positive definiteness of the Hessian, i.e. strong convexity of f, would imply, that H acts as a Lyapunov function for (1). However, this will not be the case in general and therefore a more refined argument has to be given, in order to ensure global convergence to the critical points of f. Note that the equilibria of (1) are precisely the points $(x_\infty, \dot{x}_\infty) \in TM$, where x_∞ is a critical point of f on M and $\dot{x}_\infty = 0$. The following result clarifies the convergence properties of the Newton–Euler–Lagrange flow.

Theorem 1. *Let $f : M \to \mathbb{R}$ be a smooth function with compact sublevel sets. Assume that there exists no nonconstant geodesic $x : I \to M$ on M such that*

$$\mathrm{grad\ f(x(t))} = 0 \quad \forall \mathrm{t} \in \mathrm{I}.$$

(This condition is e.g. satisfied if f has only isolated critical points.) Then the solutions $x(t)$ of (1) exist for all t and converge for $t \to \infty$ to the set of critical points of f, which are also the equilibrium points of the flow. If f has only isolated critical points, then the solutions converge to single equilibrium points.

Proof. Consider $V : TM \to \mathbb{R}$, defined by

$$V(x,\dot{x}) := f(x) + \frac{1}{2}\|\dot{x} + \mathrm{grad\ f(x)}\|^2.$$

The Lie deriviative along any solution $x(t)$ of (1) is $\dot{V} =$

$$< \mathrm{gradf(x)}, \dot{\mathrm{x}} > + < \dot{\mathrm{x}} + \mathrm{grad\ f(x)}, \nabla_{\dot{\mathrm{x}}}\dot{\mathrm{x}} + \nabla_{\dot{\mathrm{x}}}\mathrm{grad\ f(x)} >$$

using the compatibility property of the Levi-Civita connection with the metric. Moreover,

$$\nabla_{\dot{x}}\mathrm{grad\ f(x)} = \mathrm{H_f(x)\dot{x}}$$

and therefore $\dot{V} =$

$$\begin{aligned} &< \mathrm{grad\ f(x)}, \dot{\mathrm{x}} > + < \dot{\mathrm{x}} + \mathrm{grad\ f(x)}, -\mathrm{grad\ f(x)} > \\ &= -\|\mathrm{grad\ f(x)}\|^2. \end{aligned} \tag{2}$$

Thus $V(t)$ decreases along the solutions of (1). By the definition of V this shows that both $f(x)$ and $\|\dot{x} + \mathrm{grad\ f(x)}\|^2$ are bounded from above. Since f has compact sublevel sets this implies that x and $\dot{x}$ are positively bounded and therefore the solutions exist for all $t \geq 0$. Moreover, by LaSalle's invariance principle, $(x(t), \dot{x}(t))$ converges to the maximal invariant and connected subset of

$$\mathcal{X} := \{(x, y) \in TM \quad | \quad \mathrm{grad\ f(x)} = 0\}$$

of the associated first order system

$$\begin{aligned} \dot{x} &= y \\ \nabla_y y &= -\mathrm{grad\ f(x)} - \mathrm{H_f(x)y}. \end{aligned} \tag{3}$$

Suppose there exists a nontrivial solution (x, y) of this first order system that is contained in $\mathcal{X}$. Then $\mathrm{grad\ f(x(t))} = 0$ and therefore by differentiation also $\nabla_{\dot{x}}\mathrm{grad\ f(x)} = \mathrm{H_f(x)\dot{x}} = 0$. Thus $\nabla_{\dot{x}}\dot{x} = 0$ which implies that x is a geodesic. By our assumption thus x is a constant curve and therefore a critical point. This proves the claim.

The above result shows that the Newton Euler-Lagrange flow will converge to the set of critical points of f. The question arises under which conditions pointwise convergence to individual critical points happens. Of course, this will not be the case in general. The following result holds for a slightly stronger damped version than (1).

Theorem 2 ([12]). *Let M be a real analytic submanifold and f a real analytic function of M. Let $\alpha > 0$. Then the ω–limit set of any solution $x(t)$ of*

$$\nabla_{\dot{x}}\dot{x} = -\mathrm{grad\ f(x)} - \alpha\dot{\mathrm{x}} - \mathrm{H_f(x)\dot{x}}. \tag{4}$$

consists of at most one point. In particular, if f has compact sublevel sets on M, then every solution $x(t)$ of (4) converges to a single critical point of f.

We now characterize the local stability properties of the equilibrium points of (1).

Lemma 1. *The linearization of (4) at an equilibrium point* $(x_\infty, \dot{x}_\infty = 0)$ *is the linear differential equation on the tangent space* $T_{x_\infty}M$

$$\ddot{\xi} + (\alpha I + A)\dot{\xi} + A\xi = 0, \tag{5}$$

where $\xi \in T_{x_\infty}M$ *and* $A = \mathrm{H_f(x_\infty)}$.

Proof. The linearization of the Hessian operator at a critical point is just A, and at a critical point the identity

$$D\mathrm{gradf(x_\infty)} = \mathrm{H_f(x_\infty)}$$

holds. The result follows.

In order to compute the eigenvalues of the second order system we note the following lemma.

Lemma 2. *Let* $0 \le \alpha \le 1$*. The discriminant of the polynomial* $s^2 + (\alpha + \lambda)s + \lambda$ *is zero if and only if* $\lambda = \lambda_-$ *or* $\lambda = \lambda_+$*, where*

$$\lambda_\pm = 2 - \alpha \pm 2\sqrt{1-\alpha} \ge 0.$$

For $\lambda_- < \lambda < \lambda_+$ *the discriminant is negative, and the roots* s_-, s_+ *are complex with negative real part. For* $\lambda > \lambda_+$ *or if* $0 < \lambda < \lambda_-$*, the discriminant is positive and the real roots* s_-, s_+ *are negative. For* $\lambda < 0$*, one root is positive and the other negative.*

This leads to the following characterization of the spectrum of the linearization of the flows (1), (4).

Theorem 3. *Let* $0 \le \alpha \le 1$*. Let* $f : M \to \mathbb{R}$ *denote a Morse function. Then the dimensions of the positive eigenspace of the Riemannian Hessian* $\mathrm{H_f(x_\infty)}$ *and of the negative eigenspace of the linearization* B *of (1) at an equilibrium point* x_∞*, respectively, satisfy:*

$$2\mathrm{dimEig_+H_f(x_\infty)} = \mathrm{dimEig_-B}.$$

In particular, the local minima of f are precisely the local attractors of (1).

3 Second Order Principal Subspace Flows

Principal component analysis is a statistical estimation techniques for analyzing the covariance structure of stochastic signals. It is based on the estimation of the first few dominant eigenvectors of the associated covariance matrix. In the neural networks literature, see e.g. [9], a lot of algorithms have been developed to achieve that task. Their analysis is mainly based on the study of associated matrix differential equations, derived from ideas of stochastic approximation; see [4]. The most prominent of such flows for principal component analysis is that of [13], [14]. Following Oja's work, other differential equations have been proposed with similar convergence properties; see [9]. See also [15] for a rigorous convergence analysis of generalized flows for principal component analysis. Since such flows were not derived from any physical first principles, the question arises whether there exist any Hamiltonian or Lagrangian system that can be used for principal component analysis. It is the purpose of this paper to show that the answer to this question is "yes", by presenting second order flows that can be used for principal subspace analysis, i.e. flows that are capable of computing the invariant subpsace defined by the first p largest eigenvalues. We do so by describing flows on Grassmann manifolds that converge to the dominant invariant subspace of a covariance matrix A and thus perform the principal subspace analysis. See [10] for a related description of principal component flows on Stiefel manifolds (without proof details).

We now consider the closely related task of optimizing a trace function on the Grassmann manifold for principal subspace analysis. Thus let A denote a real symmetric $n \times n$ matrix, given e.g. as the covariance matrix for random data vectors $x_1, \cdots, x_N$. Let $\lambda_1 \geq \cdots \geq \lambda_n$ denote the eigenvalues of A, denoted in decreasing form. The p-dimensional principal subspace V_{pca} of A is defined as the eigenspace associated with $\lambda_1 \geq \cdots \geq \lambda_p$. The computation of V_{pca} is well known to be equivalent to a maximization problem of the trace function tr(AP) on the set rank p selfadjoint projection operators of $\mathbb{R}^n$, thus on a Grassmann manifold. Here we identify the Grassmann manifold with the isospectral set $\text{Grass}(p, \mathbb{R}^n)$ of real symmetric projection operators of rank p. Note that we can uniquely identify any projection operator P with its image space and therefore $\text{Grass}(p, \mathbb{R}^n)$ can be identified the Grassmann manifold of p-dimensional linear subspaces of $\mathbb{R}^n$. In this case one can develop the Riemannian geometry of the Grassmann manifold directly in terms of the ambient space of symmetric matrices, which leads to simple expressions for the geodesics and Newton Euler–Lagrange equations. Let $\text{S}(n)$ and $\mathfrak{so}(n)$ denote the vector spaces of $n \times n$ real symmetric and skew-symmetric matrices, respectively. We consider $\text{S}(n), \mathfrak{so}(n)$ as inner product spaces, endowed with the innner product $\langle X, Y \rangle := \text{tr}(\text{X}^\top \text{Y})$. The following result is verified by a straightforward computation, which we omit. It enables us to explicitly compute the Levi-Civita connection on the Grassmann manifold.

Proposition 1. *For any selfadjoint projection operator* $P \in \text{Grass}(p, \mathbb{R}^n)$*, the linear map*

$$\pi_P : \text{S}(n) \to \text{S}(n), \;\; X \mapsto [P, [P, X]]$$

is a self-adjoint projection onto the tangent space $T_P \text{Grass}(p, \mathbb{R}^n)$*.*

There are at least two natural Riemannian metrics defined on the Grassmannian $\text{Grass}(p, \mathbb{R}^n)$, the *induced metric* and the *normal metric*. The induced Riemannian metric on $\text{Grass}(p, \mathbb{R}^n)$ is defined by the Frobenius inner product on the tangent spaces

$$\langle X, Y \rangle := \text{tr(XY)}$$

for all $X, Y \in T_P \text{Grass}(p, \mathbb{R}^n)$, $P \in \text{Grass}(p, \mathbb{R}^n)$. The normal Riemannian metric has a somewhat more complicated definition. Since it can be shown to coincide with the induced Riemannian metric we do not discuss it here. In order to determine the geodesics, we apply Proposition 1 to compute the Levi-Civita connection ∇ on the Riemannian manifold $\mathrm{Grass}(p, \mathbb{R}^n)$. Thus for any two vector fields X, Y on $\mathrm{Grass}(p, \mathbb{R}^n)$, the Levi-Civita connection is given as

$$\nabla_X Y(P) = \pi_P \left(DY(P)X(P)\right) \quad \forall P \in \mathrm{Grass}(p, \mathbb{R}^n).$$

Therefore the geodesics on $\mathrm{Grass}(p, \mathbb{R}^n)$ are characterized as

$$\nabla_{\dot{P}} P = [P, [P, \ddot{P}]] = 0. \tag{6}$$

Since π_P is a projection operator onto the tangent space of the Grassmannian, we obtain for any smooth curve $P(t)$ on $\mathrm{Grass}(p, \mathbb{R}^n)$ the identity

$$\dot{P} = [P, [P, \dot{P}]]. \tag{7}$$

Thus, by differentiating this identity we conclude

$$\ddot{P} = [\dot{P}, [P, \dot{P}]] + [P, [P, \ddot{P}]]. \tag{8}$$

This shows

Theorem 4. *The geodesics of* $\mathrm{Grass}(p, \mathbb{R}^n)$ *are exactly the solutions of the second order differential equation*

$$\ddot{P} + [\dot{P}, [\dot{P}, P]] = 0. \tag{9}$$

It is possible to solve this quadratic matrix differential equation explicitly, by showing that any solution $P(t)$ with initial conditions $P(0) = P_0 \in \mathrm{Grass}(p, \mathbb{R}^n)$, $\dot{P}(0) = \dot{P}_0$ is given as

$$P(t) = e^{t[\dot{P}_0, P_0]} P_0 e^{-t[\dot{P}_0, P_0]}. \tag{10}$$

We have already seen that the covariant second derivative with respect to the Levi-Civita connection on $\mathrm{Grass}(p, \mathbb{R}^n)$ is given as

$$\nabla_{\dot{P}} \dot{P} = \ddot{P} + [\dot{P}, [\dot{P}, P]].$$

This leads to the following result.

Theorem 5. *The Newton Euler Lagrange equation on* $\mathrm{Grass}(p, \mathbb{R}^n)$ *is*

$$\ddot{P} + [\dot{P}, [\dot{P}, P]] + \mathrm{grad\ f(P)} + \mathrm{H_f(P)\dot{P}} = 0. \tag{11}$$

Equivalently, the associated first order system on the tangent bundle $T\mathrm{Grass}(p, \mathbb{R}^n)$ *is*

$$\begin{aligned} \dot{P} &= L \\ \dot{L} &= -[L, [L, P]] - \mathrm{grad\ f(P)} - \mathrm{H_f(P)L} = 0. \end{aligned} \tag{12}$$

Using the cost function $f(P) = \mathrm{tr(AP)}$ we obtain the following result that exhibits the Newton Euler-Lagrange equation as a principal subspace flow.

Theorem 6. *The Newton Euler-Lagrange equation for* $f(P) = \mathrm{tr(AP)}$ *on* $\mathrm{Grass}(p, \mathbb{R}^n)$ *is*

$$\ddot{P} + [\dot{P}, [\dot{P}, P]] + [P, [P, A]] + [P, [\dot{P}, A]] = 0. \qquad (13)$$

The solutions of (13) converge from any generic initial condition to the unique projection operator P_∞ *that is associated with the principal subspace* V_{pca} *of* A.

Proof. By inspection, the Riemannian gradient for the trace function tr(AP) is given as

$$\mathrm{grad\ f(P)} = [\mathrm{P}, [\mathrm{P}, \mathrm{A}]],$$

see [11]. The Riemannian Hesse operator is given as

$$\mathrm{H_f(P)L} = \nabla_{\mathrm{L}}\mathrm{grad\ f(P)}$$

and thus coincides with

$$\begin{aligned} \mathrm{H_f(P)L} &= ad_P^2([L, [P, A]] + [P, [L, A]]) \\ &= ad_P^2([L, [P, A]]) + [P, [L, A]] \\ &= [P, [L, A]] \end{aligned}$$

since it is easily seen that $ad_P^2([L, [P, A]]) = [P, [P, [L, [P, A]]]] = 0$ holds for any tangent vector L of the Grassmannian. THus the Newton Euler-Lagrange equation has the stated form.

If the eigenvalues of A are distinct, then f has only finitely many isolated critical points on the Grassmannian. Moreover, the unique local and global minimum is given as V_{pca}. Therefore, Theorem 1 implies that, for any initial condition on the Grassmannian, the flow converges to a single equilibrium point. By Theorem 3, the principal subspace V_{pca} is the only local attractor of the flow and all the other critical points have stable manifolds of positive codimension. Since the union of the stable manifolds of these unstable critical points therefore forms a set of positive codimension, its complement is everywhere dense. Thus the flow converges to V_{pca} from a generic set of initial conditions.

One can compare this flow with well-known first order flows for principal component analysis. The proper extension of Oja's flow to a principal subspace flow is

$$\dot{P} = -[P, [P, A]],$$

i.e. Brockett's double bracket gradient flow of the trace function $f(P) = \mathrm{tr(AP)}$ on the Grassmannian; c.f. [11]. Covariant differentiation of a slution of this flow yields

$$\ddot{P} + [\dot{P}, [\dot{P}, P]] + [P, [\dot{P}, A]] = 0,$$

which differs from (13) only by the double bracket term $[P, [P, A]]$.

References

1. F. Alvarez, H. Attouch, J. Bolte, and P. Redont. A second–order gradient–like dissipative dynamical system with hessian driven damping. *J.MPA*, 19:595–603, 2002.
2. F. Alvarez, J. Bolte, and O. Brahic. Hessian Riemannian gradient flows in convex programming. *SIAM J. Control and Opt.*, 43:477–501, 2004.
3. H. Attouch and P. Redont. The second–order in time continuous Newton methods. *SIAM J. Control and Opt.*, 19:595–603, 1981.
4. A. Benveniste, M. Metivier, and P. Priouret. *Adaptive Algorithms and Stochastic Approximations.* Springer Publ., Berlin, 1990.
5. A. Bloch, P. S. Krishnaprasad, J. E. Marsden, and T. S. Ratiu. Dissipation induced instabilities. *Ann. Inst. H. Poincare, Analyse Nonlineare*, 11:37–90, 1994.
6. A. Bloch, P. S. Krishnaprasad, J. E. Marsden, and T. S. Ratiu. The Euler-Poincare equations and double bracket dissipation. *Comm. Math. Physics*, 175:1–42, 1996.
7. R. W. Brockett. Differential geometry and the design of algorithms. *Proc. Symp. Pure Math*, 54:69–92, 1993.
8. R. W. Brockett. Oscillatory descent for function minimization. In B. H. M. Alber and J. Rosenthal, editors, *Current and Future Directions in Applied Mathematics*, pages 65–82. Birkhäuser, Boston, 1997.
9. K. I. Diamantaras and S. Kung. *Principal Component Neural Networks.* J. Wiley, New York, 1996.
10. U. Helmke and P.S. Krishnaprasad. Second order dynamics for principal component analysis. *3rd IFAC Workshop on Lagrangian and Hamiltonian Methods for Nonlinear Control*, Preprints, Nagoya, 2006.
11. U. Helmke and J. Moore. *Optimization and Dynamical Systems.* Springer, London, 1994.
12. C. Lageman. Convergence of gradient-like flows. submitted to *Math. Nachrichten*, 2006.
13. E. Oja. A simplified neuron model as a principal component analyzer. *J. Math. Biology*, 15:267–273, 1982.
14. E. Oja. Neural networks, principal components, and subspaces. *Int. J. Neural Systems*, 1:61–68, 1989.
15. S. Yoshizawa, U. Helmke, and K. Starkov. Convergence analysis for principal component flows. *Int. J. Appl. Math. Comput. Sci.*, 11:223–236, 2001.

Approximation of Generalized Minimizers and Regularization of Optimal Control Problems

Manuel Guerra[1] and Andrey Sarychev[2]

[1] ISEG, Technical University of Lisbon, R.do Quelhas, 6 ,1200-781, Lisboa, Portugal
mguerra@iseg.utl.pt

[2] DiMaD, University of Florence, v. C.Lombroso 6/17, Firenze, 50134, Italy
asarychev@unifi.it

Summary. An open problem, set by Yu.Orlov in his contribution to the volume "Open Problems in Mathematical Systems and Control Theory", V.Blondel, A.Megretski Eds., 2004, regards regularization of optimal control-affine problems with control-independent state-quadratic cost. It is asked whether the infima of the regularized (by adding squared L_2-norm of controls) functionals converge to the infimum of the original functional?

We show that this question can be resolved by an elementary argument. We claim that one should study minimizing sequences of the original functional, rather than its infimum. We advocate the relevance of this question, formulating it via notion of order of singularity of an optimal problem. We study this question and provide computations of the order of singularity for arbitrary singular linear-quadratic problem and also for some classes of nonlinear control-affine problems. Some open problems are set.

1 Introduction

We consider nonlinear optimal control-affine problems of the form

$$\begin{aligned} J_0^T(u(\cdot)) = \int_0^T x(t)'Px(t)dt \to \min, \\ \dot{x} = f(x) + G(x)u, \\ x(0) = x_0,\ x(T) = x_T, \end{aligned} \tag{1}$$

where $T \leq +\infty,\ P > 0$ is symmetric positive-definite matrix,

$$u = (u_1, \ldots, u_k),\ G = (g_1|g_2|\ldots|g_k),$$

and f, g_j are smooth vector fields. For $T = +\infty$ one takes $x_T = 0$.

This problem is *singular* in the sense that it lacks coercivity with respect to control parameters and therefore existence of classical minimizers is not guaranteed. In fact even in the simplest singular linear-quadratic case, where f is linear and g_j are constant, there exist so-called 'cheap' generalized minimizing controls, which belong to a space of distributions.

F. Bullo et al. (Eds.): Lag. & Hamil. Methods for Nonlin. Ctrl., LNCIS 366, pp. 269–279, 2007.
springerlink.com

We aim to characterize the *order of singularity* of problem (1) in terms of approximability of its generalized minimizers. Knowing this order would allow for construction of *proper regularizations*.

Our motivation started with an open question set in contribution of Yu.Orlov [16] to a recently published volume "Open Problems in Mathematical Systems and Control Theory", V.Blondel, A.Megretski Eds, Princeton University Press, 2004:

Question 1. Do the infima of the regularized problem

$$J_\varepsilon^T(u(\cdot)) = \int_0^{+\infty} x(t)'Px(t) + \varepsilon^2|u(t)|^2 dt$$

$$\dot{x} = f(x) + G(x)u, \tag{2}$$

$$x(0) = x_0, \ x(T) = x_T,$$

converge to the infimum of the problem (1)?

We start with (positively) answering this question by means of an elementary reasoning. Then we argue that these are not the infima of the regularized functionals but rather the elements of minimizing sequences of J_0^∞ which are worth studying.

To study them we introduce notion of order of singularity of the problem (1), which characterizes asymptotics of norms of its ε-minimizers. We provide detailed analysis of this characteristic for singular *linear-quadratic problems*, and obtain partial results in some nonlinear cases. Some open problems are set.

2 Convergence of Regularized Optimal Values

Positive answer to the Question 1 of the Introduction is contained in the following simple result.

Proposition 1. *If for finite or infinite T there exist minimizing sequences of square-integrable controls $u^m \in L_2[0,T]$ for J_0^T, then*

$$\lim_{\varepsilon\to 0} \inf J_\varepsilon^T = \inf J_0^T.$$

Proof. Without loss of generality one may think that $J_0^T(u^m) \leq \inf J_0^T + 1/m$. Let $\|u^m\|_{L_2} = \nu_m$; note that the sequence ν_m is not necessarily bounded.

For the penalized functional J_ε^T of (2):

$$\inf J_0^T \leq \inf J_\varepsilon^T \leq J_\varepsilon^T(u^m) = \\ = J_0^T(u^m) + \varepsilon\nu_m^2 \leq \inf J_0^T + 1/m + \varepsilon\nu_m^2.$$

Taking $\varepsilon_m = \nu_m^{-2}/m$ we conclude that

$$\inf J_0^T \leq \inf J_{\varepsilon_m}^T \leq J_{\varepsilon_m}^T(u^m) \leq \inf J_0^T + 2/m,$$

and hence $\lim_{\varepsilon\to 0} \inf J_\varepsilon^T = \inf J_0^T$.

Corollary 1. *Let the assumptions of the Proposition 1 hold and for each $\varepsilon > 0$, let $u_\varepsilon(\cdot)$ be a minimizer of the corresponding regularized problem (2). Then*

$$\lim_{\varepsilon \to 0} J_0^T(u_\varepsilon) = \inf J_0^T.$$

Proof

$$\inf J_0^T \leq J_0^T(u_\varepsilon) \leq J_\varepsilon^T(u_\varepsilon) = \inf J_\varepsilon^T.$$

Proposition 1 leaves some possibility of negative answer to the Question 1, whenever *all the minimizing sequences contain non square-integrable controls.* This may happen for example, if $T = +\infty$ and $f(0) \neq 0$. Still under an assumption of local stabilizability the existence of minimizing sequences of square-integrable controls in the infinite-horizon case is guaranteed.

Definition 1. *The control affine system of (1) is locally stabilizable of order $\alpha > 0$ if there exists a Lipschitzian feedback $\bar{u}(x)$ ($\bar{u}(0) = 0$), and constant C such that for each initial point x^0 from some neighborhood Ω of the origin*

$$|x(t; x^0)| \leq C|x^0|(t+1)^{-\alpha}.$$

Proposition 2. *Assume $T = +\infty$. Let the control system in (1) be locally stabilizable of order $\alpha > \frac{1}{2}$ and $\inf J_0^\infty$ be finite. Then the Question 1 of the Introduction admits positive answer.*

Sketch of the proof. Consider a minimizing sequence $\{u^{(m)}\}$, such that $J_0^\infty(u^{(m)}) \leq \inf J_0^\infty + \frac{1}{m}$, and let $\overline{u}$ denote a feedback control satisfying the conditions of Definition 1. It can be shown that there exists a sequence $\{T_m \in]0, +\infty[, m \in I\!N\}$ such that the controls

$$\tilde{u}^{(m)}(t) = \begin{cases} u^{(m)}(t) & \text{if } t \leq T_m \\ \overline{u}(x_{\overline{u}}(t)) & \text{if } t > T_m \end{cases}$$

are square integrable and satisfy $J_0^\infty(u^{(m)}) \leq \inf J_0^\infty + \frac{2}{m}$. Hence the result follows from Proposition 1.

3 Order of Singularity: Problem Setting

We set another question in relation to the problem (1). It regards asymptotics of L_2-norms $\|u^\varepsilon(\cdot)\|_{L_2}$ of ε-minimizers $\{u^\varepsilon\}$ of the functional J_0^T. If $\inf J_0^T$ is finite, then for generic data and for any minimizing sequence $u^m(\cdot)$ of controls for J_0^T the norms $\|u^m(\cdot)\|_{L_2} \stackrel{m\to\infty}{\longrightarrow} +\infty$. Therefore $J_\varepsilon^T(u^m) \stackrel{m\to\infty}{\longrightarrow} +\infty$ for any *fixed* $\varepsilon > 0$, i.e. minimizing sequences for J_0^T exhibit *"singular behavior"*.

In order to measure the degree of singularity we introduce the following quantity σ.

Definition 2. *Let $\mathcal{U}_\varepsilon$ be the set of controls which steer the control system (1) from the initial point x_0 to a point from ε-neighborhood of x_T and such that $J_0^T(u) \leq \inf J_0^T + \varepsilon$. The value*

$$\sigma = \limsup_{\varepsilon \to 0^+} \frac{\inf \left\{ \ln \|u\|_{L_2} : \ u(\cdot) \in \mathcal{U}_\varepsilon \right\}}{\ln \frac{1}{\varepsilon}}. \tag{3}$$

is called order of singularity of the problem (1).

In the rest of the paper we proceed with a study of the order of singularity.

4 The Singular Linear-Quadratic Case

In this Section we demonstrate relevance of the introduced order of singularity σ by examining singular linear-quadratic case, i.e. the problem (1) with linear controlled dynamics:

$$\dot{x} = Ax + Bu. \tag{4}$$

In this case one may consider a more general cost functional, than in (1), namely:

$$J_0^T(u) = \int_0^T x'Px + 2u'Qx + u'Ru \, d\tau, \tag{5}$$

where $R \geq 0$ is a symmetric singular matrix, and symmetric P is allowed to have negative eigenvalues. We assume $T < +\infty$.

A complete description of generalized minimizers of this problem has been provided in [8, 9]. The minimizers are proved to belong to some Sobolev space H_{-r}, with integer r related to a desingularization procedure, which reduces the initial singular problem to a regular linear-quadratic problem.

It turns out that in this case the order of singularity σ introduced by (3) is tightly connected with r and with approximability of distributions-like minimizers in Sobolev norms.

Proposition 3. *Given the problem (4)-(5) and generic boundary data x_0, x_T:*

$$\sigma = r - \frac{1}{2}.$$

For linear dynamics (4) and *positive state-quadratic cost functional* one obtains.

Corollary 2. *For the problem (1) with linear dynamics (4), cost functional $\int_0^T x'Pxdt$, $(P > 0)$ and generic boundary data, the order of singularity is*

$$\sigma = \frac{1}{2}.$$

Complete analysis is possible for arbitrary boundary data. The space $\mathbb{R}^n \times \mathbb{R}^n$ of boundary data admits a stratification. Different orders of singularity correspond to different open strata of this stratification. The dependence of σ on the boundary data is lower semicontinuous. Possible values of σ are

$$\sigma = \frac{i + 1/2}{2(j - i) - 1} : \ 0 \leq i < j \leq n. \tag{6}$$

A more extensive presentation of the study of order of singularity σ in the singular linear-quadratic case can be found in [10].

5 Singularity of Nonlinear Control-Affine Problems

While we managed to characterize the order of singularity for singular linear quadratic problems, similar questions for control-affine nonlinear problem (1) remain essentially open.

Question 2. Is the set of possible values of σ for problems (1) finite? Is the value of σ semiinteger for generic boundary data? Do all possible positive values of σ have the form (6)? Does this set of numbers correspond to a (local) stratification of the state space?

It is interesting to compare the invariant σ with the order of singularity introduced in [13, 14]). Other notions, which seem to be related to the questions, we discuss in this paper, are the one of complexity and enthropy of sub-Riemannian paths, which have been extensively studied recently ([7, 11]). Although our context is not sub-Riemannian, general approaches look similar.

We provide some partial answers for control-affine problem (1) with *positive state-quadratic cost.*

Apparently commutativity/noncommutativity of controlled vector fields affects the value of order of singularity σ a great deal. The connection between the commutativity/noncommutativity and the generalized minimizers is an established fact [15, 2, 17, 3]. Yet it is not well understood, how the Lie structure is revealed in the properties of generalized minimizers.

We make a first step by providing an upper estimate for order of singularity σ in general control-affine control-commutative case and then estimate σ for some noncommutative examples and provide general answer for driftless case.

The following assumptions of commutativity and generalized coercivity are to be made.

Assumption. For the problem (1)

i) The controlled vector fields g^j are pairwise commuting: $\forall k, \ell : \ [g^k, g^\ell] = 0$; in the following theorem we assume them to be constant.

ii_0) The drift vector field f is subquadratic at infinity, meaning that:

$$\|f(x)\| \leq \phi(\|x\|), \quad \lim_{\|x\| \to \infty} \phi(\|x\|)/\|x\|^2 = 0.$$

A stronger assumption is

ii_1) The drift vector field f and its first derivatives are subquadratic at infinity, meaning that:

$$\|f(x)\| + \|Df(x)\| \leq \phi(\|x\|), \quad \lim_{\|x\| \to \infty} \phi(\|x\|)/\|x\|^2 = 0.$$

Theorem 1 (order of singularity for input-commutative case). *Under the Assumptions i) and ii_1) for the problem (1) order of singularity:*

$$\sigma \leq 3/2.$$

6 Sketch of the Proof of the Theorem 1

First we proceed with a transformation, which appeared under the name of 'reduction' in [1] and proved to be useful for analysis of control-affine systems without apriori bounds on the value of controls.

To simplify our presentation we will assume the vector fields g^j to be constant and linearly independent. The upper estimate (1) and our method of proof both hold also without this simplification. Note that under the commutativity assumption g^j can be transformed *locally* into constant vector fields by change of variables.

For constant vector fields 'reduction' amounts to time-variant substitution of state variable

$$x = y + Gv(t),\ v(t) = \int_0^t u(s)ds.$$

This substitution leads us to the following 'reduced' problem

$$\dot{y} = f(y + Gv(t)) \tag{7}$$

$$\int_0^T (y^* Py + 2y^* PGv + v^* G^* PGv)dt \to \min . \tag{8}$$

with boundary conditions

$$y(0) = x_0,\ y(t) = x_T + GV,\ V \in I\!R^m. \tag{9}$$

This is a nonlinear optimal control problem, which is coercive with respect to the new control v, given that $v \mapsto v^* G^* PGv$ is a positive definite quadratic form and $f(y + Gv(t))$ has (at most) subquadratic growth, as $\|v\| \to \infty$.

This problem though lacks convexity with respect to v: for fixed y the set

$$\Phi(y) = \{(y^* Py + 2y^* PGv + v^* G^* PGv, f(y + Gv)),\ v \in I\!R^k\}$$

is in general nonconvex. For example for scalar v this set is a curve in $I\!R^{n+1}$.

Therefore classical minimizers of the reduced problem (7)-(9) may cease to exist, but assumptions for existence of *relaxed* minimizer (see for example [4, Ch.11], or [6, Ch.8]) are satisfied. Recall that a relaxed control can be seen as a family $t \mapsto \mu_t$ of probability measures in the space of $v \in I\!R^k$; the family $t \mapsto \mu_t$ is measurable in the weak sense with respect to $t \in [0, T]$.

The corresponding dynamic equation and the functional become

$$\dot{y} = \langle \mu_t, f(y + Gv) \rangle \tag{10}$$

$$\int_0^T \langle \mu_t, (y^* Py + 2y^* PGv + v^* G^* PGv) \rangle dt \to \min,$$

where for any $\chi(\cdot, v) :\ \langle \mu_t, \chi(\cdot, v) \rangle = \int \chi(\cdot, v) d\mu_t(v)$.

Proposition 4. *Under the Assumptions i)-ii$_0$) from the previous Section, the reduced problem (7)-(9) possesses minimizer $t \mapsto \mu_t$ in the class of relaxed controls.*

In fact basing on the latter Proposition one is able to conclude existence of generalized minimizer of the original problem (1), which belongs to the space $W_{-1,\infty}([0,T],\mathbf{P})$, where $\mathbf{P}$ is the space of probabilistic measures on the space $I\!R^k$ of control parameters. This minimizer coincides with the generalized derivative $\partial\mu_t/\partial t$.

To proceed further and to apply the Gamkrelidze Approximation Lemma ([5, 6]) we need the minimizing trajectory to be Lipschitzian. To guarantee this Lipschitzian property we reinforce the Assumption ii$_0$) to ii$_1$). Under this assumption we can proceed by using a compactification of the set of control parameters (cf. [6]) of the problem. This method has been successfully used in [18] for establishing Lipschitzian regularity of nonrelaxed minimizers in the problems with control-affine dynamics. In the relaxed case a more intricate way of compactification is needed. We will comment on this matter in much more general context on other occasion.

Under the reinforced coercivity assumption the following result holds

Proposition 5. *Under coercivity assumptions i)-ii$_1$) from the previous Section minimizing relaxed trajectories of the problem (10) are Lipschitzian, with the possible exception of abnormal relaxed minimizers.*

The rest of the proof consists of two approximation steps. At a first stage we approximate the relaxed minimizer μ_t of the reduced problem (7)-(9) by a piecewise continuous control $w_\varepsilon(\cdot)$ in such a way that the (final point of the) trajectory and the functional of (7), driven by $w_\varepsilon(\cdot)$, are ε-close to the (final point of the) trajectory and the functional of (10).

At second stage we approximate the piecewise-continuous control $w_\varepsilon(\cdot)$ by an absolutely continuous control $v_\varepsilon(\cdot) \in W_{1,2}$ whose derivative $u_\varepsilon(\cdot) = \dot{v}_\varepsilon$ becomes ε-minimizing control for the original problem (1). The asymptotics (as $\varepsilon \to +0$) of L_2-norm of $u_\varepsilon(\cdot)$, or all the same, $W_{1,2}$-norm of $v_\varepsilon(\cdot)$ provides an upper estimate for the order of singularity σ.

To proceed with the first approximation step we slightly modify the construction [6] of piecewise constant control which ε-approximates relaxed control μ_t in so-called relaxation metric. The main concern is to 'control' the number of intervals of continuity, or rather the variation of the approximating function $w_\varepsilon(\cdot)$. We prove the following

Proposition 6. *For each small $\varepsilon > 0$ there exists piecewise-constant control $w_\varepsilon(\cdot)$, with $N_\varepsilon = (n+2)[T/\varepsilon]$ intervals of constancy, such that the trajectory and the value of the functional of (7), driven by $w_\varepsilon(\cdot)$, are ε-close in $C^0[0,T] \times I\!R$ to the trajectory and the functional of (10).*

Asymptotics of the variation of this control is $O(\varepsilon^{-1})$, as $\varepsilon \to 0$.

Passing to the second step, we alter $w_\varepsilon(\cdot)$ at $N_\varepsilon = 2(n+2)[T/\varepsilon]$ intervals of lengths Δ, adjacent to the points of discontinuity of $w_\varepsilon(\cdot)$. We assume $\Delta > 0$ to be small enough, so that the corresponding intervals do not intersect. We substitute at these intervals constant pieces by linear ones, thus transforming $w_\varepsilon(\cdot)$ into (absolutely) continuous piecewise-linear function $v_\Delta(\cdot)$.

It is rather evident, that under such approximation the trajectory and the value of the cost functional are changed by the value $O(N_\varepsilon \Delta)$. As far as $N_\varepsilon = O(\varepsilon^{-1})$, as $\varepsilon \to +0$, and we want the error of approximation to be $O(\varepsilon)$, we choose

$$\Delta = \varepsilon^2. \tag{11}$$

Control $v_{\varepsilon^2}(\cdot)$ of the reduced system (7) corresponds to the control $\frac{d}{dt}v_{\varepsilon^2}(\cdot)$ for the original control-affine system of (1). This latter time-derivative is nonvanishing on $N_\varepsilon = O(\varepsilon^{-1})$ intervals of length $\Delta = \varepsilon^2$. The magnitude of $u_\varepsilon(\cdot) = \dot{v}_{\varepsilon^2}(\cdot)$ is $O(\varepsilon^{-2})$ on these intervals. Therefore

$$\|u_\varepsilon(\cdot)\|_{L_2} = \|\dot{v}_{\varepsilon^2}(\cdot)\|_{L_2} \asymp \left((\varepsilon^{-2})^2\varepsilon^2\varepsilon^{-1}\right)^{1/2}, \tag{12}$$

as $\varepsilon \to 0$, and we conclude $\sigma \leq 3/2$.

6.1 Order of Singularity for Input-Commutative Case: A Conjecture

We conjecture that the statement of the theorem, we have just proved, can be strengthened.

Conjecture. *Under the assumptions of the Theorem 1 order of singularity of the problem (1)*

$$\sigma \leq 1.$$

Our conjecture relies on Proposition 6, which is a key fragment of the proof. We trust that its statement can be reinforced: there exists a piecewise continuous control $w_\varepsilon(\cdot)$ with $\leq O(\varepsilon^{-1})$ intervals of continuity, such that the end-point of the trajectory and the value of the functional of (7) driven by $w_\varepsilon(\cdot)$ are ε^2**-close** to the end-point of the trajectory and the value of the functional of (10).

If this holds true then taking $\Delta = \varepsilon^3$ instead of (11) we arrive to the estimate $\|u_\varepsilon(\cdot)\|_{L_2} \asymp O(\varepsilon^2)$ in a computation analogous to (12). As a consequence we conclude $\sigma \leq 1$.

6.2 An Example

Example 1. Consider optimal control-affine problem

$$\dot{x} = f(x) + g^1(x)u_1, \ x = (x_1, x_2, x_3),$$
$$f = x_1\partial/\partial x_2 + \gamma(x_1)(x_1^2 - 1)\partial/\partial x_3, \ g_1 = \partial/\partial x_1,$$
$$J_0^T = \int_0^1 (x_1^2 + x_2^2 + x_3^2)dt \to \min,$$
$$x(0) = 0, \ x(1) = 0,$$

where $\gamma(x)$ is a smooth function supported at $[-2, 2] \subset I\!R$, $0 \leq \gamma(x) \leq 1$ and $\gamma(x) \equiv 1$ on $[-\frac{3}{2}, \frac{3}{2}]$.

In coordinates the dynamics of the problem is

$$\dot{x}_1 = u_1, \ \dot{x}_2 = x_1, \ \dot{x}_3 = \gamma(x_1)(x_1^2 - 1). \tag{13}$$

To estimate the infimum of this problem note that

$$0 = \int_0^1 \gamma(x_1(t))(x_1^2(t) - 1)dt \leq \int_0^1 x_1^2(t)dt - 1,$$

and hence $J_0^T \geq 1$. Now we construct a minimizing sequence of controls $u_N(\cdot)$, such that $J_0^T(u_N) \to 1$ as $N \to +\infty$.

First take the indicator function $p(t)$ of the interval $[0,1]$ and construct piecewise-constant function

$$q_N(t) = \sum_{j=0}^{2N-1} (-1)^j p(2Nt - j);$$

N being a large integer. Its intervals of constancy are of lengths $(2N)^{-1}$.

At the second moment we alter the function q_N on the intervals

$$[0, N^{-3}], [1 - N^{-3}, 1],$$
$$[j(2N)^{-1} - N^{-3}, j(2N)^{-1} + N^{-3}], \ j = 1, \ldots, 2N - 1. \tag{14}$$

We substitute on these intervals constant pieces by linear ones transforming it into piecewise-linear *continuous* function $q_N^c(t)$, whose boundary values are: $q_N^c(0) = q_N^c(T) = 0$.

This piecewise-linear (absolutely) continuous function corresponds to a piecewise-constant control $u_1(t) = \dot{q}_N^c(t)$ supported on the intervals (14). Amplitude of this control is $O(N^3)$ and its L_2-norm admits an estimate

$$\|u_1(t)\|_{L_2} \simeq ((N^3)^2 N^{-3} N)^{1/2} = N^2,$$

as $N \to +\infty$.

Taking $x_1(t) = q_N^c(t)$ and substituting it into second and third equations of (13), we conclude that the corresponding solution satisfies the conditions $x_2(1) = 0, \ x_3(1) = O(N^{-2})$, as $N \to +\infty$. Besides

$$\int_0^1 (x_1^2 + x_2^2 + x_3^2)dt - 1 = O(N^{-2}), \ \text{as } N \to +\infty.$$

Therefore the order of singularity σ is ≤ 1.

7 Non-commutative Driftless Case: General Result

It turns out that the non-commutative *driftless* case

$$J_0^T(u(\cdot)) = \int_0^T x(t)' P x(t) dt \to \min,$$

$$\dot{x} = \sum_{j=1}^{r} g^j u_j, x(0) = x_0, \ x(T) = x_T,$$

is rather simple. In this case the value of the order of singularity equals $\sigma = 1/2$ for generic boundary data, and it does not depend on the Lie structure of the system of vector fields $\{g^1, \ldots, g^r\}$.

In order to see this, consider the case when the system $\{g^1, \ldots, g^r\}$ has complete Lie rank. Then, roughly speaking, generalized optimal trajectory consists of three 'pieces': an initial 'jump', which brings it to the origin O, a constant piece $x(t) \equiv O, \ t \in]0, T[$, and a final 'jump' to the end point $x(T) = x_T$. Evidently $\inf J_0^T = 0$ and a simple homogeneity based argument shows that $\sigma \leq 1/2$.

To prove that in fact $\sigma = 1/2$, we assume $x_0 \neq O$, fix $\varepsilon > 0$ and take a control u_ε, such that

$$J_0^T(u_\varepsilon) = \int_0^T x'_{u_\varepsilon}(t) P x_{u_\varepsilon}(t) dt < \inf J_0^T + \varepsilon = \varepsilon. \tag{15}$$

Consider the set $\{x \in I\!R^n | \ x'Px \leq (1/2)x_0'Px_0\}$; let ρ be the distance from x_0 to this set. Since $x'Px$ is positive definite, one concludes from the inequality (15), that there exists $t_\varepsilon < \frac{2\varepsilon}{x_0'Px_0}$ such that $x_{u_\varepsilon}(t_\varepsilon)^* P x_{u_\varepsilon}(t_\varepsilon) < (1/2)x_0'Px_0$. Then, by Cauchy-Schwarz inequality, the control needed to achieve $x_{u_\varepsilon}(t_\varepsilon)$ from x_0 in time t_ε, must satisfy the estimate $\|u_\varepsilon\|^2_{L_2[0,T]} \geq \rho t_\varepsilon^{-1} \geq C\varepsilon^{-1}$, for some $C > 0$.

More details on the proof as well as treatment of general noncommutative case will be addressed in further publications.

Acknowledgements

The first author has been partially supported by Fundacão para a Ciência e a Tecnologia (FCT), Portugal via Center for Research in Optimization and Control (CEOC), University of Aveiro, cosponsored by European Community Fund FEDER/POCTI. The second author has been partially supported by Ministero dell'Università e della Ricerca (MIUR), Italy, via COFIN grant 2004015409-003.

References

1. Agrachev A, Sarychev A (1987), Math. USSR Sbornik, 58:15–30
2. Bressan A (1987) Rend. Semin. Univ. Padova 78:227–235
3. Bressan A, Rampazzo F (1994) J Optimization Theory Applications 81:435–457
4. Cesari L (1983) Optimization Theory. Springer, New York, Heidelberg, Berlin
5. Gamkrelidze R (1965) J. Soc. Ind. Appl. Math., Ser. A: Control. 3:106–128
6. Gamkrelidze R (1978) Principles of Optimal Control Theory. Plenum Press, New York
7. Gauthier J-P, Zakalyukin V (2006) J. Dynam. Control Syst. 12:371–404
8. Guerra M (2000) J. Dynam. Control Systems 6:265-309
9. Guerra M (2004) J. Math. Sciences 120,1:895–918.

10. Guerra M , Sarychev A (2006) Regularizations of optimal control problems for control-affine systems. In: Proc. Symp.Mathematical Theory of Networks and Systems (MTNS 2006), Kyoto, Japan
11. Jean F (2003) ESAIM Control Optim. Calc. Var. 9:485–508
12. Jurdjevic V (1997) Geometric Control Theory. Cambridge University Press
13. Kelley H , Kopp R , Moyer G (1967) Singular extremals. In: Leitmann G (ed) Topics in Optimization. Academic Press, New York
14. Krener A (1977) SIAM J. Control Optimization 15:256–293
15. Orlov Yu (1988) Theory of Optimal Systems with Generalized Controls (in Russian). Nauka, Moscow
16. Orlov Yu (2004) Does cheap control solve a singular nonlinear quadratic problem? In: Blondel V , Megretski A (eds), Unsolved Problems in Mathematical Systems and Control Theory, Princeton University Press
17. Sarychev A (1988) Differential Equations 24:1021–1031
18. Sarychev A , Torres D (2000) Applied Mathematics Optimization 41:237–254

Minimum Time Optimality of a Partially Singular Arc: Second Order Conditions

Laura Poggiolini[1] and Gianna Stefani[2]

[1] Dip. di Matematica Applicata "G. Sansone",
Università di Firenze, Italy
laura.poggiolini@math.unifi.it
[2] Dip. di Matematica Applicata "G. Sansone",
Università di Firenze, Italy
gianna.stefani@math.unifi.it

1 Introduction

This paper is part of a research project started with [2] and [3] aiming to use Hamiltonian methods to study second order conditions in optimal control. It is the opinion of the authors that a complete understanding of the problem in Hamiltonian terms will lead to a deeper insight into the synthesis problem.

We consider the minimum–time problem with fixed end–points for a control–affine system on a manifold M:

$$\text{minimize} \quad T \tag{1}$$

subject to

$$\dot{\xi}(t) = f_0(\xi(t)) + u_1 f_1(\xi(t)) + u_2 f_2(\xi(t)) \tag{2}$$

$$\xi(0) = \widehat{x}_0\,, \quad \xi(T) = \widehat{x}_f \tag{3}$$

$$u \equiv (u_1, u_2) \in U := [-1, 1]^2$$

where f_0, f_1, f_2 are C^∞ vector fields.

The aim of the authors is to give second order sufficient conditions for a reference *normal* Pontryagin extremal $(\widehat{T}, \widehat{\xi}, \widehat{u})$ to be a *local optimizer in the strong topology.*

In problems with free final time, "strong local" may have two meanings:

- $\widehat{\xi}$ is optimal for the problem with respect to a neighborhood of the graph of $\widehat{\xi}$ in $M \times \mathbb{R}$ and hence local with respect to both state and final time. We call this type of local optimality *(state, time)–local.*
- $\widehat{\xi}$ is optimal for the problem with respect to a neighborhood of the range of $\widehat{\xi}$ in M and hence local only with respect to the state. We call this type of local optimality *state–local.*

State–local optimality implies (state, time)–local optimality, but they are not equivalent, see [6].

F. Bullo et al. (Eds.): Lag. & Hamil. Methods for Nonlin. Ctrl., LNCIS 366, pp. 281–291, 2007.
springerlink.com

The authors faced the problem of state local–optimality in [6], when the extremal is bang–bang, and the problem of (time, state)–local optimality for a single–input control system in [9], when the reference trajectory is totally singular, and in [7] when it is bang–singular, i.e. it is the concatenation of a bang arc and of a singular arc. Also we should mention the papers by different authors who considered minimum time optimality: [4], [5], [8], the book [1] and the references therein.

In this paper we consider (time, state)–local optimality of a reference trajectory which is the concatenation of a bang arc and a partially singular arc on a n–dimensional manifold M, $n \geq 3$, namely we consider the case when $\widehat{u}_2$ is bang and $\widehat{u}_1$ switches from bang to singular at time $\widehat{\tau}_1 \in (0, \widehat{T})$. Without loss of generality we can assume $\widehat{u}_2 \equiv 1$ and $\widehat{u}_1 \equiv 1$ along its bang–arc, so that the time–dependent reference vector field is given by

$$\widehat{f}_t = \begin{cases} f_0 + f_1 + f_2 := f^+ & t \in [0, \widehat{\tau}_1) \\ f_0 + \widehat{u}_1(t) f_1 + f_2 := f^s_t & t \in [\widehat{\tau}_1, \widehat{T}]. \end{cases}$$

The aim of the authors is to prove that regularity conditions on the bang arc and the positivity of a suitable second variation associated to the partially singular arc are sufficient to prove optimality.

The result is proved using Hamiltonian methods, which consist in constructing a field of non–intersecting state–extremals covering a neighborhood of the graph of the reference trajectory. Here we need a non–smooth modification of the method which is explained in Section 3.

In the present case the natural second variation associated to the trajectory cannot be coercive (see Subsection 2.1) and we require

1. regularity conditions on the bang arc which are the strengthening of the necessary ones needed for the minimum–time problem with fixed end–points $\widehat{x}_0$ and
$$\widehat{x}_1 := \widehat{\xi}(\widehat{\tau}_1).$$
2. on the singular arc we assume the coercivity of a suitable second variation. This condition is a little stronger than the coercivity of the second variation of the minimum time problem with fixed end–points $\widehat{x}_1$, $\widehat{x}_f$.

2 Statement of the Result

Denote by ℓ the elements of T^*M and by $\pi \colon T^*M \to M$ the canonical projection. For notational simplicity we also denote $q = \pi\ell$. Moreover, for a possibly time–dependent Hamiltonian H_t we denote by $\overrightarrow{H}_t$ the corresponding Hamiltonian vector field and by $\mathcal{H}_t$ the flow from time $\widehat{\tau}_1$ to time t of $\overrightarrow{H}_t$. Equivalently $\mathcal{H}_t(\ell)$ can be seen as the solution at time t of the Hamiltonian system defined by H_t with initial condition $\mathcal{H}_{\widehat{\tau}_1}(\ell) = \ell$.

Recall that every possibly time–dependent vector field f_t defines a Hamiltonian function H_t on T^*M by

$$H_t(\ell) := \langle \ell\,,\; f_t(q)\rangle\,,$$

in particular we denote by $\widehat{H}_t$, H^+, H_i, the Hamiltonian functions associated to the vector fields $\widehat{f}_t$, f^+, f_i, $i = 0, 1, 2$, respectively.

Pontryagin Maximum Principle (PMP) for this problem reads as follows: *There exists* $\widehat{\lambda}\colon [0, \widehat{T}] \to T^*M$ *such that*

$$\dot{\widehat{\lambda}}(t) = \overrightarrow{\widehat{H}_t}(\widehat{\lambda}(t))\,, \quad \pi\widehat{\lambda}(t) = \widehat{\xi}(t) \tag{4}$$

$$\widehat{H}_t(\widehat{\lambda}(t)) = \max_{u\in U}\left\{\left(H_0 + \sum_{i=1}^{2} u_i H_i\right)(\widehat{\lambda}(t))\right\} = p_0 \in \{0, 1\}\,. \tag{5}$$

We assume $\widehat{\lambda}$ to be a normal extremal, i.e. $p_0 = 1$.

From PMP, introducing for $i = 0, 1, 2$ the functions

$$\varphi_i := H_i \circ \widehat{\lambda}\colon [0, \widehat{T}] \to \mathbb{R},$$

we obtain

$$\varphi_2(t) \geq 0 \text{ for any } t \in [0, \widehat{T}] \tag{6}$$

$$\varphi_1(t) \geq 0 \text{ on the bang arc} \tag{7}$$

$$\varphi_1(t) \equiv \dot{\varphi}_1(t) \equiv 0 \text{ on the singular arc} \tag{8}$$

$$\ddot{\varphi}_1(\widehat{\tau}_1^-) \geq 0\,. \tag{9}$$

We strengthen the above conditions (6)–(7)–(9) assuming that $\widehat{u}_1$ is regular degenerating on the bang arc and that $\widehat{u}_2$ is a regular bang control, namely we assume

A1 $\varphi_1(t) > 0$, $t \in [0, \widehat{\tau}_1)$ and $\ddot{\varphi}_1(\widehat{\tau}_1^-) > 0$;
A2 $\varphi_2(t) > 0$, $t \in [0, \widehat{T}]$.

We give the conditions on the singular arc fixing the control u_2 and considering the single input control system

$$\dot{\xi}(t) = \widetilde{f}(\xi(t)) + u_1(t) f_1(\xi(t)) \tag{10}$$

where $\widetilde{f}$ is defined by

$$\widetilde{f} := f_0 + f_2.$$

Introducing the Poisson parentheses $\{\cdot,\cdot\}$ between Hamiltonian functions and the Lie brackets $[\cdot,\cdot]$ between vector fields we obtain from (8) on $[\widehat{\tau}_1, \widehat{T}]$:

$$\langle\widehat{\lambda}(t)\,,\; f_1(\widehat{\xi}(t))\rangle = H_1(\widehat{\lambda}(t)) = 0 \tag{11}$$

$$\langle\widehat{\lambda}(t)\,,\; [\widetilde{f}, f_1](\widehat{\xi}(t))\rangle = \{\widetilde{H}, H_1\}(\widehat{\lambda}(t)) = 0 \tag{12}$$

$$\begin{aligned}&\langle\widehat{\lambda}(t)\,,\; \widehat{u}_1(t)[f_1, [\widetilde{f}, f_1]] + [\widetilde{f}, [\widetilde{f}, f_1]]\rangle(\widehat{\xi}(t))\\ &= \widehat{u}_1(t)\{H_1, \{\widetilde{H}, H_1\}\}(\widehat{\lambda}(t)) + \{\widetilde{H}, \{\widetilde{H}, H_1\}\}(\widehat{\lambda}(t)) \equiv 0\,.\end{aligned} \tag{13}$$

Along the u_1–singular arc we state the following regularity assumptions:

A3 (SGLC) *Strong generalized Legendre condition*

$$\{H_1, \{\widetilde{H}, H_1\}\}(\widehat{\lambda}(t)) > 0 \quad \forall t \in [\widehat{\tau}_1, \widehat{T}]\,. \tag{14}$$

A4 f_1, $[\widetilde{f}, f_1]$ are linearly independent along the range of $\widehat{\xi}_{|[\widehat{\tau}_1, \widehat{T}]}$.

Remark 1. $\{H_1, \{\widetilde{H}, H_1\}\}(\widehat{\lambda}(t)) \geq 0$ is a necessary condition for optimality, see for example [1].

Remark 2. Assumption **A4** cannot be fulfilled if $\dim M \leq 2$, and it is a generic condition if $\dim M \geq 3$. Moreover **A4** together with PMP for the normal case imply that f_1, $[\widetilde{f}, f_1]$, $[f_1, [\widetilde{f}, f_1]]$ are linearly independent near the reference trajectory.

Defining

$$\begin{aligned} \Sigma :=& \{\ell \in T^*M \colon H_1(\ell) = 0\}, \\ \mathcal{S} :=& \left\{\ell \in \Sigma \colon \{\widetilde{H}, H_1\}(\ell) = 0\,,\ \{H_1, \{\widetilde{H}, H_1\}\}(\ell) > 0\right\}, \end{aligned}$$

we obtain that the Hamiltonian function H defined near $\mathcal{S}$ by

$$H := \widetilde{H} - \frac{\{\widetilde{H}, \{\widetilde{H}, H_1\}\}}{\{H_1, \{\widetilde{H}, H_1\}\}}\, H_1$$

is such that

- $\overrightarrow{H}$ is tangent to $\mathcal{S}$,
- $\widehat{\ell}_1 := \widehat{\lambda}(\widehat{\tau}_1) \in \mathcal{S}$
- $\widehat{\lambda}(t) = \mathcal{H}_t(\widehat{\ell}_1)$ for any $t \in [\widehat{\tau}_1, \widehat{T}]$,
- $\widehat{\lambda}([\widehat{\tau}_1, \widehat{T}]) \subset \mathcal{S} \subset \Sigma$.

In order to introduce the second order condition on the u_1–singular arc define

$$\Gamma_e := \{\exp \sigma[\widetilde{f}, f_1] \circ \exp s f_1(\widehat{x}_1) \colon (s, \sigma) \in \mathbb{R}^2\}$$

and consider the following minimum time problem:

$$\begin{aligned} &\text{minimize } T \text{ subject to} \\ &\dot{\xi}(t) = f_t^s(\xi(t)) + w(t)[\widetilde{f}, f_1](\xi(t)) + \frac{w(t)^2}{2}[f_1, [\widetilde{f}, f_1]](\xi(t)) \\ &\xi(\widehat{\tau}_1) \in \Gamma_e, \quad \xi(T) = \widehat{x}_f\,. \end{aligned} \tag{15}$$

Remark that the triplet $(\widehat{T}, w \equiv 0, \widehat{\xi}_{|[\widehat{\tau}_1, \widehat{T}]})$ is a normal Pontryagin extremal with $\widehat{\lambda}_{|[\widehat{\tau}_1, \widehat{T}]}$ as adjoint covector and denote by J''_{st} the associated standard second variation. We make the following assumption:

A5 J''_{st} is coercive.

Remark 3. Notice that the natural second variation L'' of the minimum time problem with fixed end–points $\widehat{x}_1$, $\widehat{x}_f$ for system (10) is the standard second variation of the minimum time problem for system (15) with initial condition $\xi(\widehat{\tau}_1) \in \{\exp s f_1(\widehat{x}_1) \colon s \in \mathbb{R}\}$, $\xi(T) = \widehat{x}_f$, see [9].

In the case when the Hamiltonians H and $\widetilde{H}$ coincide on $\mathcal{S}$, we can give the following more explicit condition:

A5–bis Let H and $\widetilde{H}$ coincide on $\mathcal{S}$ and let $\Omega(t)$ be the fundamental solution to the variational system associated to the reference trajectory:

$$\dot{\Omega}(t) = D\widetilde{f}(\widehat{\xi}(t))\Omega(t)\,, \quad \Omega(\widehat{\tau}_1) = \mathrm{Id}\,.$$

The coercivity of J''_{st} is equivalent to:
$f_1(\widehat{\xi}(t))$ and $\Omega(t, \widehat{\tau}_1)[\widetilde{f}, f_1](\widehat{x}_1)$ are linearly independent for $t \in [\widehat{\tau}_1, \widehat{T}]$.

Remark that if $\dim M = 3$, the condition $H = \widetilde{H}$ on $\mathcal{S}$ can be assumed without loss of generality (see Remark 4.2 in [7]).

We prove the following result:

Theorem 1. *If Assumptions* **A1** *through* **A5** *hold, then* $(\widehat{T}, \widehat{u}, \widehat{\xi})$ *is a local optimizer in the strong topology.*

2.1 Necessary Conditions

The extended second variation that naturally appears is the one we get considering the sub–problem of (1)–(2)–(3) described as follows. By means of a time reparametrization we allow the switching time $\widehat{\tau}_1$ and the final time $\widehat{T}$ to move and, at the same time, we let u_1 vary on the singular arc. In this way we get a sub–problem in the fixed time–interval $[0, \widehat{T}]$. Moreover we merge the variations $v \in L^2[\widehat{\tau}_1, \widehat{T}]$ of the singular control into $\mathbb{R} \times L^2[\widehat{\tau}_1, \widehat{T}]$ by

$$v \mapsto \left(a := \int_{\widehat{\tau}_1}^{\widehat{T}} v(s)\,\mathrm{d}\,s,\ w \colon t \mapsto \int_t^{\widehat{T}} v(s)\,\mathrm{d}\,s \right)$$

and we obtain the extended second variation as a functional

$$J'' \colon \mathbb{R}^2 \times \mathbb{R} \times L^2((\widehat{\tau}_1, \widehat{T})) \to \mathbb{R}\,,$$

where $(a, w) \in \mathbb{R} \times L^2((\widehat{\tau}_1, \widehat{T}))$ is the couple of parameters we have just defined, while

$$\varepsilon := (\varepsilon_1, \varepsilon_2) \in \mathbb{R}^2$$

takes account of the variations of the lengths of the bang and of the singular arc, respectively.

To evaluate J'' denote by

$$\widehat{S}_t \colon M \to M$$

the flow from time $\widehat{\tau}_1$ to time t of the reference vector field $\widehat{f}_t$ and by g_t^1 the pull–back of f_1 from time t to time $\widehat{\tau}_1$ along the reference flow:

$$g_t^1(x) := \widehat{S}_{t*}^{-1} f_1 \circ \widehat{S}_t(x)\,.$$

Let $\beta\colon M \to \mathbb{R}$ be any function such that $\mathrm{d}\,\beta(\widehat{x}_1) = -\widehat{\lambda}(\widehat{\tau}_1)$. Using the notation

$$g \cdot \beta(x) := \langle \mathrm{D}\,\beta(x)\,,\ g(x)\rangle$$

for a vector field g, we calculate the second variation for the above described problem using the arguments in [1], [9], [7], obtaining:

$$\begin{aligned} J''[\varepsilon, a, w]^2 &= \frac{\varepsilon_1^2}{2}\, f^+ \cdot f^+ \cdot \beta(\widehat{x}_1) + \frac{a^2}{2}\, f_1 \cdot f_1 \cdot \beta(\widehat{x}_1) + \\ &+ a\,\varepsilon_1\, f^+ \cdot f_1 \cdot \beta(\widehat{x}_1) - \frac{\varepsilon_2^2}{2}\, f^s_{\widehat{\tau}_1} \cdot f^s_{\widehat{\tau}_1} \cdot \beta(\widehat{x}_1) + \varepsilon_2\, \zeta(\widehat{\tau}_1) \cdot f^s_{\widehat{\tau}_1} \cdot \beta(\widehat{x}_1) + \\ &+ \frac{1}{2}\int_{\widehat{\tau}_1}^{\widehat{T}} \left(w^2(s)\, [\dot{g}^1_s, g^1_s] \cdot \beta(\widehat{x}_1) + 2w(s)\,\zeta(s) \cdot \dot{g}^1_s \cdot \beta(\widehat{x}_1)\right) \mathrm{d}s \end{aligned} \tag{16}$$

with the linear constraint

$$\begin{aligned} &\dot{\zeta}(s) = w(s)\dot{g}^1_s(\widehat{x}_1) \quad s \in (\widehat{\tau}_1, \widehat{T}) \\ &\zeta(\widehat{\tau}_1) = \varepsilon_1 f^+(\widehat{x}_1) + \varepsilon_2 f^s_{\widehat{\tau}_1}(\widehat{x}_1) + a f_1(\widehat{x}_1), \quad \zeta(\widehat{T}) = 0\,. \end{aligned} \tag{17}$$

It is easy to see that J'' cannot be coercive. In fact, consider the identity

$$f_0 + \widehat{u}_1(\widehat{\tau}_1) f_1 + f_2 = f^+ - b f_1\,, \qquad b := 1 - \widehat{u}_1(\widehat{\tau}_1),$$

and choose $(\varepsilon, a, w) = (1, -1, -b, w \equiv 0)$. Such a choice is admissible and yields $J''(\varepsilon, a, w) = 0$.

Remark 4. J'' restricted to $\varepsilon = 0$ is the second variation L'' described in Remark 3.

3 Hamiltonian Methods

In this Section we describe the non–smooth modification of the standard Hamiltonian method used to prove sufficient conditions in optimal control. Roughly speaking the method consists in constructing a field of extremals by projecting on the state–space M the flow of the maximized Hamiltonian $H^{\max}$ emanating from a horizontal Lagrangean sub–manifold. Equivalently one can choose a Hamiltonian greater than or equal to $H^{\max}$ and coinciding with $H^{\max}$ on the range of the adjoint covector.

In this case we are not able to obtain a field of state–extremals covering a neighborhood of the graph of $\widehat{\xi}$ using a unique smooth Lagrangean sub–manifold and we proceed in the following way.

Step 1. We define a time–dependent Hamiltonian K_t, $t \in [\widehat{\tau}_1, \widehat{T}]$ and a horizontal Lagrangean sub–manifold Λ projecting on a neighborhood $\mathcal{O}$ of $\widehat{x}_1$ and defined by

$$\Lambda := \{\mathrm{d}\,\alpha(x) \colon x \in \mathcal{O}\},$$

where α is a smooth function such that $\mathrm{d}\,\alpha(\widehat{x}_1) = \widehat{\lambda}(\widehat{\tau}_1)$. We require the following properties on K_t and Λ:

1. $K_t \geq H^{\max}_{|\Sigma}$ on $[\widehat{\tau}_1, \widehat{T}]$, $K_t \circ \widehat{\lambda} = \widehat{H}_t \circ \widehat{\lambda}$, $\overrightarrow{K}_t \circ \widehat{\lambda} = \overrightarrow{\widehat{H}}_t \circ \widehat{\lambda}$, $\overrightarrow{K}_t$ is tangent to Σ;
2. $\Lambda \subset \Sigma$;
3. $\mathrm{id} \times \pi \circ \mathcal{K}\colon (t, \ell) \in [\widehat{\tau}_1, \widehat{T}] \times \Lambda \mapsto (t, \pi \circ \mathcal{K}_t(\ell)) \in [\widehat{\tau}_1, \widehat{T}] \times M$ is a local diffeomorphism.

Step 2. First we define a smooth function $\widetilde{\alpha} \geq \alpha$ on $\mathcal{O}$ such that the "half" Lagrangean sub–manifold defined by

$$\widetilde{\Lambda}^+ := \left\{ \widetilde{\ell} = \mathrm{d}\, \widetilde{\alpha}(x) \colon x \in \mathcal{O},\ f_1 \cdot \widetilde{\alpha}(x) \geq 0 \right\}$$

has the property

$$H_1(\mathcal{H}^+_t(\widetilde{\ell})) > 0 \quad \forall (t, \widetilde{\ell}) \in [0, \widehat{\tau}_1) \times \widetilde{\Lambda}^+. \tag{18}$$

Afterwards we consider

$$\Lambda^- := \{ \ell \in \Lambda \colon f_1 \cdot \widetilde{\alpha}(q) \leq 0 \}$$

and we prove that

$$\Lambda^- \cap \widetilde{\Lambda}^+ = \{ \ell = \mathrm{d}\, \alpha(x) \colon f_1 \cdot \widetilde{\alpha}(x) = 0 \} \tag{19}$$

and that

$$H_1(\mathcal{H}^+_t(\widetilde{\ell})) > 0 \quad \forall (t, \widetilde{\ell}) \in [0, \widehat{\tau}_1) \times \Lambda^-. \tag{20}$$

Remark that

$$\Lambda_{\widehat{\tau}_1} := \widetilde{\Lambda}^+ \cup \Lambda^-$$

is a continuous piecewise smooth Lagrangean sub–manifold and that

$$\mathrm{id} \times \pi \circ \mathcal{H}^+ \colon (t, \ell) \in [0, \widehat{\tau}_1] \times \Lambda_{\widehat{\tau}_1} \to (t, \pi \circ \mathcal{H}^+_t(\ell)) \in [0, \widehat{\tau}_1] \times M$$

is a local diffeomorphism.

If we can fulfill the above requirements, then we can complete the proof by lifting any admissible trajectory $\xi \colon [0, T] \to M$, to T^*M by

$$\mu(t) = \begin{cases} \mu_B(t) = \left(t, (\pi \circ \mathcal{H}^+_t)^{-1}(\xi(t))\right) & t \in [0, \widehat{\tau}_1] \\ \mu_S(t) = \left(t, (\pi \circ \mathcal{K}_t)^{-1}(\xi(t))\right) & t \in [\widehat{\tau}_1, \widehat{T}]. \end{cases}$$

Remark that we can apply this procedure only if $T \leq \widehat{T}$ and the graph of ξ belongs to a sufficiently small neighborhood of the graph of $\widehat{\xi}$.

Namely, let $p\, \mathrm{d}\, q$ be the canonical Liouville form on T^*M. Denoting

$$\Omega_B = [0, \widehat{\tau}_1] \times \Lambda_{\widehat{\tau}_1}, \quad \Omega_S = [\widehat{\tau}_1, \widehat{T}] \times \Lambda,$$

standard symplectic methods show that the one forms ω_B and ω_S defined by

$$\omega_B = \mathcal{H}^{+*} \left(p\, \mathrm{d}\, q - H^+ \, \mathrm{d}\, t \right), \quad \omega_S = \mathcal{K}^* \left(p\, \mathrm{d}\, q - K \, \mathrm{d}\, t \right)$$

are exact on Ω_B and on Ω_S, respectively.

Denoting moreover

$$\widehat{\mu}(t) = (t, \widehat{\lambda}(\widehat{\tau}_1)),\ t \in [0, \widehat{T}], \quad \ell(t) = (t, (\pi \circ \mathcal{K}_t)^{-1}(\widehat{x}_f)), \quad t \in [T, \widehat{T}]$$

we can compute

$$0 = \int_{\mu_S} \omega_S + \int_{\ell} \omega_S - \int_{\widehat{\mu}_{|[\widehat{\tau}_1, \widehat{T}]}} \omega_S + \alpha(\xi(\widehat{\tau}_1)) - \alpha(\widehat{x}_1)\,.$$

Since, by the properties of K_t, $\int_{\mu_S} \omega_S \leq 0$ and $\int_{\widehat{\mu}_{|[\widehat{\tau}_1, \widehat{T}]}} \omega_S = 0$, we obtain

$$0 \leq \int_{\ell} \omega_S + \alpha(\xi(\widehat{\tau}_1)) - \alpha(\widehat{x}_1)\,.$$

On the other hand

$$K_t((\pi \circ \mathcal{K}_t)^{-1}(\widehat{x}_f)) = K_{\widehat{\tau}_1}(\widehat{\lambda}(\widehat{\tau}_1)) = 1 + O(t - \widehat{\tau}_1)$$

so that

$$\int_{\ell} \omega_S = \int_T^{\widehat{T}} -K_t((\pi \circ \mathcal{K}_t)^{-1}(\widehat{x}_f)))\,\mathrm{d}\,t = T - \widehat{T} + o(|T - \widehat{T}|)$$

and

$$\alpha(\xi(\widehat{\tau}_1)) - \alpha(\widehat{x}_1) \geq \widehat{T} - T + o(|\widehat{T} - T|)\,.$$

Using the same symplectic argument on Ω_B we compute

$$0 = \int_{\mu_B} \omega_B - \int_{\widehat{\mu}_{|[0, \widehat{\tau}_1]}} \omega_B + \widetilde{\alpha}(\widehat{x}_1) - \begin{cases} \widetilde{\alpha}(\xi(\widehat{\tau}_1)) & \text{if } f_1 \cdot \widetilde{\alpha}(\xi(\widehat{\tau}_1)) > 0 \\ \alpha(\xi(\widehat{\tau}_1)) & \text{if } f_1 \cdot \widetilde{\alpha}(\xi(\widehat{\tau}_1)) \leq 0 \end{cases}$$

Inequalities (18) and (20) give $\int_{\mu_B} \omega_B \leq 0$ and $\int_{\widehat{\mu}_{|[0, \widehat{\tau}_1]}} \omega_B = 0$, so that, since $\widetilde{\alpha} \geq \alpha$, we finally get

$$0 \geq \alpha(\xi(\widehat{\tau}_1)) - \alpha(\widehat{x}_1) \geq \widehat{T} - T + o(|\widehat{T} - T|).$$

If T is sufficiently close to $\widehat{T}$ we can deduce $T = \widehat{T}$.

4 Proof of Theorem 1

To complete the proof of Theorem 1 we only need to show that it is possible to accomplish the requirements in Steps 1 and 2 of the previous Section.

4.1 Step 1

On the singular arc we have reduced ourselves to the single input system (10) with singular reference control, hence we can use the results in [9] where it is proved that there exists a neighborhood $\mathcal{U}$ of $\mathcal{S}$ in T^*M and a function

$$\rho\colon \mathcal{U} \to \mathbb{R}$$

such that

1. the Hamiltonian vector field $\overrightarrow{K}_t$ associated to the Hamiltonian function $K_t(\ell) := \widehat{H}_t + \frac{\rho}{2}\{\widetilde{H}, H_1\}^2$ is tangent to Σ.
2. $\rho(\ell) = \dfrac{1}{\{H_1, \{\widetilde{H}, H_1\}\}(\ell)}$ $\quad \forall \ell \in \mathcal{S}$, so that $\rho > 0$ on $\mathcal{U}$.

It is easy to see that the requirements in a) of Step 1 are fulfilled by K_t.

Consider a chart at $\widehat{x}_1$ defined in $\mathcal{O}$ and such that

$$\begin{aligned} f_1 &\equiv \frac{\partial}{\partial x_1}, \\ [\widetilde{f}, f_1](x) &= \frac{\partial}{\partial x_2} + x_1\Big(\frac{\partial}{\partial x_3} + O(x)\Big), \\ [f_1, [\widetilde{f}, f_1]](x) &= \frac{\partial}{\partial x_3} + O(x), \end{aligned}$$

this can be obtained by assumption **A4**. Moreover, in these coordinates, we get

$$\widehat{\ell}_1 = \sum_{i=3}^{n} \lambda_i \, \mathrm{d}\, x_i, \quad \lambda_3 > 0.$$

With the same proof of [9] and [7] we can show that there exists a sufficiently large b such that

$$\alpha(x) = \sum_{i=3}^{n} \left(\lambda_i x_i + \frac{b}{2} x_i^2 \right)$$

has the properties required in b) and c) of Step 1.

4.2 Step 2

Define

$$\widetilde{\alpha}(x) = \alpha(x) + \frac{b}{2} x_1^2$$

and remark that

$$f_1 \cdot \alpha(x) = 0 \;\text{ and }\; f_1 \cdot \widetilde{\alpha}(x) = b x_1$$

so that

$$\Lambda_{\widehat{\tau}_1} = \{\mathrm{d}\,\alpha(x) \colon x_1 \le 0\} \cup \{\mathrm{d}\,\widetilde{\alpha}(x) \colon x_1 \ge 0\};$$
$$\widetilde{\Lambda}^+ \cap \Lambda^- = \{\mathrm{d}\,\alpha(x) \colon x_1 = 0\}.$$

hence (19) is fulfilled. The following Lemma proves that (18)–(20) are fulfilled and hence Theorem 1 is proved.

Lemma 1. *Possibly restricting $\mathcal{O}$,*

$$H_1(\mathcal{H}_t^+(\ell)) \ge 0$$

for any $(t,\ell) \in [0,\widehat{\tau}_1] \times \Lambda_{\widehat{\tau}_1}$.

Proof. Since
$$[f_1,[\widetilde{f},f_1]] \cdot \widetilde{\alpha}(\widehat{x}_1) = [f_1,[\widetilde{f},f_1]] \cdot \alpha(\widehat{x}_1) = \lambda_3,$$
we can choose, possibly restricting $\mathcal{O}$, $\varepsilon > 0$ such that

$$[f_1,[\widetilde{f},f_1]] \cdot \widetilde{\alpha}(\exp((t-\widehat{\tau}_1)\widetilde{f})(x)) > 0\,,$$
$$[f_1,[\widetilde{f},f_1]] \cdot \alpha(\exp((t-\widehat{\tau}_1)\widetilde{f})(x)) > 0$$

for any $(t,x) \in [\widehat{\tau}_1 - \varepsilon, \widehat{\tau}_1] \times \mathcal{O}$.

With a Taylor expansion with respect to t we get

$$H_1(\mathcal{H}_t^+)(\ell)) = H_1(\ell) + (t-\widehat{\tau}_1)\left\{H^+,H_1\right\}(\ell) + \\ + \frac{(t-\widehat{\tau}_1)^2}{2}\left\{H_1,\left\{H^+,H_1\right\}\right\}(\mathcal{H}_{s(t)}^+(\ell))$$

for some $s(t) \in (t,\widehat{\tau}_1)$.

If $\ell \in \widetilde{\Lambda}^+$ then we have

$$H_1(\mathcal{H}_t^+(\ell)) = f_1 \cdot \widetilde{\alpha}(x) + (t-\widehat{\tau}_1)[\widetilde{f},f_1] \cdot \widetilde{\alpha}(x) + \\ + \frac{(t-\widehat{\tau}_1)^2}{2}[f_1,[\widetilde{f},f_1]] \cdot \widetilde{\alpha}\left(\widehat{S}_{s(t)}(x)\right) \ge x_1\left[b + (t-\widehat{\tau}_1)(1 + bx_3 + O(x))\right]$$

which is greater than or equal to 0.

If $\ell \in \Lambda^-$ then we have

$$H_1(\mathcal{H}_t^+(\ell)) = f_1 \cdot \alpha(x) + (t-\widehat{\tau}_1)[\widetilde{f},f_1] \cdot \alpha(x) + \\ + \frac{(t-\widehat{\tau}_1)^2}{2}[f_1,[\widetilde{f},f_1]] \cdot \alpha\left(\widehat{S}_{s(t)}(x)\right) \ge x_1(t-\widehat{\tau}_1)\,(1 + bx_3 + O(x))$$

which is greater than or equal to 0.

Since $H_1(\widehat{\lambda}(t)) > 0$ for any $t \in [0,\widehat{\tau}_1)$, there exists $\delta > 0$ such that $H_1(\widehat{\lambda}(t)) \ge \delta$ for any $t \in [0,\widehat{\tau}_1 - \varepsilon]$, so that we may assume

$$H_1(\mathcal{H}_t^+(\ell)) \ge \frac{\delta}{2}$$

for any $(t,\ell) \in [0,\widehat{\tau}_1 - \varepsilon] \times \Lambda_{\widehat{\tau}_1}$. This completes the proof of the lemma.

References

1. Andrei A. Agrachev and Yuri L. Sachkov. *Control theory from the geometric viewpoint.* Springer–Verlag, 2004.
2. Andrei A. Agrachev, Gianna Stefani, and PierLuigi Zezza. An invariant second variation in optimal control. *Internat. J. Control*, 71(5):689–715, 1998.
3. Andrei A. Agrachev, Gianna Stefani, and PierLuigi Zezza. Strong minima in optimal control. *Proc. Steklov Inst. Math.*, 220:4–26, 1998. translation from Tr. Mat. Inst. Steklova 220, 8-22 (1998).
4. A.V. Dmitruk. Quadratic order conditions of a local minimum for singular extremals in a general optimal control problem. In *Proceedings of Symposia in Pure Mathematics*, volume 64, pages 163–198, 1999.
5. Helmut Maurer and Nikolai P. Osmolovskii. Second order sufficient conditions for time optimal bang-bang control problems. *SIAM J. Control and Optimization*, to appear.
6. Laura Poggiolini and Gianna Stefani. State-local optimality of a bang-bang trajectory: a Hamiltonian approach. *Systems & Control Letters*, 53:269–279, 2004.
7. Laura Poggiolini and Gianna Stefani. Minimum time optimality for a bang-singular arc: second order sufficient conditions. In *44th IEEE Conference on Decision and Control and European Control Conference ECC 2005*, pages on CD–ROM. IEEE Control Systems Society, 2005.
8. Andrei V. Sarychev. Sufficient optimality conditions for Pontryagin extremals. *Syst. Control Lett.*, 19(6):451–460, 1992.
9. Gianna Stefani. Minimum-time optimality of a singular arc: second order sufficient conditions. In *Proceeding of CDC04*, 2004.

Hamiltonian Engineering for Quantum Systems

Sonia G. Schirmer

Dept of Applied Mathematics & Theoretical Physics, University of Cambridge, Cambridge, CB3 0WA, United Kingdom
sgs29@cam.ac.uk

Summary. We describe different strategies for using a semi-classical controller to engineer Hamiltonians for quantum systems to solve control problems such as quantum state or process engineering and optimization of observables.

1 Introduction

Extending control to the quantum domain, i.e., to physical systems whose behavior is not governed by classical laws but dominated by quantum effects, has become an important area of research recently. It is also an essential prerequisite for the development of novel technologies such as quantum information processing, as well as new applications in quantum optics, quantum electronics, or quantum chemistry. Choreographing the behavior of interacting quantum particles is a rather difficult task in general, for a variety of reasons, including the destructive effects of uncontrollable interactions with the environment and measurement backaction, both of which lead to loss of coherence, for instance. Yet, many problems in quantum optics, quantum electronics, atomic physics, molecular chemistry, and quantum computing can be reduced to quantum state or process engineering, and optimization of observables, which can be solved using Hamiltonian engineering techniques. In this paper we outline and compare different strategies for accomplishing this in various quantum settings.

2 Control System Model

A control system must obviously consist of a system to be controlled and a controller. In quantum control the former is always quantum-mechanical. The latter can be either quantum or classical, and it may seem natural to choose another quantum system as the controller. Indeed, this is useful for some applications, for instance in quantum optics [14]. The main challenge in general though, and the focus of this paper, is control at the interface between the quantum world and the classical world we experience, i.e., control of systems that obey the laws of quantum physics using semi-classical sensors and actuators that interact with the quantum system but accept classical input (such as different settings of the classical control switches of the laboratory equipment) and return classical information (See Fig. 1).

F. Bullo et al. (Eds.): Lag. & Hamil. Methods for Nonlin. Ctrl., LNCIS 366, pp. 293–304, 2007.
springerlink.com

This model is applicable to many quantum control settings. For example, the quantum system could be an ensemble of molecules involved in a chemical reaction, subject to laser pulses produced by an actuator consisting of a laser source and pulse shaping equipment, and a detector that might consist of a mass spectrometer to identify the reaction products. It could be a solid-state system such as an ensemble of quantum dots representing qubits, with actuators and sensors consisting of control electrodes and single-electron transistors, respectively. It could be an ensemble of molecules with nuclear spins subject to actuators generating magnetic and radio-frequency fields, and sensors that detect the magnetization of the sample, etc.

3 System Dynamics

In the setting defined above the state of the system can be represented either by a Hilbert space vector $|\Psi(t)\rangle$, or more generally, a density operator $\rho(t)$, i.e., a positive trace-one operator acting on the system's Hilbert space $\mathcal{H}$. Although the full Hilbert spaces of most quantum systems are hardly finite or separable—even the Hilbert space of the hydrogen atom is not separable if the full continuum of ionized states is included, we are usually only interested in controlling the dynamics of a finite-dimensional subspace of the system's full Hilbert space (e.g., a subset of bound states of an atom), and full control of an infinite-dimensional system is generally not possible in any case [see e.g. [12]]. We will therefore assume in this paper that the Hilbert spaces of interest are finite-dimensional, and thus all operators have matrix representations, etc.

Neglecting decoherence and the effect of measurements for a moment, the evolution of the quantum system to be controlled is governed by the Schrodinger equation

$$\frac{d}{dt}|\Psi(t)\rangle = -\frac{i}{\hbar}H[\mathbf{f}(t)]|\Psi(t)\rangle, \tag{1}$$

or the (equivalent) quantum Liouville equation

$$\frac{d}{dt}\rho(t) = -\frac{i}{\hbar}\left[H[\mathbf{f}(t)], \rho(t)\right], \tag{2}$$

where $[A, B] = AB - BA$ is the commutator. $\hbar = h/2\pi$, where h is the Planck constant, but we will often choose units such that $\hbar = 1$. Thus, the dynamics is determined by the Hamiltonian $H[\mathbf{f}(t)]$, which is an operator acting on the system's Hilbert space $\mathcal{H}$, and depends on a set of (classical) control fields $\mathbf{f}(t) = (f_1(t), \ldots, f_M(t))$ produced by the actuators.

The main difference between Eqs (1) and (2) is that the former applies only to pure-state systems, while the latter applies equally to pure and mixed-state systems (quantum ensembles). Moreover, unlike the Schrodinger equation, the quantum Liouville equation can be generalized for systems subject to decoherence or weak measurements by adding (non-Hamiltonian) superoperators

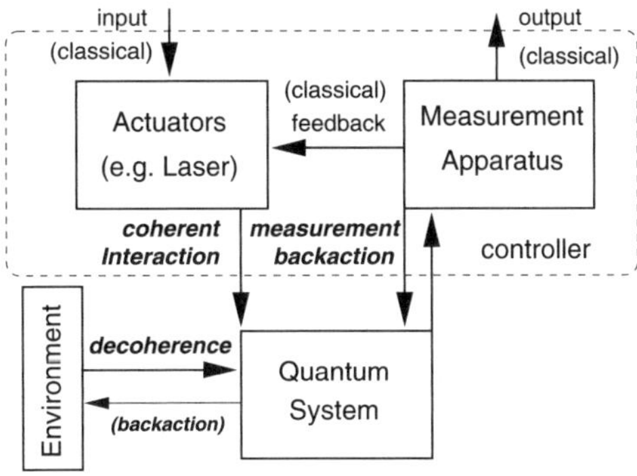

Fig. 1. Semi-classical Quantum Control Model

[operators acting on Hilbert space operators such as $\rho(t)$] to account for the contributions of measurements and dissipation to the dynamics of the system

$$\dot{\rho}(t) = \mathcal{L}_H[\rho(t)] + \mathcal{L}_M[\rho(t)] + \mathcal{L}_D[\rho(t)], \tag{3}$$

where $\mathcal{L}_H[\rho(t)] = -i\,[H[\mathbf{f}(t)], \rho(t)]$. The main difference between $\mathcal{L}_M$ and $\mathcal{L}_D$ is conceptual: the former depends on the measurements performed, i.e., the configuration of the sensors, which we can control, the latter on usually uncontrollable interactions of the system with the environment.

Due to the physically prescribed evolution equations for quantum systems, quantum control—at least in the classical controller model described—is fundamentally non-linear, even in the case of open-loop control. In the special case when the system's interaction with both sensors and the environment is negligible (which is obviously only possible for open-loop control), the system's dynamics is completely determined by a Hamiltonian operator $H[\mathbf{f}(t)]$, and the main task of quantum control is to find effective ways to engineer this Hamiltonian to achieve a desired objective. For simplicity we will restrict our attention here mostly to Hamiltonian engineering in this case, although many of the techniques are still useful in the more there general case when the system is subject to measurements and/or uncontrollable interactions with the environment [Eq. (3)].

4 Quantum Control Objectives

Although the objectives vary depending on the application, most problems in quantum control can be reduced to quantum process engineering, quantum state engineering or optimization of observables. The first, quantum process engineering, involves finding a Hamiltonian $H[\mathbf{f}(t)]$ such that

$$\exp_+\left[-i\int_{t_0}^{t_F} H[\mathbf{f}(t)]\,dt\right] = U, \tag{4}$$

where U is a desired target process, and the subscript $+$ indicates that the exponential must be interpreted as a time-ordered exponential with positive time-ordering due to the time-dependent nature of the Hamiltonian. This problem is of particular interest in quantum computing, where the target processes are quantum logic gates. In general, not every process can be implemented for a given system, and the set of reachable processes is determined by the dynamical Lie group associated with the set of Hamiltonians $\{H[\mathbf{f}(t)] : \mathbf{f}(t) \in \mathcal{A}\}$, where $\mathcal{A}$ is the set of admissible controls that can be produced by the actuators and satisfy physical constraints, e.g., on the field strength, etc. For example, for a Hamiltonian system a necessary (minimum) requirement is that U be a *unitary* operator (acting on the system's Hilbert space), and other constraints may restrict the set of reachable operators further. However, if the target process is reachable then there are usually many controls $\mathbf{f}(t)$ and corresponding Hamiltonians $H[\mathbf{f}(t)]$, which give rise to the same process.

The second problem, quantum state engineering, requires finding a Hamiltonian $H[\mathbf{f}(t)]$ such that (a) $|\Psi(t_F)\rangle = |\Psi_1\rangle$, $|\Psi(t_0)\rangle = |\Psi_0\rangle$, and $|\Psi(t)\rangle$ satisfies the dynamical equation (1), or (b) $\rho(t_F) = \rho_1$, $\rho(t_0) = \rho_0$, and $\rho(t)$ satisfies the dynamical equation (2) [or (3) in the general case], where ρ_0 $[|\Psi_0\rangle]$ and ρ_1 $[|\Psi_0\rangle]$ represent the initial and target state, respectively.

The final problem, optimization of observables, requires finding a Hamiltonian such that the expectation value (or ensemble average) of an observable A, $\mathrm{Tr}[A\rho(t_F)]$, assumes a maximum or minimum at a certain time t_F, given that the state $\rho(t)$ evolves according to Eq. (2) and satisfies an initial condition $\rho(t_0) = \rho_0$. Problems of this type arise frequently in atomic and molecular physics and chemistry, where the observables of interest can range from the position or momentum of a particle, to the dipole moment or the vibrational energy of a molecular bond, etc., but they are also relevant in quantum computing where we wish to maximize the gate fidelity or the projection of the system onto subspaces that are robust with regard to decoherence, etc.

While the second problem may appear much simpler than the first, it can be shown that for a *generic quantum ensemble*, i.e., a quantum ensemble described by a density operator ρ with a maximum number of distinct eigenvalues, the problems of quantum state and process engineering are essentially equivalent, up to a usually unobservable (and hence insignificant) global phase factor. Similarly, the third problem is equivalent to quantum process engineering if the initial state of the system is a generic quantum ensemble and the target observable is represented by an operator A with distinct eigenvalues (occurring with multiplicity 1). It should be noted, however, that this equivalence does not hold for pure-state systems, which are always represented by density operators of rank 1, and for which process engineering is in general a much harder problem than the others. See e.g. [9]

5 Hamiltonian Engineering

Having shown how many problems in quantum control can be reduced to Hamiltonian engineering problems, we shall now consider various strategies for finding and implementing control Hamiltonians for quantum systems, and their advantages and drawbacks. So far, we have not made any assumptions about the structure of the Hamiltonian $H[\mathbf{f}(t)]$ or the nature of the control fields $\mathbf{f}(t)$ applied, both of which depend on the specific physical systems considered. While we will try to avoid being too specific some additional assumptions are necessary.

We can always partition $H[\mathbf{f}(t)]$ into a system part H_S, which describes the system's intrinsic dynamics and is independent of the controller, and a control part $H_C[\mathbf{f}(t)]$. Although $H_C[\mathbf{f}(t)]$ can depend on the control functions $f_m(t)$ in a nonlinear fashion, in many situations assuming a linear dependence on the field components $f_m(t)$

$$H_C[\mathbf{f}(t)] = \sum_{m=1}^{M} f_m(t) H_m, \tag{5}$$

is a reasonably good approximation. Furthermore, while in some applications such as solid-state architectures with multiple control electrodes, there are naturally multiple independent fields, in many applications there is only a single effective control field such as the electromagnetic field induced by a laser pulse, a maser or radio-frequency field, for instance, and thus the control Hamiltonian simplifies further $H_C[\mathbf{f}(t)] = f(t)H_1$.

In the control-linear case (5), a necessary and sufficient condition for being able to engineer any (unitary) process up to a global phase factor is that the Lie algebra generated by iH_S and iH_m, $m = 1, \ldots, M$ is either $u(N)$ or $su(N)$, where N is the dimension of the relevant Hilbert space $\mathcal{H}$ [1, 10]. As outlined in the previous section, this is also a necessary and sufficient condition for quantum state or observable controllability, at least for generic quantum ensembles and observables. Most quantum systems can be shown to be controllable, but constructive control can be challenging.

5.1 Geometric Control Techniques

Given a set of (independent) control Hamiltonians H_m, $m = 1, \ldots, M$, which is complete in that the iH_m generate the entire Lie algebra, the simplest general strategy we can pursue is to expand the (unitary) target process U into a product of elementary (complex) rotations $\exp(icH_m)$ in a cyclic, iterating pattern

$$U = \prod_{k=1}^{K} \left[\prod_{m=1}^{M} \exp(ic_{km} H_m) \right] \tag{6}$$

and determine the constants c_{km} in the expansion using Lie group decomposition methods, from which we can derive suitable values for the field strengths f_{mk} and control pulse lengths t_{mk} via the relation $c_{mk} = \int_0^{t_{mk}} f_{mk}(t)\, dt$, which reduces to $c_{mk} = f_{mk} t_{mk}$ for piecewise constant fields.

If the H_m are orthogonal, $\mathrm{Tr}[H_m H_n] = \text{const.}\, \delta_{mn}$, then there are well-developed geometric techniques to solve this type of problem. Explicit solutions have been developed especially for many problems involving spin$-1/2$ particles, which are of interest in nuclear magnetic resonance (NMR) applications (see, e.g. [2]), and even constructive algorithms for the generation of arbitrary unitary operators on n-qubits via $SU(2^n)$ decompositions have been proposed [8]. For non-qubit systems the problem is harder but some explicit decomposition algorithms for simple N-level systems have also been proposed (e.g. [7]).

In practice, however, most physical systems are subject to internal system dynamics in addition to the control-induced dynamics. This leads to a drift term H_S, which is usually non-trivial and cannot be turned off. In principle, we can include this term in the expansion (6) by replacing the control Hamiltonians H_m by $H_S + H_m$, for example. However, even if the H_m are orthogonal, the new effective Hamiltonians in the expansion are usually not, which significantly complicates the problem. Moreover, even if we find a decomposition for a given problem, it may not be physically realizable as the c_{km} could be negative. Since the control coefficient corresponding to H_S is fixed ($f_S = 1$), implementing such a rotation would require letting the system evolve for negative times, which is usually impossible (except possibly if the evolution of the system is periodic). Decomposition techniques for non-orthogonal Hamiltonians that take these constraints into account exist for $SU(2)$ [6] but even in this simple case the resulting pulse sequences are much more complicated and for higher-dimensional systems even the minimum number of pulses necessary to generate a desired unitary transformation given non-orthogonal Hamiltonians is generally not known.

The usual way to circumvent the problem of drift is by transforming to a rotating frame (RF). For instance, we can define

$$|n(t)\rangle = e^{-itE_n}|n\rangle, \tag{7}$$

where $\{|n\rangle : n = 1, \ldots, N = \dim \mathcal{H}\}$ is a basis of $\mathcal{H}$ consisting of eigenstates $|n\rangle$ of H_S with eigenvalues (energies) E_n, $H_S|n\rangle = E_n|n\rangle$. Setting $U_S(t) = \exp(-itH_S)$ the dynamics in the rotating frame is governed by the new (interaction picture) Hamiltonian

$$H_C'[\mathbf{f}(t)] = U_S(t)^\dagger H_C[\mathbf{f}(t)] U_S(t). \tag{8}$$

Thus we have transformed away the drift term but the previously time-independent Hamiltonians H_m in the control-linear approximation are now time-dependent

$$H_m' = U_S(t)^\dagger H_m U_S(t). \tag{9}$$

However, for certain applications we can decompose the control fields into several distinct frequency components. An especially useful decomposition is

$$f(t) = \sum_{n,n'>n} A_{nn'}(t)\cos(\omega_{nn'}t + \phi_{nn'}), \tag{10}$$

where $\omega_{nn'} = E_{n'} - E_n$ is the transition frequency between states $|n'\rangle$ and $|n\rangle$, $A_{nn'}(t)$ are "amplitude functions" and $\phi_{nn'}$ constant phases. If the system is (a) strongly regular, i.e., $\omega_{nn'} \neq \omega_{mm'}$ unless $(n, n') = (m, m')$, and (b) the transition frequencies $\omega_{nn'}$ are sufficiently well separated, then choosing control fields of the form (10) with amplitude functions $A_{nn'}(t)$ that vary slowly compared to $\omega_{nn'}^{-1}$, enables us to address individual transitions via frequency selective pulses, and decompose the control Hamiltonian as

$$H'_C = \sum_{n,n'>n} A_{nn'}(t)\cos(\omega_{nn'}t + \phi_{nn'})\, U_S(t)^\dagger H_{nn'} U_S(t), \tag{11}$$

where $H_{nn'} = d_{nn'}[|n\rangle\langle n'| + |n'\rangle\langle n|]$ and $d_{nn'}$ is the dipole moment (or coupling strength) of the transition. Inserting $U_S(t) = \sum_{n=1}^{N} e^{-i\omega_n t}|n\rangle\langle n|$ and $2\cos(\omega_{nn'}t + \phi_{nn'}) = e^{+i(\omega_{nn'}t+\phi_{nn'})} + e^{-i(\omega_{nn'}t+\phi_{nn'})}$, we obtain after simplification

$$H'_C = \frac{1}{2}\sum_{n,n'>n} A_{nn'}(t)\left[e^{i\phi_{nn'}}|n\rangle\langle n'| + e^{-i\phi_{nn'}}|n'\rangle\langle n|\right] \tag{12}$$
$$+A_{nn'}(t)\left[e^{-i(2\omega_{nn'}t+\phi_{nn'})}|n\rangle\langle n'| + e^{i(2\omega_{nn'}t+\phi_{nn'})}|n'\rangle\langle n|\right]$$

If each control pulse $f_{nn'}(t)$ is much longer than $\omega_{nn'}^{-1}$, then the contribution of the terms oscillating with frequency $2\omega_{nn'}$ will average to zero over the length of the pulse. Hence we can drop these terms and simplify Eq. (12) to

$$H_C^{RWA} = \sum_{n,n'>n} \Omega_{nn'}(t) H_{nn'}(\phi_{nn'}) \tag{13}$$

where $\Omega_{nn'} = \frac{1}{2}A_{nn'}(t)d_{nn'}$, $H_{nn'}(\phi_{nn'}) = x_{nn'}\cos\phi_{nn'} + y_{nn'}\sin\phi_{nn'}$, and $x_{nn'} = |n\rangle\langle n'| + |n'\rangle\langle n|$. $y_{nn'} = i(|n\rangle\langle n'| - |n'\rangle\langle n|)$. The resulting rotating wave approximation (RWA) Hamiltonian (13) is drift-free with time-independent components $H_{nn'}$, as desired.

The RF and RWA are ubiquitous in physics, and the RWA control Hamiltonian is the starting point for the design of control schemes using geometric techniques in many applications. As the derivation shows, however, it relies on the validity of several assumptions such as strong regularity of the system, control dynamics much slower than the intrinsic system dynamics (control pulses must be much longer than the oscillation periods $2\pi/\omega_{nn'}$ for (10) and RWA to make sense), and the negligibility of any off-resonant excitation. For an analysis of the validity of geometric control schemes for atomic and molecular systems see e.g. [9].

In practice there are often further complications. For example, for multipartite systems with fixed coupling, the drift term H_S usually consists of two parts: the internal Hamiltonian governing the intrinsic dynamics of the component systems, and a coupling term governing the interaction between the parts. Since the time-scales of the internal dynamics and coherent oscillations induced by the weak coupling terms are often vastly different, the rotating frame transformation (7) must be modified in this case to obtain a useful RWA Hamiltonian. Given n two-level systems (e.g., spin-1/2 particles) with internal Hamiltonian $H_S^{(k)} = \frac{1}{2}\omega_k\sigma_z^{(k)}$, where $\sigma_z^{(k)}$ is an n-factor tensor product whose kth factor is $\sigma_z = \mathrm{diag}(1, -1)$ and all others the identity $I = \mathrm{diag}(1, 1)$, one usually transforms into a multiply-rotating frame given by $U_S(t) = \exp[-it\sum_k H_S^{(k)}]$. This results in a simplified RWA Hamiltonian with a residual drift term, which is in practice often neglected by assuming in addition that the control pulses are sufficiently strong (hard) so that the influence of the drift term is negligible and the control Hamiltonians can be assumed to be effectively orthogonal for the decomposition. Thus, now the control pulses at once must be hard (and fast) enough so that inter-qubit coupling is negligible, but slow enough so that the RWA remains valid and off-resonant excitation negligible.

5.2 Optimal Strategies for Fast Control

If we wish to achieve control on time-scales comparable to the intrinsic dynamics of the system, or wish to control systems with too many or insufficiently distinct transition frequencies, as is the case for complex molecular systems in chemistry for example, then a different approach is required. A promising alternative is to formulate the control problem in terms of optimization of an objective functional such as the distance from a target state or target process, the expectation value of an observable or the gate fidelity, subject to the constraint that the dynamical evolution equations (1), (2) or (3) be satisfied.

A particularly flexible approach is to formulate the control problem as a variational problem by defining a functional $\mathcal{J} = \mathcal{A} - \mathcal{D} - \mathcal{C}$, which incorporates the objective $\mathcal{A}$, the dynamical constraints $\mathcal{D}$ and the costs $\mathcal{C}$ and depends on variational trial functions, and to use variational calculus to find necessary and sufficient conditions for an extremum. For instance, a very popular choice in physical chemistry involves setting $\mathcal{A} = \mathrm{Tr}[A\rho(t_F)]$, where A is a (Hermitian) operator representing the objective (e.g., A could be the projector $|\Psi\rangle\langle\Psi|$ onto a target state $|\Psi\rangle$), with a dynamical constraint functional given by

$$\mathcal{D} = \int_{t_0}^{t_F} \mathrm{Tr}\left[A_v(t)\left(\dot{\rho}_v + \mathcal{L}_{\mathbf{f}(t)}[\rho_v(t)]\right)\right] dt \tag{14}$$

where $\mathcal{L}_{\mathbf{f}(t)}[\rho_v(t)] = \mathcal{L}_{H[\mathbf{f}(t)]}[\rho_v(t)] + \mathcal{L}_M[\rho_v(t)] + \mathcal{L}_D[\rho_v(t)]$ and $\rho_v(t)$ and $A_v(t)$ are variational trial functions (Lagrange multipliers) for the state (density operator) and observable, respectively, and a cost term related to the control field energies

$$\mathcal{C} = \sum_{m=1}^{M} \frac{\lambda_m}{2} \int_{t_0}^{t_F} |f_m(t)|^2 \tag{15}$$

where λ_m are penalty weights. Setting the independent variations of $\mathcal{J}$ with regard to A_v, ρ_v and f_m to 0, a necessary condition for $\mathcal{J}$ to have an extremum, then leads to a set of coupled differential equations with mixed boundary conditions, the Euler-Lagrange equations, which can be solved numerically to obtain a solution for the control fields $f_m(t)$ and the corresponding trajectories for the state $\rho(t)$ and the observable $A(t)$.

The key advantages of this approach are that we can deal with complex (and even non-linear) control Hamiltonians (no RWA or other approximations required), and are in fact not limited to Hamiltonian systems at all, and a wide range of costs or trade-offs can be taken into account. A drawback is that the Euler-Lagrange equations are almost always nontrivial and can only be solved

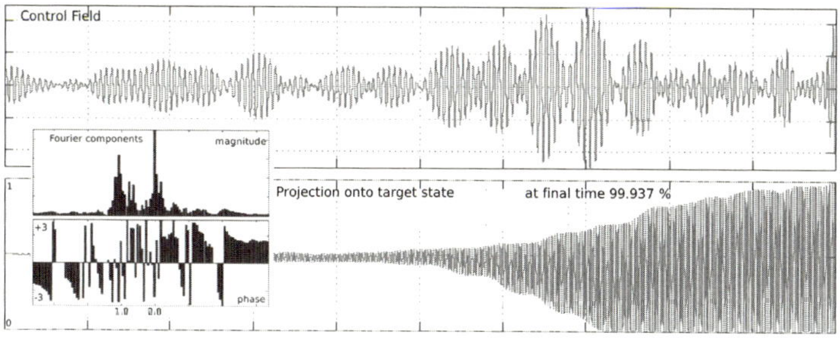

Fig. 2. Optimally shaped pulse and corresponding projection of system state $\rho(t)$ onto target state, $\mathrm{Tr}[A\rho(t)]$ with $A = |\psi_{\mathrm{Bell}}\rangle\langle\psi_{\mathrm{Bell}}|$, for preparation of Bell state $|\psi_{\mathrm{Bell}}\rangle = \frac{1}{\sqrt{2}}(|00\rangle + |11\rangle)$ for a two-qubit system with fixed Heisenberg coupling governed by $H_S = 0.9\sigma_z^{(1)} + \sigma_z^{(2)} + 0.1(\sigma_x \otimes \sigma_x + \sigma_y \otimes \sigma_y + \sigma_z \otimes \sigma_z)$ and $H_C[f(t)] = f(t)(\sigma_x^{(1)} + \sigma_x^{(2)})$ and $\rho_0 = |00\rangle\langle 00|$. The projection onto target state exhibits oscillatory behavior in the stationary frame as expected for a non-stationary target state, but grows in magnitude and reaches close to 100% at the target time. The Fourier decomposition of the pulse (inset) shows peaks at the single qubit resonance frequencies 1.8 and 2.0, as expected.

using numerical techniques. However, efficient algorithms with good convergence properties exist for a large class of problems (see e.g. [4]), and promising results have been obtained for various applications, especially laser control of molecular systems using ultra-fast (sub-picosecond) pulses, a regime well outside the realm of applicability for frequency-selective geometric control schemes. An example of a shaped pulse obtained for the problem of Bell state preparation for system of two coupled spins is given in Fig 2.

5.3 Robust Control Through Adiabatic Passage

On the opposite end of the spectrum are adiabatic techniques. Rather than applying control fields to induce (fast) transitions between various states of the system by absorption or emission of field quanta (usually photons), adiabatic techniques rely on (slow) continuous deformation of the energy surfaces by strong, slowly varying control fields, and adiabatic following of the system's state, making use of the eigenstate decomposition of the Hamiltonian

$$H[\mathbf{f}(t)] = \sum_{n=1}^{N} \epsilon_n(t) |\Psi_n(t)\rangle\langle\Psi_n(t)|. \tag{16}$$

For a Hamiltonian with a control-induced time-dependence the eigenvalues $\epsilon_n(t)$ and corresponding eigenstates $|\Psi_n(t)\rangle$ vary in time. The idea of adiabatic passage is that if we start in a particular initial state $|\Psi_0\rangle$, which will usually be an eigenstate of the system's intrinsic Hamiltonian H_S, and then *slowly* switch on and vary suitable control fields $\mathbf{f}(t)$, then rather than inducing spontaneous transitions to other states, the state $|\Psi(t)\rangle$ is going to remain an eigenstate of the Hamiltonian and follow the path determined by the control fields via Eq. (16). Adiabatic passage is the basis for many control schemes in atomic physics [11], most notably STIRAP.

The simplest example is population transfer in a three-level Λ-system (see Fig. 3) simultaneously driven by two fields $f_m(t) = A_m(t)\cos(\omega_m t)$, $m = 1, 2$, that resonantly excite the $1 \to 2$ and $2 \to 3$ transitions, respectively. In this case the RWA Hamiltonian (13) simplifies to

$$H_C^{RWA} = -\begin{bmatrix} 0 & \Omega_1(t) & 0 \\ \Omega_1(t) & 0 & \Omega_2(t) \\ 0 & \Omega_2(t) & 0 \end{bmatrix} \tag{17}$$

with $\Omega_m = d_{m,m+1}A_m/2$ for $m = 1, 2$. Setting $\Omega = \sqrt{\Omega_1^2 + \Omega_2^2}$ and $\theta(t) = \arctan[\Omega_1(t)/\Omega_2(t)]$, it is easy to verify that the eigenstates of this Hamiltonian are $|\Psi_\pm(t)\rangle = \Omega_1|1\rangle \pm \Omega|2\rangle + \Omega_2|3\rangle$ for $\lambda_\pm = \pm\Omega$, respectively, and $|\Psi_0(t)\rangle = \cos\theta(t)|1\rangle - \sin\theta(t)|3\rangle$ for $\lambda_0 = 0$. Thus, if the system is initially in state $|1\rangle$, then applying control fields such that $\Omega_1(t)/\Omega_2(t)$ changes (sufficiently slowly) from 0 at t_0 to ∞ at t_F, results in adiabatic passage of the system from state $|1\rangle = |\Psi_0(0)\rangle$ to $|3\rangle = |\Psi_0(t_F)\rangle$ as $\theta(t)$ goes from 0 to $\pi/2$. STIRAP is based on the realization that we can achieve this simply by applying two overlapping Gaussian pulses in a *counter-intuitive* sequence, i.e, so that $\Omega_2(t)$ starts and ends before $\Omega_1(t)$ as shown in Fig. 2. Since the upper level $|2\rangle$ is decoupled, i.e., not populated during this process, the transfer is robust against decay from the excited state, which is often a major limiting factor in the control of atomic or molecular systems, although it must be noted that the scheme is sensitive with regard to other forms of decoherence etc.

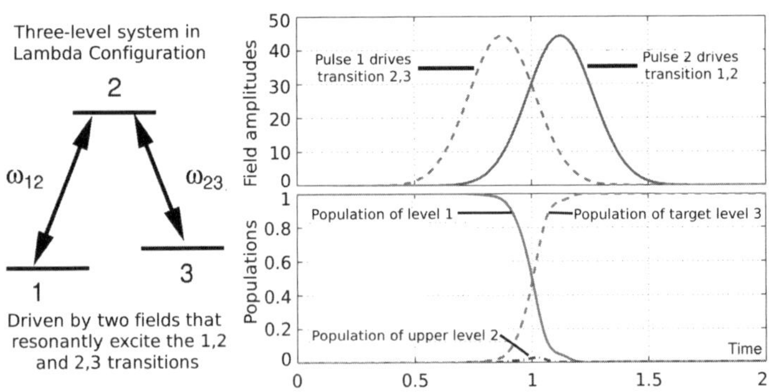

Fig. 3. STIRAP pulse sequence (top) for robust population transfer from state $|1\rangle$ to state $|3\rangle$ (bottom) for three-level Λ system (inset)

5.4 Learning Control Via Closed-Loop Experiments

A potential problem with all of the techniques described so far is their reliance on a model of the system. Unfortunately, such models are not always available, especially for complex systems. While improved techniques for quantum control system identification will hopefully eventually allow us to overcome this problem, an alternative which has already been successfully demonstrated in the laboratory (see e.g. [5]) are experimental closed-loop learning techniques [3]. The former must not be confused with real-time quantum feedback control [13]. While both schemes rely on (classical) information gained from measurements, the former involve repeated experiments on many copies of the system, while the latter rely on continuous observation of the same system via weak measurements (e.g. homodyne detection or passive photodetectors).

Closed-loop learning techniques are essentially adaptive, using feedback to guide an evolutionary process. There are many variations but the basic strategy is simple. We define an objective functional, usually called fitness function, and select an initial set (population) of control fields. Each of these is then applied to an identical *copy* of the system, and the observable measured to determine the fitness of the field. Then a new generation of control fields is computed from a subset of (mostly) well-performing fields using a set of predefined rules for mutations and crossovers, and the experiments are repeated with the new generation of fields until we have arrived at a population of fields with a sufficiently high fitness.

Although there are some disadvantages to this approach (requirement of many identical copies of the system or the ability to efficiently re-initialize the system in the same state after each experiment, need for high experimental duty cycles due to slow convergence, etc.), closed-loop learning techniques have proved useful in the laboratory for a wide range of systems, and incorporating feedback from such closed-loop learning experiments in some form is likely to be essential for quantum control to succeed in the laboratory. A particularly promising avenue may be adaptive system identification strategies based on feedback from closed-loop experiments.

This work was supported by the Cambridge-MIT Institute's Quantum Technology Project and an EPSRC Advanced Fellowship.

References

1. Albertini F, D'Alessandro D (2001) Notions of controllability for quantum-mechanical systems electronic preprint `http://arxiv.org/abs/quant-ph/0106128`
2. D'Alessandro D (2000) Algorithms for quantum control based on decompositions of Lie groups. In: Proceedings of the 39th IEEE Conference on Decision and Control. IEEE New York, pages 1074–1075
3. Judson R S, Rabitz H (1992) Teaching lasers to control molecules. Phys. Rev. Lett. 68: 1500
4. Maday Y, Turinici G (2003) New formulations of monotonically convergent quantum control algorithms. J. Chem. Phys. 118(18): 8191

5. Pearson B J *et al.* (2001) Coherent control using adaptive learning algorithms. Phys. Rev. A 63: 063412
6. Ramakrishna V *et al.* (2000) Quantum control by decompositions of SU(2). Phys. Rev. A 62: 053409
7. Ramakrishna V *et al.* (2000) Explicit generation of unitary transformations in a single atom or molecule. Phys. Rev. A 61: 032106
8. Sa Earp H A, Pachos J K (2005) A constructive algorithm for the cartan decomposition of $SU(2^n)$. J. Math. Phys. 46: 1
9. Schirmer S G *et al.* (2002) Constructive control of quantum systems using factorization of unitary operators. J. Phys. A 35: 8315–8339
10. Schirmer S G, Leahy J V, Solomon A I (2002) Degrees of controllability for quantum systems and applications to atomic systems. J. Phys. A 35: 4125
11. Shore B W (1990) Theory of coherent atomic excitation. John Wiley & Sons, New York
12. Tarn T J, Clark J W, Lucarelli D J (2000) Controllability of quantum-mechanical systems with continuous spectra. In: Proceedings of the 39th IEEE Conference on Decision and Control. IEEE, New York, pages 2803–2809
13. Wiseman H M (1994) Quantum theory of continuous feedback. Phys. Rev. A 49: 2133
14. Yanagisawa M, Kimura H (2003) Transfer function approach to quantum control. IEEE Trans. Autom. Control 48: 2107 and 2121

Stability Analysis of 2-D Object Grasping by a Pair of Robot Fingers with Soft and Hemispherical Ends

Morio Yoshida[1], Suguru Arimoto[2], and Ji-Hun Bae[3]

[1] Department of Robotics, Ritsumeikan University, 1-1-1 Nojihigashi, Kusatsu, Shiga, 525-8577 Japan
rm156950@se.ritsumei.ac.jp
[2] Department of Robotics, Ritsumeikan University, 1-1-1 Nojihigashi, Kusatsu, Shiga, 525-8577 Japan
arimoto@se.ritsumei.ac.jp and also Nagoya, RIKEN
[3] Pohang Institute of Intelligent Robotics, San 31, Hyoja-Dong, Nam-Gu, Pohang, Gyeongbuk 790-784, S. Korea
joseph@postech.ac.kr

1 Introduction

It is said that "human hand" is an agent of the brain. This might attract attention from many prominent robot engineers and researchers who eventually attempted to design multi-fingered robot hands that mimic human hands. In the history of development of multi-fingered robot hands (see the literature [1] $\sim$ [5]), a variety of sophisticated robot hands designed and make are indeed reported. However, most of them have not yet been used widely in practice such as assembly tasks and other automation lines in place of human hands. The most important reason of this must be owing to the high cost of manufacturing such multi-fingered hands with many joints together with expensive sensing devices such as tactile and/or force sensors, which can not redeem human potentials of flexibility and versatility in execution of a variety of tasks. In fact, multi-fingered robot hands were used only in open-loop control(see [2]) and the importance of sensory feedback was not discussed in the literature until around the year of 2000 (see [12]). This paper firstly introduces a mathematical model of full dynamics of planar but vertical motion a rigid object grasped by a pair of two and three d.o.f fingers with soft and deformable tips whose shape is hemispherical. The behavior of the soft finger tips is lumped-parameterized by assuming that the soft material is distributively composed of massless springs with spring constant k (stiffness constant per unit area) and dampers in parallel.

Secondly, motivated from the previous result concerning "blind grasping" in the case of rigid finger-ends (see [11]), a class of control signals is proposed, which can be constructed easily by using only physical parameters of fingers and measurement data on finger joints. It is shown theoretically that such a control

F. Bullo et al. (Eds.): Lag. & Hamil. Methods for Nonlin. Ctrl., LNCIS 366, pp. 305–316, 2007.
springerlink.com

signal renders the closed-loop dynamics asymptotically stable on an equilibrium manifold satisfying the force/torque balance in a dynamic sense. This shows that a pair of robot fingers can grasp a thing securely in a blind manner, that is, without knowing object kinematics or using external sensings such as tactile or visual sensing. Numerical simulation results are also given to verify the theoretical results.

2 Dynamics of PINCHING

Firstly let us derive dynamics of pinch motion by a pair of two and three DOF (Degrees-of-Freedom) fingers with soft tips (see Fig. 1). In this setup, symbols O and O′ denote first joint centers of the left and right fingers respectively, point O also denotes the origin of Cartesian coordinates fixed at the base frame, and $O_{c.m.}$ denotes the center of mass of the object whose position is expressed in terms of $\boldsymbol{x} = (x, y)^{\mathrm{T}}$ of the Cartesian coordinates. Symbols O_1 and O_2 denote centers of

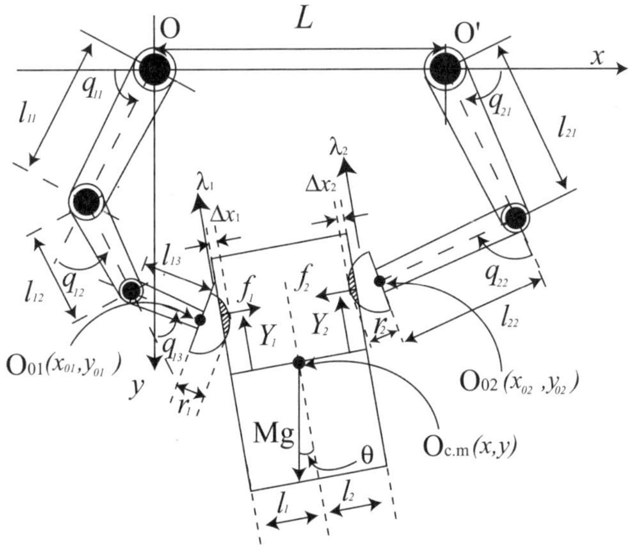

Fig. 1. Two robot fingers pinching an object with parallel flat surfaces under the gravity effect

area contacts whose Cartesian coordinates are described as $\boldsymbol{x}_1 = (x_1, y_1)^{\mathrm{T}}$ and $\boldsymbol{x}_2 = (x_2, y_2)^{\mathrm{T}}$ respectively and O_{01} and O_{02} denote centers of hemispherical soft finger tips which are expressed in terms of Cartesian coordinates as $\boldsymbol{x}_{0i} = (x_{0i}, y_{0i})^{\mathrm{T}}$,(i=1,2) respectively. Next let us denote the Y-component of center O_1 of area-contact of the left finger in terms of Cartesian coordinates (X, Y) fixed

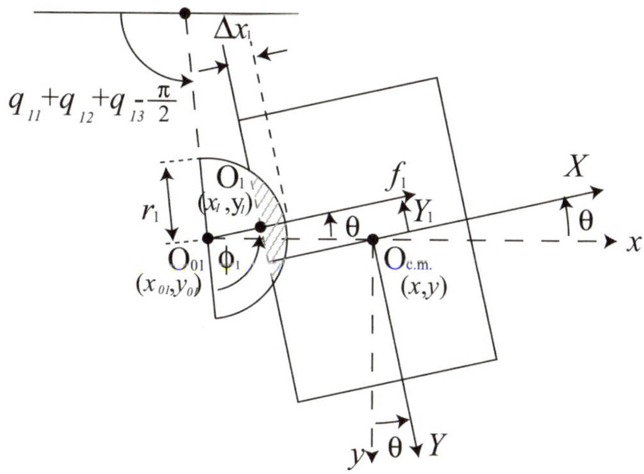

Fig. 2. Definition of physical variables

at the object (see Figs.1 and 2) by Y_1 and that of the right finger by Y_2. Other symbols L, l_i, and l_{ij} are defined in Fig.1. Then, obviously it follows that

$$\Delta x_i = r_i + l_i + (-1)^i (\boldsymbol{x} - \boldsymbol{x}_{0i})^{\mathrm{T}} \boldsymbol{r}_X, \; i = 1, 2 \tag{1}$$

$$\boldsymbol{x}_i = \boldsymbol{x}_{0i} - (-1)^i (r_i - \Delta x_i) \boldsymbol{r}_X, \; i = 1, 2 \tag{2}$$

where $\Delta x_i (i = 1, 2)$ denote the maximum displacements of deformation arised at the centers of area-contact respectively and $\boldsymbol{r}_X = (\cos\theta, -\sin\theta)^{\mathrm{T}}$. Similarly, it follows that

$$\boldsymbol{x} = \boldsymbol{x}_1 + l_1 \boldsymbol{r}_X - Y_1 \boldsymbol{r}_Y = \boldsymbol{x}_2 - l_2 \boldsymbol{r}_X - Y_2 \boldsymbol{r}_Y \tag{3}$$

from which it follows that

$$Y_i = (\boldsymbol{x}_{0i} - \boldsymbol{x})^{\mathrm{T}} \boldsymbol{r}_Y, \; i = 1, 2 \tag{4}$$

where $\boldsymbol{r}_Y = (\sin\theta, \cos\theta)^{\mathrm{T}}$. Since velocities of O_1 (or O_2) (the center of the left (or right) hand contact area, see Fig.2) in terms of finger-end coordinates and in terms of object coordinates $\mathrm{O}_{\mathrm{c.m.}} - XY$ are equal, it follows that

$$-(r_i - \Delta x_i) \frac{\mathrm{d}}{\mathrm{d}t} \left(\frac{3\pi}{2} - (-1)^i \theta - \boldsymbol{q}_i^{\mathrm{T}} \boldsymbol{e}_i \right) = \frac{\mathrm{d}}{\mathrm{d}t} Y_i, \quad i = 1, 2 \tag{5}$$

Equation (5) is of the form of total differentials. Hence, it is reasonable to introduce Lagrange multiplies λ_i in such a way that

$$0 = \lambda_i \left\{ \frac{\mathrm{d}Y_i}{\mathrm{d}t} + (r_i - \Delta x_i) \frac{\mathrm{d}}{\mathrm{d}t} \phi_i \right\}, \qquad i = 1, 2 \tag{6}$$

$$\phi_i = \frac{3}{2}\pi - (-1)^i \theta - \boldsymbol{q}_i^{\mathrm{T}} \boldsymbol{e}_i, \qquad i = 1, 2 \tag{7}$$

where c_1 and c_2 are a constant of integration.

According to the lumped-parametrization of contact forces caused by defomation of finger-tip material (see[6] and Appendix A), the reproducing force $\bar{f}_i(\Delta x_i)$ arising in the direction normal to the object surfaces at the center $\mathrm{O_i}$ of contact area is characterized as

$$\bar{f}_i(\Delta x_i) = k_i \Delta x_i^2, \qquad i = 1, 2 \tag{8}$$

with stiffness parameter $k_i > 0[\mathrm{N/m^2}]$. Furthermore, we assume that lumped parametrized viscous forces also arise from distributed viscosity of the finger-tip material, which are accompanied with reproducing forces in such a way that

$$f_i\left(\Delta x_i, \Delta \dot{x}_i\right) = \bar{f}_i\left(\Delta x_i\right) + \xi_i\left(\Delta x_i\right)\Delta \dot{x}_i, \qquad i = 1, 2 \tag{9}$$

where $\xi_i(\Delta x_i)$ is a positive scalar function increasing with increase of Δx_i. Then the total potential energy of reproducing forces and the total kinetic energy can be given as

$$P = P_1 + P_2 - Mgy + \sum_{i=1}^{2} \int_0^{\Delta x_i} \bar{f}_i(\xi)\mathrm{d}\xi \tag{10}$$

$$K = \frac{1}{2}\left\{ \sum_{i=1,2} \dot{\boldsymbol{q}}_i^{\mathrm{T}} H_i(\boldsymbol{q}_i)\dot{\boldsymbol{q}}_i + M \parallel \dot{\boldsymbol{x}} \parallel^2 + I\dot{\theta}^2 \right\} \tag{11}$$

where $H_i(i = 1, 2)$ denote the inertia matrices of the fingers $i = 1, 2$, I the inertia moment of the object, M the mass of the object, P_i the potential energy for finger i, and $-Mgy$ denotes that of the object. Finally, the Lagrange equation of motion of the overall system can be derived by applying Hamilton's principle described as

$$\int_{t_0}^{t_1}\left[\delta\left(K-P\right) - \sum_{i=1,2} \frac{\partial \frac{1}{2}\left\{\xi_i(\Delta x_i)\Delta \dot{x}_i^2\right\}}{\partial \Delta \dot{x}_i}\delta \Delta x_i + \sum_{i=1,2} \boldsymbol{u}_i^{\mathrm{T}}\delta \boldsymbol{q}_i\right]\mathrm{d}t$$
$$= -\sum_{i=1,2} \lambda_i \left\{\frac{\partial Y_i}{\partial \boldsymbol{z}} + (r_i - \Delta x_i)\frac{\partial \phi_i}{\partial \boldsymbol{z}}\right\} \tag{12}$$

where $\boldsymbol{z} = (\boldsymbol{q}_1^{\mathrm{T}}, \boldsymbol{q}_2^{\mathrm{T}}, x, y, \theta)^{\mathrm{T}}$, which results in

$$H_i(\boldsymbol{q}_i)\ddot{\boldsymbol{q}}_i + \left(\frac{1}{2}\dot{H}_i + S_i\right)\dot{\boldsymbol{q}}_i - (-1)^i f_i J_{0i}^{\mathrm{T}} \boldsymbol{r}_X$$
$$+\lambda_i\left\{(r_i - \Delta x_i)\,\boldsymbol{e}_i - J_{0i}^{\mathrm{T}}\boldsymbol{r}_Y\right\} + g_i(\boldsymbol{q}_i) = \boldsymbol{u}_i \tag{13}$$

$$M\ddot{\boldsymbol{x}} - (\boldsymbol{r}_X, \boldsymbol{r}_Y)\left(f_1 - f_2, -\lambda_1 - \lambda_2\right)^{\mathrm{T}} - (0, Mg)^{\mathrm{T}} = \boldsymbol{0} \tag{14}$$

$$I\ddot{\theta} - f_1 Y_1 + f_2 Y_2 + l_1\lambda_1 - l_2\lambda_2 = 0 \tag{15}$$

where $J_{0i}^{\mathrm{T}} = \partial \boldsymbol{x}_{0i}^{\mathrm{T}}/\partial \boldsymbol{q}_i$. It is obvious that the input-ouput pair $\{\boldsymbol{u} = (\boldsymbol{u}_1, \boldsymbol{u}_2)^{\mathrm{T}}, \dot{\boldsymbol{q}} = (\dot{\boldsymbol{q}}_1, \dot{\boldsymbol{q}}_2)^{\mathrm{T}}\}$ concerning the dynamics of eqs.(13) to (15) satisfies the equation

$$\int_0^t (\dot{\boldsymbol{q}}_1^{\mathrm{T}}\boldsymbol{u}_1 + \dot{\boldsymbol{q}}_2^{\mathrm{T}}\boldsymbol{u}_2)\mathrm{d}\tau = E(t) - E(0) - \int_0^t \sum_{i=1,2} \xi\left(\Delta x_i(\tau)\right)\Delta \dot{x}_i^2(\tau)\mathrm{d}\tau \tag{16}$$

where $E = K + P$.

3 Control Signals for Blind Grasping

Motivated from the analysis in the case of rigid rolling contacts between rigid finger-ends and a rigid object (see [12]), we propose the following control signal that should exert torques on finger joints:

$$\boldsymbol{u}_i = \boldsymbol{g}_i(\boldsymbol{q}_i) - c_i\dot{\boldsymbol{q}}_i + (-1)^i \frac{f_d}{r_1+r_2} J_{0i}^{\mathrm{T}} \begin{pmatrix} x_{01} - x_{02} \\ y_{01} - y_{02} \end{pmatrix} - \frac{\hat{M}g}{2}\left(\frac{\partial y_{0i}}{\partial \boldsymbol{q}_i}\right) - r_i \hat{N}_i \boldsymbol{e}_i, \qquad i = 1,2 \tag{17}$$

where

$$\hat{M} = \hat{M}(0) + \int_0^t \frac{g\gamma_M^{-1}}{2} \sum_{i=1,2} \left(\frac{\partial y_{0i}}{\partial \boldsymbol{q}_i}\right)^{\mathrm{T}} \dot{\boldsymbol{q}}_i \, \mathrm{d}\tau = \hat{M}(0) + \frac{g\gamma_M^{-1}}{2}\left(y_{01}(t) + y_{02}(t) - y_{01}(0) - y_{02}(0)\right) \tag{18}$$

$$\hat{N}_i = \gamma_{N_i}^{-1} \int_0^t \left(r_i \boldsymbol{e}_i^{\mathrm{T}} \dot{\boldsymbol{q}}_i\right) \mathrm{d}\tau = \gamma_{N_i}^{-1} r_i \boldsymbol{e}_i^{\mathrm{T}} \left(\boldsymbol{q}_i(t) - \boldsymbol{q}_i(0)\right) \ (i = 1,2) \tag{19}$$

and γ_M and $\gamma_{Ni} (i = 1,2)$ are positive constants.

4 Theoretical Proof of Feasibility of Blind Grasping

First, define

$$\begin{cases} \Delta f_i = f_i + (-1)^i \frac{Mg}{2} \sin\theta + \frac{f_d}{r_1+r_2}\left(\boldsymbol{x}_{01} - \boldsymbol{x}_{02}\right)^{\mathrm{T}} \boldsymbol{r}_X \\ \Delta\lambda_i = \lambda_i - \frac{Mg}{2}\cos\theta + (-1)^i \frac{f_d}{r_1+r_2}\left(\boldsymbol{x}_{01} - \boldsymbol{x}_{02}\right)^{\mathrm{T}} \boldsymbol{r}_Y \\ N_i = (-1)^i \frac{f_d}{r_1+r_2}\left(\boldsymbol{x}_{01} - \boldsymbol{x}_{02}\right)^{\mathrm{T}} \boldsymbol{r}_Y - \frac{Mg}{2}\cos\theta \end{cases} \tag{20}$$

$$\begin{cases} S = -f_d\left(1 - \frac{\Delta x_1 + \Delta x_2}{r_1+r_2}\right)(Y_1 - Y_2) - \frac{Mg}{2}N \\ N = (Y_1 + Y_2)\sin\theta - (l_1 - l_2)\cos\theta \end{cases} \tag{21}$$

Note that from eq.(4) and eq.(1)

$$\begin{cases} \left(\boldsymbol{x}_{01} - \boldsymbol{x}_{02}\right)^{\mathrm{T}} \boldsymbol{r}_Y = Y_1 - Y_2 \\ -\left(\boldsymbol{x}_{01} - \boldsymbol{x}_{02}\right)^{\mathrm{T}} \boldsymbol{r}_X = l_1 + l_2 + r_1 + r_2 - (\Delta x_1 + \Delta x_2) \end{cases} \tag{22}$$

Next define

$$f_0 = f_d\left\{1 + \frac{l_1 + l_2 - \Delta x_1 - \Delta x_2}{r_1 + r_2}\right\} \tag{23}$$

Differently from the case of rigid finger-ends [11], f_0 is not a constant but dependent on the magnitude of $\Delta x_1 + \Delta x_2$. Nevertheless, it is possible to find $\Delta x_{di} (i = 1,2)$ for a given $f_d > 0$ so that they satisfy

$$\bar{f}_i(\Delta x_{di}) = \left(1 + \frac{l_1 + l_2 - \Delta x_{d1} - \Delta x_{d2}}{r_1 + r_2}\right) f_d, \qquad i = 1, 2 \tag{24}$$

because $\bar{f}_i(\Delta x)$ is of the form of $\bar{f}_i(\Delta x) = k_i \Delta x^2$ [6]. Then, by substituting eq.(16) into eq.(12) and referring to eqs.(19) to (21), we obtain the closed-loop dynamics of the overall fingers-object system in the following way:

$$H_i(\boldsymbol{q}_i)\ddot{\boldsymbol{q}}_i + \left(\frac{1}{2}\dot{H}_i + S_i\right)\dot{\boldsymbol{q}}_i - (-1)^i \Delta f_i J_{0i}^{\mathrm{T}} \boldsymbol{r}_X$$

$$-\Delta\lambda_i \boldsymbol{r}_{\lambda i} - \Delta M g \frac{\partial y_{0i}}{\partial \boldsymbol{q}_i} - r_i \Delta N_i \boldsymbol{e}_i = 0 \tag{25}$$

$$M\ddot{\boldsymbol{x}} - (\Delta f_1 - \Delta f_2)\boldsymbol{r}_X - (\Delta\lambda_1 + \Delta\lambda_2)\boldsymbol{r}_Y = 0 \tag{26}$$

$$I\ddot{\theta} - \Delta f_1 Y_1 + \Delta f_2 Y_2 + l_1 \Delta\lambda_1 - l_2 \Delta\lambda_2 + S = 0 \tag{27}$$

where $\Delta N_i = \hat{N}_i - (1 - \Delta x_i / r_i) N_i (i = 1, 2)$ and

$$\boldsymbol{r}_{\lambda i} = -\left\{(r_i - \Delta x_i)\boldsymbol{e}_i - J_{0i}^{\mathrm{T}} \boldsymbol{r}_Y\right\} \tag{28}$$

Then, it is important to note that along a solution to the equations of (25) to (27) under the constraints of eq.(5) the following energy relation is satisfied:

$$\frac{\mathrm{d}}{\mathrm{d}t} W = \sum_{i=1,2} -\left\{c_i \|\dot{\boldsymbol{q}}_i\|^2 + \xi(\Delta x_i)\Delta \dot{x}_i^2\right\} \tag{29}$$

where

$$W = K + \Delta P + \frac{f_d}{2(r_1 + r_2)}(Y_1 - Y_2)^2 + \frac{\gamma_M}{2}\Delta M^2$$

$$+ \sum_{i=1,2} \frac{\gamma_{Ni}}{2}\hat{N}_i^2 + \frac{Mg}{2}\left\{(y_{01} + y_{02} - 2y)\right\} \tag{30}$$

$$\frac{y_{01} + y_{02}}{2} - y = \frac{Y_1 + Y_2}{2}\cos\theta - \frac{1}{2}\left\{(l_1 - l_2)\right.$$

$$\left. + (r_1 - r_2) - (\Delta x_1 - \Delta x_2)\right\}\sin\theta \tag{31}$$

$$\Delta P = \sum_{i=1,2} \int_0^{\delta x_i} \left\{\bar{f}_i(\Delta x_{di} + \xi) - \bar{f}_i(\Delta x_{di})\right\} \mathrm{d}\xi \tag{32}$$

where $\delta x_i = \Delta x_i - \Delta x_{di}$. Now, it is convenient to define

$$\Delta\boldsymbol{\lambda} = \left(\Delta\bar{f}_1, \Delta\bar{f}_2, \Delta\lambda_1, \Delta\lambda_2, \frac{\Delta M}{2} g, \Delta N_1, \Delta N_2, \frac{S}{r_3}\right)^{\mathrm{T}} \tag{33}$$

$$A = \begin{bmatrix} -J_1^{\mathrm{T}} \boldsymbol{r}_X & 0 & \boldsymbol{r}_{\lambda 1} & 0 \\ 0 & J_2^{\mathrm{T}} \boldsymbol{r}_X & 0 & \boldsymbol{r}_{\lambda 2} \\ r\cos\theta & -r\cos\theta & -r\sin\theta & -r\sin\theta \\ -r\sin\theta & r\sin\theta & -r\cos\theta & -r\cos\theta \\ Y_1 & -Y_2 & -l_1 & l_2 \end{bmatrix}, \quad D = \begin{bmatrix} -\frac{\partial y_{01}}{\partial \boldsymbol{q}_1} & -r_1 \boldsymbol{e}_1 & 0 & 0 \\ -\frac{\partial y_{02}}{\partial \boldsymbol{q}_2} & 0 & -r_2 \boldsymbol{e}_2 & 0 \\ 0 & 0 & 0 & 0 \\ 0 & 0 & 0 & 0 \\ 0 & 0 & 0 & r_3 \end{bmatrix} \tag{34}$$

where $\Delta \bar{f}_i = \Delta f_i - \xi_i(\Delta x_i)\Delta \dot{x}_i$. Then, the closed-loop dynamics of (25), (26), and (27) can be expressed in the following unified matrix-vector form:

$$H\ddot{\bar{z}} + \left(\frac{1}{2}\dot{H} + S\right)\dot{\bar{z}} + C\dot{\bar{z}} - [A, D]\Delta\boldsymbol{\lambda} + \sum_{i=1,2} \xi_i(\Delta x_i)\Delta \dot{x}_i \left(\frac{\partial \Delta x_i}{\partial \bar{\boldsymbol{z}}}\right) = 0 \tag{35}$$

where $\bar{\boldsymbol{z}} = \left(\boldsymbol{q}_1^{\mathrm{T}}, \boldsymbol{q}_2^{\mathrm{T}}, r^{-1}\boldsymbol{x}^{\mathrm{T}}, \theta\right)^{\mathrm{T}}$,

$$\begin{cases} H = \mathrm{diag}(H_1, H_2, r^2 M I_2, I), \ S = \mathrm{diag}(S_1, S_2, 0, 0, 0) \\ C = \mathrm{diag}(c_1 I_3, c_2 I_2, 0, 0, 0) \end{cases} \tag{36}$$

a positive scale factor r is introduced to balance numerical values of coefficients among motion equations in terms of $\dot{\boldsymbol{x}}$ with the physical unit of force [N] and rotational motion equations of $\dot{\boldsymbol{q}}_i$ and $\dot{\theta}$ with the unit of torque [Nm], and $r_3 > 0$ is also an appropriate scale factor. Next, define

$$p_1 = \sum_{j=1}^{3} q_{ij}, \quad p_2 = \sum_{j=1}^{2} q_{ij} \tag{37}$$

and note that $(Y_1 - Y_2)^2$ and $\hat{N}_i^2$ are quadratic in θ, p_1, and p_2. It follows from the definition of ΔP that ΔP is a positive definite function in δx_1 and δx_2. Hence, it is easy to check that W has a minimum W_m under the constraints of eq.(5). This means that $\|\dot{\bar{z}}\|$ is bounded and thereby it is possible to show that $\|\Delta\boldsymbol{\lambda}\|$ is bounded from eq.(35) and constraints of eq.(5). Thus, $\ddot{\bar{z}}$ becomes bounded and thereby $\dot{\bar{z}}$ becomes uniformly continuous in t. Since $\dot{\boldsymbol{q}}_i(t)(i = 1, 2)$ and $\Delta \dot{x}_i$ are in $L^2(0, \infty)$ from (29), Barbalat's lemma implies that $\dot{\boldsymbol{q}}_i(t) \to 0$ and $\Delta \dot{x}_i(t) \to 0$ as $t \to \infty$, which means that $\dot{\theta}(t) \to 0$ as $t \to \infty$ from constraints of eq.(5). Since the matrix $[A, D]$ is of an 8×8 squared matrix and nonsingular as easily checked, it follows that $\Delta\boldsymbol{\lambda}(t) \to 0$ as $t \to \infty$. Thus, as $t \to \infty$ the force/torque balance is established in a dynamic sense.

The proof presented above has been rather sketchy owing to limitation of given pages, but it can be ascertained by carrying out numerical simulations.

Further, it should be remarked that dynamics of the overall fingers-object system depicted in Fig.1 is redundant in degrees-of-freedom. In fact, the dimension of the generalized position coordinates is eight and there are two holonomic constraints concerning rolling. Therefore, the total d.o.f of the overall system is six. Then, the force/torque balance is realized through specification of physical values of Δx_1, Δx_2, λ_1, λ_2, and the magnitude of $y - (y_{01} + y_{02})/2$. Thus, one d.o.f is redundant. Actually, blind grasping can be realized when each robot finger has the two d.o.f, though in this case $\Delta\boldsymbol{\lambda}(t)$ in (35) converges to some non-zero constant vector $\Delta\boldsymbol{\lambda}_\infty$ as $t \to \infty$. The details of the discussions including exponential convergence of $\Delta\boldsymbol{\lambda}(t)$ to zero (in a redundant case) or to some constant $\Delta\boldsymbol{\lambda}_\infty$ (in the non-redundant case) as $t \to \infty$ must be omitted in this paper due to the page limitation.

5 Numerical Simulation of 2-D Case

We carried out computer simulation for an object with non-parallel flat surfaces as shown in Fig.1. Control signals as defined in eq.(17) are applied to dynamics of the overall finger-object systemthat are expressed as Lagrange's equation of motion given in eqs.(13) to (15). In the simulation, the constraints of eq.(5) can be ensured by using Baumgarte's method called the CSM (Constraint Stabilization Method). Physical parameters of the fingers-object system are given in Table 1 and physical gains in control signals are given in Talbe 2. Transient responses of principal physical variables appearing in the closed-loop dynamics are shown in Fig.3 from (a) to (j). It is seen from Fig.3 that except $Y_1 - Y_2$ and θ all magnitudes of constraint forces f_i and $\lambda_i (i = 1, 2)$ converge to their correspponding target values respectively, that is, $\Delta \bar{f}_i$ and $\Delta\lambda_i$ converge to zero as $t \to \infty$. Other variables ΔM, $\Delta N_i (i = 1, 2)$, and S converge to zero as $t \to \infty$, too. It should be also noted that $Y_1 - Y_2$ and θ also converge to some constant values as seen from Fig.3 (a) and (b). At the initial position, we set $\Delta x_1 = 0$ and $\Delta x_2 = 0$ and therefore $f_1 = 0$ and $f_2 = 0$. As seen from (c) and(d) of Fig.3, $\Delta \bar{f}_i (i = 1, 2)$ are around -2.5[N] at $t = 0$ because of $f_1 = f_2 = 0$ at $t = 0$. However, once $f_i > 0$ just after $t > 0$, $f_i(\Delta x_i, \Delta \dot{x}_i)(i = 1, 2)$ are kept to be positive forever, which means that contacts between finger-ends and the object are maintained throughout movements of the overall system.

Table 1. Physical parameters

link length	length [m]	link mass	weight [kg]	link inertia moment	inertia moment [kgm^2]
$l_{11} = l_{21}$	0.065	$m_{11} = m_{21}$	0.045	$I_{11} = I_{21}$	1.584×10^{-5}
l_{12}	0.039	m_{12}	0.025	I_{12}	3.169×10^{-6}
l_{13}	0.026	m_{13}	0.015	I_{13}	8.450×10^{-7}
l_{22}	0.065	m_{22}	0.040	I_{22}	1.408×10^{-5}
radius of fingertip	length [m]	base length	length [m]	object width	length [m]
$r_1 = r_2$	0.010	L	0.063	$l_1 = l_2$	0.0015
object height	length [m]	object inertia moment	inertia moment [m]	stiffness of fingertip	stiffness [N/m^2]
h	0.050	I	1.333×10^{-5}	$k_1 = k_2$	3.000×10^{5}
viscosity of fingertip			viscosity [Ns/m^2]		
$c_{\Delta 1} = c_{\Delta 2}$			1000		

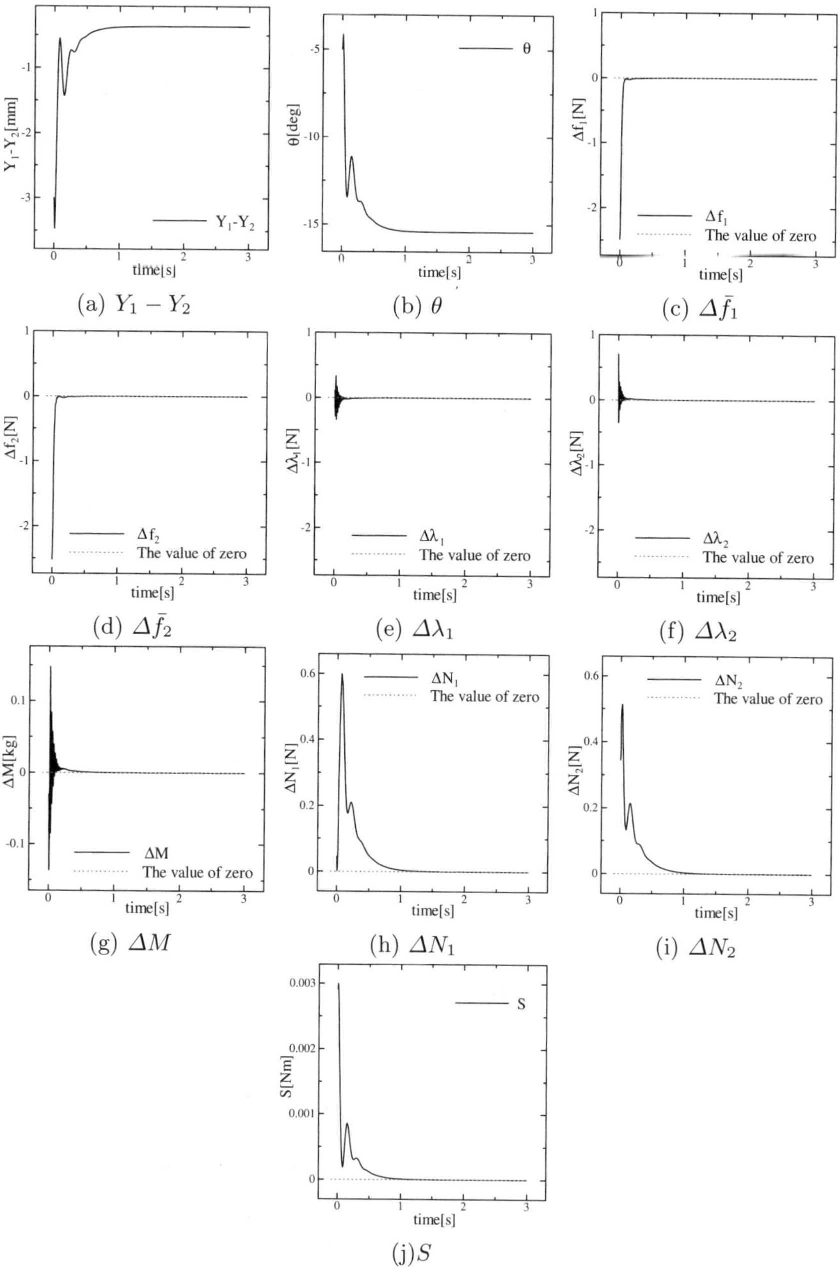

(a) $Y_1 - Y_2$ (b) θ (c) $\Delta\bar{f}_1$

(d) $\Delta\bar{f}_2$ (e) $\Delta\lambda_1$ (f) $\Delta\lambda_2$

(g) ΔM (h) ΔN_1 (i) ΔN_2

(j)S

Fig. 3. Transient responses of physical variables

Table 2. Parameters of Control Signals & Initial Value of Estimator

internal force	force [N]	damping coefficient	coefficient [Nms/rad]
f_d	1.0	$c_1 = c_2$	0.003
regressor gain	gain [mrad/N]	regressor gain	gain [m^2/kgs^2]
γ_M	0.01	$\gamma_{N1} = \gamma_{N2}$	0.001
initial value of object estimate		mass [kg]	
$M(0)$		0.010	

6 Conclusions

This paper shows that there exists a class of control signals constructed from using only finger kincmatics and measurement data of finger joint angles, which enable a pair of multi-d.o.f fingers with soft finger-tips to grasp a rigid object securely and manipulate it towards an equilibrium state of force/torque balance. This shows that even a pair of robot fingers can grasp an object securely in a blind manner like human grasp it even if they close their eyes. A sketchy proof of convergence of solution trajectories of the closed-loop dynamics toward an equilibrium state satisfying the force/torque balance is given and computer simulation results verify the validity of the theoretical prediction.

References

1. M. R. Cutkovsky (1985) Robotic Grasping and Fine Manipulation. Kluwer Academic, Dordrecht, Netherlands,
2. K. B. Shimoga (1996) Robot grasp synthesis algorithms: A survey. Int. J. of Robotics Research,15,3:230–266
3. J. D. Crisman, C. Kanojia, I. Zeid (1996) Graspar: A flexible, easily controllable robotic hand. IEEE Robotics and Automation Magazine,June:32–38
4. A. M. Okamura, N. Smaby, M. R. Cutkovsky (2000) An overview of dexterous manipulation. Proc. of the 2000 IEEE Robotics and Automation, SanFrancisco, CA, April 25-28:255–262
5. R. M. Murray, Z. Li, S. S. Sastry (1994) A Mathematical Introduction to Robotic Manipulation. CRC Press, Boca Raton and Tokyo
6. S. Arimoto, P.T.A. Nguyen, H.-Y. Han, Z. Doulgeri (2000) Dynamics and control of a set of dual fingers with soft tips. Robotica,18,1:71–80
7. K. Tahara, M. Yamaguchi, P.T.A. Nguyen, H.-Y. Han, S. Arimoto (2000) Stable grasping and posture control for a pair of robot fingers with soft tips. Proc. of the Int. Conf. on Machine Automation(ICMA2000), Osaka, Japan, Sept. 27-29:33–38
8. J. P. LaSalle (1960) Some extensions of Lyapunov's second method. IRE Trans. on Circuit Theory,7:520–527
9. S. Arimoto, K. Tahara, M. Ymaguchi, P.T.A Nguyen, H.-Y. Han (2001) Principle of superposition for controlling pinch motions by means of robot fingers with soft tips. Robotica,19,1:21–28
10. S. Arimoto, P.T.A. Nguyen (2001) Principle of superposition for realizing dexterous pinching motion of a pair of robot fingers with soft-tips. IEICE Trans. on Fundamentals,E84-A,1:39–47

11. S. Arimoto, R. Ozawa, M. Yoshida (2005) Two-dimensional stable blind grasping under the gravity effect. Proc. of the 2005 IEEE Int. Conf. on Robotics and Automation:1208–1214
12. S. Arimoto (2004) Intelligent control of multi-fingered hands. Annual Review in Control,28,1,75–85
13. D.J. Montana (1992) Contact stability for Two-Fingered Grasps. IEEE Trans. on Robotics and Automation,8,4:421–430

Appendix A

As shown in Fig.A, a narrow strip with width $r\mathrm{d}\theta$ and radius $r\sin\theta$ in the contact area produces a reproducing force in the direction to the center O_{0i} of the curvature of the hemisphere with the magnitude

$$k(2\pi r\sin\theta)\mathrm{d}\theta \times r(\cos\theta - \cos\theta_0)\cos\theta \tag{A-1}$$

where k denotes the stiffness parameter of the soft and defomable material per unit area, $(2\pi r^2\sin\theta)\mathrm{d}\theta$ is the area of the narrow circular strip as shown in Fig.A, and

$$\begin{aligned}\frac{r(\cos\theta - \cos\theta)}{\cos\theta} &= \left(r - \frac{r - \Delta x}{\cos\theta}\right) \\ &= r - \frac{r\cos\theta_0}{\cos\theta}\end{aligned} \tag{A-2}$$

denotes the length of deformation at angle θ. Since the total reproducing force with magnitude (A-1) generated from the narrow circular strip contributes to the direction Δx (the arrow denoted by f in Fig.A) by $\cos\theta$, the total reproducing force can be expressed as the integral

$$\begin{aligned}\bar{f} &= \int_0^{\theta_0} 2\pi k r^2 \sin\theta(\cos\theta - \cos\theta_0)\mathrm{d}\theta \\ &= \pi k r^2(1 - \cos\theta_0)^2 = \pi k \Delta x^2\end{aligned} \tag{A-3}$$

This means that the reproducing force produced by the deformed area can be approximately expressed by an increasing function of Δx (the maximum length of displacement). It should be noted that the moment $M = (M_x, M_y, M_z)^{\mathrm{T}}$ around O_{0i} becomes

$$\overrightarrow{O_{0i}P} \times k\overrightarrow{PQ}\mathrm{d}S \tag{A-4}$$

where $\mathrm{d}S = r^2\sin\theta\mathrm{d}\theta\mathrm{d}\phi$ (which denotes an infinitesimally small area shown in Fig.8). It is evident to see that the integral of eq.(A-4) over ϕ in$[0, 2\pi]$ times θ in $[0, \theta_0]$ vanishes due to the symmetry of sinusoidal functions $\sin\phi$ and $\cos\phi$ appearing in eq.(A-4). Thus, the moment acting around the point O_{0i} caused by overall deformations of the soft material becomes zero.

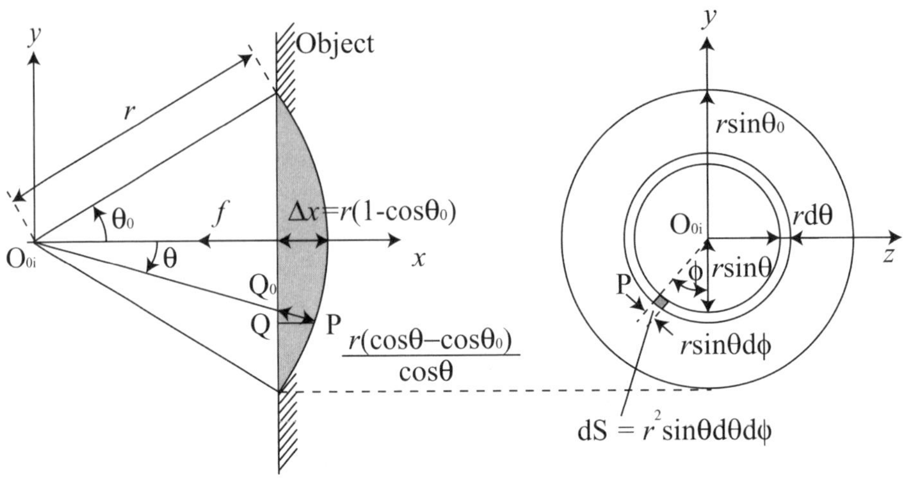

Fig. A. Geometric relations related to lumped-parametrization of finger-tip stress distribution into a single reproducing force focused on the center of curvature O_{0i}

As to the lumped-parametrization of distributed viscous forces, a similar argument can be applied. It can be concluded that the viscous force acting to the point O_{0i} can be expressed in the form of $\xi_i(\Delta x_i)\Delta\dot{x}_i$, where $\xi_i(\Delta x_i)$ is an increasing function of Δx_i.

Intrinsic Control Designs for Abstract Machines

Robert Fuentes[1], Gregory Hicks[2], and Jason Osborne[3]

[1] Raytheon Missile Systems, 1151 E. Hermans Road, Bldg. 805, Tucson, AZ 85706
rjfuentes@ieee.org

[2] General Dynamics Advanced Information Systems, 14700 Lee Road, Chantilly, VA 20151
gregory.hicks@gd-ais.com

[3] North Carolina State University, 245 Harrelson Hall, Raleigh, NC 27695
jmosborn@ncsu.edu

Summary. A geodesic-based formulation of geometric PD tracking control for fully actuated mechanical systems is derived from the Riemannian metric. The region of stability is determined directly from the magnitude of the system's injectivity radius and, for a restricted set of control problems, the locus of cut points about a desired reference point in the manifold. Exponential stability is proven for controlled motion along a geodesic, yielding a particularly simple, yet elegant, methodology for control design.

1 Introduction

Many recent advances in control theory have evolved from the machinery of differential geometry and topology. Specifically, regarding the theory of control design as addressed in [1, 2, 3], it has clearly been demonstrated that Riemannian geometry can be employed to *naturally* approach a variety of control design problems and obtain a unique perspective on commonly used control logic. Of particular interest to us is the notion of the geodesic spring introduced by Bullo, Murray, and Sarti in [1].

In [1], F. Bullo et.al. introduce a geodesic based proportional–derivative (PD) fixed point tracking controller on S^2. The generalization of this design philosophy, completed in [2], creates a geometric tracking control law rooted in Riemannian geometry. The resulting control architecture is shown to hinge upon inexorably linked geometric notions of configuration and velocity error that ensure closed loop stability.

As the geodesic spring paradigm of [1] seemed to be the original impetus for [2], it would be natural and thus highly desirable to return to a geodesic setting for tracking on Riemannian manifolds. Indeed, that is the purpose of this writing. We are particularly interested in demonstrating how a machine's intrinsic geometry (i.e. the geometry induced by the kinetic energy tensor) can be exploited within the general framework of [1]. In particular, we demonstrate

F. Bullo et al. (Eds.): Lag. & Hamil. Methods for Nonlin. Ctrl., LNCIS 366, pp. 317–328, 2007.
springerlink.com

that free motion is the glue that binds geometrically natural representations of configuration and velocity error. As a result, the ensuing control design explicitly realizes the aforementioned paradigm.

This paper is organized as follows. Section 2.1 makes note of the needed background material and notational conventions. The control system and a glimpse of the framework introduced in [2] is summarized in Section 2.2. The main contributions are provided in Section 3. In particular we prove, as a corollary to the main control result of Bullo and Murray, an exponential stability result that makes use of intrinsic geometric implements. In addition, the conditions under which our control law results in geodesic spring-damper dynamics are explained.

2 Background

2.1 Intrinsic Geometry

Assuming a knowledge of intrinsic geometry, we begin by setting the notation for the requisite notions from that subject. Specific details regarding the subject matter in this brief exposition can be found in [4].

- Q a connected, complete Riemannian manifold with metric, g
- A smooth curve σ in Q on $[a, b]$ is called a segment connecting $q = \sigma(b)$ and $r = \sigma(a)$ with $\mathcal{S}(r, q)$ the set of all segments connecting q and r.
- The length of a segment is

$$l^{\sigma}(r, q) = \int_a^b \sqrt{g\left(\frac{\mathrm{d}\sigma}{\mathrm{d}u}, \frac{\mathrm{d}\sigma}{\mathrm{d}u}\right)}\, du$$

- The Riemannian manifold's intrinsic distance is defined by

$$\mathrm{dist}(r, q) = \inf_{\sigma \in \mathcal{S}(q, r)} l^{\sigma}(r, q)\ .$$

- Let ∇ be the Levi-Civita connection with $\frac{D}{du}$ the covariant derivative, wrt ∇, of vector fields on smooth curves; The geodesic equations are given by $\frac{D}{du}\gamma' = 0$
- $\exp : TQ \to Q$ the exponential mapping
- For q in a normal neighborhood of $r \in Q$,

$$\mathrm{dist}(r, q) = l^{\gamma}(r, q) \tag{1}$$

  where γ is the unique geodesic from r to q
- $\sigma_{r \to q} : T_r Q \to T_q Q$ the point–wise, smooth linear parallel transport mapping defined by

$$\nabla_{\sigma'} V = 0 \tag{2}$$

- $C(r)$ the cut-locus of $r \in Q$
- The injectivity radius of r, $\mathrm{Inj}(r)$ and the injectivity radius of Q, $\mathrm{Inj}(Q)$:

$$\mathrm{Inj}(r) = \inf_{q \in C(r)} \mathrm{dist}(r, q) \ \text{ and } \ \mathrm{Inj}(Q) = \inf_{r \in Q} \mathrm{Inj}(r)$$

- If f is a mapping between manifolds then denote $f_y(x) \triangleq f(x,y)|_{y \text{ fixed}}$ and $f_x(y) \triangleq f(x,y)|_{x \text{ fixed}}$

2.2 Dynamics and Control

Let us begin here by describing the potential-free (post compensated), fully actuated, simple, mechanical, control system or more simply, an *abstract machine.* An abstract machine is a mathematical construct of the form $[Q, g, \mathcal{F}]$, where *i)* Q is a smooth n–dimensional manifold called the configuration manifold and *ii)* g is a Riemannian metric on Q and *iii)* $\mathcal{F}$ is a full rank control co-distribution. The abstract machine dynamics are given by

$$g^\flat \frac{D\dot{\gamma}}{dt} = u_i \mathcal{F}_i \,. \tag{3}$$

We now highlight the control design framework of [2, 3], beginning with the generalized concept of position error.

Definition 1. [Configuration Error] *Let P be a subbundle of $Q \times Q$ with connected fibre F such that $\pi(P) = Q$ and* $\dim(F) = \dim(Q)$. *A smooth function $\varphi\colon P \to \mathbb{R}$ is a* Configuration Error Function *if for each fixed $r \in Q$, $\varphi_r(q)$ is smooth, symmetric, proper, and bounded from below and φ satisfies i) $\varphi(r,r) = 0$ and ii)* $\mathrm{d}\varphi_r(q)|_{q=r} = 0$ *and iii)* $\mathrm{Hess}\, \varphi_r(q)|_{q=r}$ *is positive definite.*

As shown in [3], one can define the *L-sublevel set of* φ_r as $\varphi_r^{-1}(\leq L) = \varphi^{-1}((-\infty, L])$ and the connected component of the L-sublevel set $\varphi_r^{-1}(\leq L)$ containing $x_0 \in Q$ as $\varphi_r^{-1}(\leq L, x_0)$. Let

$$L_{\mathrm{reg}}(\varphi_r(q), r) \triangleq \sup \left\{ L \in \mathbb{R} \,|q \in \varphi_r^{-1} (\leq L, r) \setminus \{r\} \Rightarrow \mathrm{d}\varphi_r(q) \neq 0 \right\}$$
$$L_{\mathrm{reg}}(\varphi, Q) \triangleq \inf \left\{ L_{\mathrm{reg}}(\varphi_r(q), r) \mid r \in Q \right\}$$

A configuration error function φ is *uniformly quadratic* if $L_{\mathrm{reg}}(\varphi, Q) > 0$ and if, for all $L \in (0, L_{\mathrm{reg}}(\varphi, Q))$ and $r \in Q$, there exist two constants $0 < a \leq b$ such that for all $q \in \varphi_r^{-1}(B_L(0)) \backslash \{r\}$, we have

$$0 < a \, \|\mathrm{d}\varphi_r(q)\|_g^2 \leq \varphi(r,q) \leq b \, \|\mathrm{d}\varphi_r(q)\|_g^2 \tag{4}$$

where $\|\cdot\|_g$ denotes the induced or operator norm w.r.t. the norm associated to the inner product g.

Next we need to describe the generalized notion of velocity error. This requires the concept of *transport map* $\mathcal{T}$. A transport map is simply a construct through which one can compare velocities which reside in different fibers of the tangent bundle TQ. Thus, given a curve $(r(t), q(t)) \in P$ one may define a velocity error in T_qQ by

$$\dot{e} = \dot{q} - \mathcal{T}_{r \to q} \dot{r}$$

where $\mathcal{T}_{r \to q}$ denotes the evaluation of $\mathcal{T}$ at $p = (r, q)$. The formal definition of a transport mapping is as follows.

Definition 2. [TRANSPORT MAP] *A* Transport Mapping *is a smooth bitensor field*[1] $\mathcal{T}$ *on the previously defined subbundle* P *of* $Q \times Q$ *satisfying: i)* $\mathcal{T}_{r\to q} \in \mathrm{GL}(T_rQ, T_qQ)$ *and ii)* $\mathcal{T}_{r\to r} = \mathrm{id}$.

A given transport mapping is said to be *compatible with configuration error* φ if it satisfies the relation

$$\mathrm{d}\varphi_q(r) = -\mathcal{T}^*_{r\to q}\left(\mathrm{d}\varphi_r(q)\right) .$$

Finally, we conclude this section by stating a main theorem of [2].

Theorem 1 (Bullo and Murray). *Consider the control system in equation* (3)*, and let* $\{r(t), t \in \mathbb{R}_+\}$ *be a twice differentiable reference trajectory. Let* φ *be an error function on* $Q \times Q$*,* $\mathcal{T}$ *a transport map satisfying the compatibility condition, and* K_d *a dissipation function.*

If the control input is defined as $F = F_{\mathrm{PD}} + F_{\mathrm{FF}}$ *with*

$$\begin{aligned} F_{\mathrm{PD}} &= -\mathrm{d}\varphi_r(q) - K_d\dot{e} \\ F_{\mathrm{FF}} &= g^\flat(q)\left((\nabla_{\dot q}\mathcal{T}_{r\to q})\dot r + \frac{d}{dt}|_{q\ fixed}(\mathcal{T}_{r\to q}\dot r)\right) \end{aligned}$$

then the curve $q(t) = r(t)$ *is stable with Liapunov function*

$$W_{total}(q, \dot q, r, \dot r) = \varphi(r, q) + \frac{1}{2}g(\dot e, \dot e) .$$

In addition, if the error function φ *satisfies the quadratic assumption with a constant* L*, and if the boundedness assumptions*

$$d_1 \le \inf_{q\in Q} \|K_d(q)\|_g \le \sup_{q\in Q} \|K_d(q)\|_g \le d_2$$

$$\sup_{(r,q)\in Q\times Q} \|\nabla\mathcal{T}_{r\to q}\|_g < \infty$$

$$\sup_{(r,q)\in Q\times Q} \|\nabla \mathrm{d}\varphi_r(q)\|_g < \infty$$

$$\sup_{t\in\mathbb{R}} \|\dot r\|_g < \infty$$

where $0 < d_1 \le d_2$*, hold, then the curve* $q(t) = r(t)$ *is locally exponentially stable with Liapunov function* W_{total} *from all initial conditions* $(q(0), \dot q(0))$ *such that*

$$W_{total}(0) < L .$$

[1] See [5] (pg. 48) for bitensors (or alternatively two-point tensors) in classical physics coordinate notation. Alternatively, a bitensor field is a map $(r,q) \mapsto \phi(r,q)$, $(r,q) \in P \subseteq Q \times Q$,with $\phi(r,q)$ interpreted as an element of $\mathrm{GL}(T_rQ, T_qQ)$. The bitensor approach to transport mappings is completely equivalent to the vector bundle mapping approach of [3].

3 Intrinsic Tracking Control

Theorem 1 does not tell one how to actually select the configuration error φ and a compatible transport mapping $\mathcal{T}$. In our view and in the spirit of [1], intrinsic geometry provides a natural pair. In this section we show, as a corollary to Theorem 1, that the function

$$\varphi(r,q) = \frac{1}{2}\mathrm{dist}(r,q)^2 \tag{5}$$

and the traditional parallel transport along length minimizing geodesics, form a compatible configuration error-transport mapping pair.

To this end, we begin by specifying a submanifold P of $Q \times Q$ on which equation (1) holds and $l^\gamma(r,q)$ depends smoothly on $p = (r,q) \in P$. Assume that $\mathrm{Inj}(Q)$ is non-zero and consider the fibered submanifold of TQ defined by

$$BQ_{\mathrm{Inj}(Q)} \triangleq \left\{ v \in TQ : \sqrt{g(v,v)} < \mathrm{Inj}(Q) \right\}.$$

This is, in fact, a subbundle of TQ.

We now describe an embedding of $BQ_{\mathrm{Inj}(Q)}$ into $Q \times Q$. The embedded space is the appropriate domain for (5).

Proposition 1. *$f\colon BQ_{\mathrm{Inj}(Q)} \to Q \times Q$ defined by $f(v) = (\pi(v), \exp(v))$ is an embedding.*

Proof. The map f bijective and has local form $F(\mathbf{r},\mathbf{v}) = (\mathbf{r}, \mathrm{Exp}(\mathbf{r},\mathbf{v}))$. exp is a differentiable function and thus all partials of Exp exist and are smooth. TF is a triangular block matrix with diagonal $(I, T\mathrm{Exp}_r)$ and thus is injective so long as $T\mathrm{Exp}_r$ is. Given any fixed $r \in Q$, if $v_r \in B_{\mathrm{Inj}(Q)}(0)$, then $\exp_r(v)$ is not a conjugate point of r and so $T\mathrm{Exp}_{\mathbf{r}}(\mathbf{v})$ is injective. By the inverse function theorem, f is a local diffeomorphism and the result follows.

Given a Riemannian manifold Q with non-zero injectivity radius, we define the bundle $EQ \triangleq F(BQ_{\mathrm{Inj}(Q)})$ with fiber over each point r diffeomorphic to a normal neighborhood centered at r of radius $\mathrm{Inj}(Q)$.

Corollary 1. *Let Q be a manifold with $\mathrm{Inj}(Q) > 0$. Then for $p = (r,q) \in EQ$, there exists a unique (unit speed), length minimizing geodesic γ between r and q that is differentiably dependent on r and q. As a consequence, equation (1) holds and depends smoothly on p.*

This corollary allows us to now make some preparatory calculations. Consider a curve $p : \mathbb{R} \to EQ : u \mapsto (r(u), q(u))$ which does not touch the diagonal of $Q \times Q$. Composing this curve with l^γ provides a length function with domain, codomain $\mathbb{R}$ parameterized by u, which we shall also denote by l^γ. To which l^γ we refer should be clear by context. Being it the case that $l^\gamma : EQ \to \mathbb{R}$ is smooth and the curve p is smooth, $l^\gamma : \mathbb{R} \to \mathbb{R}$ is smooth and we may take its derivative with respect to the parameter u.

Let $\eta(s,u)$ be the family of length minimizing geodesics parameterized by arc length, s, between $r(u)$ and $q(u)$. Select some fixed parameter value and call it u^*. Taking $\gamma(s)$ to be the minimizing geodesic between $r = r(u)|_{u^*}$ and $q = q(u)|_{u^*}$ we see that $\eta(s,u)$ is a geodesic variation of $\gamma(s)$ with boundary values $\eta(s,u)|_{s=l^\gamma(u)} = q(u)$ and $\eta(s,u)|_{s=0} = r(u)$. That $\eta(s,u)$ depends smoothly on the boundary values $q(u)$ and $r(u)$ follows from corollary 1.

One finds $\frac{dl^\gamma}{du}$ to be related to the partials of η at u^* through the metric by the first variation formula, which is stated in the following lemma.

Lemma 1. [Gauss] *Let the curve p and the geodesic variation $\eta(s,u)$ be as described in the above discussion. Then*

$$\left.\frac{dl^\gamma}{du}\right|_{u^*} = g\left(\left.\frac{dq}{du}\right|_{u^*}, \left.\frac{\partial\eta}{\partial s}\right|_{(l^\gamma(u^*),u^*)}\right) - g\left(\left.\frac{dr}{du}\right|_{u^*}, \left.\frac{\partial\eta}{\partial s}\right|_{(0,u^*)}\right)$$

This lemma proves to be invaluable as it is the foundation of the following vital result.

We know by corollary 1 that l^γ is smooth on EQ and thus the differentials $\mathrm{d}l^\gamma_q$ and $\mathrm{d}l^\gamma_r$ exist. Using the Gauss lemma, we can calculate them.

Corollary 2. *Let the circumstances be those given in Lemma 1. Then the differential of the length function with respect to q and to r are, respectively*

$$\mathrm{d}l^\gamma_r(q) = g^\flat(q)\frac{d\gamma}{ds}(l^\gamma(r,q)) \tag{6}$$

$$\mathrm{d}l^\gamma_q(r) = -g^\flat(r)\frac{d\gamma}{ds}(0) \tag{7}$$

where, in the right members of these equations, γ is the unique geodesic segment connecting q to r.

Proof. Consider the vector v_q and let $\sigma(u) \in v_q$. Using the curve σ we may create, in accord with the above discussions, a curve $p(u) = (r, \sigma(u)) \in EQ$ and an associated geodesic variation $\eta(s,u)$ in which one endpoint, namely r, is fixed. It follows from Lemma 1 that

$$\begin{aligned}
[\mathrm{d}l^\gamma_r](v_q) &= v_q(l^\gamma_r) \\
&= \left.\frac{d}{du}l^\gamma_r(u)\right|_{u=0} \\
&= g\left(v_q, \frac{\partial\eta}{\partial s}(l^\gamma(0),0)\right) \\
&= g\left(v_q, \frac{d\gamma}{ds}(l^\gamma(r,q))\right) \\
&= \left[g^\flat(q)\,\frac{d\gamma}{ds}(l^\gamma(r,q))\right](v_q)
\end{aligned}$$

Since such is the case for all $v_q \in T_qQ$ the result follows. Equation (7) follows along exactly the same lines.

The following series of results ensure that the main objective of this paper is met.

Theorem 2 (Intrinsic Configuration Error). *Assuming* Inj$(Q) > 0$, *the function*

$$\varphi(r,q) = \frac{1}{2}(l^{\gamma}(r,q))^2 \tag{8}$$

is a uniformly quadratic tracking error function with $L_{\text{reg}} = \frac{1}{2}(\text{Inj}(Q))^2$.

Proof. The proof that this mapping is tracking error function is indeed simple if we make use of a normal chart (U,ψ) centered at r. In such a chart we have the representation $\Phi_r(q_1,\ldots,q_n) = \frac{1}{2}(q_1^2 + \ldots + q_n^2)$. The local expression of the differential is then $(\mathrm{d}\Phi_r)_{\psi(q)} = \begin{bmatrix} q_1 \cdots q_n \end{bmatrix}$. So $(\mathrm{d}\Phi_r)_{\psi(r)=0} = 0$ and r is a critical point of φ_r. This is a geometric property and does not depend on the local representation. Since r is a critical point of φ_r, the Hessian of this function Hess φ_r is coordinate independent and is positive definite provided the matrix $\left[\frac{\partial^2 \Phi_r}{\partial q_i \partial q_j}\right]$ is positive definite for any chart whatsoever we should choose. Using normal coordinates we obtain the identity matrix which completes the discussion concerning the conjecture that φ_r is a tracking error function.

We now demonstrate that φ is uniformly quadratic with $L_{\text{reg}} = \frac{1}{2}(\text{Inj}(Q))^2$. Let $r \in Q$. For a given geodesic γ emanating from r, φ is defined only within the injectivity radius and $\mathrm{d}\varphi_r \neq 0$ for every interval $[0,L]$ with $L < \frac{1}{2}(\text{Inj}(Q))^2$. Hence $L_{\text{reg}}(\varphi_r(q),r) = \frac{1}{2}(\text{Inj}(Q))^2$ for each $r \in Q$. Taking the infimum over all $r \in Q$, we have $L_{\text{reg}}(\varphi, Q) = \frac{1}{2}(\text{Inj}(Q))^2$.

Now, using the Cauchy-Schwarz inequality, we have, suppressing the evaluation of $\frac{d\gamma}{ds}$ at $l^{\gamma}(r,q)$,

$$\begin{aligned}
|\mathrm{d}\varphi_r(q)v_q|^2 &= (l^{\gamma})^2 g\left(\frac{d\gamma}{ds}, v_q\right)^2 \\
&\leq (l^{\gamma})^2 g\left(v_q, v_q\right) .
\end{aligned}$$

Taking the square root of both sides and then the supremum over $|v_q| = 1$, it is clear that $\|\mathrm{d}\varphi_r(q)\|_g$ can be no greater than l^{γ}. If we take $v_q = \frac{d\gamma}{ds}|_{s=l^{\gamma}(r,q)}$, the supremum is reached and thus the result. Taking $a = b = \frac{1}{2}$ in inequality (4) completes the discussion.

Lemma 2. *The parallel transport of two vectors $v_r, w_r \in T_rQ$ from a point r to another point q along a curve $\gamma \subset Q$, denoted $\gamma_{r\to q}$, preserves the length of each vector and the inner product between the transport of v_r, w_r in the Riemannian metric $g(\cdot,\cdot)$.*

Theorem 3 (Intrinsic Transport Map). *Parallel transport along length minimizing geodesics is a transport mapping on EQ.*

Proof. Writing the parallel transport equations out in coordinates we get the IVP

$$\frac{\mathrm{d}v^k}{\mathrm{d}u} + \Gamma^k_{ij} v^j \frac{\mathrm{d}\sigma_i}{\mathrm{d}u} = 0 \qquad k = 1, \ldots, n$$

with $v^k(0) = v^k_0$. Thus we get a linear (possibly nonautonomous) system and a unique solution with infinite continuation is assured. The solution at u^* is determined by the associated state transition matrix $\Phi(u^*, 0)$. So we see that, within a chart, parallel transport is, point–wise, a smooth linear mapping with local representation Φ. This conclusion, however, remains true even if we have need of involving multiple charts along the way from r to q. It is, therefore, clear that $\sigma_{r \to q}$ describes a bitensor field over EQ.

By the definition of parallel transport, $\sigma_{r \to r} = \mathrm{id}$. Further, Lemma 2 indicates that parallel transport preserves frames. Thus, we also know that $\sigma_{r \to q} \in GL(T_r Q, T_q Q)$. Should we show that $\sigma_{r \to q}$ is smooth, then we may conclude that it is a transport mapping. We now take the argument up.

The geodesic field on TQ is smooth. So, by the theory of ordinary differential equations, the geodesics depend smoothly on the initial data. Further, since f is a diffeomorphism of $BQ_{\mathrm{Inj}(Q)}$ onto EQ, the length minimizing geodesics depend smoothly on the boundary data $(r, q) \in EQ$. Thus the field described by equation (2) is smoothly parameterized by the boundary data $(r, q) \in EQ$. Hence, calling upon the theory of ordinary differential equations once more, we determine that the solutions $V(u; v_r, (r, q))$ depend smoothly on the initial condition v_r and the boundary data (r, q). Therefore, $\sigma_{r \to q}$ takes smooth fields to smooth sections of $Q \times TQ$. That is, for $Y \in \mathcal{X}(Q)$, $V(1; Y(r), (r, q))$ is a smooth section of $Q \times TQ$.

Theorem 4 (Compatibility). *The parallel transport along length minimizing geodesics and the tracking error function* (8) *are* compatible *in the sense that*

$$\mathrm{d}\varphi_q(r) = -\gamma^*_{r \to q} \left(\mathrm{d}\varphi_r(q) \right) \ .$$

Proof. Since $\mathrm{d}\varphi_r(q) = l^\gamma(r, q) \mathrm{d}l^\gamma_r(q)$, $d\varphi_q(r) = l^\gamma(r, q) \mathrm{d}l^\gamma_q(r)$, and $\gamma^*_{r \to q}$ is a linear operator, we need only show

$$\mathrm{d}(l^\gamma_q(r)) = -\gamma^*_{r \to q} \left(\mathrm{d}(l^\gamma_r(q)) \right) .$$

Consider an arbitrary vector $v_r \in T_r Q$. Computing directly:

$$\begin{aligned}
[\mathrm{d}l^\gamma_q(r)](v_r) &\overset{eq.(7)}{=} \left[-g^\flat(r) \frac{d\gamma}{ds}(0) \right] (v_r) \\
&= -g \left(\frac{d\gamma}{ds}(0), v_r \right) \\
&\overset{Lm.2}{=} -g \left(\gamma_{r \to q} \frac{d\gamma}{ds}(0), \gamma_{r \to q} v_r \right) \\
&= -g \left(\frac{d\gamma}{ds}(l^\gamma(r, q)), \gamma_{r \to q} v_r \right) \\
&= \left[-\gamma^*_{r \to q} \left(g^\flat(q) \left. \frac{d\gamma}{ds} \right|_{l^\gamma(r,q)} \right) \right] (v_r) \\
&\overset{eq.(6)}{=} \left[-\gamma^*_{r \to q} \left(\mathrm{d}l^\gamma_r(q) \right) \right] (v_r) .
\end{aligned}$$

This calculation holds for all $v_r \in T_r Q$ and so the result.

We thus have the following generalization of [1], or an instance of Theorem 1, in which the intrinsic geometry of an abstract machine creates a closed-loop system analogous to a geodesic mass–spring–damper.

Corollary 3. *Suppose that one has an abstract machine with dynamics given by equation* (3)*. Further assume that the machine's configuration manifold Q has a positive injectivity radius. Let the control force be defined as $F = F_{\mathrm{FB}} + F_{\mathrm{FF}}$ where*

$$\begin{aligned} F_{\mathrm{FB}} &= -\alpha \, \mathrm{d}\varphi_r(q) - \beta \, g^\flat(q)\dot{e} \\ &= -\alpha \, g^\flat(q) \, \mathrm{grad}\, \varphi_r - \beta \, g^\flat(q)\dot{e} \\ &= -\alpha \, \mathrm{dist}(r,q) \, g^\flat(q) \, \frac{d\gamma}{ds} - \beta \, g^\flat(q)\dot{e} \\ \dot{e} &= \dot{q} - \gamma_{r\to q}\dot{r} \\ F_{\mathrm{FF}} &= g^\flat(q) \left(\frac{\mathrm{D}}{\mathrm{d}t} \gamma_{r\to q}\dot{r} \right) \end{aligned}$$

where $\varphi(r,q)$ is the tracking error function given in equation (8) *and γ is the unique minimizing geodesic between $q(t)$ and $r(t)$ at time t. Let*

$$V(t) = \frac{1}{2} \, \alpha \, \mathrm{dist}^2(r(t), q(t)) + \frac{1}{2} \, g\,(\dot{e}(t), \dot{e}(t))$$

be the Liapunov candidate. Then, $r(t)$ is stable and should the boundedness conditions of Theorem 1 be satisfied of $P = EQ$, the curve $r(t)$ is locally exponentially stable from all initial conditions satisfying

$$\mathrm{dist}^2(r(0), q(0)) + \frac{1}{\alpha} \, g\,(\dot{e}(0), \dot{e}(0)) < \mathrm{Inj}(Q)^2. \tag{9}$$

In the next theorem, we show that this particular design induces the mass–spring–damper paradigm *explicitly* when the reference trajectory and mechanical system initial condition belongs to $T\mathrm{Im}(\gamma)$. We shall also see that corollary 3 has another advantage.

Theorem 5. *Let $r(t)$ be a smooth curve with image contained within a geodesic path $\gamma \subset Q$, where Q is a manifold with positive injectivity radius. For any initial condition of the mechanical system satisfying i) Inequality 9, ii) $q(0) \in Im(\gamma)$ and iii) $\dot{q}(0) \in T_{q(0)}Im(\gamma)$ the closed loop solution q to equation* (3) *under the control law of corollary* (3) *satisfies* 1) *$q(t)$ lies in the trace of γ for all $t > 0$ and* 2) *$q(t) = r(t)$ is exponentially stable with rate of convergence dictated by α and β.*

Proof. Choosing a base point $o \in \mathrm{Im}(\gamma)$ and a unit tangent vector $v_o \in T_o\mathrm{Im}(\gamma)$, parameterize the curve γ by the associated signed arc length measure s so that $\gamma(s) \triangleq \gamma(s, v_o)$. Let the reference trajectory be parameterized by, say $s_r(t)$, so that $r(t) = \gamma(s_r(t))$. Now, let us assume momentarily or "guess" that the signed arc length difference $\Delta s(t)$, where $\mathrm{dist}(r, q) = |\Delta s(t)|$, between the points $r(t)$ and $q(t)$ satisfies the mass–spring–damper differential equation:

$$\Delta\ddot{s} = -\alpha\Delta s - \beta\Delta\dot{s}\,. \tag{10}$$

We show that the solution to the closed loop dynamic system under the control action of corollary 3 is

$$q(t) = \gamma(s_r(t) + \Delta s(t))\,.$$

During our computations we will suppress the functional dependence on t. Taking derivatives, we have the velocity vectors

$$\dot{q} = \left.\frac{d\gamma}{ds}\right|_{(s_r+\Delta s)} (\dot{s}_r + \Delta\dot{s}) \;\text{ and }\; \dot{r} = \left.\frac{d\gamma}{ds}\right|_{s_r} \dot{s}_r.$$

The parallel transport of $\dot{r}(t)$ to the base point $q(t)$ is given by

$$\gamma_{r\to q}\dot{r} = \gamma_{r\to q}\left.\frac{d\gamma}{ds}\right|_{s_r} \dot{s}_r = \left.\frac{d\gamma}{ds}\right|_{(s_r+\Delta s)} \dot{s}_r$$

and so $\dot{e} = \dot{q} - \gamma_{r\to q}\dot{r} = \left.\frac{d\gamma}{ds}\right|_{(s_r+\Delta s)} \Delta\dot{s}$. Taking the covariant derivative of the velocity error vector, we have, suppressing the evaluation of $\frac{d\gamma}{ds}$ at $s_r + \Delta s$,

$$\begin{aligned}\frac{\mathrm{D}\dot{e}}{\mathrm{d}t} &= \frac{\mathrm{D}}{\mathrm{d}t}\left(\frac{d\gamma}{ds}\Delta\dot{s}\right) = \frac{\mathrm{D}}{\mathrm{d}t}\left(\frac{d\gamma}{ds}\right)\Delta\dot{s} + \frac{d\gamma}{ds}\Delta\ddot{s} \\ &= \frac{\mathrm{D}}{\mathrm{d}s}\left(\frac{d\gamma}{ds}\right)(\dot{s}_r + \Delta\dot{s})\,\Delta\dot{s} + \frac{d\gamma}{ds}\Delta\ddot{s} = \frac{d\gamma}{ds}\Delta\ddot{s}\end{aligned}$$

as $\frac{\mathrm{D}}{\mathrm{d}s}\frac{d\gamma}{ds} = 0$ due to the fact that γ is a geodesic. Substituting (10), we have

$$\begin{aligned}\frac{\mathrm{D}\dot{e}}{\mathrm{d}t} &= \left.\frac{d\gamma}{ds}\right|_{(s_r+\Delta s)} (-\alpha\Delta s - \beta\Delta\dot{s}) \\ &= -\alpha\,\mathrm{grad}\varphi_r(q) - \beta\,\dot{e}\end{aligned}$$

Under the operator $g^\flat(q)$ we arrive at the closed loop dynamics given by the control law of Corollary 3. By uniqueness of solutions we find that our "guess" of the mass–spring–damper dynamics in equation (10) was correct. Furthermore, note that the Liapunov function presented in Corollary 3 takes the form

$$\begin{aligned}V(t) &= \frac{\alpha}{2}\Delta s^2 + \frac{1}{2}g(\dot{e},\dot{e}) \\ &= \frac{\alpha}{2}\Delta s^2 + \frac{1}{2}\Delta\dot{s}^2\left|\frac{d\gamma}{ds}\right|^2 \\ &= \frac{\alpha}{2}\Delta s^2 + \frac{1}{2}\Delta\dot{s}^2\end{aligned}$$

on the geodesic. This is a Liapunov function we're used to seeing for the mass-spring damper.

For this restricted case, we can design the arc length and arc length rate to be over-damped, critically damped, or under-damped through judicious choice of positive constants α and β. Hence, this is a very practical design tool for those interested in tuning the performance of an abstract machine. Imbedded in this geometric control methodology is a second order linear, time-invariant ODE in kinetic energy distance. It is seen that the relative kinetic energy dissipates as the mechanical system's trajectory converges to the reference trajectory.

Another interesting consequence unfolds if we further restrict to the set point control problem with zero initial conditions.

Corollary 4. *Let* $S(r) \triangleq Q \setminus C(r)$. *Let* $\dot{r}(t) = \dot{q}(0) = 0$. *Then there exists a control law such that* $q(t)$ *exponentially converges to* r *for all* $q(0) \in S(r)$.

Proof. Choose α and β as to ensure the ODE in equation (10) is over-damped. The definitions for arc length and parallel transport can be extended along each geodesic up until its intersection with the cut locus. The arc length is strictly decreasing and the trajectory length can never leave the geodesic between r and $q(0)$. The convergence is exponential using the same proof argument of Theorem 5.

4 Conclusions

Since the definitions and control design in [2] are heavily influenced by Riemannian geometry, it should come as little surprise that natural distance and parallel transport should fit within the context of any self-consistent control theory so modeled. The abstraction of position and velocity error described in [2, 3] illustrates the role of compatibility in stabilization; in other words, position and velocity errors should be linked by geometry to ensure stability. Furthermore, geodesic lines can provide that link and serve as an intrinsic basis for control. Thus, this contribution provides a simple, yet elegant, solution that directly addresses the geodesic spring paradigm in control.

Acknowledgements

This research was supported by the Air Force Office of Scientific Research and the Air Force Research Laboratory. The authors were also respectively supported by Raytheon Missile Systems; General Dynamics, SRS; and NSF (grant DMS-0306017). We thank these fine organizations for their patronage.

References

1. Bullo, F., Murray, R.M., Sarti, A.: Control on the sphere and reduced attitude stabilization. In: Symposium on Nonlinear Control Systems. Volume 2., Tahoe City, CA (1995) 495–501

2. Bullo, F., Murray, R.: Tracking for fully actuated mechanical systems: a geometric framework. Automatica **35** (1999) 17–34
3. Bullo, F., Lewis, A.: Geometric Control of Mechanical Systems. Springer, New York (2005)
4. DoCarmo, M.: Riemannian Geometry. Birkhäuser, Boston (1992)
5. Synge, J.: Relativity: The General Theory. North-Holland Publishing Company, Amsterdam (1971)

Shape Control of a Multi-agent System Using Tensegrity Structures

Benjamin Nabet and Naomi Ehrich Leonard

Mechanical and Aerospace Engineering, Princeton University, Princeton, NJ 08544 USA
{bnabet,naomi}@princeton.edu

Summary. We present a new coordinated control law for a group of vehicles in the plane that stabilizes an arbitrary desired group shape. The control law is derived for an arbitrary shape using models of tensegrity structures which are spatial networks of interconnected struts and cables. The symmetries in the coupled system and the energy-momentum method are used to investigate stability of relative equilibria corresponding to steady translations of the prescribed rigid shape.

1 Introduction

We address shape control of a group of mobile agents or vehicles in the plane. Shape refers to the geometry, configuration or formation of the group and is invariant under translation and rotation of the group as a whole. In [1], shape coordinates are defined based on Jacobi coordinates and a law to control small formations on Jacobi shape space is derived using a control Lyapunov function. In this paper, we use models of *tensegrity structures* to synthesize and analyze the shape dynamics of a group.

Shape control and, more generally, collective control of multi-agent systems have applications in a variety of engineering problems. One specific application motivating this research is the control of a fleet of autonomous underwater vehicles (AUVs) recently used for an adaptive sampling experiment in Monterey Bay, CA (AOSN), see e.g. [2, 3]. For the design of mobile sensors carrying out sampling or searching tasks, the configuration of the group can be critical. Depending on the field that is being surveyed, smaller or larger formations might be more efficient, and certain shapes of the group might be preferable for estimating field parameters such as gradients or higher-order derivatives from noisy measurements made by the mobile sensors, see e.g., the problem of generating a contour plot with a mobile sensor network [4].

We present a constructive method to stabilize an arbitrary planar shape for a group of n vehicles using virtual tensegrity structures. We model each of the n vehicles as a particle, moving in the plane, under the influence of a control force. The control forces are designed as if the particle group forms a tensegrity

F. Bullo et al. (Eds.): Lag. & Hamil. Methods for Nonlin. Ctrl., LNCIS 366, pp. 329–339, 2007.
springerlink.com

structure in which particles are treated like nodes and connections between particles simulate struts or cables. Stabilization of relative equilibria corresponding to the desired shape in steady translation is investigated using symmetries in the multi-agent system together with the energy-momentum method.

Tensegrity structures [5] are geometric structures formed by a combination of *struts* (in compression) and *cables* (in tension) which we classify together more generally as *edges*. The edges of a tensegrity structure meet at *nodes*. A generic combination of cables and struts will not be in equilibrium; if the corresponding structure were physically built, it would collapse. We define tensegrity structures as only those structures that are in equilibrium. The artist Kenneth Snelson [6] built the first tensegrity structure, and Buckminster Fuller [7] coined the term tensegrity by combining the words tension and integrity.

Tensegrity structures have been widely studied with different motivations and approaches. For instance, there is growing interest in tensegrities in the context of designing structures whose shape can be adjusted and controlled. Models of the forces and formalization of the notion of stability for tensegrity structures were proposed by Connelly through an energy approach in [8, 9, 10, 11]. Tensegrity structures have also been used to model biological systems such as proteins, [12] or cellular structure, [13]. It is known [5] that the shape of a tensegrity structure can be changed substantially with little change in the potential energy of the structure. This motivates us in part to use tensegrity structures as a model for shape control of a group of vehicles.

In this paper, we define a tensegrity structure that realizes any arbitrary desired shape. Each vehicle, modelled as a particle, is identified with one node of the tensegrity structure. The edges of the tensegrity structure correspond to communications and direction of forces between the vehicles. If an edge is a cable, the force is attractive; if the edge is a strut then the force is repulsive. The magnitude of the forces depends on the tensegrity structure parameters and the relative distance between the vehicles associated with the edge. In this setting, it is possible to see a tensegrity structure as an undirected graph with the interconnection between nodes weighted by the magnitude of the force. This allows us to use the formalism and results from algebraic graph theory. We note that because we use *virtual* tensegrity structures our model cannot impose the constraints that physical struts only increase in length and cables only decrease in length; an important consequence is the need for a nonlinear model that isolates the desired equilibrium shape. In Sections 2 and 3 of this paper we discuss different models for the forces. In Section 4 we present a systematic method to generate any shape. In Section 5 we investigate the stability of the generated shapes.

2 Linear Force Model

In this section we describe the simplest way of modelling the forces induced by the two types of edges of a tensegrity structure. We then find the relationship between the choice of cables, struts and parameters for the corresponding model and the equilibria. We model cables as springs with zero rest length and struts as springs with zero rest length and with a negative spring constant [8, 9]. Hence if we consider two nodes i, j we have

$$\overrightarrow{f}_{i\to j} = \omega_{ij}(\overrightarrow{q_i} - \overrightarrow{q_j}) = -\overrightarrow{f}_{j\to i}, \tag{1}$$

where $\overrightarrow{f}_{i\to j} \in \mathbb{R}^2$ is the force applied to node j as a result of the presence of node i. Here, $\overrightarrow{q_i} = (x_i, y_i) \in \mathbb{R}^2$ is the position vector of node i and ω_{ij} is the spring constant of the edge ij. The spring constant ω_{ij} is positive if ij is a strut, negative if ij is a cable and zero if there is no connection between the nodes i and j. We call ω_{ij} the *stress* of the edge ij. The absolute position of the structure in the plane is given by a vector $\mathbf{q} \in \mathbb{R}^{2n}$ which we call a *placement*. Let $\mathbf{x} = (x_1, \ldots, x_n)^T$ and $\mathbf{y} = (y_1, \ldots, y_n)^T$, then $\mathbf{q} = \begin{pmatrix}\mathbf{x}\\ \mathbf{y}\end{pmatrix}$. The potential energy of a tensegrity structure is

$$E_\omega(\mathbf{q}) = \frac{1}{2}\sum_{i=1}^{n}\sum_{j=i+1}^{n} \omega_{ij}\|\overrightarrow{q_j} - \overrightarrow{q_i}\|^2. \tag{2}$$

We write $\sum\limits_{i<j}$ to represent $\sum\limits_{i=1}^{n}\sum\limits_{j=i+1}^{n}$. We note that this potential increases as we stretch the cables or shrink the struts. Using cartesian coordinates in the plane, equation (2) becomes

$$E_\omega(\mathbf{q}) = \frac{1}{2}\Big(\sum_{i<j}\omega_{ij}(x_j - x_i)^2 + \sum_{i<j}\omega_{ij}(y_j - y_i)^2\Big). \tag{3}$$

The equilibria of the system are the critical points of the potential (3). We rewrite (3) to more easily calculate the critical points.

Using notations from algebraic graph theory, we consider the undirected graph $G = (V, E)$, where V is the set of nodes and E the set of edges. Let d_j be the degree of node j, then the Laplacian L of the graph G is the $n \times n$ matrix defined by

$$L_{ij} = \begin{cases} d_j & \text{if } i = j \\ -1 & \text{if } (i,j) \in E \\ 0 & \text{otherwise.} \end{cases} \tag{4}$$

In our setting communications are not identical from one edge to the other, but rather are weighted by the spring constants ω_{ij}.

Our goal is to solve for and stabilize a tensegrity structure. To do this we solve for the weights ω_{ij}. A tensegrity structure can then be viewed as an undirected graph for which we define the weighted Laplacian Ω by

$$\Omega_{ij} = \begin{cases} \sum_{j=1}^{n}\omega_{ij} & \text{if } i = j \\ -\omega_{ij} & \text{if } i \neq j. \end{cases}$$

This matrix Ω introduced by Connelly (but derived in a different way) in [8] is called the *stress matrix*. The stress matrix is an $n \times n$ symmetric matrix and the

n-dimensional vector $\mathbf{1} = \begin{pmatrix}1 \cdots 1\end{pmatrix}^T$ is in the kernel of Ω (the last property is true for all Laplacians). Using this matrix, we can rewrite the potential (3) as

$$E_\omega(\mathbf{q}) = \frac{1}{2}\mathbf{q}^T(\Omega \otimes I_2)\mathbf{q},$$

where I_2 is the 2×2 identity matrix and $\Omega \otimes I_2$ is the $2n \times 2n$ block diagonal matrix $\begin{pmatrix}\Omega & 0 \\ 0 & \Omega\end{pmatrix}$. It is now easy to see that the critical points, and hence the equilibria, are given by

$$\mathbf{q}^T(\Omega \otimes I_2) = 0. \tag{5}$$

Using the fact that the stress matrix is symmetric, we see that a placement $\mathbf{q}^e$ is an equilibrium if and only if $\mathbf{x}^e$ and $\mathbf{y}^e$ are in the kernel of Ω. Recall that $\mathbf{1}$ is in the kernel of Ω. Assuming that the nodes are not all in a line, $\mathbf{x}^e, \mathbf{y}^e$ and $\mathbf{1}$ are linearly independent. We can conclude that with this model, a combination of cables and struts will have an equilibrium if and only if $\text{rank}(\Omega) \leq n - 3$. We assume from now on that $n \geq 4$. By choosing the stresses of the edges of the structure so that $\text{rank}(\Omega) = n - 3$, the kernel of the stress matrix Ω is exactly three dimensional, and we can prescribe the shape of the equilibrium.

However, we cannot prescribe the size of the equilibrium configuration. Indeed if $\ker(\Omega) = \text{span}\{\mathbf{x}^e, \mathbf{y}^e, \mathbf{1}\}$ then the structure described by $\mathbf{q}^e = (\alpha\mathbf{x}^e, \beta\mathbf{y}^e)$ is also an equilibrium $\forall \alpha, \beta \in \mathbb{R}$. For real tensegrities this is not a problem because the cable and strut constraints preclude the existence of any but the original equilibrium. In the virtual setting, however, where we cannot impose the constraints, we get a continuum of equilibria which is not desirable. For example, if we prescribe the tensegrity to be a square, it will be the case that not only all squares but also all rectangles will be equilibria. In the next section we exploit the simple equation (5), derived using the linear model (1), that determines the tensegrity shape as a function of the parameters ω_{ij}. We propose a nonlinear model for the forces along edges that isolates a tensegrity, fixing both shape and size.

3 Nonlinear Force Model

In the previous section we chose to model the forces between a pair of nodes with a linear function of the relative distance between the nodes. We now model the forces along the edges as nonlinear springs with finite, nonzero rest length. Cables will always be longer than their rest length and struts will always be shorter than their rest length. We consider two nodes i, j and we define

$$\overrightarrow{f}_{i \to j} = \alpha_{ij}|\omega_{ij}|\frac{r_{ij} - L_{ij}}{r_{ij}}(\overrightarrow{q}_i - \overrightarrow{q}_j). \tag{6}$$

Here $r_{ij} = \|\overrightarrow{q}_i - \overrightarrow{q}_j\|$ is the relative distance between nodes i and j, L_{ij} is the rest length of the spring that models the edge ij, ω_{ij} is the spring constant from model (1) and α_{ij} is a scalar parameter that fixes the spring constant of model (6) for the edge ij.

The corresponding potential energy is

$$E_\omega(\mathbf{q}) = \frac{1}{2}\sum_{i<j} \alpha_{ij}|\omega_{ij}|(r_{ij} - L_{ij})^2. \tag{7}$$

The equilibria of this system can be found by solving for the critical points of the potential (7). After some manipulation, we find that the critical points of (7) are given by

$$\sum_{j=1}^{n} \alpha_{ij}|\omega_{ij}|(\overrightarrow{q_j} - \overrightarrow{q_i})(1 - \frac{L_{ij}}{r_{ij}}) = 0, \;\; i = 1, \ldots, n.$$

¿From this set of equations, we can define an analogue of the stress matrix Ω of (2). The new stress matrix is not a constant matrix. Rather it depends on the relative distances between pairs of nodes. We define

$$\widetilde{\omega}_{ij}(\mathbf{x}, \mathbf{y}) = \alpha_{ij}|\omega_{ij}|(1 - \frac{L_{ij}}{r_{ij}}) \tag{8}$$

to be the stress of the edge ij. The entries of the new stress matrix are given by

$$\widetilde{\Omega}_{ij}(\mathbf{x}, \mathbf{y}) = \begin{cases} \sum_{k=1}^{n} \widetilde{\omega}_{ik}(\mathbf{x}, \mathbf{y}) & \text{if } i = j \\ -\widetilde{\omega}_{ij}(\mathbf{x}, \mathbf{y}) & \text{if } i \neq j. \end{cases}$$

We note that the vector $\mathbf{1}$ is also in the kernel of $\widetilde{\Omega}$, $\forall\ (\mathbf{x}, \mathbf{y}) \in \mathbb{R}^n \times \mathbb{R}^n$. Now if we wish the placement $\mathbf{q}^e = (\mathbf{x}^e, \mathbf{y}^e)$ to be the tensegrity structure (i.e the stable equilibrium of the system), we need to pick (if possible) the parameters $\alpha_{ij}, \omega_{ij}, L_{ij}$ so that

$$\widetilde{\Omega}(\mathbf{x}^e, \mathbf{y}^e)\mathbf{x}^e = 0$$
$$\widetilde{\Omega}(\mathbf{x}^e, \mathbf{y}^e)\mathbf{y}^e = 0.$$

As a first step we choose parameters α_{ij} and L_{ij} for all i, j so that $\widetilde{\Omega}(\mathbf{x}^e, \mathbf{y}^e) = \Omega$. If edge ij is a cable, then $\omega_{ij} > 0$, and by (8) we have $\widetilde{\omega}_{ij} = \alpha_{ij}\omega_{ij}(1 - \frac{L_{ij}}{r_{ij}})$. To make $\widetilde{\omega}_{ij}(\mathbf{x}^e, \mathbf{y}^e) = \omega_{ij}$, we make $\alpha_{ij}(1 - \frac{L_{ij}}{r_{ij}^e}) = 1$ where r_{ij}^e is the relative distance between nodes i and j for the desired placement. This last equation is solved by picking $\alpha_{ij} = 2$ and $L_{ij} = \frac{1}{2}r_{ij}^e$. If edge ij is a strut, then $\omega_{ij} < 0$, and by (8) we have $\widetilde{\omega}_{ij} = -\alpha_{ij}\omega_{ij}(1 - \frac{L_{ij}}{r_{ij}})$. We make $\alpha_{ij}(1 - \frac{L_{ij}}{r_{ij}^e}) = -1$ by picking $\alpha_{ij} = 1$ and $L_{ij} = 2r_{ij}^e$. The choice of L_{ij} and α_{ij} is not unique (in case of a strut or a cable); the effect of picking other values for parameters α_{ij} and L_{ij} is to be determined.

We show in the next section, that we can also find parameters ω_{ij} independent of parameters α_{ij} and L_{ij}, so that $\ker(\Omega) = \text{span}\{\mathbf{x}^e, \mathbf{y}^e, \mathbf{1}\}$, and such that the nonzero eigenvalues of Ω are all positive. This makes the equilibrium $\mathbf{q}^e = (\mathbf{x}^e, \mathbf{y}^e)$ an isolated minimum of the potential (modulo rigid transformations), i.e., our choices ensure that we have the right combination of struts and cables to make $\mathbf{q}^e = (\mathbf{x}^e, \mathbf{y}^e)$ a tensegrity structure.

4 Construction of the Constant Stress Matrix

In this section we solve the following problem: given a desired placement $\mathbf{q}^e = (\mathbf{x}^e, \mathbf{y}^e)$, find stresses ω_{ij} such that $\ker(\Omega) = \text{span}\{\mathbf{x}^e, \mathbf{y}^e, \mathbf{1}\}$ and the nonzero eigenvalues of Ω are positive.

We know that Ω is symmetric, hence it has only real eigenvalues and can be diagonalized using an orthonormal basis. As mentioned previously, if we do not consider the case when all the nodes are in a line, then $\mathbf{x}^e, \mathbf{y}^e$ and $\mathbf{1}$ are linearly independent. We can complete these three vectors with $n - 3$ others so that we have a basis of $\mathbb{R}^n$. Then if we apply the Gram-Schmidt procedure to those vectors, we get an orthonormal basis $(\mathbf{v}_1, \ldots, \mathbf{v}_n)$ for $\mathbb{R}^n$ that satisfies

$$\text{span}\{\mathbf{v}_1, \mathbf{v}_2, \mathbf{v}_3\} = \text{span}\{\mathbf{x}^e, \mathbf{y}^e, \mathbf{1}\}.$$

Now we define the $n \times n$ diagonal matrix D with diagonal elements $(0, 0, 0, 1, \ldots, 1)$ and the orthonormal $n \times n$ matrix $\Lambda = \begin{pmatrix} \mathbf{v}_1 & \cdots & \mathbf{v}_n \end{pmatrix}$. If we compute $\Lambda D \Lambda^T$ we have a symmetric positive semi-definite matrix with its kernel equal to $\text{span}\{\mathbf{x}^e, \mathbf{y}^e, \mathbf{1}\}$. Setting $\Omega = \Lambda D \Lambda^T$ determines the values of stresses ω_{ij} that make the placement $\mathbf{q}^e = (\mathbf{x}^e, \mathbf{y}^e)$ a tensegrity structure. The effect of choosing smaller or larger positive eigenvalues for Ω (set here to 1) is to be determined.

5 Relative Equilibrium Stability

In this section we use the energy-momentum method to look at the stability of the relative equilibrium corresponding to the tensegrity structure modelled by (6) in steady translation. The *energy-momentum method* is a technique for proving stability of relative equilibria [14]. For simple mechanical systems, we have the following setting: a configuration space Q, a symplectic manifold $P = T^*Q$ with a symplectic action of a Lie group G on P, an equivariant momentum map $\mathbf{J} : P \mapsto \mathfrak{g}^*$ and a G-invariant Hamiltonian $H : P \mapsto \mathbb{R}$. Here $\mathfrak{g}^*$ is the dual of the Lie algebra $\mathfrak{g}$ of G. If the Hamiltonian vector at the point $\mathbf{z}^e \in P$ points in the direction of the group orbit through $\mathbf{z}^e$, then the point is called a *relative equilibrium*. Let $\mu = \mathbf{J}(\mathbf{z}^e)$.

Theorem 1. Relative Equilibrium Theorem [14] *$\mathbf{z}^e$ is a relative equilibrium if and only if there is a $\xi \in \mathfrak{g}$ such that $\mathbf{z}^e$ is a critical point of the augmented Hamiltonian $H_\xi(\mathbf{z}) := H(\mathbf{z}) - \langle \mathbf{J} - \mu, \xi \rangle$.*

Definition 1 [14]. *Let S be a subspace of $T_{\mathbf{z}^e}P$ such that $S \subset \ker\mathbf{DJ}(\mathbf{z}^e)$ and S is transverse to the G_μ-orbit within $\ker\mathbf{DJ}(\mathbf{z}^e)$, where $G_\mu = \{g \in G \mid g \cdot \mu = \mu\}$, $\mu \in \mathfrak{g}^*$ and $g \cdot \mu$ is the coadjoint action of G on $\mathfrak{g}^*$.*

Theorem 2. Energy Momentum Theorem
[14]. *If $\delta^2 H_\xi(\mathbf{z}^e)$ is definite on the subspace S, then $\mathbf{z}^e$ is G_μ-orbitally stable in $\mathbf{J}^{-1}(\mu)$ and G-orbitally stable in P.*

For our system, the configuration space is $Q = (\mathbb{R}^2)^n$, $\mathbf{q} = (\mathbf{x}, \mathbf{y}) \in Q$ and an element in the cotangent bundle $\mathbf{z} \in T^*Q$ can be written as $\mathbf{z} = (\mathbf{q}, \mathbf{p}) = (\mathbf{x}, \mathbf{y}, \mathbf{p}_x, \mathbf{p}_y)$. Assuming unit mass nodes, the system kinetic energy is $1/2(\|\dot{\mathbf{x}}\|^2 + \|\dot{\mathbf{y}}\|^2)$ and $\mathbf{p}_x = \dot{\mathbf{x}}$ and $\mathbf{p}_y = \dot{\mathbf{y}}$. The Hamiltonian of the system is given by

$$H(\mathbf{z}) = \frac{1}{2}(\|\mathbf{p}_x\|^2 + \|\mathbf{p}_y\|^2) + \frac{1}{2}\sum_{i<j} \alpha_{ij}|\omega_{ij}|(r_{ij} - L_{ij})^2.$$

Noting the fact that the potential energy of the system only depends on the relative distances between nodes, we have that the Hamiltonian of the system is invariant under the following action of the Lie group $SE(2)$ on Q:

$$g.\mathbf{q} = (\cos\theta\mathbf{x} - \sin\theta\mathbf{y} + T_x\mathbf{1}, \sin\theta\mathbf{x} + \cos\theta\mathbf{y} + T_y\mathbf{1}), \tag{9}$$

where $g = (\theta, T_x, T_y) \in SE(2)$. Let $g(t) \in SE(2)$ such that $g(0) = (0, 0, 0)$ and $\dot{g}(0) = \xi = (\omega, V_x, V_y) \in \mathfrak{se}(2) = \mathfrak{g}$. The infinitesimal generator corresponding to the action (9) is

$$\begin{aligned} \xi_Q(q) &= \left.\frac{d}{dt}\right|_{t=0} g(t) \cdot (\mathbf{x}, \mathbf{y}) \\ &= (-\omega\mathbf{y} + V_x\mathbf{1}, \omega\mathbf{x} + V_y\mathbf{1}). \end{aligned}$$

The momentum map $\mathbf{J} : T^*Q \mapsto \mathfrak{g}^*$ is given by the formula [15]

$$\langle \mathbf{J}(\mathbf{x}, \mathbf{y}, \mathbf{p}_x, \mathbf{p}_y), \xi \rangle = \langle (\mathbf{p}_x, \mathbf{p}_y), \xi_Q(q) \rangle. \tag{10}$$

¿From equation (10), we get

$$\mathbf{J}(\mathbf{z}) = \begin{pmatrix} \langle \mathbf{x}, \mathbf{p}_y \rangle - \langle \mathbf{y}, \mathbf{p}_x \rangle \\ \sum p_{x_i} \\ \sum p_{y_i} \end{pmatrix}. \tag{11}$$

The components of the momentum map are the total angular momentum of the tensegrity about the origin and the total linear momenta in the x and y directions.

Let $\mathbf{q}^e = (\mathbf{x}^e, \mathbf{y}^e)$ correspond to a tensegrity as designed in the previous sections. Let $\xi = (\omega, V_x, V_y) \in \mathfrak{se}(2)$. By Theorem 1, the relative equilibria $\mathbf{z}^e = (\mathbf{x}^e, \mathbf{y}^e, \mathbf{p}_x^e, \mathbf{p}_y^e) \in \mathbb{R}^{4n}$ satisfy

$$\frac{\partial H_\xi}{\partial \mathbf{x}}(\mathbf{z}^e) = -\omega\mathbf{p}_y^e + \Omega\mathbf{x}^e = \mathbf{0} \tag{12}$$

$$\frac{\partial H_\xi}{\partial \mathbf{y}}(\mathbf{z}^e) = \omega\mathbf{p}_x^e + \Omega\mathbf{y}^e = \mathbf{0} \tag{13}$$

$$\frac{\partial H_\xi}{\partial \mathbf{p}_x}(\mathbf{z}^e) = \mathbf{p}_x^e + \omega\mathbf{y}^e - V_x\mathbf{1} = \mathbf{0} \tag{14}$$

$$\frac{\partial H_\xi}{\partial \mathbf{p}_y}(\mathbf{z}^e) = \mathbf{p}_y^e - \omega\mathbf{x}^e - V_y\mathbf{1} = \mathbf{0}. \tag{15}$$

By design $\Omega \mathbf{x}^e = \Omega \mathbf{y}^e = \mathbf{0}$. Choosing $\xi = (0, V_x, V_y)$ and $(\mathbf{p}_x^e, \mathbf{p}_y^e) = (V_x \mathbf{1}, V_y \mathbf{1})$, we satisfy equations (12)-(15). Therefore $\mathbf{z}^e = (\mathbf{x}^e, \mathbf{y}^e, V_x \mathbf{1}, V_y \mathbf{1})$ is a relative equilibria of the system.

Next we compute $\delta^2 H_\xi(\mathbf{z}^e)$, the second variation of H_ξ evaluated at the relative equilibrium:

$$\delta^2 H_\xi(\mathbf{z}^e) = \begin{pmatrix} \Omega + L_{\omega x}(\mathbf{q}^e) & L_{\omega xy}(\mathbf{q}^e) & 0_n & 0_n \\ L_{\omega xy}(\mathbf{q}^e) & \Omega + L_{\omega y}(\mathbf{q}^e) & 0_n & 0_n \\ 0_n & 0_n & I_n & 0_n \\ 0_n & 0_n & 0_n & I_n \end{pmatrix} \tag{16}$$

where the ijth element of each matrix is

$$L_{\omega x}(i,j) = \begin{cases} -\alpha_{ij}|\omega_{ij}|\frac{(x_i-x_j)^2 L_{ij}}{r_{ij}^3} & \text{if } i \neq j \\ \sum_{j=1 j\neq i}^{n} \alpha_{ij}|\omega_{ij}|\frac{(x_i-x_j)^2 L_{ij}}{r_{ij}^3} & \text{if } i = j \end{cases}$$

$$L_{\omega y}(i,j) = \begin{cases} -\alpha_{ij}|\omega_{ij}|\frac{(y_i-y_j)^2 L_{ij}}{r_{ij}^3} & \text{if } i \neq j \\ \sum_{j=1 j\neq i}^{n} \alpha_{ij}|\omega_{ij}|\frac{(y_i-y_j)^2 L_{ij}}{r_{ij}^3} & \text{if } i = j \end{cases}$$

and

$$L_{\omega xy}(i,j) = \\ \begin{cases} -\alpha_{ij}|\omega_{ij}|\frac{(x_i-x_j)(y_i-y_j) L_{ij}}{r_{ij}^3} & \text{if } i \neq j \\ \sum_{j=1 j\neq i}^{n} \alpha_{ij}|\omega_{ij}|\frac{(x_i-x_j)(y_i-y_j) L_{ij}}{r_{ij}^3} & \text{if } i = j. \end{cases}$$

We first show that this matrix is positive semi-definite. This is equivalent to proving that the top left $2n$ by $2n$ block in (16) given by

$$K = \begin{pmatrix} \Omega + L_{\omega x}(\mathbf{q}^e) & L_{\omega xy}(\mathbf{q}^e) \\ L_{\omega xy}(\mathbf{q}^e) & \Omega + L_{\omega y}(\mathbf{q}^e) \end{pmatrix} \tag{17}$$

is positive semi-definite. Recall that we have designed Ω to be positive semi-definite. The matrices $L_{\omega x}$ and $L_{\omega y}$ are symmetric and have all their off diagonal terms negative. The diagonal terms are positive so that each row sums to 0. In other words, $L_{\omega x}$ and $L_{\omega y}$ are diagonally dominant symmetric matrices. From the Gersgorin theorem [16] those two matrices are positive semi-definite. We assert (and will prove in a later publication) that $\begin{pmatrix} L_{\omega x} & L_{\omega xy} \\ L_{\omega xy} & L_{\omega y} \end{pmatrix} \geq 0$. This implies that $K \geq 0$ and $\delta^2 H_\xi(\mathbf{z}^e) \geq 0$.

Lemma 1. *The kernel of* $\delta^2 H_\xi(\mathbf{z}^e)$ *is equal to*

$$\mathit{span}\left\{ \begin{pmatrix} \mathbf{1} \\ \mathbf{0} \\ \mathbf{0} \\ \mathbf{0} \end{pmatrix}, \begin{pmatrix} \mathbf{0} \\ \mathbf{1} \\ \mathbf{0} \\ \mathbf{0} \end{pmatrix}, \begin{pmatrix} -\mathbf{y}^e \\ \mathbf{x}^e \\ \mathbf{0} \\ \mathbf{0} \end{pmatrix} \right\}.$$

Proof: See the appendix.

We now investigate if the vectors in the kernel of $\delta^2 H_\xi(\mathbf{z}^e)$ are in $\mathcal{S}$ (given by Definition 1). First we compute $\mathbf{DJ}(\mathbf{z}^e)$ to be

$$\mathbf{DJ}(\mathbf{z}^e) = \begin{pmatrix} V_y\mathbf{1}^T & -V_x\mathbf{1}^T & -\mathbf{y}^{eT} & \mathbf{x}^{eT} \end{pmatrix}.$$

Next we compute G_μ. In our case, $G = SE(2)$ and $\mu \in \mathfrak{se}(2)^*$. By (11), we have $\mu = \mathbf{J}(\mathbf{z}^e) = (\mu_1, nV_x, nV_y)$. Let $g = (\theta, T_x, T_y) \in SE(2)$ then the coadjoint action $g \cdot \mu$ is

$$\begin{aligned} &Ad^*_{(\theta,T_x,T_y)^{-1}}(\mu_1, nV_x, nV_y) \\ &= (\mu_1 - R(\theta)\mathbf{V}\cdot \mathbb{J}\mathbf{T}, R(\theta)\mathbf{V}), \end{aligned}$$

$$R(\theta) = \begin{pmatrix} \cos\theta & -\sin\theta \\ \sin\theta & \cos\theta \end{pmatrix}, \quad \mathbb{J} = \begin{pmatrix} 0 & 1 \\ -1 & 0 \end{pmatrix},$$

$$\mathbf{T} = \begin{pmatrix} T_x \\ T_y \end{pmatrix}, \quad \mathbf{V} = \begin{pmatrix} nV_x \\ nV_y \end{pmatrix}.$$

If $\mathbf{V} \neq 0$ then

$$G_\mu = \{g \in SE(2) \mid R(\theta) = I \text{ and } V_x T_y = V_y T_x\}.$$

Having computed $\mathbf{DJ}(\mathbf{z}^e)$ and G_μ, we can now compute S as follows [14]. Let

$$\mathcal{V} = \{\delta\mathbf{q} \in T_{\mathbf{q}^e}Q \mid \langle \delta\mathbf{q}, \xi_Q(\mathbf{q}^e)\rangle = 0 \ \forall \xi \in \mathfrak{g}_\mu\}. \tag{18}$$

Then,

$$\mathcal{S} = \{\delta\mathbf{z} \in \ker \mathbf{DJ}(\mathbf{z}^e) \mid T\pi_Q \cdot \delta\mathbf{z} \in \mathcal{V}\} \tag{19}$$

where $\pi_Q : T^*Q \to Q$ is the projection. This leads to $\delta\mathbf{z} = (\delta\mathbf{q}, \mathbf{0}, \mathbf{0}) \in \mathcal{S}$ if and only if

$$\begin{pmatrix} V_x\mathbf{1}^T & V_y\mathbf{1}^T \end{pmatrix} \begin{pmatrix} \delta\mathbf{x} \\ \delta\mathbf{y} \end{pmatrix} = 0 \tag{20}$$

$$\begin{pmatrix} V_y\mathbf{1}^T & -V_x\mathbf{1}^T \end{pmatrix} \begin{pmatrix} \delta\mathbf{x} \\ \delta\mathbf{y} \end{pmatrix} = 0. \tag{21}$$

Since there exist $\alpha, \beta \in \mathbb{R}$ such that

$$\delta\mathbf{q} = \begin{pmatrix} \delta\mathbf{x} \\ \delta\mathbf{y} \end{pmatrix} = \alpha \begin{pmatrix} \mathbf{1} \\ \mathbf{0} \end{pmatrix} + \beta \begin{pmatrix} \mathbf{0} \\ \mathbf{1} \end{pmatrix} + \begin{pmatrix} -\mathbf{y}^e \\ \mathbf{x}^e \end{pmatrix} \tag{22}$$

satisfies both (20)-(21), then $\delta\mathbf{z} = (\delta\mathbf{q}, \mathbf{0}, \mathbf{0}) \in \mathcal{S}$ is also in the kernel of $\delta^2 H_\xi(\mathbf{z}^e)$.

Thus, the Energy Momentum Theorem does not provide a conclusive result on the stability of the relative equilibrium $\mathbf{z}^e$. However, we note that $\delta\mathbf{z} = (\delta\mathbf{q}, \mathbf{0}, \mathbf{0})$ with $\delta\mathbf{q}$ given by (22) corresponds to a rotation of the tensegrity about its center of mass. Any rotated tensegrity moving with constant velocity is just another relative equilibrium. Indeed there is a continuum of relative equilibria for a given tensegrity shape moving at constant velocity, parameterized by the orientation of the tensegrity. Simulations restricted to a constant momentum surface with $\mu = (0, nV_x, nV_y)$ reveal no rotational drift. In the full space, simulations exhibit drift only in the SE(2) directions, i.e., translational and rotational drift. Of particular note, there are solutions in the full space, near the relative equilibrium studied, that correspond to a rotating and translating formation.

6 Final Remarks

We present a new coordinated control law for a group of vehicles in the plane that creates an arbitrary desired group shape. The control law is derived for an arbitrary shape using tensegrity structures modelled by (6). The symmetries in the coupled system and the energy-momentum method are used to investigate stability of relative equilibria corresponding to steady translations of the prescribed rigid shape. The energy momentum method alone does not provide conclusive results; however, the relative equilibrium does appear to be stable when the dynamics are restricted to the constant momentum surface. The analysis led to the discovery of solutions in the full space corresponding to rotating and translating formations; One way of dealing with these rotating solutions, seems to add damping in the dynamics. Simulations of the system with damping seem to suggest that the relative equilibrium we studied becomes stable even when we do not restrain ourselves to the constant momentum surface. The formation rotating and translating is not anymore an equilibrium.

Acknowledgements

This work was supported in part by the Office of Naval Research grants N00014–02–1–0826 and N00014–04–1–0534 and by NSF Research grant CMS-0625259.

References

1. F. Zhang, M. Goldgeier, and P. S. Krishnaprasad. Control of small formations using shape coordinates. *Proc. 2003 IEEE Int. Conf. Robotics Aut.*, pages 2510–2515, 2003.
2. N.E. Leonard, D. Paley, F. Lekien, R. Sepulchre, D. M. Fratantoni, and R. Davis. Collective motion, sensor networks and ocean sampling. *Proceedings of the IEEE*, Special Issue on Networked Control Systems, 2006. To appear.
3. E. Fiorelli, N. E. Leonard, P. Bhatta, D. Paley, R. Bachmayer, and D. M. Fratantoni. Multi-AUV control and adaptive sampling in Monterey Bay. In *Proc. IEEE Workshop on Multiple AUV Operations*, 2004. To appear, IEEE J. Oceanic Engineering.
4. F. Zhang and N. Leonard. Generating contour plots using multiple sensor platforms. In *Proc. of 2005 IEEE Symposium on Swarm Intelligence*, pages 309–314, 2005.

5. R.E. Skelton, J.W. Helton, R. Adhikari, J.P. Pinaud, and W. Chan. An introduction to the mechanics of tensegrity structures. In *The Mechanical Systems Design Handbook.* CRC Press, 2001.
6. K. Snelson. Continuous tension, discontinuous compression structures. *U.S. Patent 3,169,611*, 1965.
7. R. Buckminster Fuller. Tensile-integrity structures, *U.S. Patent 3,063,521*, 1962.
8. R. Connelly. Rigidity and energy. *Invent. Math.*, 66:11–33, 1982.
9. R. Connelly. Generic global rigidity. *Discrete Comput. Geom.*, 33:549–563, 2005.
10. R. Connelly. Tensegrity structures: Why are they stable? In *Rigidity Theory and Applications*, pages 47–54. Plenum Press, 1999.
11. R. Connelly and W. Whiteley. Second-order rigidity and prestress stability for tensegrity frameworks. *SIAM J. Discrete Math.*, 9(3):453–491, 1996.
12. G. Zanotti and C. Guerra. Is tensegrity a unifying concept of protein folds? *FEBS Letters*, 534:7–10, 2003.
13. D.E Ingber. Cellular tensegrity: Defining new rules of biological design that govern the cytoskeleton. *Journal of Cell Science*, 104:613–627, 1993.
14. J.E. Marsden. *Lectures on Mechanics.* Cambridge University Press, 2004. Third ed.
15. J.E. Marsden and T.S Ratiu. *Introduction to Mechanics and Symmetry.* Springer-Verlag, 1999. Second ed.
16. R.A. Horn and C.R. Johnson. *Matrix Analysis.* Cambridge University Press, 1985.

APPENDIX: Proof of Lemma 1

Proving Lemma 1 is equivalent to proving that the kernel of K is exactly equal to the span of

$$\{\mathbf{w}_1, \mathbf{w}_2, \mathbf{w}_3\} = \left\{ \begin{pmatrix} \mathbf{1} \\ \mathbf{0} \end{pmatrix}, \begin{pmatrix} \mathbf{0} \\ \mathbf{1} \end{pmatrix}, \begin{pmatrix} -\mathbf{y}^e \\ \mathbf{x}^e \end{pmatrix} \right\}.$$

We can rewrite K as

$$K = K_1 + K_2 = \begin{pmatrix} \Omega & 0_n \\ 0_n & \Omega \end{pmatrix} + \begin{pmatrix} L_{\omega x} & L_{\omega xy} \\ L_{\omega xy} & L_{\omega y} \end{pmatrix}.$$

Since K_1, K_2 are symmetric positive semi-definite,

$$\mathbf{q} \in \ker(K) \iff \mathbf{q} \in \ker(K_1) \text{ and } \mathbf{q} \in \ker(K_2).$$

By design $\{\mathbf{w}_1, \mathbf{w}_2, \mathbf{w}_3\}$ are in the kernel of K_1. By direct computation, they are also in the kernel of K_2. We now show that the kernel of K_2 is exactly equal to the span of $\{\mathbf{w}_1, \mathbf{w}_2, \mathbf{w}_3\}$. Since we exclude the case of all nodes in a line, the span of $\{\mathbf{w}_1, \mathbf{w}_2, \mathbf{w}_3\}$ is three dimensional. We complete $\{\mathbf{w}_1, \mathbf{w}_2, \mathbf{w}_3\}$ with $2n - 3$ vectors $\{\mathbf{w}_4, \ldots, \mathbf{w}_{2n}\}$ from the canonical basis of $\mathbb{R}^{2n}$ making sure that we have $2n$ linearly independent vectors. It is then easy to check that

$$K_2 \mathbf{v}_i \neq 0 \quad \forall i \geq 4$$

and so the kernel of K_2 and therefore the kernel of K is exactly spanned by $\{\mathbf{w}_1, \mathbf{w}_2, \mathbf{w}_3\}$. □

A Family of Pumping-Damping Smooth Strategies for Swinging Up a Pendulum

Javier Aracil[1], Francisco Gordillo[1], and Karl J. Åström[2]

[1] Escuela Superior de Ingenieros, Universidad de Sevilla. Camino de los Descubrimientos s/n. Sevilla-41092. Spain
{aracil,gordillo}@esi.us.es

[2] Department of Automatic Control LTH. Lund University. Box 118 SE-221 00 Lund. Sweden
kja@control.lth.se

Summary. In this paper we present some additional results regarding the pumping-damping strategy for swinging up a pendulum introduced in [3]. Here, the family of energy functions is enlarged and the corresponding pumping-damping functions are proposed giving rise to new smooth controllers that swing up and stabilize the pendulum. Furthermore, a generalization of the stability criterion is introduced for this larger class of controllers.

Keywords: Pendulum, Shaping Hamiltonians, Swing-up, Energy management.

1 Introduction

The family of the inverted pendula has attracted the attention of control researchers in recent decades as a benchmark for testing and evaluating a wide range of classical and contemporary nonlinear control methods. As it is well known the inverted pendulum displays two main problems: swinging up the pendulum to the upright position and stabilizing it in this position once it is reached. The classical solution to this problem is given in [4], and is based on energy considerations. However, this solution is a hybrid one. A main challenge for the control community has been to find a single smooth law that copes with the global problem of leading the pendulum from the hanging position (or any other initial state, both in position and velocity) to the upright one. A solution to that problem is given in [2, 3]. In the present paper we generalize the results of [3] to a larger class of energy functions and pumping-damping strategies.

We consider the simplest version of the pendulum [4]: the control action is the acceleration of the pivot and, thus, a 2-dimensional model is used. Here, we return to the idea of [3]. First, an energy shaping control law is designed in such a way that: 1) the closed-loop energy presents a minimum at the desired position; and 2) the energy shaping controller is globally defined. Since the chosen target energy has other minima different than the desired equilibrium, a combination of energy dissipation (damping) and injection (pumping) is needed in order to

F. Bullo et al. (Eds.): Lag. & Hamil. Methods for Nonlin. Ctrl., LNCIS 366, pp. 341–352, 2007.
springerlink.com

globally stabilize the origin. To that end an oval closed curve circumscribing about the region where pumping is needed was introduced in [3] for a quite simple case. The resultant law is smooth, no commutations are needed, and the origin of the final closed-loop system is almost-globally asymptotically stable. In this paper, we enlarge the family of controllers based on these ideas, giving several examples of control laws that fulfil the control objective and we broaden also the pumping-damping strategies with the new concept of circumscribing ovals.

The paper is organized as follows. Section 2 presents a family of suitable target energy functions for the pendulum, as well as the pumping-damping idea. In Section 3, several examples of controllers belonging to the family are presented. In Section 4 a stability criterion which generalices the one in [3] is introduced. The paper closes with a Section of conclusions.

2 Energy Shaping and Pumping-Damping

The normalized model of the pendulum system is

$$\begin{aligned} \dot{x}_1 &= x_2 \\ \dot{x}_2 &= \sin x_1 - u \cos x_1, \end{aligned} \tag{1}$$

being x_1 the angular position of the pendulum (the origin at the upright position) and x_2 the velocity, so it is defined on the manifold $\mathcal{S} \times R$.

To design a controller for the swing up problem the potential energy shaping method is used. For the moment we will focus in the choice of the energy function and, thus, a conservative target system will be chosen, leaving for later damping-pumping addition. For this, consider the desired system

$$\begin{aligned} \dot{x}_1 &= x_2 \\ \dot{x}_2 &= -V_d'(x_1), \end{aligned} \tag{2}$$

which is a Hamiltonian system with Hamiltonian function

$$H_d(x_1, x_2) = V_d(x_1) + \frac{x_2^2}{2}, \tag{3}$$

where V_d should have a single minimum at the desired upright position. Thus, we should impose that $V_d'(0) = 0$, $V_d''(0) > 0$, and for symmetry reasons, $V_d(x_1) = V_d(-x_1)$.

To solve the matching problem of (1) and (2) a good choice of V_d', in order to avoid the division by $\cos x_1$, is

$$V_d' = -\sin x_1 + \beta(x_1) \cos x_1, \tag{4}$$

and then, $u = \beta(x_1)$.

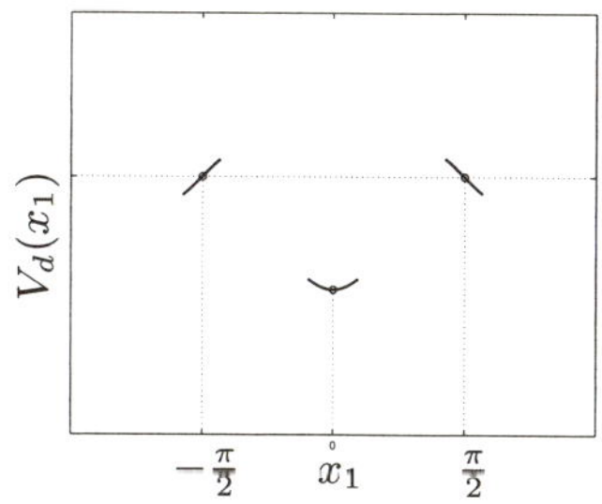

Fig. 1. Main properties shared by the V_d belonging to the family (5)

A family of functions V_d that fulfills these conditions for appropriate values of parameters a_i is given by

$$V_d = a_0 + \cos x_1 - a_2 \cos^2 x_1 - a_3 \cos^3 x_1 - \cdots , \tag{5}$$

which yields

$$V_d' = -\sin x_1 + \sin x_1 \cos x_1 (2a_2 + 3a_3 \cos x_1 + ...), \tag{6}$$

and therefore $\beta(x_1) = \sin x_1(2a_2 + 3a_3 \cos x_1 + ...)$ to match expression (5) with (4). It should be noted that function H_d given by (3) and (5) is conservative for system (1) with $u = \beta(x_1)$.

The simplest case of this family is obtained by taking $a_0 = -1/4a$, $a_2 = a$ and $a_k = 0, \forall k > 2$, which leads to

$$V_d(x_1) = \cos x_1 - a \cos^2 x_1 - \frac{1}{4a}. \tag{7}$$

and then to the feedback law

$$u = u_{es} = 2a \sin x_1, \tag{8}$$

where the notation u_{es} has been introduced in order to point out that this is an energy-shaping control law. This controller is studied in [3].

We consider now the problem of the choice of the energy variation strategy. Notice that a pure damping control law would not work. To see this, consider the shape of any V_d belonging to family (5). It can be seen that it displays the traits shown in Fig. 1. This means the there is at least a maximum at $x_1 = x_1^0$ in the interval $(0, \pi/2)$ (resp. $(-\pi/2, 0)$). These maxima will give rise to saddles in the energy function (3). Out of the interval $(-x_1^0, x_1^0)$ there must exist at least one minima (see Fig. 1). With pure damping strategies, these extra minima give rise to undesired attraction basins, which we will call in the sequel "undesirable wells" in the energy landscape, because they preclude the global nature of the stability of the equilibrium at the upright position.

To overcome the difficulty with the undesirable wells we propose a strategy that consists in pumping energy to make the trajectories to leave them. This strategy is discussed in the following.

System (1) has as passive output $y = -x_2 \cos x_1$. This means that taking as input $u = -k_a y$ the system behaves in such a way that H_d decreases. However our goal is to increase H_d in the regions where pumping of energy is needed. Therefore we should modulate the sign of k_a according needs.

If we make $u = u_{es} + u_{pd}$ (where u_{pd} stands for pumping-damping) in (1), the instantaneous energy variation is given by

$$\dot{H}_d = -x_2 u_{pd} \cos x_1 . \tag{9}$$

We take $k_a = bF(x_1, x_2)$ where $b > 0$ and function F is negative in the region of the state space where we want to pump energy into the system and positive where we desire to damp the system. Then, $u_{pd} = bx_2 F(x_1, x_2) \cos x_1$, and therefore the control law is

$$u = \underbrace{\beta(x_1)}_{u_{es}} + \underbrace{bx_2 F(x_1, x_2) \cos x_1}_{u_{pd}} . \tag{10}$$

The first term of this controller can be interpreted as a nonlinear spring and we can therefore call it the "spring term". This term makes the pendulum to behave conservative. The second one is the pumping-damping term. In the next Section the problem of choosing appropriate F functions for different V_d is discussed.

3 Different Choices of the Energy Function

In this Section three concrete cases for the energy function V_d are considered. All the cases are based on the same idea: first, an V_d belonging to the family (5) and with a minimum at the origin is chosen; then a control law is defined with the following structure:

$$u = u_{es} + bx_2 F(x_1, x_2) \cos x_1, \tag{11}$$

where u_{es} shapes the energy in order to obtain the corresponding

$$H_d(x_1, x_2) = V_d + \frac{x_2^2}{2}, \tag{12}$$

and $F(x_1, x_2) = 0$ is an appropriated oval that circumscribes about the boundaries of the undesirable wells introduced by the shape of V_d. Parameter $b > 0$ defines the amount of energy that is pumped or damped.

3.1 A Case with Three Wells

Consider the following potential energy function that belongs to the family (5)

$$V_{d_A}(x_1) = -\frac{\cos 3x_1}{3} - \frac{1}{3} = \cos x_1 - \frac{4}{3}\cos^3 x_1 - \frac{1}{3}, \tag{13}$$

which is shown in Fig. 2. It can be seen that it has three minima. The central one corresponds to the desired position and the other two are undesirable. The

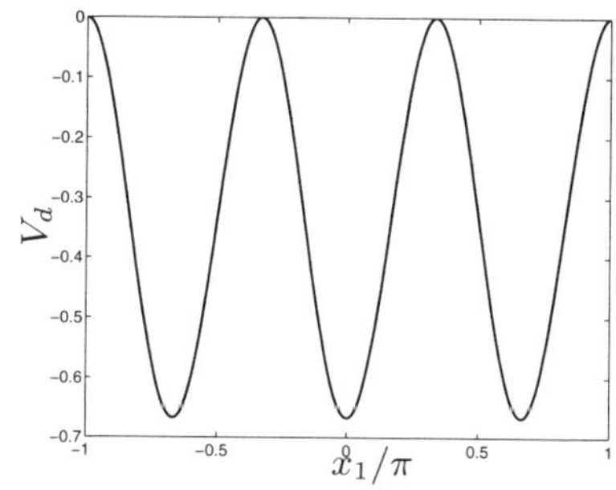

Fig. 2. Graph of $V_{d_A}(x_1)$

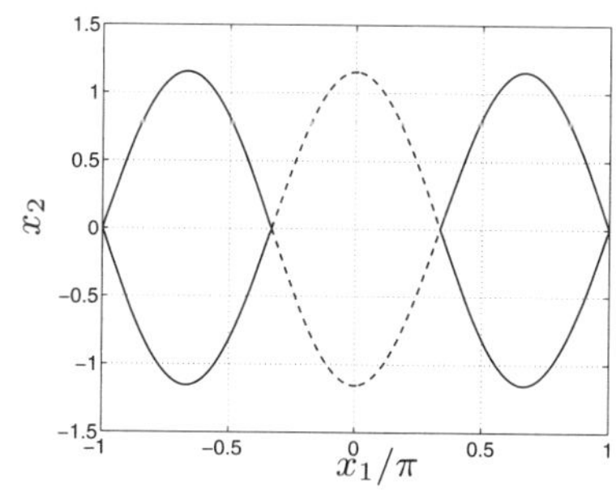

Fig. 3. Curve $\frac{x_2^2}{2} + V_{d_A}(x_1, x_2) = 0$. In solid the boundaries of the undesirable wells.

constant term in this function has been chosen so that the undesirable wells are bounded by $H_{d_A} = V_{d_A} + x_2^2/2 = 0$. These boundaries are shown in Fig. 3.

V_{d_A} of Eq. (13) yields

$$u_{es_A} = 4\cos x_1 \sin x_1 = 2\sin 2x_1, \tag{14}$$

that accomplishes the matching goal.

Note in Fig. 3 that neither of the undesirable wells contains the hanging position, which is an unstable equilibrium of saddle type. This fact induces a difference with respect to the case studied in [3]: Here the equilibrium $(x_1, x_2) = (\pi, 0)$ is a saddle even in absence of pumping-damping, while in [3] this equilibrium is a center for $b = 0$ and unstable for $b \neq 0$.

In order to find a curve that circumscribes about the undesirable wells we choose a family of curves of the form

$$\frac{x_2^2}{2} + \alpha_0 + \alpha_1 \cos x_1 + \alpha_2 \cos^2 x_1 = 0. \tag{15}$$

Imposing that the curve passes through the points $(\pi/3, 0)$ and $(\pi, 0)$ and that the curve is tangent to $x_2^2/2 + V_{d_A} = 0$ we obtain:

$$F_A(x_1, x_2) = \frac{x_2^2}{2} - 1 + \cos x_1 + 2\cos^2 x_1 \tag{16}$$

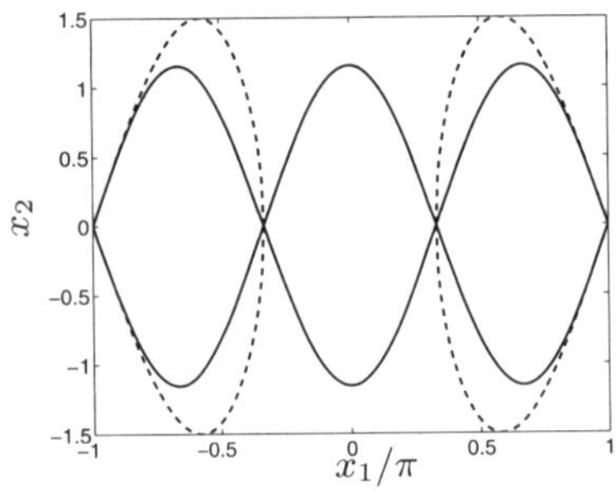

Fig. 4. Shape of $F_A = 0$ (dashed) circumscribing about the boundaries of the undesirable wells associated with the curve $x_2^2/2 + V_{d_A}(x_1) = 0$ (solid)

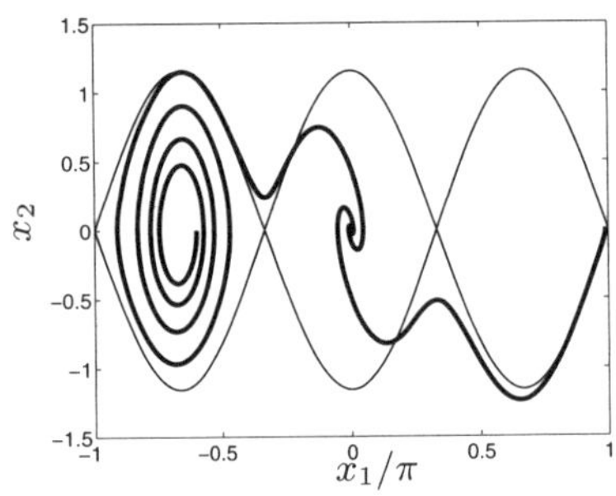

Fig. 5. Results of two simulations corresponding to controller (17) with $b = 0.4$

The curve $F_A = 0$ is represented in Fig. 4. With respect to [3], a term in $\cos^2 x_1$ has been added to shape appropriately the oval.

Therefore, we propose a controller such as (11) that now becomes

$$u = 2a \sin 2x_1 + b x_2 F_A(x_1, x_2) \cos x_1, \tag{17}$$

with F_A given by (16).

Simulations show the good performance of this controller (Fig. 5). If the initial condition is the hanging position at rest, this controller does not need energy injection (as does the one in [3]) for leaving this hanging position, as shown in the simulation that appears on the right of Fig. 5.

Remark 1. Other choices of circumscribing curves are possible. For example (Fig. 6)

$$\tilde{F}_A(x_1, x_2) = \cos x_1 - \frac{1}{2} + \frac{2x_2^2}{3}, \tag{18}$$

which corresponds to $\alpha_2 = 0$, and remembers the ovals used in [3]. Simulations show that this controller works satisfactorily. However, it will be seen below that in this case the stability criteria proposed in the current paper does not guaranty stability. This fact justifies the inclusion of the term $\cos^2 x_1$ in F.

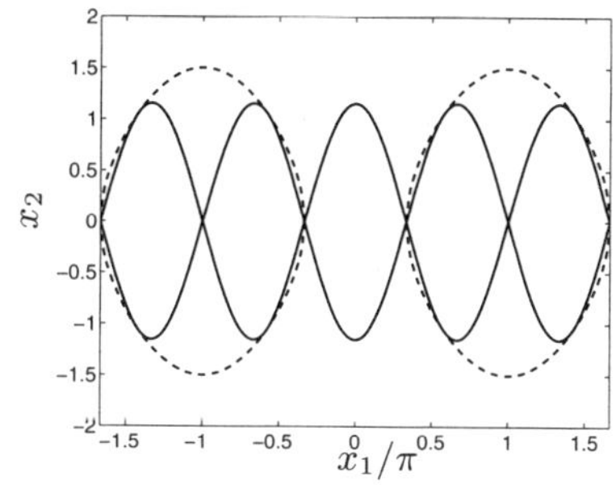

Fig. 6. Curves $x_2^2/2 + V_{d_A}(x_1, x_2) = 0$ (solid) and $\tilde{F}_A = 0$ (dashed)

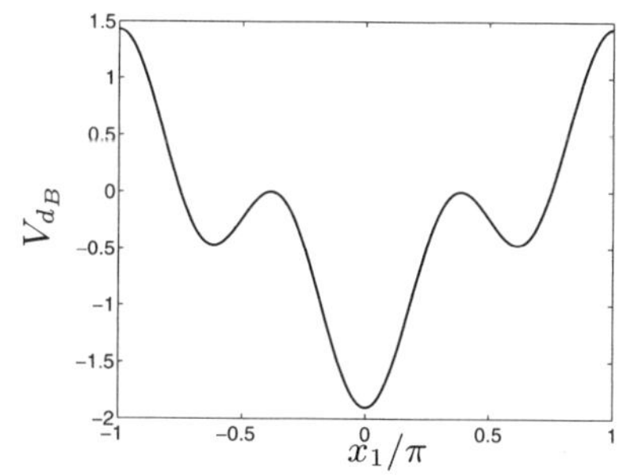

Fig. 7. Graph of $V_{d_B}(x_1)$

3.2 A Case with Asymmetrical Undesirable Wells

Another interesting V_d (represented in Fig. 7) is obtained by the choice $a_0 = -\sqrt{2}/6$, $a_3 = 8/3$ and $a_{2k+1} = 0, \forall k \neq 0$ in (5), which gives [1]

$$V_{d_B}(x_1) = \cos x_1 - \frac{8}{3}\cos^3 x_1 - \frac{\sqrt{2}}{6}. \tag{19}$$

As before, the term a_0 has been chosen so that the undesirable wells are bounded by $V_{d_B} = 0$. In this last case the matching yields

$$u_{es_B} = 4 \sin 2x_1. \tag{20}$$

The shape of V_{d_B} is shown in Fig. 7. It displays two undesirable wells that are asymmetrical, in the sense that the maximum values at both sides of the undesirable wells are different.

The asymmetry of the undesirable wells leads to the fact that, in absence of pumping and damping, their boundaries are no more heteroclinic orbits but homoclinic ones (Fig. 8). Figure 8 shows that the closed curve defined by

$$\frac{x_2^2}{2} - \frac{4+\sqrt{2}}{6} + \frac{1+2\sqrt{2}}{3}\cos x_1 + \frac{2\sqrt{2}+8}{3}\cos^2 x_1 = 0, \tag{21}$$

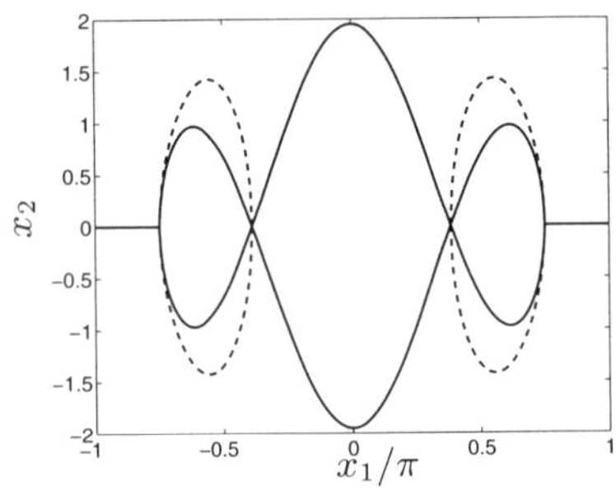

Fig. 8. Curves $x_2^2/2 + V_{d_B}(x_1, x_2) = 0$ (solid) and $F_B = 0$ (dashed)

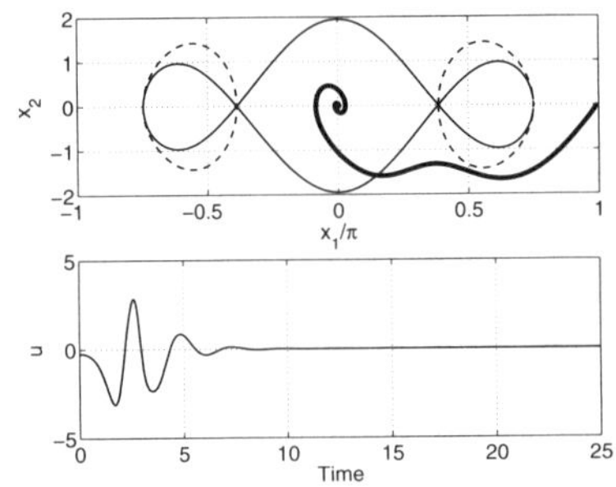

Fig. 9. Results of a simulation corresponding to the case (20)–(21) with $b = 0.4$

circumscribes about the two undesirable wells. It is worthy to note that in this case the circumscribing curve is split in two, one for each well.

As before, simulations show the good performance of this controller (Fig. 9).

3.3 A More Sophisticated Case

Many other V_d belonging to the same family can be conceived. Recalling (5), this expression can be rewritten as

$$V_d = a_0 + \cos x_1 - \cos^2 x_1 (a_2 + a_3 \cos x_1 + \cdots ,)$$

therefore

$$V_d = a_0 + \cos x_1 - \cos^2 x_1 f(\cos x_1)$$

where $f(y)$ has to be chosen such that V_d has the appropriate shape (a minimum at the desired upright position) and it has a series expansion of the form

$$f(y) = b_0 + b_1 y + b_2 y^2 + \cdots$$

For instance, $f(y) = \exp(y)(1 + y)$ yields

$$V_{d_C} = \cos x_1 - \cos^2 x_1 \exp(\cos x_1)(1 + \cos x_1) - 0.1497, \quad (22)$$

which is shown in Fig. 10. With V_{d_C} and applying the same procedure as before

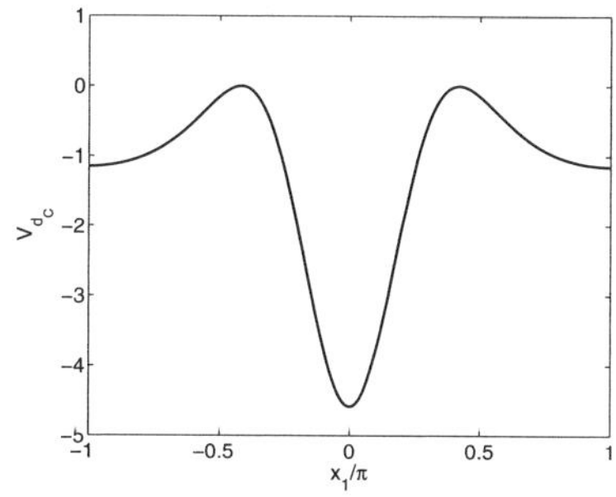

Fig. 10. Graph of $V_{d_C}(x_1)$

$$
\begin{aligned}
u = {} & \sin x_1 \exp(\cos x_1)(2 + 4\cos x_1 + \cos x_1^2) \\
& + b x_2 F_C(x_1, x_2) \cos x_1, \qquad (23)
\end{aligned}
$$

where a reasonable oval circumscribing about the only undesirable well is given by

$$F_C(x_1, x_2) = 2x_2^2 + 4\cos x_1 - 1. \qquad (24)$$

Figure 12 shows the good results for two simulations of the system with the resultant control law.

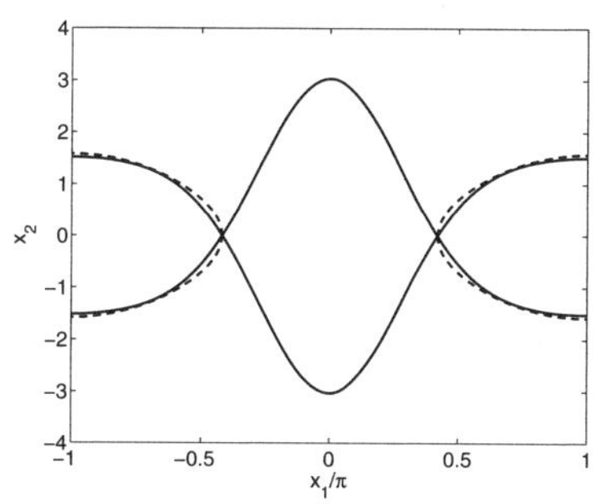

Fig. 11. Curves $x_2^2/2 + V_{d_C}(x_1)$=0 (solid) and $F_C = 0$ (dashed)

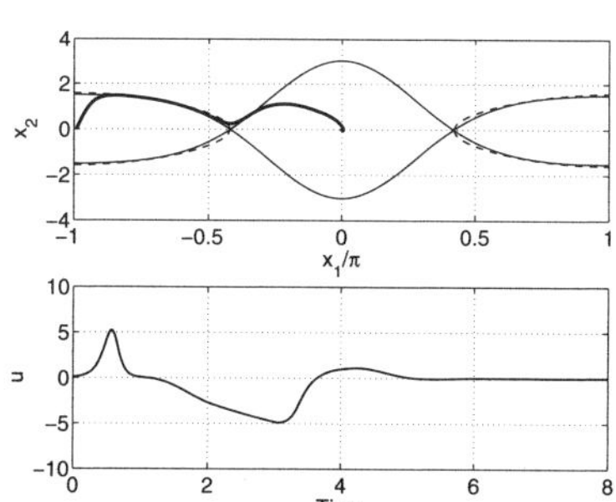

Fig. 12. Example of trajectory (top) and control signal (bottom) showing the good behavior of controller (23-24) with $b = 1.75$

4 A Stability Criterion

The stability criterion presented in [3] for the concrete case studied there, is generalized here for the whole family of energy functions (5) considered in the current paper. For each $V_d(x_1, x_2)$ having a minimum at the origin, the corresponding $F(x_1, x_2)$ is chosen in such a way that it circumscribes about the undesirable wells. Once function F is chosen, the following functions can be defined:

- $\varphi_H(x_1)$ (with $x_1 \in [0, \pi]$) as the x_2 coordinate of the upper curve defined by $H(x_1, x_2) = V(x_1) + x_2^2/2 = 0$. For the values of x_1 where no value of x_2 satisfies this last expression, take $\varphi_H(x_1) = 0$.
- $\varphi_F(x_1)$ as the x_2 coordinate of the upper curve defined by $F(x_1, x_2) = 0$. This curve will only be defined in some set Ω. Let $\bar{\Omega}$ be the complement set of Ω in $[0, \pi]$, that is $\bar{\Omega} = [0, \pi] \backslash \Omega$.
-
$$\begin{aligned}\Phi = & \int_{\bar{\Omega}} \varphi_H(x_1) \cos^2(x_1) F(x_1, \varphi_H(x_1)) dx_1 \\ & + \int_{\Omega} \varphi_F(x_1) \cos^2(x_1) F(x_1, \varphi_H(x_1)) dx_1 \end{aligned} \tag{25}$$

Theorem 1. *Consider system (1) with control law (11) and F defined in such a way that $F = 0$ circumscribes about the boundaries of the undesirable wells. The origin is almost-globally asymptotically stable if $\Phi > 0$.*

Proof. : Only a sketch of the proof is given here[1]. Due to the pumping mechanism inside the ovals, it is easy to see that trajectories go out of the undesirable wells. Consider a trajectory once out of any undesirable well. It can reach the desired well or turn around the manifold $S \times R$. In the first case, asymptotic stability is guaranteed by Lyapunov theory. In the second case, we can compute the energy balance through a 2π turn. From Eq. (9) the energy change along such a turn is given by

$$\Delta H_d(x_1, x_2) = -\int_{-\pi}^{\pi} b x_2 \cos^2 x_1 F(x_1, x_2) dx_1,$$

where ΔH_d is the net energy loss along a 2π cycle of x_1. It can be shown that $\Delta H_d > \Phi$, because Φ, according to its definition in (25), gives a conservative measure of the bounds of the net energy loss during a 2π cycle (for details, see [3]). This measure is conservative in the sense that energy injection is overestimated, while damping is underestimated. In consequence $\Phi > 0$ means that the dissipation is higher than energy injection, and so the trajectory in the state space shrinks until it reaches the central well where stability is guaranteed by Lyapunov theory.

[1] We are considering only the case where, in the definition of Φ, $\Omega = [0, \pi]$ and consequently $\bar{\Omega} = \emptyset$. The case where $\bar{\Omega} \neq \emptyset$ is a little more involved but leads to the same conclusions.

Table 1 shows the corresponding values of Φ for the examples given above. As was said before, stability is not guaranteed for the third case in this table, since for this case $\Phi < 0$, in spite of the fact that the system seems to be stable by simulation. This fact shows the conservativeness of the criterion. In any case, the corresponding controller was included just for illustrative purposes since it is clearly outperformed by the ones corresponding to line 2 in Table 1. For the rest of the proposed controllers Theorem 1 guarantees stability. An open problem is the performance comparison of the different choices for potential energy functions as well as for the circumscribing ovals.

Table 1. Values of Φ for the examples of Sect 3

$V_d(x_1)$	$F(x_1, x_2)$	Φ
V_d in [3] with $a = 1$	F in [3]	0.0598
V_{d_A} Eq. (13)	F_A Eq. (16)	1.9846
V_{d_A} Eq. (13)	$\tilde{F}_A$ Eq. (18)	−0.4481
V_{d_B} Eq. (19)	F_B Eq. (21)	7.0034
V_{d_C} Eq. (22)	F_C Eq. (24)	28.3871

5 Conclusions

In this paper a further elaboration of the approach to the swing up problem of the inverted pendulum introduced in [3] is presented. Te approach is based in a two step procedure. First a feedback law is applied such that the desired equilibria becomes stable although not asymptotically stable. However, this introduces more equilibria in the closed loop system. Then, the damping term is added to introduce damping around the desired equilibrium to change it from being stable to being asymptotically stable; and moreover, negative damping (energy pumping) is introduced in the region around the undesirable equilibria to make them unstable. In this way a single controller is obtained that swings up the pendulum and stabilizes it from almost all initial conditions (positions and velocities). In the present paper the results of the previous one are extended to a wider class of energy functions. Furthermore, a new concept of circumscribing ovals is introduced that considerably deepens the approach. For these energy functions and circumscribing ovals a stability criterion is stated that generalices the one of [3]. Four different controllers based on this approach are included in the paper.

Acknowledgments

The Spanish authors have been supported under the MCyT-FEDER grant DPI2003-00429.

References

1. José Ángel Acosta. *Nonlinear control of underactuated systems.* PhD thesis, University of Seville, 2004. In Spanish.
2. J. Aracil and F. Gordillo. The inverted pendulum: A challenge for nonlinear control (in Spanish). *Revista Iberoamericana de Automática e Informática Industrial*, 2(2):8 – 19, 2005.
3. K.J. Åström, J. Aracil, and F. Gordillo. A new family of smooth strategies for swinging up a pendulum. In *16th IFAC World Congress*, Prague, 2005.
4. K.J. Åström and K. Furuta. Swinging up a pendulum by energy control. *Automatica*, 36:287–295, 2000.

An Energy-Shaping Approach with Direct Mechanical Damping Injection to Design of Control for Power Systems

Xiaohong Jiao[1], Yuanzhang Sun[2], and Tielong Shen[3]

[1] Institute of Electrical Engineering, Yanshan University, Qinhuangdao 066004, China
jiaoxh@tsinghua.edu.cn
[2] Department of Electrical Engineering, Tsinghua University, Beijing 100084, China
yzsun@tsinghua.edu.cn
[3] Department of Mechanical Engineering, Sophia University, Tokyo 102-8554, Japan
tetu-sin@sophia.ac.jp

1 Introduction

Recently, the investigation of design approaches to nonlinear control, which can thoroughly exploit the structure and the properties of the physical systems, has been given much attention due to the designed controllers with relatively simple form and effective operation. And it has been shown that for the mechanical and electromechnical systems, by incorporating the Hamiltonian structure into passivity-based control design techniques, the controllers based on physical considerations can be obtained [1],[2],[3]. This design idea has been successfully realized in the control of power systems and a lot of achievements have been gotten (see e.g. [1]-[9]). In [4], the fact that the power system with excitation control is a Hamiltonian structure is exploited by selecting a storage function constructed by system potential and kinetic energy and an appropriate passivating output on the basis of physical arguments. Along this research line, the excitation control problem for multi-machine power systems has been investigated in [5],[6]. Moreover, an adaptive L_2 disturbance attenuation controller based on Hamiltonian structure has been provided for power systems in [7].

However, the selected storage function does not necessarily have a minimum at the desired equilibrium, or the stability region at the desired equilibrium is not large enough for the requirement of the transient stability of power system. In this case, it is required by injecting damping and pre-feedback to shape the energy function with the modified function such that the closed-loop system has dissipative Hamiltonian structure and the stability region is enlarged. This technique is called interconnection and damping assignment passivity-based control (IDA-PBC) in [3]. The IDA-PBC design method is applied to the single excitation loop problem of power system in [8] and the excitation and governor dual

F. Bullo et al. (Eds.): Lag. & Hamil. Methods for Nonlin. Ctrl., LNCIS 366, pp. 353–364, 2007.
springerlink.com

loop control problem in [9]. It should be noted that in IDA-PBC design technique, the energy shaping is achieved by modifying the pattern of energy transfer between the mechanical and the electrical components of the system and assigning the damping in the electrical dynamic description, not directly assigning the mechanical damping in the swing equation. While, the damping coefficient in the swing equation just is an important factor influencing the transient stability of the power system.

Motivated by this intuition, in this paper a novel energy-shaping design method, called an energy-shaping with direct mechanical damping injection approach, is provided for enhancing the transient stability of power systems. The main feature of this method is that a compensation damping is directly injected into the mechanical damping in the swing equation. The adjusted parameters of the designed controller have directly corresponding relation to the physical meaning of the mechanical damping and electrical damping of power system, and the mechanical damping and electrical damping are assigned separately by selecting an appropriate output regulation variable. The remaining of the paper is organized as follows. The problem is descried in section 2. In section 3, a direct mechanical damping assignment method is presented to design an excitation controller. The design idea of direct mechanical damping injection is also extended to the adaptive excitation controller for power system with transmission line faults and parameters uncertainties. Simulation results for four cases with different three phase short circuit fault and the unknown perturbation in the mechanical power are given in section 4. A comparison with the results via the controller designed in [8] is also provided. Concluding remarks follow in final section.

2 Problem Description

For the clarification of the design idea, a single-machine infinite-bus system with the excitation control is considered, which is described as [10]

$$\begin{cases} \dot{\delta} = \omega_s \omega_r \\ \dot{\omega}_r = \dfrac{1}{M}\{-D\omega_r + P_m - P_e(t)\} \\ \dot{E}'_q = \dfrac{1}{T_{d0}}\left\{-\dfrac{X_{d\Sigma}}{X'_{d\Sigma}}E'_q + \dfrac{X_d - X'_d}{X'_{d\Sigma}}V_s\cos\delta + u_f\right\} \end{cases} \tag{1}$$

where $P_e(t)$ is the active electrical power delivered by the generator, which is given by

$$P_e(t) = \frac{V_s}{X'_{d\Sigma}}E'_q \sin\delta - \frac{V_s^2}{2}\frac{X_q - X'_d}{X_{q\Sigma}X'_{d\Sigma}}\sin 2\delta$$

$\delta(t)$ denotes the rotor angle of the generator, $\omega_r(t)$ the relative angular speed of the generator, ω_s the synchronous rotating speed. $E'_q(t)$ is the q axis transient potential, V_s the voltage at the infinite bus. P_m is the turbine mechanical input power. D is the damping constant, M the inertia constant. $X_{d\Sigma} = X_d + X_T + X_L$

is the total reactance which takes into account the direct axis reactance of the generator X_d, the reactance of the transmission line X_L and the reactance of the transformer X_T. $X'_{d\Sigma} = X'_d + X_T + X_L$ with X'_d denoting the direct axis transient reactance of the generator. $X_{q\Sigma} = X_q + X_T + X_L$ with X_q denoting the generator q axis transient reactance. T_{d0} is time constant of the field winding. u_f is the control input of the SCR amplifier of the generator.

Due to the excitation control considered only, the mechanical power P_m is assumed to be a constant in the nominal model (1). If $(\delta_s, \omega_{rs}, E'_{qs})$ denotes the operating equilibrium of the system, it satisfies the following condition

$$\omega_{rs} = 0, \quad P_{es} = a_1 E'_{qs} \sin\delta_s - a_2 \sin 2\delta_s = P_m, \quad u_{fs} = bE'_{qs} - c\cos\delta_s \tag{2}$$

where

$$a_1 = \frac{V_s}{X'_{d\Sigma}}, \quad a_2 = \frac{V_s^2}{2}\frac{X_q - X'_d}{X_{q\Sigma} X'_{d\Sigma}}, \quad b = \frac{X_{d\Sigma}}{X'_{d\Sigma}}, \quad c = \frac{X_d - X'_d}{X'_{d\Sigma}} V_s$$

u_{fs} is the static value of the excitation control. For the simplification of description, the system (1) can be rewritten as the following form

$$\begin{cases} \dot{\delta} = \omega_s \omega_r \\ \dot{\omega}_r = \dfrac{1}{M}\{-D\omega_r + P_m - P_{e0} - a_1(E'_q - E'_{qs})\sin\delta\} \\ \dot{E}'_q = \dfrac{1}{T_{d0}}\{-bE'_q + c\cos\delta + u_f\} \end{cases} \tag{3}$$

where $P_{e0} = a_1 E'_{qs}\sin\delta - a_2 \sin 2\delta$.

Note that $\delta_s \in (0, \frac{\pi}{2})$ and there is another unstable equilibrium point $(\delta_u, 0, E'_{qs})$ with $\sin\delta_u = \sin\delta_s$ for the system (1), which may be close to the stable one making the stability region very small. From the characteristic of the system (3) it can be seen that if there exists an excitation controller not only compensating the damping of the electrical dynamics (namely increasing the coefficient b) but directly assigning the mechanical damping (increasing the coefficient D), the system will have better dynamic properties and large stability margin. And in this case, the tuning parameters of the controller will have directly corresponding relation to the desired mechanical damping and electrical damping of the system.

Therefore, an objective of this paper is to provide a design method with direct mechanical damping injection to excitation controller $u_f = \alpha(\delta, \omega_r, E'_q)$ such that the resulting closed-loop system has large asymptotical stability region at the desired equilibrium $(\delta_s, \omega_s, E'_{qs})$.

In addition, it is probable in practice that there exist short circuits fault in the transmission line and perturbation in the mechanical power. When the short circuits fault occurs in the transmission line, the transmission reactance X_L varied causes the variation of parameters a_1, a_2, b and c. It can be seen from (2) that the variation of the system parameters will cause the equilibrium to shift, so that the generator may be driven to be unstable, loss of synchronous. When the parameter uncertainties are considered, let $\theta_1 = a_1,\ \theta_2 = a_2,\ \theta_3 = b,\ \theta_4 = c$, then, the system (1) can be rewritten as the following

$$\begin{cases} \dot{\delta} = \omega_s\omega_r \\ \dot{\omega}_r = \dfrac{1}{M}\{P_m - D\omega_r - \theta_1 E'_q \sin\delta + \theta_2 \sin 2\delta\} \\ \dot{E}'_q = \dfrac{1}{T_{d0}}\left\{-\theta_3 E'_q + \theta_4 \cos\delta + u_f\right\} \end{cases} \tag{4}$$

Therefore, the other objective of this paper is to solve the problem of the transient stabilization for the system (4) with the unknown parameters θ_i, $(i = 1, 2, 3, 4, 5)$, $\theta_5 = P_m$, which can be formulated as follows: with help of the method of energy-shaping with direct mechanical damping injection, to design an adaptive excitation controller

$$\begin{cases} u_f = \alpha(\delta, \omega_r, E'_q, \hat{\theta}) \\ \dot{\hat{\theta}} = \beta(\delta, \omega_r, E'_q, \hat{\theta}) \end{cases} \tag{5}$$

such that the resulting closed-loop system is Lyapunov stable at the only equilibrium $(\delta_s, \omega_s, E'_{qs}, \hat{\theta}_s)$ and $\delta \to \delta_s$, $\omega \to \omega_s$, $E'_q \to E'_{qs}$ as $t \to \infty$, where $\hat{\theta}$ is the estimate of $\theta = [\theta_1\ \ \theta_2\ \ \theta_3\ \ \theta_4\ \ P_m]^T$.

3 Design with Direct Mechanical Damping Injection

First, an energy-shaping with direct mechanical damping injection approach will be presented to derive an excitation controller for the system (3).

One output regulation variable y is defined

$$y = -a_1(E'_q - E'_{qs})\sin\delta + K_D\omega_r \tag{6}$$

with the tuning parameter $K_D > 0$, then, the subsystem (δ, ω_r) of (3) is equivalent to the following form

$$\begin{cases} \dot{\delta} = \omega_s\omega_r \\ \dot{\omega}_r = \dfrac{1}{M}\{P_m - P_{e0} - (D + K_D)\omega_r + y\} \end{cases} \tag{7}$$

It is seen that K_D is directly added into the term of the mechanical damping, tuning K_D is equivalent to directly assigning the mechanical damping D. And the dynamics of the output variable y is described as the following form

$$\begin{aligned} \dot{y} = & -\frac{a_1 \sin\delta}{T_{d0}}(-bE'_q + c\cos\delta + u_f) - a_1 \Delta E'_q \omega_s\omega_r \cos\delta \\ & + \frac{K_D}{M}\{P_m - P_{e0} - (D + K_D)\omega_r + y\} \end{aligned} \tag{8}$$

with $\Delta E'_q = E'_q - E'_{qs}$.

It is well known [10] that for the second order swing equation of power system the classical transient energy function consists of the kinetic and potential energy of the system, i.e.

$$W(\delta, \omega_r) = \frac{1}{2} M \omega_s \omega_r^2 + \int_{\delta_s}^{\delta} (P_e - P_m) d\delta \tag{9}$$

Thus, for the power system with excitation control law, a candidate of storage function is constructed as the following form

$$H(\delta, \omega_r, E'_q) = \frac{1}{2} M \omega_s \omega_r^2 + \int_{\delta_s}^{\delta} (P_{e0} - P_m) d\delta + \frac{1}{2} y^2 \tag{10}$$

Then, let the system state variable $x = [\delta \ \omega_r \ y]^T$, the partial differential of H along the state variable x can be calculated as follows:

$$\begin{cases} \dfrac{\partial H}{\partial \delta} = P_{e0} - P_m \\ \dfrac{\partial H}{\partial \omega_r} = M \omega_s \omega_r \\ \dfrac{\partial H}{\partial y} = y \end{cases} \tag{11}$$

Motivated by the design idea of the passivity-based control with Hamiltonian system structure, the objective is to find a excitation control law such that the resulting closed-loop system has the Hamiltonian structure with Hamiltonian function H as (10), i.e. the resulting closed-loop system can be represented as the form

$$\dot{x} = [J(x) - R(x)] \frac{\partial H}{\partial x}(x) \tag{12}$$

where $H(x)$ is the desired total energy function, which has a minimum at the desired equilibrium $x_s = (\delta_s, \omega_{rs}, E'_{qs})$, $J(x) = -J^T(x)$ and $R(x) = R^T(x) \geq 0$ are some desired interconnection and damping matrices, respectively. Thus, choose the interconnection and damping matrices $J(x)$, $R(x)$ as

$$J(x) = \begin{bmatrix} 0 & \dfrac{1}{M} & 0 \\ -\dfrac{1}{M} & 0 & \dfrac{1}{M} \\ 0 & -\dfrac{1}{M} & 0 \end{bmatrix}, \quad R(x) = \begin{bmatrix} 0 & 0 & 0 \\ 0 & \dfrac{D + K_D}{M^2 \omega_s} & 0 \\ 0 & 0 & K_y \end{bmatrix} \tag{13}$$

with the tuning parameter $K_y > 0$, then, the subsystem (δ, ω_r) (7) can be represented as

$$\begin{cases} \dot{\delta} = \dfrac{1}{M} \dfrac{\partial H}{\partial \omega_r} \\ \dot{\omega}_r = -\dfrac{1}{M} \dfrac{\partial H}{\partial \delta} - \dfrac{D+K_D}{M^2 \omega_s} \dfrac{\partial H}{\partial \omega_r} + \dfrac{1}{M} \dfrac{\partial H}{\partial y} \end{cases} \tag{14}$$

and the dynamics of y is described as

$$\dot{y} = -\frac{1}{M}\frac{\partial H}{\partial \omega_r} - \frac{1}{M}\frac{\partial H}{\partial y} = -\omega_s\omega_r - K_y y \tag{15}$$

It is noted that tuning K_y can directly play a role to improve the dynamic property of $y(E_q')$.

Further, from comparison (8) with (15), it follows that the excitation control law can be chosen as

$$\begin{aligned} u_f = bE_q' - c\cos\delta + \frac{T_{d0}}{a_1 \sin\delta}\{\omega_s\omega_r - a_1 \Delta E_q'\omega_s\omega_r\cos\delta \\ + \frac{K_D}{M}[P_m - P_{e0} - (D+K_D)\omega_r + y] + K_y y\Big\} \end{aligned} \tag{16}$$

such that the model (12) is matched with an energy function of the form (10).

Therefore, the following conclusion can be reached:

Proposition 1. *For the system (1), if the state feedback excitation controller is designed as (16) with the tuning parameters K_D and K_y for the mechanical damping and electrical damping, the resulting closed loop system takes the form (12), (13) and is asymptotically stable at the desired equilibrium $x_s = (\delta_s, \omega_{rs}, E_{qs}')$, $J(x) = -J^T(x)$ with Lyapunov function H as form (10).*

Proof. From the above analysis, it follows that with the choices of $J(x)$, $R(x)$ in (13) the closed loop system consisting of (1) with (16) matches the model (12). Hence, along the trajectories of (12), the time derivative of the Hamiltonian function H constructed as the form (10) satisfies the following equality

$$\dot{H}(\delta, \omega_r, E_q') = -\frac{\partial^T H}{\partial x}R(x)\frac{\partial H}{\partial x} = -(D+K_D)\omega_s\omega_r^2 - K_y y^2 \tag{17}$$

H plays the role of Lyapunov function, which ensures the Lyapunove stability at the point $(\omega_r = 0, y = 0)$.

Furthermore, from LaSalle invariance principle, it follows that $\omega_r \to 0$ and $y \to 0$ as $t \to \infty$. Then, by the definition of y, $E_q' \to E_{qs}'$ as $t \to \infty$.

On the other hand, from the dynamical description of (1) with (16), it follows that $P_m = a_1 E_{qs}' \sin\delta - a_2 \sin 2\delta$ as $t \to \infty$. This means that $\delta \to \delta_s$ from the equilibrium condition. Hence, the result is concluded that the closed-loop system is asymptotically stable at the equilibrium $(\delta_s, 0, E_{qs}')$ as $t \to \infty$. □

Remark 1. It should be noted that the distinguishing feature of the proposed design method is the mechanical damping compensation is feeded back to the excitation loop and the mechanical and electrical damping can be separately arbitrarily assigned.

Remark 2. Based on the above design idea, by modifying the output regulation variable

$$y = -\hat{\theta}_1(E_q' - E_{qs}')\sin\delta + K_D\omega_r \tag{18}$$

an adaptive excitation controller can be designed as

$$u_f = \hat{\theta}_3 E'_q - \hat{\theta}_4 \cos\delta + \frac{T_{d0}}{\hat{\theta}_1 \sin\delta}\Big\{ \frac{K_D}{M}[\hat{P}_m - \hat{\theta}_1 E'_q \sin\delta + \hat{\theta}_2 \sin 2\delta - D\omega_r] \\ + \omega_s\omega_r - \dot{\hat{\theta}}_1 \Delta E'_q \sin\delta - \hat{\theta}_1 \Delta E'_q \omega_s\omega_r \cos\delta + K_y y \Big\} \tag{19}$$

$$\dot{\hat{\theta}}_1 = -\frac{1}{r_1}\left(\omega_s\omega_r + \frac{K_D}{M} y\right) E'_q \sin\delta, \quad \dot{\hat{\theta}}_2 = \frac{1}{r_2}\frac{K_D}{M} y \sin 2\delta \tag{20}$$

$$\dot{\hat{\theta}}_3 = \frac{1}{r_3 T_{d0}} y \hat{\theta}_1 E'_q \sin\delta, \quad \dot{\hat{\theta}}_4 = -\frac{1}{2r_4 T_{d0}} y \hat{\theta}_1 \sin 2\delta, \quad \dot{\hat{P}}_m = \frac{1}{r_5}\frac{K_D}{M} y \tag{21}$$

with the Lyapunov function constructed as

$$V(\delta, \omega_r, E'_q, \hat{\theta}) = H(\delta, \omega_r, E'_q) + \frac{1}{2}(\theta - \hat{\theta})^T \Gamma (\theta - \hat{\theta}) \tag{22}$$

such that the closed loop system is locally Lyapunov stable at the unknown equilibrium $(\delta_s, \omega_{rs}, E'_{qs})$ $(\delta_s < \frac{\pi}{2})$, and $\delta \to \delta_s$, $\omega_{rs} \to 0$, $E'_q \to E'_{qs}$ as $t \to \infty$.

4 Simulation Research and Results

The physical parameters of the power system employed for the simulation research are given as follows:

$$\omega_s = 1, \quad D = 0.1, \quad M = 7, \quad T_{d_0} = 8, \quad V_s = 0.995, \\ X_d = 1.8, \quad X'_d = 0.3, \quad X_q = 1.76, \quad X_T = 0.15, \quad X_{l1} = 0.5, \quad X_{l2} = 0.93.$$

and $X_L, X_{d\Sigma}, X'_{d\Sigma}, X_{q\Sigma}$ can be calculated as

$$X_L = 0.3252, \quad X_{d\Sigma} = 2.2752, \quad X'_{d\Sigma} = 0.7752, \quad X_{q\Sigma} = 2.2352$$

The fault considered in simulations is a symmetrical three phase short circuit fault which occurs on one of the transmission lines. The following four cases with different fault sequences and the unknown perturbation in the mechanical power are discussed.

Case 1. Temporary Fault. The system is in a pre-fault steady state, a short circuit fault occurs at $t = 0.5s$ on the transmission line X_{l2},, the fault is removed by opening the breakers of the faulted line at $t = 0.7s$, the transmission line is restored at $t = 2s$ and the system is in a post-fault state.

Case 2. Permanent Fault. The system is in a pre-fault steady state, a short circuit fault occurs at $t = 0.5s$ on the transmission line X_{l2}, the fault is removed by opening the breakers of the faulted line at $t = 0.7s$, the system is in a post-fault state.

Case 3. Temporary Fault and Unknown Perturbation. The system is in a pre-fault steady state, a short circuit fault occurs at $t = 0.5s$ on the transmission line X_{l1}, the fault is removed by opening the breakers of the faulted line at $t = 0.7s$,

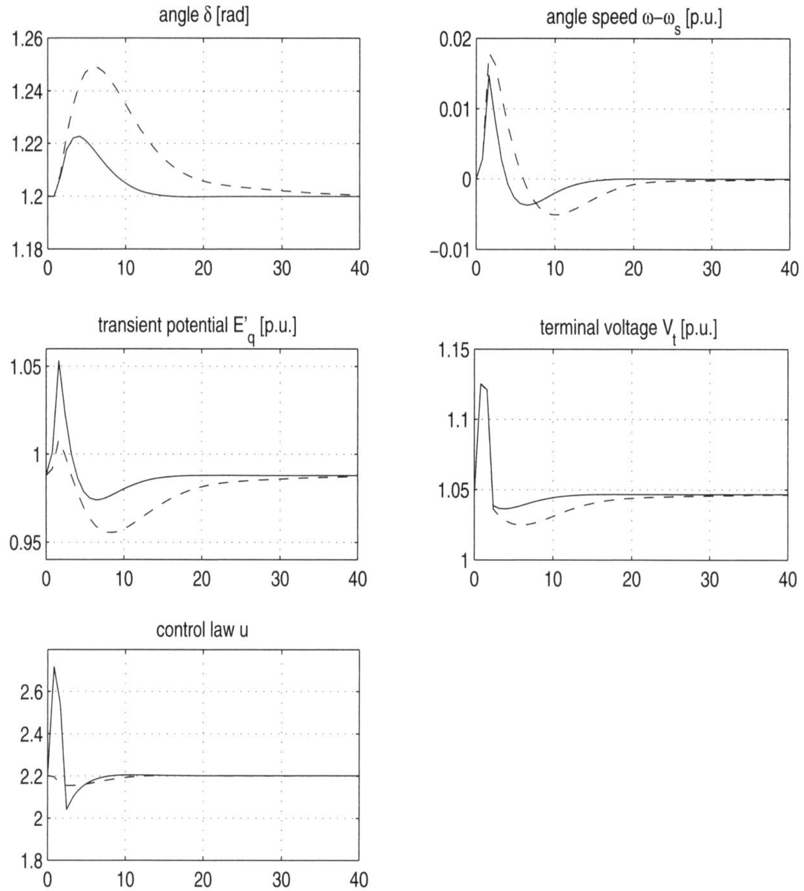

Fig. 1. Responses for case 1 with temporary fault in transmission line

the transmission line is restored at $t = 2s$ and the system is in a post-fault state. Meanwhile, there exists a perturbation in the mechanical power between $t = 0.5s$ and $t = 1s$.

Case 4. Temporary Fault and Turbine Failure. The system is in a pre-fault steady state, a short circuit fault occurs at $t = 0.5s$ on the transmission line X_{l1}, the fault is removed by opening the breakers of the faulted line at $t = 0.7s$, the transmission line is restored at $t = 2s$ and the system is in a post-fault state. Meanwhile, sudden turbine failure happens at $t = 0.5s$ and the fault is not recovered so that the mechanical power abruptly change to an unknown value.

By employing the proposed excitation control law, we obtained the following simulation results, which are shown as the solid curves in Fig.1-4. In the simulation, the tuning parameters of the proposed controller are chosen as $K_D = 4, \quad K_y = 2$. Moreover, for comparison, the results via the controller

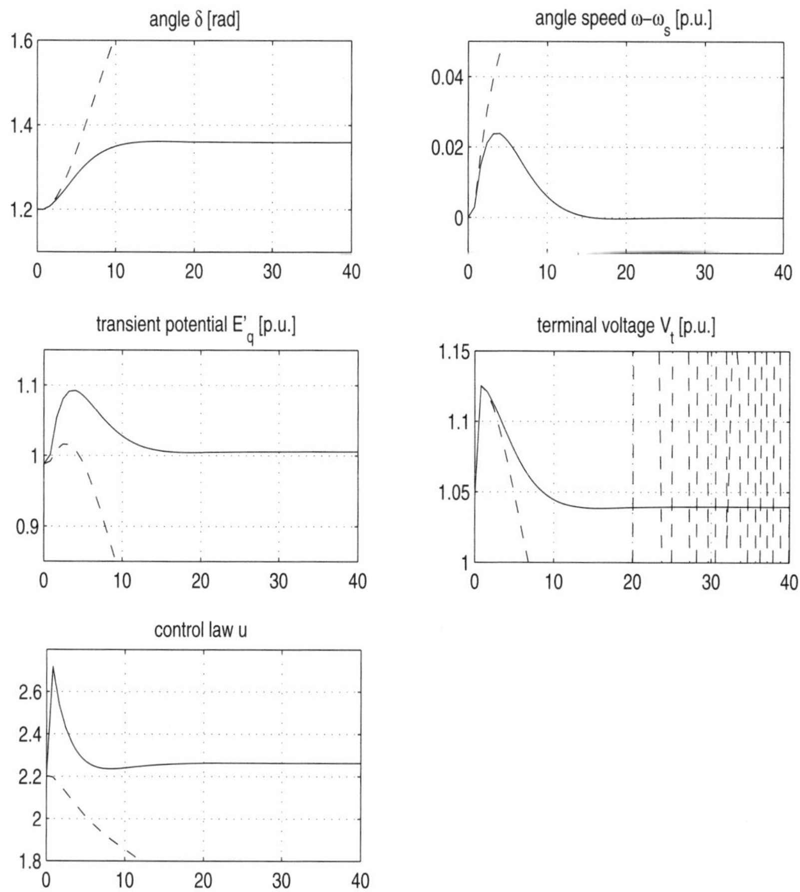

Fig. 2. Responses for case 2 with permanent fault in transmission line

designed in [8] are also given as the dash curves in Fig.1-4. The Hamiltonian function and the excitation controller are designed in [8] as follows:

$$\begin{aligned} H_2(\delta,\omega_r,E'_q) = {} & \frac{1}{2}M\omega_s\omega_r^2 + P_m(\delta_s-\delta) + a_1E'_q(\cos\delta_s-\cos\delta) \\ & + \frac{1}{2}a_2(\cos 2\delta - \cos 2\delta_s) + \frac{a_1 b}{2c}(E'_q - E'_{qs})^2 + \Phi(\delta, E'_q) \\ & + \alpha_1 a_1\left\{\delta\cos\delta_s - \sin\delta + \frac{b}{c}\delta(E'_q - E'_{qs} + \frac{\alpha_1}{2}\delta)\right\} \end{aligned} \tag{23}$$

with

$$\Phi(\delta, E'_q) = -\frac{\alpha_1 a_1 b}{c}\delta_s[\alpha_1(\delta-\delta_s) + (E'_q - E'_{qs})] + \frac{\gamma}{2\alpha_1^2}[\alpha_1(\delta-\delta_s) + (E'_q - E'_{qs})]^2$$

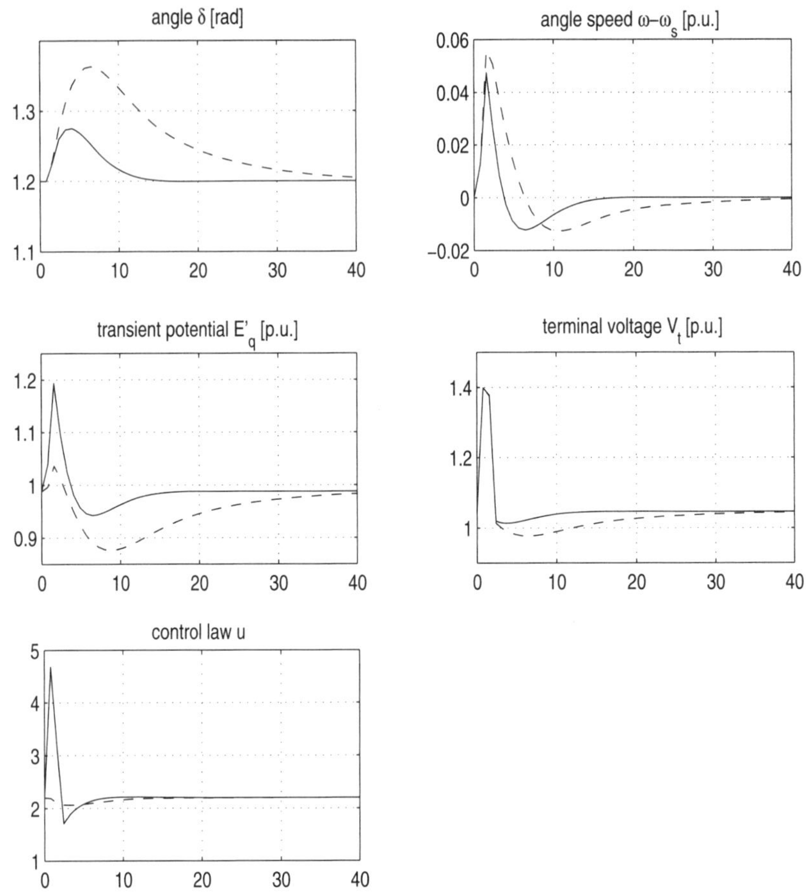

Fig. 3. Responses for case 3 with temporary fault and unknown perturbation

$$u_{f2}=bE'_{qs}-c\cos\delta_s-\frac{1}{T'_{d0}}\Big\{ra_1(\cos\delta_s-\cos\delta)+(r\alpha_2+\frac{\gamma}{\alpha_1^2}\frac{c}{a_1T'_{d0}})(E'_q-E'_{qs})$$
$$+\alpha_1\omega_s\omega_r+(r+\frac{c}{a_1T'_{d0}})\alpha_1\alpha_2(\delta-\delta_s)\Big\} \quad (24)$$

where the tuning parameters $r>0$, $\alpha_1>0$ and $\alpha_2=\frac{\gamma}{\alpha_1^2}+\frac{a_1b}{c}$. In the simulation, the tuning parameters of the proposed controller are chosen as $r=0.4$, $\alpha_1=1$, $\gamma=15$.

From the simulation results shown in Fig.1-4, it can be concluded that

1. The excitation controller proposed in this paper can effectively improve transient stability of the power system. The system can keep transiently stable even in the case where a large sudden variation occurs in the transmission line and there exists a large amount of the uncertainty in the parameters of power

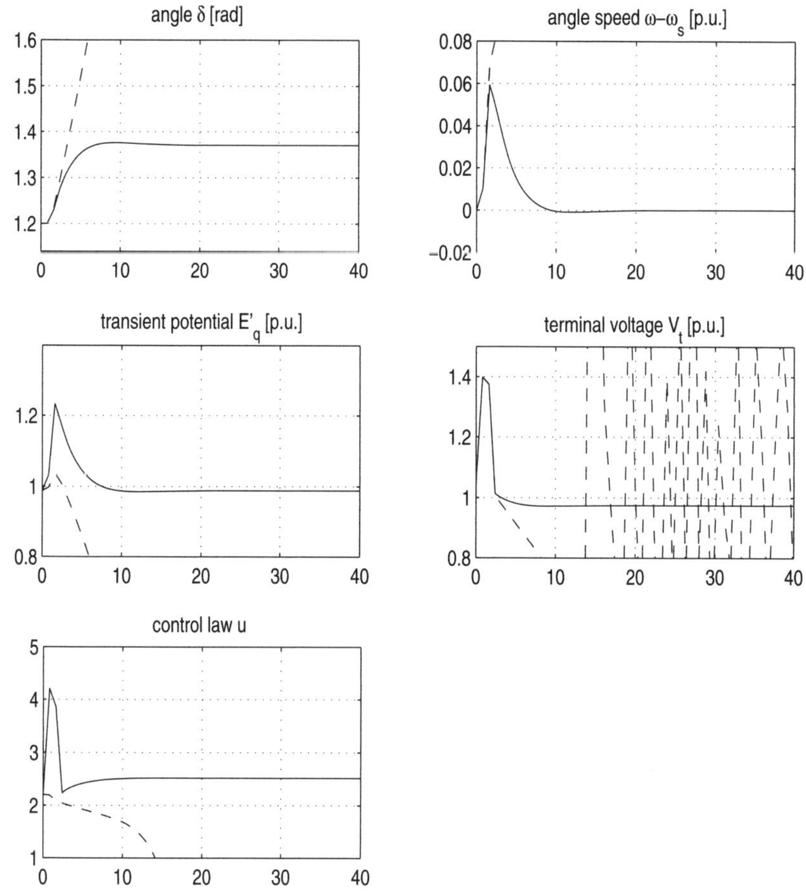

Fig. 4. Responses for case 4 with temporary fault and turbine failure

system, such as in case 2 and case 4. While under case 2 and case 4, the system forced by the excitation controller designed in [8] is unstable.

2. The excitation controller proposed in this paper can achieve better dynamic properties, i.e. the control law rendering the system to converge quickly to a stable equilibrium point, irrespective of the operating equilibrium point of the system and the fault sequence.

5 Conclusions

A novel energy-shaping with direct mechanical damping injection design method is provided for control of power systems. The main feature of this method is that the mechanical damping and electrical damping are separately assigned so that the tuning parameters of the proposed controller have directly corresponding relation to the assigned coefficients of the mechanical damping and electrical damping. The theoretical analysis and simulation results all show that the excitation controller designed by the proposed method can enlarge the stability region of the closed-loop equilibrium and improves the dynamical properties of the system.

References

1. Fujimoto K, Sugie T (2001) Canonical Transformation and Stabilization of Generalized Hamiltonian Systems. Systems and Control Letters 42:217-227
2. Ortega R, van der Schaft A, Mareels I, Maschke B (2001) Putting Energy back in Control. IEEE Control System Magazine 21:18-33
3. Ortega R, van der Schaft A, Maschke B, Escobar G (2003) Interconnect And Damping Assignment Passivity-Based Control of Port-controlled Hamiltonian Systems. Automatica 38:585-596
4. Sun Y, Song YH, Li X (2000) Novel Energy-Based Lyapunov Function for Controlled Power Systems. IEEE Power Engineering Review 55-57
5. Sun Y, Shen T, Ortega R, Liu Q (2001) Decentralized Controller Design for Multimachine Power Systems Based on the Hamiltonian Structure. Proceedings of the 40th IEEE CDC, FL USA, Dec. 4-7
6. Wang Y, Cheng D, Li C, Ge Y (2003) Dissipative Hamiltonian Realization and Energy-Based L_2 Disturbance Attenuation Control of Multimachine Power Systems. IEEE Trans. Autom. Control 48:1428-1433
7. Shen T, Ortega R, Lu Q, Mei S, Tamura K (2003) Adaptive L_2 Disturbance Attenuation of Hamiltonian Systems with Parametric Perturbation and Application to Power Systems. Asian Journal of Control 5:143-152
8. Galaz M, Ortega R, Bazanella AS, and Stankovic AM (2003) An Energy-shaping Approach to the Design of Excitation Control of Synchronous Generators. Automatica 39:111-119
9. Shen T, Sun Y, Ortega R, Mei S (2005) Energy-Shaping Control of Synchronous Generators with Exciter-Governor Dual Control Loops. International Journal of Control 78:100-111
10. Lu Q, Sun Y, Mei S (2000) Nonlinear Control Systems and Power System Dynamics. Kluwer Academic Publishers, Boston

Remarks on Quadratic Hamiltonians in Spaceflight Mechanics

Bernard Bonnard[1], Jean-Baptiste Caillau[2], and Romain Dujol[2]

[1] Institut de mathématiques de Bourgogne, CNRS, Dijon, France
bernard.bonnard@u-bourgogne.fr
[2] ENSEEIHT-IRIT, CNRS, Toulouse, France
{caillau, dujol}@n7.fr

Summary. A particular family of Hamiltonian functions is considered. Such functions are quadratic in the moment variables and arise in spaceflight mechanics when the averaged system of energy minimizing trajectories of the Kepler equation is computed. An important issue of perturbation theory and averaging is to provide integrable approximations of nonlinear systems. It turns out that such integrability properties hold here.

Keywords: Controlled Kepler equation, averaging, quadratic Hamiltonians, Riemannian problems.

1 Introduction

As explained in [4], the energy minimizing trajectories of the controlled Kepler equation [7] can be approximated by trajectories of an *averaged* system. The coplanar single-input system we consider is

$$\ddot{q} = -\mu \frac{q}{r^3} + u \frac{\dot{q}}{|\dot{q}|} \tag{1}$$

for the energy performance index $\int_0^{t_f} |u|^2 \to \min$, where q is the position vector in $\mathbf{R}^2$ and r the radius $(q_1^2 + q_2^2)^{1/2}$. Except in §5 where the real system is considered for the numerical computations, the gravitation constant μ will be normalized to one in the text. In accordance with (1), the thrust is directed along the speed $\dot{q}$—that is *tangential*—, and if we restrict ourselves to bounded trajectories, the state space in coordinates $(q, \dot{q})$ is the four-dimensional manifold $Q = \{r > 0,\ \dot{q}^2/2 - 1/r < 0\}$. Like the system with two inputs, this single-input model is shown to be controllable and is physically important since it is interpretated as the limit of cone-constrained problems where the control has to remain in a cone directed by the velocity of the spacecraft. In modern applications such as low-thrust orbit transfer [7], controls are very small and act as perturbations not only in (1), but also in the Hamiltonian system provided by Pontryagin maximum principle and describing the extremals with respect to the energy criterion. Such a point of view is standard in celestial mechanics, see for

F. Bullo et al. (Eds.): Lag. & Hamil. Methods for Nonlin. Ctrl., LNCIS 366, pp. 365–373, 2007.
springerlink.com

instance [9]. In the framework of Hamiltonian systems, one defines the averaged dynamical system [1] to retrieve the action of the perturbation up to first order.

The averaging process is presented in §2. Then we recall in §3 how a Riemannian control problem can be associated with a Hamiltonian such as the averaged one, which is quadratic in the moment variable p. Averaging is intended to provide an integrable approximation of the perturbation of an integrable system, here the coplanar Kepler equations. Integrability of the canonical equations is addressed in §4, first for a two-dimensional subsystem, then for the full system with three degrees of freedom. We end by providing in §5 energy minimizing trajectories obtained by continuation from the averaged system.

2 Averaged System

Defining the feedback $u' = u/|v|$ where v stands for $\dot{q}$, the system (1) is written as an affine single-input control one,

$$\dot{x} = F_0(x) + u'F_1(x), \tag{2}$$

where x is the state $(q, \dot{q})$, and where

$$F_0 = -\frac{q}{r^3}\frac{\partial}{\partial v}, \quad F_1 = v\frac{\partial}{\partial v}.$$

Computing Lie brackets of the two vector fields up to length three, one checks that [5]

$$\begin{aligned}
[F_0, F_1] &= -v\frac{\partial}{\partial q} - \frac{q}{r^3}\frac{\partial}{\partial v}, \\
[F_1, [F_0, F_1]] &= -F_0, \\
[F_0, [F_0, F_1]] &= \frac{2q}{r^3}\frac{\partial}{\partial q} + \frac{2}{r^5}\left[(2q_1^2 - q_2^2)v_1 + 3q_1q_2v_2\right]\frac{\partial}{\partial v_1} \\
&\quad + \frac{2}{r^5}\left[3q_1q_2v_1 + (2q_2^2 - q_1^2)v_2\right]\frac{\partial}{\partial v_2}.
\end{aligned}$$

Hence, F_0, F_1, $[F_0, F_1]$ and $[F_0, [F_0, F_1]]$ form a frame, and the system, whose drift F_0 is periodic, is controllable [8]. So as to perform averaging, we change coordinates and replace the cartesian ones by an angle, the *longitude* l, together with three first integrals of the unperturbed motion: $x = (l, n, e, \omega)$ where n is the *mean movement*, e the *eccentricity*, and ω the *argument of pericenter* (see, *e.g.*, [9]). The mean movement, the eccentricity and the argument of pericenter define the geometry of the osculating elliptic orbit, while the longitude represents the position on the ellipse. In these coordinates, $Q = \{n > 0,\ e < 1\}$, and

$$\dot{l} = \frac{n[1 + e\cos(l - \omega)]^2}{(1 - e^2)^{3/2}},$$

in order that trajectories can be reparameterized by the cumulated longitude. Before doing so, we consider the system (2) with performance index

$$\int_0^{t_f} u^2 dt = \int_0^{t_f} u'^2 |v|^2 dt \to \min,$$

fixed endpoints, and free final time t_f. More precisely, the final cumulated longitude, $l \in \mathbf{R}$, is fixed to l_f. For l_f big enough, there are admissible trajectories. Indeed, since the system is controllable, there are trajectories such that $l(t_f) = l_f \bmod 2\pi$. Then, setting the control to zero, n, e and ω remain unchanged while the longitude is increased until it reaches the desired l_f.

Proposition 1. *The problem of the minimization of energy with fixed cumulated longitude and free final time admits no abnormal extremals.*

Proof. The Hamiltonian of the problem is $H = p^0 u'^2 v^2 + H_0 + u' H_1$. For an abnormal extremal, p^0 is zero, and $H_1 = 0$ because of the maximization condition of Pontryagin maximum principle. Then, by differentiation, $\{H_0, H_1\} = \{H_0, \{H_0, H_1\}\} + u'\{H_1, \{H_0, H_1\}\} = 0$. Since the final time is free, $H = 0$ so that H_0 is also zero. As $[F_1, [F_0, F_1]]$ is colinear to F_0, $\{H_1, \{H_0, H_1\}\}$ is also zero. Using the fact that F_0, F_1, $[F_0, F_1]$ and $[F_0, [F_0, F_1]]$ form a frame, we get the contradiction. □

It is shown in [4] that, up to a renormalization, the averaged Hamiltonian associated to normal extremals of the system reparameterized by longitude is

$$H = \frac{1}{2n^{5/3}} \left[n^2 p_n^2 + \frac{4}{9} \frac{(1-e^2)^{3/2}}{1+(1-e^2)^{1/2}} p_e^2 + \frac{4}{9} \frac{1-e^2}{1+(1-e^2)^{1/2}} \frac{p_\omega^2}{e^2} \right], \tag{3}$$

which defines a quadratic form of full rank with respect to the moment p. The underlying Riemannian problem is presented in the next section.

3 Associated Riemannian Metric

The averaged system can be seen as a rotating deformable solid. The first and second coordinates define the geometry of the solid, an ellipse of given eccentricity e and semi-major axis a ($a^3 n^2 = 1$)—actually, up to a homothety, the geometry is defined by e alone—, while the third one, ω, fixes the angle of rotation around its center. As we are now going to see, the system is associated with the Riemannian problem whose distribution is

$$\begin{aligned} \dot{n} &= u_1 n^{1/6}, \\ \dot{e} &= u_2 \frac{g(e)^{1/2}}{n^{5/6}}, \\ \dot{\omega} &= u_3 \frac{k(e)^{1/2}}{n^{5/6}}, \end{aligned}$$

so that the cyclicity of ω implies that there is coupling between the deformation and the rotation: the geometry acts on the rotation, not the converse.

The averaged Hamiltonian is clearly normalized to

$$H(q,p) = \frac{1}{2}x^{\alpha}\left(x^2p_x^2 + g(y)p_y^2 + k(y)p_z^2\right) \tag{4}$$

setting $\alpha = -5/3$, $g(y) = (4/9)(1-y^2)^{3/2}/[1+(1-y^2)^{1/2}]$, and $k(y) = [4/(9e^2)](1-y^2)/[1+(1-y^2)^{1/2}]$.

Remark 1. In the case of two controls, when the direction of the thrust is not prescribed anymore, the averaged Hamiltonian obtained in [3] also belongs to the same class, still with $\alpha = -5/3$.

For α not zero, the Hamiltonian (4) defines a quadratic form in p parameterized by $q = (x, y, z)$ in $Q = \{x > 0\}$, and this form can be written as a sum of squares [3]. The task is straightforward here since we have *orthogonal coordinates* [2], in order that

$$H = \frac{1}{2}\sum_{i=1}^{3} P_i(q,p)^2$$

with $P_i = \langle p, F_i(q)\rangle$, $i = 1,\ldots,3$, $F_1 = x^{1+\alpha/2}\partial/\partial x$, $F_2 = x^{\alpha/2}g^{1/2}(y)\partial/\partial y$, and $F_3 = x^{\alpha/2}k^{1/2}(y)\partial/\partial z$. Hence, H can be seen as the Hamiltonian associated with the Riemannian problem [6] with dynamics $\dot{q} = \sum_{i=1}^{3} u_iF_i(q)$, $u = (u_1, u_2, u_3)$ in $\mathbf{R}^3$, and criterion $\int_0^{t_f} |u|^2 dt \to \min$ with prescribed final time (again denoted t_f for the sake of simplicity). Writing the dynamics $\dot{q} = B(q)u$, one has $|u|^2 = \left((BB^T)^{-1}(q)\dot{q}|\dot{q}\right)$ and the Riemannian metric is

$$ds^2 = \frac{1}{x^{\alpha}}\left(\frac{dx^2}{x^2} + \frac{dy^2}{g(y)} + \frac{dz^2}{k(y)}\right). \tag{5}$$

As z is a cyclic coordinate of the Hamiltonian, the system can be restricted to the two-dimensional subspace $Q_0 = Q \cap \{z = 0\}$. We start by studying integrability on this subspace.

4 Integrability

We first compute a normal form of the metric.

Lemma 1. *A normal form for the metric (5) of the full system is*

$$ds^2 = du^2 + u^2\left(dv^2 + \frac{dw^2}{l(v)}\right).$$

Proof. Consider the change of coordinates defined by $u = -2/(\alpha x^{\alpha/2})$, $v = \varphi(y)$, $w = z$, where φ is the quadrature

$$\varphi = \frac{|\alpha|}{2}\int \frac{dy}{g^{1/2}(y)}.$$

Letting $l = (4/\alpha^2)k \circ \varphi^{-1}$, one gets the desired expression. □

Corollary 1. *A normal form of the Hamiltonian is*

$$H = \frac{1}{2}p_u^2 + \frac{1}{u^2}H_2$$

with $H_2 = (1/2)(p_v^2 + l(v)p_w^2)$.

Restricted to $Q_0 = Q \cap \{w = 0\}$ and $\{p_w = 0\}$, the Hamiltonian becomes $H = (1/2)(p_u^2 + \frac{1}{u^2}p_v^2)$ and the metric is in *polar form*, $ds^2 = du^2 + u^2dv^2$. It is clearly isometric to the flat metric in dimension two so that, in coordinates $u\cos v$, $u\sin v$, the proposition hereafter holds.

Proposition 2. *The geodesics of the two-dimensional subsystem are straight lines.*

Using the complex notation $c = c_1t + c_2$ to parameterize such lines, we are able to write the geodesics in the original coordinates, $x = [4/(\alpha^2|c|^2)]^{1/\alpha}$, $y = \varphi^{-1}(\arg(c))$. We address now integrability of the system in dimension three.

Our first step is to reduce the analysis to the study of the *Liouville metric* [2] $ds^2 = dv^2 + dw^2/l(v)$. To this end, we have the following lemma.

Lemma 2. *The squared coordinate* u^2 *is a polynomial of degree two in* t.

Proof. The canonical equations in (u, p_u) are $\dot{u} = \partial H/\partial p_u = p_u$ and $\dot{p}_u = -\partial H/\partial u = 2H_2/u^3$. As a result, $d(up_u)/dt = p_u^2 + 2H_2/u^2$ that is equal to twice the Hamiltonian. Then, $d^2u^2/dt^2 = 4H$, whence the conclusion. □

The main result follows.

Proposition 3. *The full three-dimensional system is integrable by quadratures.*

Proof. The two variables u, p_u are computed thanks to the previous lemma. If we reparameterize the system in the remaining terms according to the time change $d\tau = dt/u^2$, we obtain the canonical equations of the auxiliary Hamiltonian $H_2 = (1/2)(p_v^2 + l(v)p_w^2)$ with two degrees of freedom. Since w is a cyclic coordinate, p_w is a first integral in involution with H_2 which is then integrable by Liouville theorem. □

Remark 2. The metric associated with H_2, $ds^2 = dv^2 + dw^2/l(v)$, is a Liouville metric, that is a metric of the form $(f(x) + g(y))(dx^2 + dy^2)$ since

$$ds^2 = l^{-1}(v)\left[(l^{1/2}dv)^2 + dw^2\right].$$

As such, it is known to be integrable [2]. Moreover, the relevant quadrature is obviously deduced from the canonical equations:

$$\dot{v}^2 + l(v)p_w^2 = \text{constant}.$$

Remark 3. Liouville integrability is also obtained by noting that H, H_2 and p_w are three independent first integrals in involution.

In summary, the three quadratures used to integrate the whole system are:

$$\varphi = \frac{|\alpha|}{2} \int \frac{dy}{g^{1/2}(y)},$$
$$\tau = \int \frac{dt}{u^2(t)},$$
$$\psi = \int \frac{dv}{(1 - bl(v))^{1/2}},$$

where u^2 is a degree two polynomial in t, $l = k \circ \varphi^{-1}$ and b a constant. We end the paper by expliciting some of this computations for the application considered.

5 Numerical Results

According to the previous sections, the whole system is integrable and the two-dimensional subsystem[1] is flat. The first quadrature can be explicitly computed,

$$\varphi(e) = \frac{5}{4}\mathrm{arcsin}[1 - 2(1 - e^2)^{1/2}], \tag{6}$$

as well as the associated coordinates on Q_0.

Namely, in dimension two,

$$n = n = (5|c|/6)^{6/5}, \tag{7}$$

$$e = \pm \left[1 - \frac{(1 - \sin((4/5)\arg c))^2}{4}\right]^{1/2}, \tag{8}$$

where $c = c_1 t + c_2$ is, as in §4, a complex polynomial of degree one in t. Equation (8) is multiform because the quadrature (6) defines a diffeomorphism either of $]-1, 0[$ or $]0, 1[$ to $]-5\pi/8, 5\pi/8[$. Even in the two-dimensional case, contact with the boundary of Q_0 may occur, either with the parabolic boundary $\{e = 1\}$, or with $\{n = 0\}$ (see [3] for a discussion in the two-input situation). We do not touch this point here and restrict ourselves to complete geodesics. Clearly then, when $t \to \infty$, $n \to n_\infty = \infty$ (that is $a \to 0$, a semi-major axis) and $e \to e_\infty = \pm[1 - (1/4)\,(1 - \sin((4/5)\arg_\infty))^2]^{1/2}$, where $\arg_\infty = \arg c_1$ when c_1 is not zero, $\arg c_2$ otherwise (stationary case). This asymptotic behaviour is summarized in the last proposition.

Table 1. Boundary conditions

Variable	Initial cond.	Final cond.
n	$5.2475e-1$ h^{-1}	$2.6251e-1$ h^{-1}
e	0.75	0
ω	0 rad	0 rad

[1] This subsystem is important in practice since it corresponds to transfers towards circular orbits.

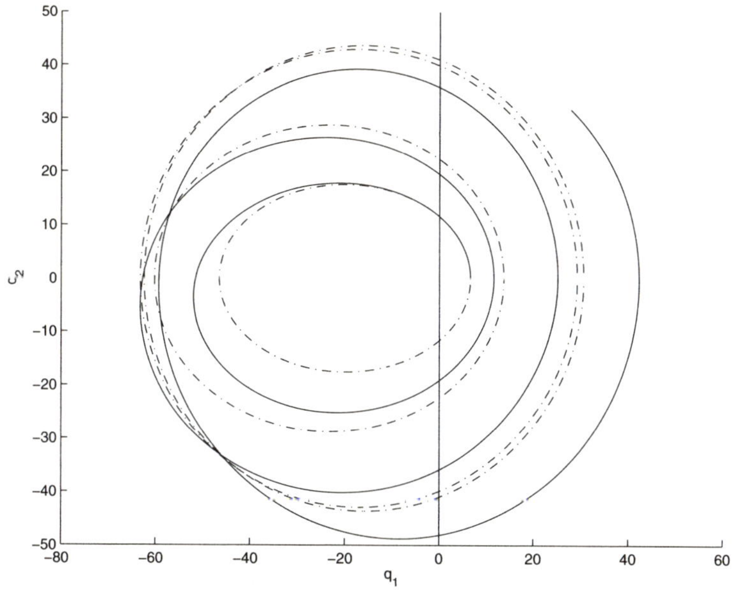

Fig. 1. Energy minimizing transfer towards the geostationary orbit, final time t_f = 19.290 hours. The trajectory is the solid line that osculates the dashed intermediary orbits of the averaged system.

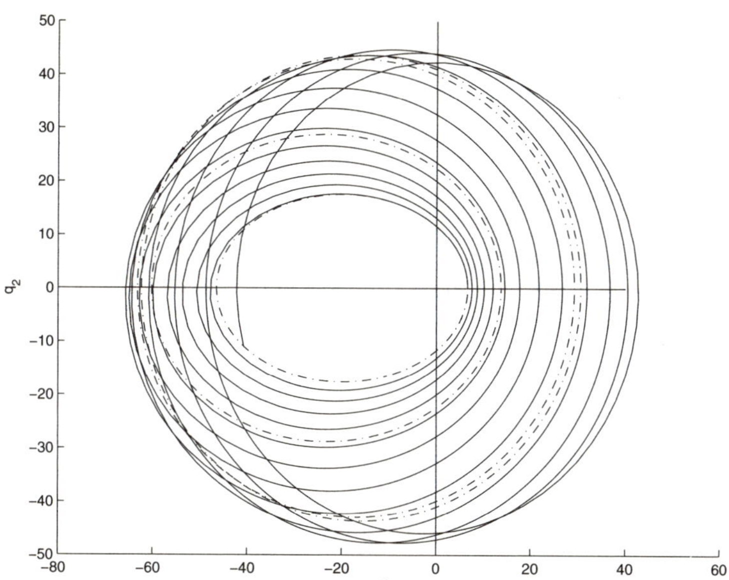

Fig. 2. Energy minimizing transfer towards the geostationary orbit, final time t_f = 77.160 hours (solid line)

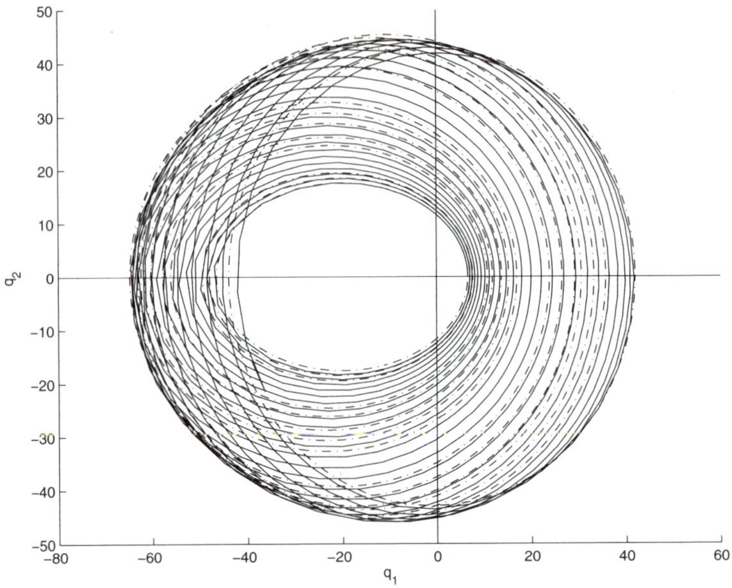

Fig. 3. Energy minimizing transfer towards the geostationary orbit, final time t_f = 154.32 hours (solid line)

Proposition 4. *The (non-stationary) complete trajectories of the two-dimensional system converge to a collision with a limit value of the eccentricity.*

We end the section with numerical results. Using the boundary conditions of table 1, the analytical solution of the two-dimensional averaged subsystem is used to initialize the computation of energy minimizing trajectories of Kepler equation (1) by a standard shooting method. The gravitation constant is the Earth constant, $\mu = 5165.8620912\ \mathrm{Mm}^3{\cdot}\mathrm{h}^{-2}$, and the target is the geostationary orbit (the initial orbit around the Earth is taken low and eccentric). Results are given for different values of the fixed final time at figures 1 to 3. As the transfer time increases, there are more and more revolutions that osculate intermediate orbits of the averaged motion with a growing accuracy, thus illustrating convergence towards the averaged problem.

References

1. V. I. Arnold. *Mathematical Methods of Classical Mechanics.* Springer-Verlag, New-York, 1978.
2. A. V. Bolsinov and A. T. Fomenko. *Integrable geodesics flows on two dimensional surfaces.* Kluwer, New-York, 2000.
3. B. Bonnard and J.-B. Caillau. Riemannian metric of the averaged energy minimization problem in orbital transfer with low thrust. *Ann. Inst. H. Poincaré Anal. Non Linéaire*, 25(3):395–411,2007.

4. B. Bonnard, J.-B. Caillau, and R. Dujol. Energy minimization of single-input orbit transfer by averaging and continuation. *Bulletin des Sciences Mathématiques*, 130(8):707–719, 2007.
5. B. Bonnard, J.-B. Caillau, and E. Trélat. Geometric optimal control of elliptic keplerian orbits. *Discrete and Continuous Dynamical Systems B*, 5(4):929–956, 2005.
6. B. Bonnard and M. Chyba. *Singular trajectories and their role in optimal control.* Springer-Verlag, Paris, 2003.
7. J.-B Caillau, J. Gergaud, and J. Noailles. Minimum time control of the kepler equation. In V. Blondel and A. Megretski, editors, *Unsolved Problems in Mathematical Systems and Control Theory*, pages 89–92. Princeton University Press, 2004.
8. V. Jurdjevic. *Geometric control theory.* Cambridge University Press, 1996.
9. J. Kevorkian and J. D. Cole. *Perturbation methods in applied mathematics.* Springer-Verlag, New-York, 1981.

Controlling a Submerged Rigid Body: A Geometric Analysis

Monique Chyba, Thomas Haberkorn, Ryan N. Smith, and George R. Wilkens

University of Hawaii at Manoa, Mathematics Department and Ocean and Resources Engineering Department, Honolulu, HI 96822
mchyba@math.hawaii.edu, haberkor@math.hawaii.edu, ryan@ore.hawaii.edu, grw@math.hawaii.edu

Summary. In this paper we analyze the equations of motion of a submerged rigid body. Our motivation is based on recent developments done in trajectory design for this problem. Our goal is to relate some properties of singular extremals to the existence of decoupling vector fields. The ideas displayed in this paper can be viewed as a starting point to a geometric formulation of the trajectory design problem for mechanical systems with potential and external forces.

Keywords: Submerged Rigid Body, Decoupling Vector Fields, Singular Trajectories.

1 Introduction

A very important class of control systems, even though they are non-generic, is the class of controlled mechanical systems. Examples of these systems include the planar rigid body with a single variable direction thruster, the snakeboard, and underwater vehicles; see for instance [2, 9, 11]. We are interested in trajectory design problems for such systems.

All the computations in this paper are carried out on the following system: a controlled submerged rigid body. This application is particularly well adapted to our analysis for several reasons. First, there is a clear practical motivation coming from the recent trend to build autonomous underwater vehicles. The authors are currently working on implementing their techniques on a real vehicle, see [8]. The other reasons are more mathematically oriented. An underwater vehicle can be modeled as a simple mechanical control system, i.e. coming from a Lagrangian of the form kinetic minus potential energy, with dissipative forces. The uncontrolled dissipative forces reflect the damping of the system, and the restoring forces are represented in the potential forces. The corresponding system, when neglecting the external forces, has been extensively studied and thus serves as a good starting point for our analysis, see [3] and references therein. Clearly, the external forces introduce additional challenges for the trajectory design problem.

In [6, 7], we develop a numerical algorithm to compute time efficient trajectories for a controlled submerged rigid body that can be implemented on a real

F. Bullo et al. (Eds.): Lag. & Hamil. Methods for Nonlin. Ctrl., LNCIS 366, pp. 375–385, 2007.
springerlink.com

vehicle. The motivation and discussion of this algorithm are application oriented. We call our algorithm *the Switching Time Parametrization Algorithm* (STPA). These papers are application oriented. In [8] we describe experiments on a real vehicle based on this algorithm. The experiments are performed at the University of Hawaii in collaboration with the Autonomous System Laboratory from the College of Engineering. A major goal is to establish a mathematical formulation of the STPA based on the differential geometric properties of the system; such as the notion of decoupling vector fields. This paper initiates a discussion in that direction.

The key notions involved in the discussions of this paper are the ones of decoupling vector fields for invariant systems on a Lie group, namely the configuration space of a rigid body SE(3), and the notion of singular extremals coming from optimal control. Our analysis is based on [1, 3, 4] for the notion of decoupling vector fields and on [5, 9] for the properties of singular extremals in our application.

The outline of this paper is as follows. In Section 2 we introduce the equations of motion for a controlled submerged rigid body. Section 3 introduces the definition of decoupling vector fields for an invariant system on a Lie group, and we recall the results for a controlled rigid body when neglecting the external forces. Section 4 is concerned with the application of the maximum principle to our situation and the properties of singular extremals. Our main contribution is in Section 5 where we relate the two notions introduced before.

2 Equations of Motion for Underwater Vehicles

Due to subtle differences in notation and reference frames, we include a short derivation of the equations of motion for a controlled rigid body immersed in a real fluid. By *real fluid* we mean an ideal fluid which is not inviscid. Notice that a real fluid is assumed to be irrotational but that from a practical point of view a viscous fluid is rotational: the definition of a real fluid used in this paper is introduced for theoretical reasons. This concise derivation is based on the equations found in [10].

We will identify the position and orientation of the rigid body, with respect to the inertial frame, with an element (b, R) of the Special Euclidean group of order 3; $(b, R) \in \mathrm{SE}(3)$. We take $b = (x, y, z)^t \in \mathbb{R}^3$ to denote the position vector for the origin of the body frame, and $R \in \mathrm{SO}(3)$ as the rotation matrix describing the orientation of the body. The translational and angular velocities in the body-fixed frame are denoted by $\nu = (u, v, w)^t$ and $\Omega = (p, q, r)^t$ respectively. It follows that the kinematic equations for a rigid body are given by

$$\dot{b} = R\nu \text{ and } \dot{R} = R\widehat{\Omega}, \tag{1}$$

where the operator $\widehat{\ }: \mathbb{R}^3 \to \mathfrak{so}(3)$ is defined by $\hat{y}\, z = y \times z$; $\mathfrak{so}(3)$ being the space of 3×3 skew-symmetric matrices. By letting p be the total translational momentum, and π the total angular momentum in the inertial frame, and P, Π be the respective quantities in the body-fixed frame, we get that $p = RP$ and

$\pi = R\Pi + \hat{b}p$. Differentiating the later two expressions, we express the dynamic equations of motion with the following:

$$\dot{P} = \widehat{P}\,\Omega + F \tag{2}$$

$$\dot{\Pi} = \widehat{\Pi}\Omega + \widehat{P}\,\nu + \sum_{i=1}^{k}(R^t(x_i - b)) \times R^t f_i + \tau \tag{3}$$

where $F = R^t(\sum_{i=1}^{k} f_i)$, $\tau = R^t(\sum_{i=1}^{l} \tau_i)$ represent the external forces and torques in the body-fixed frame, and x_i is the vector from the origin of the inertial frame to the line of action of the force f_i. Next, we compute the total kinetic energy of the system in order to express equations (2) and (3) in terms of linear and angular velocities. To this end, we note that the kinetic energy of the body, T_{body}, is given by

$$T_{body} = \frac{1}{2}\begin{pmatrix} v \\ \Omega \end{pmatrix}^t \begin{pmatrix} mI_3 & -m\hat{r}_{CG} \\ m\hat{r}_{CG} & J_b \end{pmatrix}\begin{pmatrix} v \\ \Omega \end{pmatrix} \tag{4}$$

where m is the mass of the rigid body, I_3 is the 3×3-identity matrix and r_{C_G} is a vector that denotes the location of the body's center of gravity with respect to the origin of the body-fixed frame. The matrix J_b represents the body moments of inertia. Based on Kirchhoff's equations we have that the kinetic energy of the fluid, T_{fluid}, is given by:

$$T_{fluid} = \frac{1}{2}\begin{pmatrix} v \\ \Omega \end{pmatrix}^t \begin{pmatrix} M_f & C_f^t \\ C_f & J_f \end{pmatrix}\begin{pmatrix} v \\ \Omega \end{pmatrix} \tag{5}$$

where M_f, J_f and C_f are respectively referred to as the added mass coefficients, the added moments of inertia coefficients and the added cross-terms. These coefficients depend on the fluid's density and on the vehicle's shape. By adding the above relations as shown, the total kinetic energy is given by:

$$T_{body} + T_{fluid} = T = \frac{1}{2}\begin{pmatrix} v \\ \Omega \end{pmatrix}^t \mathbb{I}\begin{pmatrix} v \\ \Omega \end{pmatrix}, \tag{6}$$

$$\mathbb{I} = \begin{pmatrix} mI_3 + M_f & -m\hat{r}_{CG} + C_f^t \\ m\hat{r}_{CG} + C_f & J_b + J_f \end{pmatrix} \tag{7}$$

where $\mathbb{I} = \begin{pmatrix} \mathbb{I}_{11} & \mathbb{I}_{12} \\ \mathbb{I}_{12}^t & \mathbb{I}_{22} \end{pmatrix}$. Which we can simplify into:

$$T = \frac{1}{2}(\nu^t\,\mathbb{I}_{11}\,\nu + 2\Omega^t\,\mathbb{I}_{12}^t\,\nu + \Omega^t\,\mathbb{I}_{22}\,\Omega). \tag{8}$$

Here we add the assumption that b, the origin of the body-fixed frame, is located at C_G; equivalently, $\hat{r}_{C_G} = 0$. We assume the body has three planes of symmetry. Hence choosing the body axes to coincide with the principal axes of inertia implies that J_b, M_f and J_f are diagonal, while C_f is zero. This leads to $P = (mI_3 + M_f)\nu = M\nu$ and $\Pi = (J_b + J_f)\Omega = J\Omega$. It follows from equations (2) and (3) that $M\dot{\nu} = M\nu \times \Omega + F$ and $J\dot{\Omega} = J\Omega \times \Omega + M\nu \times \nu + \sum_{i=1}^{k}(R^t(x_i - b)) \times R^t f_i + \tau$, where the terms $M\nu \times \Omega$, $J\Omega \times \Omega$ and $M\nu \times \nu$ account for the Coriolis and centripetal effects. The Coriolis and centripetal effects seen in the above equations can also be expressed in terms of ∇; the Levi-Civita affine-connection for the Riemannian metric induced by the kinetic energy T. Explicitly, if $\gamma(t) = (b(t), R(t))$ is a curve in SE(3), and $\gamma'(t) = (\nu(t), \Omega(t))$ is its pseudo-velocity (1), then

$$\nabla_{\gamma'}\gamma' = \begin{pmatrix} \dot{\nu} + M^{-1}(\Omega \times M\nu) \\ \dot{\Omega} + J^{-1}(\Omega \times J\Omega + \nu \times M\nu) \end{pmatrix}. \tag{9}$$

Let us now discuss the external forces acting on a submerged rigid body. We assume that the vehicle is neutrally buoyant, which means that the buoyancy force and the gravitational force cancel each other. Since the origin of the body-fixed frame is C_G, the only restoring force acting on the vehicle is the torque from the buoyancy force induced upon listing, $r_{C_B} \times R^t \rho g \mathcal{V} k$. Here r_{C_B} is the vector from C_G to C_B, where ρ is the fluid density, g the acceleration of gravity, $\mathcal{V}$ the volume of fluid displaced by the rigid body and k the unit vector pointing in the direction of gravity. Additional hydrodynamic forces experienced by the rigid body are due to drag effects. We make the assumption that we have a drag force $D_\nu(\nu)$ and a drag momentum $D_\Omega(\Omega)$ quadratic in the velocities and we neglect the off-diagonal terms. We summarize our computations in the following definition.

Definition 1. *Under our assumptions, the equations of motion, in the body-fixed frame, for a rigid body submerged in a real fluid are given by:*

$$\begin{aligned} M\dot{\nu} &= M\nu \times \Omega + D_\nu(\nu)\nu + F \\ J\dot{\Omega} &= J\Omega \times \Omega + M\nu \times \nu \\ &\quad + D_\Omega(\Omega)\Omega - r_{C_B} \times R^t \rho g \mathcal{V} k + \tau \end{aligned} \tag{10}$$

where M accounts for the mass of the rigid body and the added mass coefficients, J accounts for the body moments of inertia and the added moments of inertia coefficients. The matrices $D_\nu(\nu), D_\Omega(\Omega)$ represent respectively the drag force and momentum, while the restoring force acting on the body is due to the torque induced by the buoyancy force. Finally, $F = (f_1, f_2, f_3)^t$ and $\tau = (\tau_1, \tau_2, \tau_3)^t$ account for the control.

Section 3 of this paper deals with the geometric properties of the system. In that section the domain of control is assumed to be unbounded, thus allowing us to identify some important geometric structures. Once these structures are identified, we will discard the unbounded assumption since the controls represent forces with limited power, such as thrusters. Subsequent sections will, for simplicity, assume the domain of control to be $\mathcal{F} = \{f \in \mathbb{R}^3 \mid -1 \leq f_{1,2,3} \leq 1\}$, $\mathcal{T} = \{\tau \in \mathbb{R}^3 \mid -1 \leq \tau_{1,2,3} \leq 1\}$. An admissible control is a measurable bounded function $\varphi : [0, T] \to \mathcal{U} = \mathcal{F} \times \mathcal{T}$.

A final remark concerns the equations of motion of a rigid body moving in the air. In this case, the dissipation term due to the drag and the restoring forces are negligible. Moreover, since the density of air is much smaller than that of water we can neglect the terms due to the added mass tensor. We are then left with the well known equations of motion of a rigid body moving in the air. These equations can serve as a first approximation when considering a real fluid.

3 Affine-Connection Control Systems and Decoupling Vector Fields

Together, equations (1) and (10) form a first-order affine control system on the tangent bundle $T\,\mathrm{SE}(3)$ that represents the second-order *forced affine-connection control system* on SE(3)

$$\nabla_{\gamma'}\gamma' = \begin{pmatrix} M^{-1}(D_\nu(\nu)\nu + F) \\ J^{-1}(D_\Omega(\Omega)\Omega - r_{C_B} \times R^t \rho g \mathcal{V} k + \tau) \end{pmatrix}. \tag{11}$$

In the absence of a restoring force $r_{C_B} \times R^t \rho g \nabla k$, a drag momentum $D_\Omega(\Omega)\Omega$, and a drag force $D_\nu(\nu)\nu$ the equations of motion (11) represent a *left-invariant affine-connection control system* on the Lie group SE(3),

$$\nabla_{\gamma'}\gamma' = \begin{pmatrix} M^{-1}F \\ J^{-1}\tau \end{pmatrix}. \tag{12}$$

Definition 2. *An* affine-connection control system *on a manifold Q is determined by an affine-connection, ∇, and a constant-rank distribution $\mathcal{Y} \subset TQ$. A* trajectory *for the system is a curve $\gamma : [0,T] \to Q$ such that $\gamma' : [0,T] \to TQ$ is absolutely continuous, $\gamma'(0) = 0 \in T_{\gamma(0)}Q$, and $\nabla_{\gamma'(t)}\gamma'(t) \in \mathcal{Y}_{\gamma(t)}$ for almost all $t \in [0,T]$.*

A common presentation for such a system is $\nabla_{\gamma'(t)}\gamma'(t) = \sum_{a=1}^{k} u^a(t) Y_a(\gamma(t))$, where $u^1(t), \ldots, u^k(t)$ are measurable controls and $Y_1, \ldots, Y_k$ are independent vector-fields on Q that span $\mathcal{Y}$.

Just as equation (11) on SE(3) is equivalent to equations (1) and (10) on $T\,\mathrm{SE}(3)$, an affine-connection control system on Q is equivalent to an affine control system on TQ. The equivalence is realized via the *geodesic spray* of an affine-connection and the *vertical lift* of tangent vectors to Q.

Definition 3. *Let $v \in T_qQ \subset TQ$, then the* vertical lift *at v is a map* $\mathrm{vlft}_v : T_qQ \to T_vTQ$. *For $w \in T_qQ$, we define* $\mathrm{vlft}_v(w) = \frac{d}{dt}(v + tw)|_{t=0}$. *In components,* $\mathrm{vlft}_v(w) = \begin{pmatrix} 0 \\ w \end{pmatrix} \in T_vTQ$.

Definition 4. *The* geodesic spray *of ∇ is the vector field S, on TQ, that generates geodesic flow. Specifically, for $v \in T_qQ$, $S(v) = \frac{d}{dt}\gamma_v'(t)|_{t=0}$ where γ_v is the unique ∇-geodesic such that $\gamma_v(0) = q$ and $\gamma_v'(0) = v$.*

In the special case of our Levi-Civita connection (9),

$$S(b,R,\nu,\Omega)=\begin{pmatrix}\nu\\ \Omega\\ -M^{-1}(\Omega\times M\nu)\\ -J^{-1}(\Omega\times J\Omega+\nu\times M\nu)\end{pmatrix}.$$

For this presentation of $S(b,R,\nu,\Omega)$, the components are expressed relative to the standard left-invariant basis of vector fields on $T\,\mathrm{SE}(3)$ rather than coordinate vector fields. Equation (1) can be used to recover expressions for $\dot{b}$ and $\dot{R}$.

Given an affine-connection control system $\nabla_{\gamma'}\gamma' = \sum_{a=1}^{k} u^a(t)Y_a(\gamma(t))$ on Q, we associate to it the following affine control system: $\Upsilon'(t) = S(\Upsilon(t)) + \sum_{a=1}^{k} u^a(t)\,\mathrm{vlft}(Y_a)$ on TQ. In [3, p224] the authors show that trajectories for the affine-connection control system on Q map bijectively to trajectories for the affine control system on TQ whose initial points lie on the zero-section. The bijection maps the trajectory $\gamma : [0,T]\to Q$ to the trajectory $\Upsilon = \gamma' : [0,T] \to TQ$.

Definition 5. *A* decoupling vector field *for an affine-connection control system is a vector field V on Q having the property that every reparametrized integral curve for V is a trajectory for the affine-connection control system. More precisely, let $\gamma : [0,S] \to Q$ be a solution for $\gamma'(s) = V(\gamma(s))$ and let $s : [0,T] \to [0,S]$ satisfy $s(0) = s'(0) = s'(T) = 0$, $s(T) = S$, $s'(t) > 0$ for $t \in (0,T)$, and $(\gamma\circ s)' : [0,T] \to TQ$ is absolutely continuous. Then $\gamma\circ s : [0,T] \to Q$ is a trajectory for the affine-connection control system.*

A necessary and sufficient condition for V to be a decoupling vector field is that both V and $\nabla_V V$ are sections of $\mathcal{Y}$ [3, p. 426]. Notice that if $\mathcal{Y} = TQ$ then every vector field is a decoupling vector field, and if $\mathcal{Y} = \mathrm{Span}\{Y\}$ then V is a decoupling vector field if and only if both V and $\nabla_V V$ are multiples of Y.

Decoupling vector fields for an *under-actuated* system (12) are analyzed in [4]. In the under-actuated setting, decoupling vector fields are found by solving a system of homogeneous quadratic polynomials in several variables. For our model, the control forces $F = (f_1, f_2, f_3)^t$ and $\tau = (\tau_1, \tau_2, \tau_3)^t$ are unconstrained; the model is a *fully-actuated* affine-connection control system. In this case there are no quadratic polynomials to solve and every left-invariant vector field is a decoupling vector field. Something interesting can occur, however, when we seek *single-input subsystems that admit a decoupling vector field.* Let us start with a single-input affine-connection control system, where the input vector field is $Y = (F,\tau)^t$. As mentioned above, to be decoupling for $\mathcal{Y} = \mathrm{Span}\,\{Y\}$ a vector field V, as well as $\nabla_V V$, must be a multiple of Y.

Proposition 1. *Let ∇ be the affine connection (9) and consider the single input affine-connection control system $\nabla_{\gamma'}\gamma' = u(t)Y(\gamma(t))$, where $Y = (F,\tau)^t$. Then $V = (F,\tau)^t$ is a decoupling vector field for Y if and only if V is a multiple of $(e_i, e_i)^t$, where e_1, e_2, e_3 is the standard basis for $\mathbb{R}^3$, or all but one of its components are zero.*

Proof. The conditions for V to be decoupling eventually reduce to finding an eigenvector belonging to a real eigenvalue for the following matrix:

$$\begin{pmatrix} -\hat{\tau} & 0 \\ -\widehat{F} & -\hat{\tau} \end{pmatrix} \begin{pmatrix} M & 0 \\ 0 & J \end{pmatrix} \begin{pmatrix} F \\ \tau \end{pmatrix} = \lambda \begin{pmatrix} M & 0 \\ 0 & J \end{pmatrix} \begin{pmatrix} F \\ \tau \end{pmatrix}. \tag{13}$$

It is equivalent to $-\hat{\tau}MF = \lambda MF$ and $-\hat{F}MF - \hat{\tau}J\tau = \lambda J\tau$. Since λ is real and $\hat{\tau}$ is skew-symmetric, $\lambda = 0$. If F is zero, we have that $\hat{\tau}J\tau = \tau \times J\tau = 0$. We see that $V = (0, \tau)^t$ will be a decoupling vector field for $Y = (0, \tau)^t$ provided τ is an eigenvector for J. Similarly, when $\tau = 0$, then we must have $\lambda = 0$ and $\widehat{F}MF = F \times MF = 0$. We see that $V = (F, 0)^t$ will be a decoupling vector field for $Y = (F, 0)^t$ provided F is an eigenvector for M. Since M and J are diagonal matrices, the eigenvectors are precisely those for which all but one of $\{f_1, f_2, f_3, \tau_1, \tau_2, \tau_3\}$ are zero. Assume now $F \neq 0$. It follows that $MF = \mu\tau$ where μ is some constant and $-\hat{F}MF - \hat{\tau}J\tau = 0$. This last equality is equivalent to $\tau \times (J\tau - \mu F) = 0$ or in other words $J\tau - \mu F = \alpha\tau$ where α is a constant. By using the two previous relations between F and τ, we have that $(J - \mu^2 M^{-1})\tau = \alpha\tau$. Since $J - \mu^2 M^{-1}$ is diagonal, τ is an eigenvector if all but one of its component are zero. The same is deduced them for F.

We note that if we consider the case when all but one of the $\{f_1, f_2, f_3, \tau_1, \tau_2, \tau_3\}$ are zero, these decoupling vector fields are exactly the *pure motions* for a rigid body moving in air, and were investigated in [7].

4 Singular Extremals

Our goal is to establish a relation between singular extremals arising in optimal control and the conditions for a vector field to be decoupling. To this end, we first recall the notion of singular extremals.

Consider the minimum time problem for a controlled submerged rigid body. We assume the domain of control to be as stated in Section 2. Since the necessary conditions of the maximum principle are local, we assume the equations of motion expressed in local coordinates found in [7].

We introduce ψ, a triple of Euler angles for $R \in \mathrm{SO}(3)$; $\eta = (b, \psi)$, local coordinates for $(b, R) \in \mathrm{SE}(3)$; $\chi = (\eta, \nu, \Omega)$, local coordinates for $T\,\mathrm{SE}(3)$; and let $\chi_0 = \chi(0)$ and $\chi_T = \chi(T)$ be the initial and final states for our submerged rigid body. We let Θ be the 3×3 matrix for which $\dot{\psi} = \Theta\Omega$ and, as always, $\dot{b} = R\nu$. Note first, that the equations of motion derived in the previous section can be written as an affine control system:

$$\dot{\chi}(t) = Y_0(\chi(t)) + \sum_{i=1}^{6} Y_i(t)\varphi_i(t) \tag{14}$$

where $\varphi = (F, \tau)^t$ is the control and the drift Y_0 is given by

$$Y_0 = \begin{pmatrix} R\nu \\ \Theta\Omega \\ M^{-1}[M\nu \times \Omega + D_\nu(\nu)\nu] \\ J^{-1}a_{41} \end{pmatrix} \tag{15}$$

with $a_{41} = J\Omega \times \Omega + M\nu \times \nu + D_\Omega(\Omega)\Omega - r_{C_B} \times R^t \rho g\|$. Notice that when neglecting the external forces, the drift is the geodesic spray. The input vector fields are given by:

$$Y_i = \text{vlft}(\mathbb{I}_i^{-1}) = \begin{pmatrix} 0 \\ 0 \\ \mathbb{I}_i^{-1} \end{pmatrix} \tag{16}$$

with $\mathbb{I}_i^{-1}$ being the ith column of the matrix $\mathbb{I}^{-1} = \begin{pmatrix} M^{-1} & 0 \\ 0 & J^{-1} \end{pmatrix}$.

Assume that there exists an admissible time-optimal control $\varphi : [0, T] \to \mathcal{U}$, such that the corresponding trajectory $\chi = (\eta, \nu, \Omega)$ is a solution of the equations of motion, in Section 2, and steers the body from χ_0 to χ_T. The Maximum Principle, see [12], implies that there exists an absolutely continuous vector $\lambda = (\lambda_\eta, \lambda_\nu, \lambda_\Omega) : [0, T] \to \mathbb{R}^{12}$, $\lambda(t) \neq 0$ for all t, such that the following conditions hold almost everywhere:

$$\dot{\eta} = \frac{\partial H}{\partial \lambda_\eta}, \dot{\nu} = \frac{\partial H}{\partial \lambda_\nu}, \dot{\Omega} = \frac{\partial H}{\partial \lambda_\Omega}, \tag{17}$$

$$\dot{\lambda}_\eta = -\frac{\partial H}{\partial \eta}, \dot{\lambda}_\nu = -\frac{\partial H}{\partial \nu}, \dot{\lambda}_\Omega = -\frac{\partial H}{\partial \Omega} \tag{18}$$

where the Hamiltonian function H is given by

$$\begin{aligned} H(\chi, \lambda, \varphi) = {} & \lambda_\eta^t (R\nu, \Theta\Omega)^t + \lambda_\nu^t M^{-1}[M\nu \times \Omega + D_\nu(\nu)\nu + F] \\ & + \lambda_\Omega^t J^{-1}[J\Omega \times \Omega + M\nu \times \nu + D_\Omega(\Omega)\Omega - r_{C_B} \times R^t \rho g \nabla k + \tau]. \end{aligned}$$

Additionally, the maximum condition

$$H(\chi(t), \lambda(t), \varphi(t)) = \max_{\gamma \in \mathcal{U}} H(\chi(t), \lambda(t), \gamma)$$

holds for almost every t. The maximum of the Hamiltonian is constant along the solutions of (17), (18) and must satisfy $H(\chi(t), \lambda(t), \varphi(t)) = \lambda_0$, $\lambda_0 \geq 0$. A triple (χ, λ, φ) that satisfies the Maximum Principle is called an extremal, and the vector function λ is called the adjoint vector.

The component ϕ_i, $i = 1, \ldots, 6$, of the vectors $\lambda_\nu^t M^{-1}$ and $\lambda_\Omega^t J^{-1}$ are called the switching functions. The maximum condition, along with the control domain $\mathcal{U}$ and the form of the equations of motion, is equivalent almost everywhere to $\varphi_i(t) = -1$ if $\phi_i(t) < 0$ and $\varphi_i(t) = +1$ if $\phi_i(t) > 0$ for $i = 1, \ldots, 6$. Notice that from our assumptions on the fluid and the body, the matrices M and J are diagonal. It follows that the switching functions are, modulo a constant, the last six components of the adjoint vector, and are absolutely continuous. Clearly, the zeros of the switching functions determine the structure of the extremals. We are interested in the case when a switching function is identically zero on a non trivial interval $[t_1, t_2]$. We say that the extremal is singular in the corresponding control on this interval. To compute a singular control φ_i we must differentiate the switching function ϕ_i.

We finish the section with these computations. By construction, we have that $\phi_i(t) = \lambda^t(t)Y_i$. It is then a standard fact that the derivative $\dot{\phi}_i$ along an extremal is given by:

$$\dot{\phi}_i(t) = \lambda^t(t)[Y_0, Y_i](\chi(t)) + \sum_{j=1}^{6} \lambda^t(t)[Y_j, Y_i](t)\varphi_j(t) \tag{19}$$

where [,] denotes the Lie bracket of vector fields. Since the vector fields Y_i are constants, their Lie brackets are zero and the switching functions have absolutely continuous derivatives: $\dot{\phi}(t) = \lambda^t(t)[Y_0, Y_i](\chi(t))$. Differentiating once more, we obtain

$$\ddot{\phi}_i(t) = \lambda^t(t) \operatorname{ad}^2_{Y_0} Y_i(\chi(t)) + \sum_{j=1}^{6} \lambda^t(t)[Y_j, [Y_0, Y_i]](\chi(t))\varphi_j(t)$$

Proposition 2. *For a rigid body moving in the air, the following holds:*

$$[Y_i, [Y_0, Y_i]](\chi) \equiv 0, \quad i = 1, \ldots, 6. \tag{20}$$

In a real fluid, the previous Lie bracket is not zero but satisfies:

$$[Y_i, [Y_0, Y_i]](\chi) \in \operatorname{Span}\{Y_i\}, \quad i = 1, \ldots, 6. \tag{21}$$

Proof. A first remark is that only quadratic terms with respect to the velocities in Y_0 can produce a non zero component for $[Y_j, [Y_0, Y_i]](\chi(t))$. Hence we have that $[Y_j, [Y_0, Y_i]](\chi) \in \operatorname{Span}\{Y_1, \ldots, Y_6\}$. Moreover, it is an easy verification that under our assumptions neither $M\nu \times \Omega$, $J\Omega \times \Omega$ nor $M\nu \times \nu$ contains terms of the form $u^2, v^2, w^2, p^2, q^2, r^2$. We can conclude that for a rigid body moving in air (or more generally when we neglect the external forces) the Lie brackets $[Y_i, [Y_0, Y_i]](\chi)$ are identically zero for $i = 1, \ldots, 6$. In a real fluid, this deduction is not possible since the drag terms, as assumed in this paper, will produce a non zero component. However, it can be verified that in a real fluid the only non zero component of$[Y_i, [Y_0, Y_i]](\chi)$ is a multiple of Y_i. Note that the restoring forces play no role in the value of this Lie bracket.

Equations (20) and (21) can be interpreted in terms of the order of singular extremals. Along a φ_i-singular arc, let q be such that $\frac{d^{2q}}{dt^{2q}}\phi_i$ is the lowest order derivative in which φ_i appears explicitly with a nonzero coefficient. We define q as the order of the singular control φ_i. This definition assumes the well known result that a singular control φ_i first appears explicitly in an even order derivative of ϕ_i, see [13].

Proposition 3. *Let χ be an extremal that is singular for the component φ_i of the control while the other ones are bang. Then, for a submerged rigid body moving in air or in a real fluid, the singular control is of order at least 2.*

Proof. Let us assume χ is an extremal that is singular for the component φ_i of the control while the other ones are constant, *i.e.* the function ϕ_i is identically zero along the extremal. We can compute the singular control from equation (20) as long as the term $\lambda^t[Y_i,[Y_0,Y_i]](\chi)$ is non zero. From Proposition 2, this Lie bracket is zero for a rigid body moving in air and a multiple of Y_i for a real fluid. Since along the extremal $\phi_i = \lambda^t Y_i = 0$ the result is true in a real fluid as well. This means that we must compute at least the fourth derivative of the switching function to obtain the singular control as a feedback.

5 Main Observation

The main result consists of an observation that suggests directions for further study. The observation concerns a possible relationship between singular extremals of order greater than 1 and decoupling vector fields. As seen above, for a rigid body moving in the air the order of the singular control φ_i is greater than 1 since it satisfies the condition $[Y_i,[Y_0,Y_i]](\chi) \equiv 0$. Note that this also implies that Y_i is a decoupling vector field for the single-input affine control system where every control is set to 0 except for φ_i. This last result was already proved in Proposition 1 using an ad hoc method without referring to singular extremals or differential geometric properties of the system. To see the relation between equation 20 and Proposition 1 we note that Y_i is the vertical lift of the control $(F,\tau) = (0,\ldots,1,\ldots,0)$, where the 1 is in the ith position. We also note that for a rigid body moving in the air Y_0 (14) is the geodesic spray for the Levi-Civita affine connection (9). The final ingredient is the easily verifiable equation $[\mathrm{vlft}(X),[S,\mathrm{vlft}(X)]] = \mathrm{vlft}(2\nabla_X X)$. Hence $[Y_i,[Y_0,Y_i]](\chi) \equiv 0$ says that $\nabla_{Y_i} Y_i = 0$, where we abuse notation by using Y_i to also denote the vector field whose vertical lift is Y_i. So Y_i being auto-parallel implies that it is decoupling and that φ_i has order greater than one. This establishes a relationship between the condition that V is a decoupling vector field if both V and $\nabla_V V$ are sections of $\mathcal{Y}$ and Lie bracket properties of singular extremals for the time minimal problem.

For a forced affine-connection control system, such as a submerged rigid body in a real fluid, we should be somewhat cautious since the notion of a decoupling vector field is not clearly defined. Singular extremals, however, certainly are and the condition $[Y_i,[Y_0,Y_i]](\chi) \in \mathrm{Span}\{Y_i\}$ still holds in our case. This perhaps suggests a proper generalization and characterization of the notion of a decoupling vector field for forced affine-connection control systems.

6 Thanks

The authors would like to warmly thank A.D. Lewis for very illuminating discussions and suggestions.

References

1. F. Bullo. Trajectory design for mechanical systems: from geometry to algorithms. *European Journal of Control*, 10(5):397–410, 2004.
2. F. Bullo, N.E. Leonard, and A.D. Lewis. Controllability and motion algorithms for underactuated lagrangian systems on lie groups. *Institute of Electrical and Electronics Engineers. Transactions on Automatic Control*, 45(8):1437–1454, 2000.
3. F. Bullo and A.D. Lewis. *Geometric Control of Mechanical Systems*. Number 49 in Texts in Applied Mathematics, New York-Heidelberg-Berlin, 2004.
4. F. Bullo and K.M. Lynch. Kinematic controllability for decoupled trajectory planning in underactuated mechanical systems. *IEEE Trans. Robotics and Automation*, 17(4):402–412, 2001.
5. M. Chyba and T. Haberkorn. Designing efficient trajectories for underwater vehicles using geometric control theory. *Proceedings of the 24th International Conference on Offshore Mechanics and Artic Enginnering*, 2005.
6. M. Chyba and T. Haberkorn. Optimization techniques for autonomous underwater vehicles: a practical point of view. *Proceedings of the 17th International Symposium on Mathematical Theory of Networks and Systems*, 2006.
7. M. Chyba, T. Haberkorn, R.N. Smith, and S.K. Choi. Implementable efficient time and energy consumption trajectories design for an autonomous underwater vehicle. *6th Conf. Computer and IT Applications in the Maritime Industries (COMPIT), Cortona*, pages 443–457, 2007.
8. M. Chyba, T. Haberkorn, R.N. Smith, S. Zhao, and S.K. Choi. Towards practical implementation of time optimal trajectories for underwater vehicles. *Proceedings of the 25th International Conference on Offshore Mechanics and Artic Enginnering*, 2006.
9. M. Chyba, N.E. Leonard, and E.D. Sontag. Singular trajectories in multi-input time-optimal problems: Application to controlled mechanical systems. *J. of Dynamical and Control Systems*, 9:73–88, 2003.
10. N. E. Leonard. Stability of a bottom-heavy underwater vehicle. *Automatica*, 33:331–346, 1997.
11. A. D. Lewis, J.P. Ostrowski, R.M. Murray, and J.W. Burdick. Nonholonomic mechanics and locomotion: the snakeboard example. *Proceedings of the IEEE International Conference on Robotics and Automation, Institute of Electrical and Electronics Engineers*, pages 2391–2400, 1994.
12. L. S. Pontryagin et al. *The mathematical theory of optimal processes*, volume 55 of *International series of monographs in pure and applied mathematics*. Macmillan, 1964. Translated by D.E. Brown.
13. H.M. Robbins. A generalized Legendre-Clebsch condition for the singular cases of optimal control. *IBM Journal of Research and Development*, 11(4):361–372, 1967.

Explicit Structured Singular Value Analysis of Manipulators with Passivity Based Control

Satoru Sakai[1] and Kenji Fujimoto[2]

[1] Chiba University, Department of Electronics and Mechanical Engineering, Yayoi-cho, Inage-ku, Chiba-shi, 263-0043, Japan
satorusakai@faculty.chiba-u.jp

[2] Nagoya University, Graduate School of Engineering, Furo-cho, Chikusa-ku, Nagoya-shi, Aichi 464-8603, Japan
fujimoto@nagoya-u.ac.jp

1 Introduction

Hamiltonian control systems [11, 6] are the systems described by Hamilton's canonical equations which represent general physical systems. Recently port-controlled Hamiltonian systems are introduced as a generalization of Hamiltonian systems [3]. They can represent not only ordinary mechanical, electrical and electro-mechanical systems, but also nonholonomic systems [4]. The special structure of physical systems allows us to utilize the passivity which they innately possess and a lot of fruitful results were obtained so far. These methods are so called passivity based control [8].

One of the advantages of passivity based control is robustness against physical parameter variations. Utilizing the intrinsic passive property of physical systems, it is easy to stabilize the system with few dynamic parameters. In addition, some optimal gains are clarified to improve transient behavior in the cases of a class of mechanical Hamiltonian systems [10] and redundant manipulators [9].

On the other hand, when we control a robotic manipulators in practice, these discussion can be useful only before the settling time. After the manipulator endpoint converses to the neighborhood of the desired point in free space, the endpoint interacts with the environment which is not free space any more. This means that the endlink mass is not constant and the perturbation of the mass is unknown in the case of unstructured environment. The discussion after the settling time is required in the case of robotic pick-and-place manipulators apart from the other Hamiltonian systems, such as magnetic levitation systems and vehicle systems.

In order to discuss this matter, the concept of the structured singular value is needed. The structured singular value is a well-known measure of not only the stability margin for structured uncertainty but also robust performance.

However, unfortunately, the structured singular value can not be expressed explicitly and not be calculated exactly in general.

F. Bullo et al. (Eds.): Lag. & Hamil. Methods for Nonlin. Ctrl., LNCIS 366, pp. 387–396, 2007.
springerlink.com

In this paper, we solve this problem and derive the explicit and exact structured singular value of robotic manipulators. In Section II, we focus on a dynamics with endlink mass perturbation after the settling time. In Section III, as a main result of this paper, we derive the explicit and exact structured singular value by using the structural property of the dynamics, In Section IV, quantitatively, we analyze the robust stability of passivity based control. In Section V, we conclude this paper.

2 Port Hamiltonian Systems

2.1 Port Hamiltonian Systems [12]

A port Hamiltonian system with a Hamiltonian H is a system with the following state-space realization.

$$\begin{cases} \dot{x} = J(x,t)\frac{\partial H}{\partial x}(x,t)^T + g(x,t)u \\ y = g(x,t)^T\frac{\partial H}{\partial x}(x,t)^T \end{cases} \tag{1}$$

where $u, y \in \mathbf{R}^m$, $x \in \mathbf{R}^n$ and J is skew-symmetric, i.e. $J = -J^T$. Port Hamiltonian systems are natural generalization of physical systems. The following properties of such systems are known.

Lemma 1. [12] *Consider the system (1). Suppose Hamiltonian H is lower bounded and satisfies $\partial H/\partial t \leq 0$. Then the system is passive with respect to the storage function H, and the following feedback renders $(u, y) \to 0$. Furthermore, if the system is zero-state detectable, then the following feedback renders the system asymptotically stable*

$$u = -Cy \tag{2}$$

where $C > 0$ is any positive definite matrix.

The zero-state detectability, which is assumed in Lemma 1, does not always hold for general systems. In such a case, the stabilization method by generalized canonical transformation [1] is useful.

2.2 Dynamics After Settling

The following Hamiltonian system is considered here.

$$\begin{cases} \begin{bmatrix} \dot{q} \\ \dot{p} \end{bmatrix} = \begin{bmatrix} 0 & I \\ -I & 0 \end{bmatrix} \begin{bmatrix} \frac{\partial H}{\partial q}^T \\ \frac{\partial H}{\partial p}^T \end{bmatrix} + \begin{bmatrix} 0 \\ I \end{bmatrix} u \\ y \ = \frac{\partial H}{\partial p}(q,p)^T \end{cases} \tag{3}$$

with the Hamiltonian

$$H = \frac{1}{2}p^T M(q)^{-1} p \tag{4}$$

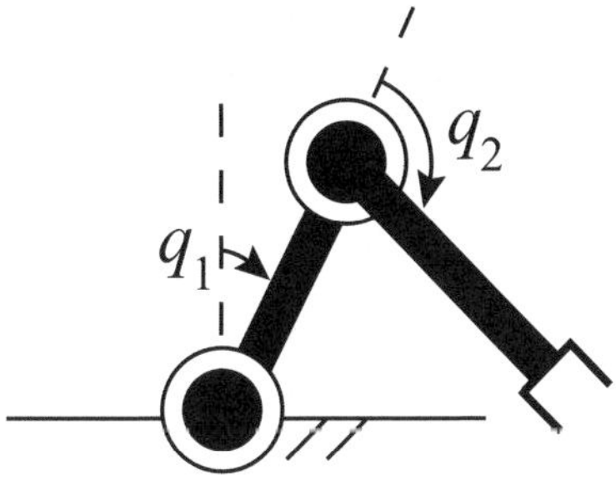

Fig. 1. Two-link manipulator

where $M(q)$ is mass matrix which makes this system equivalent to equations of motion of robotic manipulators and $q, p \in \mathbf{R}^n$ are position and momentum, respectively.

By using the stabilization procedure based on the generalized canonical transformation, the following controller renders $(q, p) \to (q_d, 0)$.

$$u = -Cy + \frac{\partial U}{\partial q}^T. \tag{5}$$

where q_d is the desired position and U is a positive definite function with $U = 0$ at $q = q_d$.

Here, we assume the moment of inertia is linear with respect to the link mass (homogeneous link) and describe the case of $n = 2$ shown in Fig.1 for simplicity. In addition, let $U(q) = -(K/2)(q - q_d)^2$ and K and C be diagonal without lose of generality. From the following process in this paper, we can see there are few difficulties to extend the results to the more general cases.

After the settling time, the closed-loop system of (6), (7) and (8), which is nothing but the classical Hamiltonian system,

$$\begin{bmatrix} \dot{q} \\ \dot{p} \end{bmatrix} = \begin{bmatrix} 0 & I \\ -I & -C \end{bmatrix} \begin{bmatrix} \frac{\partial (H+U)}{\partial q}^T \\ \frac{\partial (H+U)}{\partial p}^T \end{bmatrix} \equiv f_{cl}(q, p) \tag{6}$$

with the Hamiltonian

$$H = \frac{1}{2} p^T \begin{bmatrix} a + c + 2b\cos(q_2) & c + b\cos(q_2) \\ c + b\cos(q_2) & c \end{bmatrix}^{-1} p \tag{7}$$

can be approximated as the following linearized system around the desired point.

$$\dot{x} = \frac{\partial f_{cl}}{\partial x} x \equiv A_{cl} x = \begin{bmatrix} 0 & I_2 \\ -A_p K & -A_p C \end{bmatrix} x \tag{8}$$

where $x = (q, p)^T$ and

$$A_p = \begin{bmatrix} \frac{c}{-ac+b^2\cos^2(q_{d2})} & \frac{-(c+b\cos(q_{d2}))}{-ac+b^2\cos^2(q_{d2})} \\ \frac{-(c+b\cos(q_{d2}))}{-ac+b^2\cos^2(q_{d2})} & \frac{a+c+2b\cos(q_{d2})}{-ac+b^2\cos^2(q_{d2})} \end{bmatrix}. \tag{9}$$

Remark. A_{cl} is NOT constant because the parameters in A_p depend on the perturbation of the endlink mass m. For example, in the case of two homogeneous rectangular links with mass m_i and moment of inertia I_i, these parameters are

$$\begin{cases} a = I_1 + m_1 r_1^2 + m l_1^2 \\ b = m l_1 r_2 \\ c = I_2 + m r_2^2 \end{cases} \tag{10}$$

where l_i is link length and r_i is the distance form the joints to the enter of mass. See [5] for the details of these physical parameters.

The above stabilization procedure is equivalent to the energy shaping and damping injection [7]. However, we emphasize the canonical transformation approach for two reasons. First, we can clearly see that the passivity based control (5) preserves Hamiltonian structure and the controlled manipulators are nothing but new manipulators with spring and damping (6). This gives an important interpretation to the following results in this paper. Second, in our next work, we will extend the following results to tracking and dynamic output feedback stabilization cases which are the results due to the canonical transformation approach.

Strictly speaking, the above linearization procedure should be after the Legendre transformation: $(q, p) \to (q, \dot{q})$, because the state p is the momentum and depends on the perturbation of the endlink mass m. However results in this paper are the same because Coriolis and centrifugal forces in $\dot{p}$ are high-order nonlinear terms.

3 Exact Structured Singular Value

In this section, as the main result of this paper, we derive an exact and explicit expression of the structured singular value of the robotic manipulators with endlink mass perturbation after the settling time.

3.1 Structured Singular Value

Consider the loop shown in Fig.2. For $M \in \mathbf{C}^{n\times n}$, structured singular value μ is defined as

$$\mu_{\mathbf{\Delta}}(M) = \frac{1}{\min\{\bar{\sigma}(\Delta) | \Delta \in \mathbf{\Delta},\ \det(I - M\Delta) = 0\}} \tag{11}$$

unless no $\Delta \in \mathbf{\Delta}$ makes $I - M\Delta$ singular, in which case $\mu_{\mathbf{\Delta}} := 0$, where

$$\mathbf{\Delta} = \{\mathrm{diag}(\delta_1 I_{r1}, ..., \delta_S I_{rS}, \Delta_1, ..., \Delta_F) \text{ s.t. } \delta_i \in \mathbf{C}, \Delta_j \in \mathbf{C}^{m_j \times m_j}\},$$
$$\sum_{i=1}^{S} r_i + \sum_{j=1}^{F} m_j = n. \tag{12}$$

The following theorem and lemma are fundamental results of robust linear control theory.

Theorem 1. *Let $b > 0$. The loop shown in Fig.2 is well-posed and internally stable for all $\Delta(\bullet) \in \mathcal{M}(\mathbf{\Delta})$ with $\|\Delta\|_\infty < 1/b$ if and only if*

$$\sup_{\omega \in \mathrm{R}} \mu_{\mathbf{\Delta}}(M(j\omega)) \leq b \tag{13}$$

where ω is frequency and $\mathcal{M}(\mathbf{\Delta})$ is the set of all block diagonal and stable rational transfer functions that have block structures such as $\mathbf{\Delta}$.

Lemma 2. *The structured singular value is related to the following linear algebra quantities as*

$$\rho(M) \leq \mu(M) \leq \bar{\sigma}(M)) \tag{14}$$

where $\rho(M)$ is the spectrum radius and $\bar{\sigma}(M)$ is the largest singular value. The first equality holds at $(S, F) = (1, 0)$ and the second equality holds at $(S, F) = (0, 1)$.

Unfortunately, in general, the structured singular value can not be expressed explicitly though the lower bound and the upper bounds in (14) can. Furthermore, the difference between these bounds can be large arbitrary, so these bounds can not be used for the estimation. In many practical cases, this difficulty is avoided by numerical methods, even though we can not yet get the exact structured singular value due to non-convex optimization problem.

3.2 Exact Structured Singular Value of Robotic Manipulators

In this subsection, we give an explicit and exact structured singular value of the robotic manipulation system with endlink mass perturbation after the settling time.

Theorem 2. *Suppose the system in (12) has the following the endlink mass perturbation,*

$$m = \bar{m} + \delta_m \tag{15}$$

where $\bar{m}$ is the nominal endlink mass and δ_m is the corresponding perturbation. Then, the structured singular value is explicitly expressed as

$$\mu_{\mathbf{\Delta}}(M) = \max_i \left| \lambda_i \left(\left[\begin{array}{c|c} A & B \\ \hline C & D \end{array} \right] \right) \right| \tag{16}$$

where $\lambda(\bullet)$ is eigenvalue, $\Delta = \delta_m I$, $a = a_1 m + a_2$, $b = b_1 m$, $c = c_1 m$,

$$A = \begin{bmatrix} 0 & I \\ -KA_s & -CA_s \end{bmatrix} \tag{17}$$

$$B = \begin{bmatrix} \frac{1}{\bar{m}} I & 0 & 0 & \frac{1}{\bar{m}} I & B_l I & 0 & 0 & B_l I & 0 \\ 0 & \frac{1}{\bar{m}} I & B_s K & B_s C & 0 & B_l I & 0 & 0 & 0 \end{bmatrix}$$

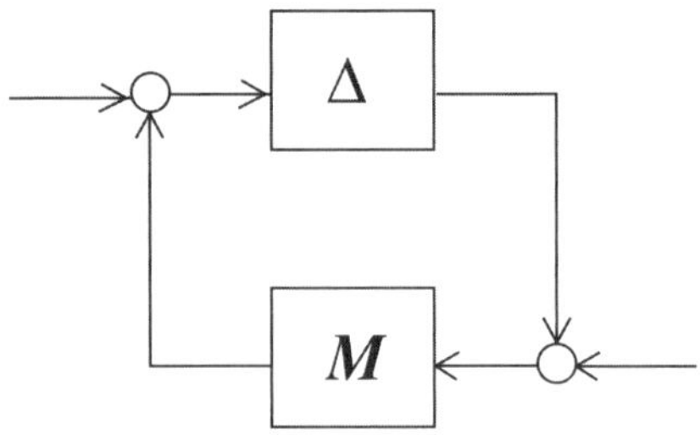

Fig. 2. Closed-loop system with structured uncertainty

$$C = \begin{bmatrix} -A \\ I \\ -\bar{m}A \\ \bar{m}I \\ -\bar{m}^2A \\ \bar{m}^2I \\ -\bar{m}^3A \\ \bar{m}^3I \end{bmatrix} \tag{18}$$

$$D=\begin{bmatrix} -\frac{1}{\bar{m}}I & A_t & -B_lI & -B_t & 0\,0\,0\,0 \\ 0 & 0 & 0 & 0 & 0\,0\,0\,0 \\ 0 & \bar{m}A_t & -\bar{m}B_lI & -B_t & 0\,0\,0\,0 \\ 0 & I & 0 & 0 & 0\,0\,0\,0 \\ 0 & \bar{m}^2A_t & c_1a_1I & -\bar{m}B_t & 0\,0\,0\,0 \\ 0 & \bar{m}I & 0 & I & 0\,0\,0\,0 \\ 0 & \bar{m}^3A_t & \bar{m}c_1a_1I & -\bar{m}^2B_t & I\,0\,0\,0 \\ 0 & \bar{m}^2I & 0 & \bar{m}I & 0\,I\,0\,0 \end{bmatrix} \tag{19}$$

Before the proof of this theorem, we discuss the guideline. In this proof, we focus on two special structures of the closed-loop systems. First, the uncertainty of this robotic manipulation system is structured, that is, the perturbation exist in only the endlink mass, while the whole system is physically modeled. This means that if we can pull out the perturbation as $\Delta = \delta_m I$, the first equality of (14) holds.

Now assume

$$A_{cl} = \bar{A}_{cl} + \delta_m \tilde{A}_{cl}. \tag{20}$$

In this case, intuitively, it is easy to give the corresponding $M(s)$ as

$$\left[\begin{array}{c|c} \bar{A}_{cl} + \tilde{A}_{cl} & I \\ \hline \tilde{A}_{cl} & 0 \end{array}\right] = \tilde{A}_{cl}(sI - (\bar{A}_{cl} + \tilde{A}_{cl}))^{-1}. \tag{21}$$

However the elements of A_{cl} in (9) are not affine of m and (20) does not hold. It is not trivial but difficult to derive the corresponding $M(s)$ to Δ. So, we need more detail structural property of the closed-loop system.

Proof of Theorem 2

We focus on a structural property that the parameters in (9) are affine of the endlink mass:

$$(a\ b\ c) = m(a_1\ b_1\ c_1) + (a_2\ 0\ 0), \tag{22}$$

due to homogeneous links. The elements of A_{cl} are written in rational polynomial form. This fact implies that the corresponding descriptor form of (9) makes the elements polynomial form with respect to m as follows,

$$E_{des}\dot{x} = A_{des}x \tag{23}$$

where

$$\begin{aligned} E_{des} &= \{(-a_1c_1 + b_1^2\cos^2(q_{d2}))m^2 - a_2c_1m\}I_4 \equiv NI_4, \\ A_{des} &= NA_{cl} \end{aligned} \tag{24}$$

In this case, the following relations hold [2],

$$\begin{aligned} [E_{des}\ A_{des}] &= F_u\left(\Delta,\ \bar{M}\right) \\ &= F_u\left(\Delta,\ S\left(\begin{bmatrix} 0 & I \\ P & P_o \end{bmatrix},\ \tilde{M}\right)\right) \end{aligned}$$

where $S(\bullet,\bullet)$ is the star product, $F_u(\bullet,\bullet)$ is the lower linear fractional transformation and

$$\begin{aligned} P &= I, \\ P_o &= \bar{m}I, \\ \Delta &= \delta_m I. \end{aligned} \tag{25}$$

Now, δ_ms of A_{cl} are successfully pulled out and the corresponding M is given as follows,

$$M = \left[\begin{array}{c|c} E_0^{-1}A_0 & E_0^{-1}\bar{M}_{21} \\ \hline -\bar{M}_{12a}E_0^{-1}A_0 + \bar{M}_{12b} & \bar{M}_{11} - \bar{M}_{12a}E_0^{-1}\bar{M}_{21} \end{array}\right]$$

where E_0, A_0 are the corresponding nominal values of (23) and

$$\bar{M} = \begin{bmatrix} \bar{M}_{11} & [\bar{M}_{12a}\ \bar{M}_{12b}] \\ \bar{M}_{21} & \bar{M}_{22} \end{bmatrix}. \tag{26}$$

From here, it is a straightforward calculation to confirm

$$M = \left[\begin{array}{c|c} A & B \\ \hline C & D \end{array}\right] \tag{27}$$

with (17), (18), (18) and (19).

Recall that Δ is the repeated scalar block with only one scalar and the first equality in (14) holds. That is, the structured singular value is expressed exactly and explicitly. (Q.E.D.)

Remark. From the proof procedure, it is clear that this result can be easily extended to the more general cases such as $n \geq 3$ with the other function U, such as that of gravity compensator. These cases are omitted here only because of the space limitation.

Theorem 2 is a very fundamental result in the field of robotics. First, based on Theorem 2, we can study many problems after the settling time, such as robust performance in the presence of the perturbation. Second, the structured singular value with the passivity based control is interpreted as the structured singular value of manipulator itself, because the passivity based control (5) preserves the Hamiltonian structure as we see in Section II.

4 Robust Stability Analysis

In this section, as a demonstrative application of Theorem 2, we quantitatively analyze the robust stability of the passivity based control. As we mentioned in Section I, the passivity based control is qualitatively said to be robust and there are some papers of experimental comparison between the conventional control (e,g. computed torque control) and the passivity based control. However, there are few papers to discuss the robust stability of the passivity based control quantitatively and theoretically .

In general, *unmodeled dynamics* exists in high-frequency region. However, in Theorem 2, the direct term D of transfer function matrix $M(s)$, is not zero, that is, $M(s)$ is not strictly proper. This means that the endlink mass perturbation can lose the robust stability in the high-frequency region.

Now, we focus the following quantity $M(\infty) = D$ based on the continuity of the structured singular value, $\mu : \mathbf{C}^{n \times n} \to \mathbf{R}$.

Theorem 3. *Consider the structured singular value (16) of robotic manipulation system (9) after the settling time. Then, the structured singular value is given as*

$$\mu_{\mathbf{\Delta}}(M(\infty)) = \begin{cases} 1/\bar{m}, & \cos^2(q_{d2}) \leq r \\ \left|1/(\bar{m} + a_2 c_1/(a_1 c_1 - b_1^2 \cos^2(q_{d2}))\right|, & \text{otherwise.} \end{cases} \tag{28}$$

with

$$r = c_1(2a_1 + (a_2/\bar{m}))/(2b_1^2).$$

The size of square matrix D, that is, the order of the characteristic equation of D is larger than four. Generally we can not derive analytical eigenvalues of the corresponding large matrices, because we can not solve the m-th ($m > 4$) order algebra equations. So, again, we need to focus on structural properties of robotic manipulators.

Proof of Theorem 3
The structure of D is not triangular but has the same property:

$$\det(D) = \det([d_{ij}]) = \Pi_{i=1}^{n} d_{ii}. \tag{29}$$

This is confirmed as the following. First, by using the following equation, in repeat,

$$\det(D) = \det(D_{ii})\det(D_{jj} - D_{ji}D_{jj}^{-1}D_{ij}), \tag{30}$$

$$(i,j) = \begin{cases}(1,2) & (\det(D_{22}) \neq 0) \\ (2,1) & (\det(D_{11}) \neq 0)\end{cases}, \tag{31}$$

(30) is reduced to

$$\det(D) = \det\left(\frac{1}{\bar{m}}I\right)\det\left(-\bar{m}B_l I\right)\det\left(0\right). \tag{32}$$

Then, from the structure of $B_l I$, the property (29) is confirmed.

The eigenvalues of matrices with the property (29) is equal to the diagonal components d_{ii}. Now, we can explicitly derive the spectrum of D as

$$\left\{0,\ -\frac{1}{\bar{m}},\ -\frac{1}{\bar{m} + a_2c_1/(a_1c_1 - b_1^2\cos^2(q_{d2}))}\right\} \tag{33}$$

The remaining parts, such as a calculation of r, are trivial and the proof is straightforward calculation. (Q.E.D.)

Remark. It is shown that the structured singular value (28) is the same as that of computed torque controls. That is, the robust stability is independent of the identified parameter errors in high frequency region. However, in low frequency region, or the steady-state, the robust stability of computed torque controls depends on the identified parameters $\hat{a}, \hat{b}, \hat{c}$ of (10) as

$$\mu_{\boldsymbol{\Delta}}(M(0)) = \max\left\{\frac{\hat{b}^2\cos^2(q_{d2})-\hat{a}\hat{c}}{4k_1(-3k_2+\cos^2(q_{d2}))},\ \frac{2\hat{a}\hat{c}-(2\hat{b}^2+\hat{a}\hat{c})\cos^2(q_{d2})+\hat{b}^2\cos^4(q_{d2})}{4k_1(-3k_2+\cos^2(q_{d2}))}\right\} \tag{34}$$

apart from that of passivity based controls:

$$\mu_{\boldsymbol{\Delta}}(M(0)) = \frac{1}{4k_1(-3k_2+\cos^2(q_{d2}))}, \tag{35}$$

where the detail of k_1 and k_2 are omitted for their uniqueness.

It is an interesting result that the structured singular value (28) does not depend on the controller gains K and C and determined by the only link parameter, even though $M(\infty)$ in (4) depends on the gains. This fact seems to point out "an importance of link design (mechanical design) more than controller design in passivity based control", all the more because the passivity based control utilizes physical properties of link dynamics more than the computed torque control.

5 Conclusion

In this paper, we derive the explicit and exact structured singular value of robotic manipulators after the settling time. We derive the exact and explicit structured singular value by using structural properties. Furthermore, quantitatively, we analyze the robust stability of the passivity based control.

References

1. K. Fujimoto and T. Sugie. Canonical transformation and stabilization of generalized hamiltonian systems. *Systems & Control Letters*, 42(3):217–227, 2001.
2. M. Kawanishi and T. Sugie. Analysis/synthesis based on exact expression of physical parameter variations. In *Proc. of 3rd European Control Conference*, volume 1, pages 159–164, 1995.
3. B. M. J. Maschke and A. J. van der Schaft. Port-controlled Hamiltonian systems: modeling origins and system-theoretic properties. In *IFAC Symp. Nonlinear Control Systems*, pages 282–288, 1992.
4. B. M. J. Maschke and A. J. van der Schaft. A Hamiltonian approach to stabilization of nonholonomic mechanical systems. In *Proc. 33rd IEEE Conf. on Decision and Control*, pages 2950–2954, 1994.
5. R. Murray, Zexiang Li, and S. Sastry. *A Mathematical Introduction to Robotic Manipulation*. CRC Press, London, 1994.
6. H. Nijmeijer and A. J. van der Schaft. *Nonlinear Dynamical Control Systems*. Springer-Verlag, New York, 1990.
7. R. Ortega and Eloisa Garcia-Canseco. Interconection and damping assignment passivity-based control: A survey. *European Journal of Control*, pages 1–27, 2004.
8. R. Ortega, A. Loria, P. J. Nicklasson, and H. Sira-Ramirez. *Passivity-based Control of Euler-Lagrange Systems*. Springer-Verlag, London, 1998.
9. M. Sekimoto S. Arimoto and R. Ozawa. Natural resolution of ill-posedness of inverse kinematics for redundant robots:a challenge to bernstein degrees-of-freedom problem. *Advanced Robotics*, 19(4):401–434, 2005.
10. S. Sakai and K. Fujimoto. Dynamic output feedback stabilization of a class of nonholonomic hamiltonian systems. In *Proc. IFAC World Congress 2005*, pages 1967–1970, 2005.
11. A. J. van der Schaft. Stabilization of Hamiltonian systems. *Nonl. An. Th. Math. Appl.*, 10:1021–1035, 1986.
12. A. J. van der Schaft. *L_2-Gain and Passivity Techniques in Nonlinear Control*. Springer-Verlag, London, 2000.

Appendix

$$
\begin{aligned}
A_s &= \frac{-1}{m_e}\begin{bmatrix} -c_1 & b_1\cos(q_{d2})+c_1 \\ b_1\cos(q_{d2})+c_1 & \frac{a_2+\bar{m}(a_1+c_1+2b\cos(q_{d2}))}{\bar{m}m_e} \end{bmatrix} \\
A_t &= \begin{bmatrix} 0 & I \\ -KA_l & -CA_l \end{bmatrix} \\
A_l &= \frac{-1}{m_e}\begin{bmatrix} -c_1 & b_1\cos(q_{d2})+c_1 \\ b_1\cos(q_{d2})+c_1 & \frac{a_2+\bar{m}(a_1+c_1+2b_1\cos(q_{d2}))}{c_1a_1-b_1^2\cos^2(q_{d2})} \end{bmatrix} \\
B_s &= \frac{1}{\bar{m}m_e}\begin{bmatrix} c_1 & -(b_1\cos(q_{d2})+c_1) \\ -(b_1\cos(q_{d2})+c_1) & a_1+c_1+2b\cos(q_{d2}) \end{bmatrix} \\
B_t &= \begin{bmatrix} 0 & B_l \\ 0 & 0 \end{bmatrix} \\
B_l &= \frac{a_1c_1-b_1^2\cos^2(q_{d2})}{\bar{m}m_e} \\
m_e &= a_2c_1+\bar{m}(a_1c_1-b_1^2\cos^2(q_{d2})).
\end{aligned}
$$

Author Index

Printing: Mercedes-Druck, Berlin
Binding: Stein+Lehmann, Berlin

Lecture Notes in Control and Information Sciences

Edited by M. Thoma, M. Morari

Further volumes of this series can be found on our homepage: springer.com

Vol. 366: Bullo F.; Fujimoto K. (Eds.)
Lagrangian and Hamiltonian Methods for Nonlinear Control 2006
398 p. 2007 [978-3-540-73889-3]

Vol. 365: Bates D.; Hagström M. (Eds.)
Nonlinear Analysis and Synthesis Techniques for Aircraft Control
360 p. 2007 [978-3-540-73718-6]

Vol. 364: Chiuso A.; Ferrante A.; Pinzoni S. (Eds.)
Modeling, Estimation and Control
356 p. 2007 [978-3-540-73569-4]

Vol. 363: Besançon G. (Ed.)
Nonlinear Observers and Applications
224 p. 2007 [978-3-540-73502-1]

Vol. 362: Tarn T.-J.; Chen S.-B.; Zhou C. (Eds.)
Robotic Welding, Intelligence and Automation
562 p. 2007 [978-3-540-73373-7]

Vol. 361: Méndez-Acosta H.O.; Femat R.; González-Álvarez V. (Eds.):
Selected Topics in Dynamics and Control of Chemical and Biological Processes
320 p. 2007 [978-3-540-73187-0]

Vol. 360: Kozlowski K. (Ed.)
Robot Motion and Control 2007
452 p. 2007 [978-1-84628-973-6]

Vol. 359: Christophersen F.J.
Optimal Control of Constrained Piecewise Affine Systems
190 p. 2007 [978-3-540-72700-2]

Vol. 358: Findeisen R.; Allgöwer F.; Biegler L.T. (Eds.): Assessment and Future Directions of Nonlinear Model Predictive Control
642 p. 2007 [978-3-540-72698-2]

Vol. 357: Queinnec I.; Tarbouriech S.; Garcia G.; Niculescu S.-I. (Eds.):
Biology and Control Theory: Current Challenges
589 p. 2007 [978-3-540-71987-8]

Vol. 356: Karatkevich A.:
Dynamic Analysis of Petri Net-Based Discrete Systems
166 p. 2007 [978-3-540-71464-4]

Vol. 355: Zhang H.; Xie L.:
Control and Estimation of Systems with Input/Output Delays
213 p. 2007 [978-3-540-71118-6]

Vol. 354: Witczak M.:
Modelling and Estimation Strategies for Fault Diagnosis of Non-Linear Systems
215 p. 2007 [978-3-540-71114-8]

Vol. 353: Bonivento C.; Isidori A.; Marconi L.; Rossi C. (Eds.)
Advances in Control Theory and Applications
305 p. 2007 [978-3-540-70700-4]

Vol. 352: Chiasson, J.; Loiseau, J.J. (Eds.)
Applications of Time Delay Systems
358 p. 2007 [978-3-540-49555-0]

Vol. 351: Lin, C.; Wang, Q.-G.; Lee, T.H., He, Y.
LMI Approach to Analysis and Control of Takagi-Sugeno Fuzzy Systems with Time Delay
204 p. 2007 [978-3-540-49552-9]

Vol. 350: Bandyopadhyay, B.; Manjunath, T.C.; Umapathy, M.
Modeling, Control and Implementation of Smart Structures 250 p. 2007 [978-3-540-48393-9]

Vol. 349: Rogers, E.T.A.; Galkowski, K.; Owens, D.H.
Control Systems Theory and Applications for Linear Repetitive Processes 482 p. 2007 [978-3-540-42663-9]

Vol. 347: Assawinchaichote, W.; Nguang, K.S.; Shi P.
Fuzzy Control and Filter Design for Uncertain Fuzzy Systems
188 p. 2006 [978-3-540-37011-6]

Vol. 346: Tarbouriech, S.; Garcia, G.; Glattfelder, A.H. (Eds.)
Advanced Strategies in Control Systems with Input and Output Constraints
480 p. 2006 [978-3-540-37009-3]

Vol. 345: Huang, D.-S.; Li, K.; Irwin, G.W. (Eds.)
Intelligent Computing in Signal Processing and Pattern Recognition
1179 p. 2006 [978-3-540-37257-8]

Vol. 344: Huang, D.-S.; Li, K.; Irwin, G.W. (Eds.)
Intelligent Control and Automation
1121 p. 2006 [978-3-540-37255-4]

Vol. 341: Commault, C.; Marchand, N. (Eds.)
Positive Systems
448 p. 2006 [978-3-540-34771-2]

Vol. 340: Diehl, M.; Mombaur, K. (Eds.)
Fast Motions in Biomechanics and Robotics
500 p. 2006 [978-3-540-36118-3]

Vol. 339: Alamir, M.
Stabilization of Nonlinear Systems Using Receding-horizon Control Schemes
325 p. 2006 [978-1-84628-470-0]

Vol. 338: Tokarzewski, J.
Finite Zeros in Discrete Time Control Systems
325 p. 2006 [978-3-540-33464-4]

Vol. 337: Blom, H.; Lygeros, J. (Eds.)
Stochastic Hybrid Systems
395 p. 2006 [978-3-540-33466-8]

Vol. 336: Pettersen, K.Y.; Gravdahl, J.T.; Nijmeijer, H. (Eds.)
Group Coordination and Cooperative Control
310 p. 2006 [978-3-540-33468-2]

Vol. 335: Kozłowski, K. (Ed.)
Robot Motion and Control
424 p. 2006 [978-1-84628-404-5]

Vol. 334: Edwards, C.; Fossas Colet, E.; Fridman, L. (Eds.)
Advances in Variable Structure and Sliding Mode Control
504 p. 2006 [978-3-540-32800-1]

Vol. 333: Banavar, R.N.; Sankaranarayanan, V.
Switched Finite Time Control of a Class of Underactuated Systems
99 p. 2006 [978-3-540-32799-8]

Vol. 332: Xu, S.; Lam, J.
Robust Control and Filtering of Singular Systems
234 p. 2006 [978-3-540-32797-4]

Vol. 331: Antsaklis, P.J.; Tabuada, P. (Eds.)
Networked Embedded Sensing and Control
367 p. 2006 [978-3-540-32794-3]

Vol. 330: Koumoutsakos, P.; Mezic, I. (Eds.)
Control of Fluid Flow
200 p. 2006 [978-3-540-25140-8]

Vol. 329: Francis, B.A.; Smith, M.C.; Willems, J.C. (Eds.)
Control of Uncertain Systems: Modelling, Approximation, and Design
429 p. 2006 [978-3-540-31754-8]

Vol. 328: Loría, A.; Lamnabhi-Lagarrigue, F.; Panteley, E. (Eds.)
Advanced Topics in Control Systems Theory
305 p. 2006 [978-1-84628-313-0]

Vol. 327: Fournier, J.-D.; Grimm, J.; Leblond, J.; Partington, J.R. (Eds.)
Harmonic Analysis and Rational Approximation
301 p. 2006 [978-3-540-30922-2]

Vol. 326: Wang, H.-S.; Yung, C.-F.; Chang, F.-R.
H_∞ Control for Nonlinear Descriptor Systems
164 p. 2006 [978-1-84628-289-8]

Vol. 325: Amato, F.
Robust Control of Linear Systems Subject to Uncertain
Time-Varying Parameters
180 p. 2006 [978-3-540-23950-5]

Vol. 324: Christofides, P.; El-Farra, N.
Control of Nonlinear and Hybrid Process Systems
446 p. 2005 [978-3-540-28456-7]

Vol. 323: Bandyopadhyay, B.; Janardhanan, S.
Discrete-time Sliding Mode Control
147 p. 2005 [978-3-540-28140-5]

Vol. 322: Meurer, T.; Graichen, K.; Gilles, E.D. (Eds.)
Control and Observer Design for Nonlinear Finite and Infinite Dimensional Systems
422 p. 2005 [978-3-540-27938-9]

Vol. 321: Dayawansa, W.P.; Lindquist, A.; Zhou, Y. (Eds.)
New Directions and Applications in Control Theory
400 p. 2005 [978-3-540-23953-6]

Vol. 320: Steffen, T.
Control Reconfiguration of Dynamical Systems
290 p. 2005 [978-3-540-25730-1]

Vol. 319: Hofbaur, M.W.
Hybrid Estimation of Complex Systems
148 p. 2005 [978-3-540-25727-1]

Vol. 318: Gershon, E.; Shaked, U.; Yaesh, I.
H_∞ Control and Estimation of State-multiplicative Linear Systems
256 p. 2005 [978-1-85233-997-5]

Vol. 317: Ma, C.; Wonham, M.
Nonblocking Supervisory Control of State Tree Structures
208 p. 2005 [978-3-540-25069-2]

Vol. 316: Patel, R.V.; Shadpey, F.
Control of Redundant Robot Manipulators
224 p. 2005 [978-3-540-25071-5]

Vol. 315: Herbordt, W.
Sound Capture for Human/Machine Interfaces: Practical Aspects of Microphone Array Signal Processing
286 p. 2005 [978-3-540-23954-3]

Vol. 314: Gil', M.I.
Explicit Stability Conditions for Continuous Systems
193 p. 2005 [978-3-540-23984-0]

Vol. 313: Li, Z.; Soh, Y.; Wen, C.
Switched and Impulsive Systems
277 p. 2005 [978-3-540-23952-9]

Vol. 312: Henrion, D.; Garulli, A. (Eds.)
Positive Polynomials in Control
313 p. 2005 [978-3-540-23948-2]

Vol. 311: Lamnabhi-Lagarrigue, F.; Loría, A.; Panteley, E. (Eds.)
Advanced Topics in Control Systems Theory
294 p. 2005 [978-1-85233-923-4]

Vol. 310: Janczak, A.
Identification of Nonlinear Systems Using Neural Networks and Polynomial Models
197 p. 2005 [978-3-540-23185-1]